Trento

January 2009

Springer Complexity

Springer Complexity is an interdisciplinary program publishing the best research and academic-level teaching on both fundamental and applied aspects of complex systems – cutting across all traditional disciplines of the natural and life sciences, engineering, economics, medicine, neuroscience, social and computer science.

Complex Systems are systems that comprise many interacting parts with the ability to generate a new quality of macroscopic collective behavior the manifestations of which are the spontaneous formation of distinctive temporal, spatial or functional structures. Models of such systems can be successfully mapped onto quite diverse "real-life" situations like the climate, the coherent emission of light from lasers, chemical reaction-diffusion systems, biological cellular networks, the dynamics of stock markets and of the internet, earthquake statistics and prediction, freeway traffic, the human brain, or the formation of opinions in social systems, to name just some of the popular applications.

Although their scope and methodologies overlap somewhat, one can distinguish the following main concepts and tools: self-organization, nonlinear dynamics, synergetics, turbulence, dynamical systems, catastrophes, instabilities, stochastic processes, chaos, graphs and networks, cellular automata, adaptive systems, genetic algorithms and computational intelligence.

The two major book publication platforms of the Springer Complexity program are the monograph series "Understanding Complex Systems" focusing on the various applications of complexity, and the "Springer Series in Synergetics", which is devoted to the quantitative theoretical and methodological foundations. In addition to the books in these two core series, the program also incorporates individual titles ranging from textbooks to major reference works.

Editorial and Programme Advisory Board

Klaus Mainzer

Thinking in Complexity

The Computational Dynamics of Matter, Mind, and Mankind

Fifth Revised and Enlarged Edition
With 223 Figures
Including 4 Color Figures

Professor Dr. Klaus Mainzer
Lehrstuhl für Philosophie und Wissenschaftstheorie
Institut für Interdisziplinäre Informatik
Universität Augsburg
Universitätsstrasse 10
86135 Augsburg, Germany

Library of Congress Control Number: 2007925880

ISBN 978-3-540-72227-4 5th edition Springer Berlin Heidelberg New York
ISBN 3-540-00239-1 4th edition Springer Berlin Heidelberg New York

Springer is a part of Springer Science+Business Media

springer.com

Typesetting and Production: LE-TEX Jelonek, Schmidt & Vöckler GbR, Leipzig, Germany
Cover: WMXDesign, Heidelberg

SPIN 12052615 54/3180/YL - 5 4 3 2 1 0 Printed on acid-free paper

Das Ganze ist aber nur das durch seine Entwicklung sich vollendende Wesen. *

G. W. F. Hegel: Phänomenologie des Geistes (1807)

* *The whole, however, is merely the essential nature reaching its completeness through the process of its own development.*

G. W. F. Hegel: The Phenomenology of Mind (1807)

Preface to the Fifth Edition

Complexity determines the spirit of twenty-first century science. The expansion of the universe, the evolution of life, and the globalization of human economies and societies all involve phase transitions of complex dynamical systems. Complexity research is done in a growing number of disciplines in the physical and life sciences, the economic and social sciences, and the cognitive and computer sciences, yielding new results and insights. Thus, this book on "Thinking in Complexity" is inevitably associated with a learning process and must be updated, although the main principles of complexity research are highlighted, just as they have been from the very first edition. However, the fifth edition of "Thinking in Complexity" enlarges and revises nearly all of the sections of the former book, and includes completely new chapters on the evolution of computability and the emerging field of econophysics.

In Chapter 2, the methodological section on time series analysis now also considers fractals and multifractals as geometric criteria for complexity. Further on, power laws reveal the high level of complexity of all biological systems. They are important indications of the scale-free laws associated with fractal and multifractal features, which are additionally analyzed in Chapter 3. In Chapter 4, the traditional dualism of mind and matter is overcome by invoking the concept of the embodied mind, which was recently introduced in the fields of neurobiology and neuropsychology. The reason is that people learn bodily from experiences with their environment. In neuromedicine, the "Theory of Mind" (ToM) explains the awareness of one's own emotional states by specialized areas of the brain as a complex embodied process. Finally, subjective experiences (qualia) emerge through the bodily interactions of self-conscious organisms with their environment, which can be modeled by the nonlinear dynamics of complex systems.

Chapter 5, on the evolution of computability, is completely new. After historical background on Leibniz and an introduction to the basic concepts of computability and algorithmic complexity, we discuss degrees of complexity in information theory and the theory of probability. Probabilistic attractors allow the degrees of complexity of stochastic processes to be classified. Probabilistic states are typical of the quantum world. Quantum states are coded by quantum information, which is processed by quantum computers. Quantum computability leads to degrees of quantum complexity. Is the universe a giant expanding quantum computer of increasing complexity? All of the degrees of complexity of dynamical systems can also be simulated by cellular automata. New ideas in and results of organic computing are included in Chap. 6. The natural evolution of life and intelligence has become an important paradigm for computational models. They are no longer restricted to the symbolic knowledge representation associated with classical artificial intelligence (AI). The embodied life and mind encountered in natural evolution has led to research into embodied artificial intelligence, which involves robots and machines that undergo bodily experiences and learning interactions with their changing environments.

Chapter 7 on complex systems and the evolution of economies is also completely new. After some historical remarks on economic self-organization, it begins by introducing some basic concepts of nonequilibrium dynamics of complex economic systems. Econophysics is a discipline that has arisen relatively recently in which economic and financial systems are analyzed with the methods of nonlinear dynamics. This situation is reminiscent of the beginnings of modern molecular biology, when physicists such as Erwin Schrödinger (in his revolutionary book "What is Life?") rather than traditional biologists made breakthroughs in the life sciences using modern mathematical physics methods. It may be that physicists, again, will find useful economic and financial models based on mathematical methods derived from the theories of complex systems and nonlinear dynamics. Complex and global markets exhibit turbulence that appears to be remarkably similar to that observed in weather and climate dynamics, which follows typical power laws of probabilistic distributions. Stochastic processes with probabilistic attractors lead to abrupt and discontinuous events (the "Noah effect") or long-term trends (the "Joseph effect"). Time series analysis permits the detection of typical patterns of fractal and multifractal structures that can be used as warning signals for critical situations.

Chapter 8 on the evolution of human culture and society has been enlarged through the addition of sections on media and communication in the age of World Wide Web, mobile, and ubiquitous computing. These are examples of highly complex self-organizing networks, each of which is very similar to a kind of superbrain. The flow of data traffic can be characterized by phase transitions and attractors. In order to efficiently manage an increasing flood of information, we need user-friendly methods of information retrieval and personalized information systems. Adaptive ("tailored") e-learning presents a challenge to the application of communication technologies in modern knowledge societies. The philosophical and ethical messages of modern complexity research are then highlighted in Chap. 9.

Finally, I would like to thank Christian Caron for initiating and supporting this new edition. He follows in the good tradition of Springer-Verlag of focusing on "Thinking in Complexity" as a key topic for the twenty-first century.

Augsburg, March 2007 *Klaus Mainzer*

Preface to the Fourth Edition

The first edition of this book, published in 1994, began with the statement that the new science of complexity would characterize the scientific development of the 21st century. In the first decade of this century, this prediction has been confirmed by overwhelming new empirical results and theoretical insights the of physical and biological sciences, cognitive and computer sciences, and social and economic sciences. Complexity and nonlinearity are still prominent features in the evolution of matter, mind, and human society. Thus, the science of complexity still aims at explanations for the emergence of order in nature and mind and in the economy and society by common principles.

But a new engineering view has focused the exploration of complexity. On the one hand, we need new computational instruments to analyze complex data and recognize future trends. On the other hand, the principles of complex dynamics are increasingly becoming the blueprints of gene, bio, and computer technology. Life and computer sciences are growing into a new kind of complex engineering, changing the basic conditions of human life and society. Nonlinear dynamics are implemented in nonlinear computer chips of high speed and miniaturized size, which are not only distributed in our technical equipment and environment, but also in our body and brain. Robots are embodied. Nanotechnology with new materials, as well as articial life and artificial intelligence are dramatic challenges to the future of complexity science. In the age of globalization, humankind is growing along with worldwide computational networks of information and communication. But we are also endangered by the nonequilibrium phase transitions of technical, economic, and social dynamics. All these new topics are considered in supplemented parts and chapters of this enlarged and revised fourth edition.

Thus, *Thinking in Complexity* has the new subtitle *The Computational Dynamics of Matter, Mind, and Mankind*. We can actually define precise degrees of algorithmic and dynamic complexity. Basic theorems of computational dynamics have been proven recently. But, because of chaos and randomness, understanding computational dynamics does not mean predicting and determining the future in all its details. While we can gain experience with nonlinear dynamics through computer experiments, computer experiments cannot replace reality. As life is complex and random, we have to live it in order to experience it. From a philosophical point of view, this book outlines new standards of epistemology and ethical behavior, which the complex problems of nature, mind, economy, and society demand.

This new edition has been inspired by several fruitful collaborations. In 1997, the book was translated into Japanese, and I was invited to universities in Tokyo, Osaka, Nagasaki, and Fukui. In 1999, the Chinese translation brought invitations to universities in Beijing and Shanghai. During the world exposition 2000 (Expo) in Hannover, I was invited by the German Science Organizations to chair the topics "Understanding Complex Systems" and "Global Networking" during the conference "Science and Technology – Thinking the Future". During the academic year 2002/2003, I was a member of the international research group on "General Principles of Information Theory and Combinatorics" at the Center of Interdisciplinary Research in Bielefeld, Germany. The 7th IEEE International Workshop on "Cellular Neural Networks and Their Applications" (Frankfurt 2002) opened new insights into recent developments in analogic neural computers and chip technology. As a member of the editorial board of the *International Journal of Bifurcation and Chaos in Applied Sciences and Engineering*, I have the opportunity to get an interdisciplinary overview of worldwide explorations in the sciences of complexity. I especially want to thank the editor of our journal, Leon O. Chua (Department of Electrical Engineering & Computer Sciences, University of California, Berkeley), for his encouraging advice and kind invitation to Berkeley. Thanks also to Stephen Wolfram and Leon Chua for the permission to reproduce some figures of their books in Chaps. 5 and 6. Last, but not least, I would like to thank Wolf Beiglböck (Springer-Verlag) for initiating and supporting this new edition.

Augsburg, March 2003 *Klaus Mainzer*

Preface to the Third Edition

The second edition of "Thinking in Complexity", like the first edition, was also sold out in less than one year. Meanwhile, Japanese and Chinese translations of the second editon have been published. Once more I have taken the opportunity provided by a new edition to revise and extend the text.

A new Sect. 2.5 "Complex Systems and the Self-Construction of Materials" is included, in order to analyze the role of complex systems in the dramatic success of supramolecular chemistry, nanotechnology, and the technology of smart ("intelligent") materials. These topics lie at the boundary between materials science and life science. In recent years, life science and computer science have been growing together in a common field of research called "artificial life". A further new Sect. 5.5 "From Artificial Intelligence to Artificial Life" has been added, in which the role of complex systems in the field of artificial life is discussed. I also use the opportunity of the new edition to make some remarks about the relationship between the Santa Fe approach to complex systems and the methods of synergetics and order parameters which are key concepts in this book.

Research into complex systems continues world-wide. I have to thank the readers who have written friendly and inspiring letters from all over the world. Some months ago, a German Society of Complex Systems and Nonlinear Dynamics was founded. The honorable German Academy of Natural Scientists Leopoldina invited me to give a lecture on complexity for which I express my gratitude. Last but not least, I would again like to thank Wolf Beiglböck of Springer-Verlag for initiating and supporting this new edition.

Augsburg, November 1996 *Klaus Mainzer*

Preface to the Second Edition

The first edition of "Thinking in Complexity" was sold out in less than one year. Obviously, complexity and nonlinearity are "hot" topics of interdisciplinary interest in the natural and social sciences. The situation is well summarized by a quotation of Ian Stewart (Mathematics Institute, University of Warwick) who wrote a nice review of my book under the title "Emerging new science" [Nature **374**, 834 (1995)]: "Nonlinearity is not a universal answer, but it is often a better way of thinking about the problem".

I have taken the opportunity provided by a second edition to revise and extend the text. In Sect. 2.4 a supplement is included on the recent importance of conservative self-organization in supramolecular chemistry and the material sciences. Some references are given to the recent discussion of self-organization in alternative cosmologies. Some remarks are made about new results on dissipative self-organization in living cells (Sect. 3.3). The success and limitations of adaptive neural prostheses in neurotechnology are analyzed in more detail (Sect. 5.4). The last chapter is extended into an "Epilogue on Future, Science, and Ethics": After a short introduction to traditional forecasting methods, their limitations and new procedures are discussed under the constraints of nonlinearity and complexity in the natural and social sciences. In particular, the possibilities of predicting and modeling scientific and technological growth are extremely interesting for the contemporary debates on human future and ethics.

General methods of nonlinear complex systems must be developed in cooperation with the natural and social sciences under their particular observational, experimental, and theoretical conditions. Thus, I want to thank some colleagues for their helpful advice: Rolf Eckmiller (Dept. of Neuroin-formatics, University of Bonn), Hans-Jörg Fahr and Wolf Priester (Dept. of Astrophysics and Max-Planck Institute for Radioastronomy, Bonn), Hermann Haken (Institute of Theoretical Physics and Synergetics, Stuttgart), Benno Hess (Max-Planck Institute for Medical Research, Heidelberg), S. P. Kurdyumov (Keldysh Institute of Applied Mathematics, Moscow), Renate Mayntz (Max-Planck Institute for Social Sciences, Cologne), Achim Müller (Dept. of Inorganic Chemistry, University of Bielefeld). Last but not least, I would like to thank Wolf Beiglböck (Springer-Verlag) for initiating and supporting this new edition.

Augsburg, November 1995 *Klaus Mainzer*

Preface to the First Edition

Complexity and nonlinearity are prominent features in the evolution of matter, life, and human society. Even our mind seems to be governed by the nonlinear dynamics of the complex networks in our brain. This book considers complex systems in the physical and biological sciences, cognitive and computer sciences, social and economic sciences, and philosophy and history of science. An interdisciplinary methodology is introduced to explain the emergence of order in nature and mind and in the economy and society by common principles.

These methods are sometimes said to foreshadow the new sciences of complexity characterizing the scientific development of the 21st century. The book critically analyzes the successes and limits of this approach, its systematic foundations, and its historical and philosophical background. An epilogue discusses new standards of ethical behavior which are demanded by the complex problems of nature and mind, economy and society.

The "nucleus" of this book was a paper contributed to a conference on complex nonlinear systems which was organized by Hermann Haken and Alexander Mikhailov at the Center for Interdisciplinary Studies in Bielefeld, in October 1992. In December 1992, Dr. Angela M. Lahee (Springer-Verlag) suggested that I elaborate the topics of my paper into a book. Thus, I would like to express my gratitude to Dr. Lahee for her kind and efficient support and to Hermann Haken for his cooperation in several projects on complex systems and synergetics. I also wish to thank the German Research Foundation (DFG) for the support of my projects on "Computer, Chaos and Self-organization" (1990–1992: Ma 842/4-1) and "Neuroinformatics" (1993–1994: Ma 842/6-1). I have received much inspiration from teaching in a mathematical graduate program on "Complex Systems" (supported by the DFG) and an economic program on "Nonlinearity in Economics and Management" at the University of Augsburg. In 1991 and 1993, the Scientific Center of Northrhine-Westphalia (Düsseldorf) invited me to two international conferences on the cultural effects of computer technology, neurobiology, and neurophilosophy.

Last but not least, I would especially like to thank J. Andrew Ross (Springer-Verlag) for carefully reading and correcting the book as a native speaker, and Katja E. Hüther and Jutta Janßen (University of Augsburg) for typing the text.

Augsburg, June 1994 *Klaus Mainzer*

Contents

1 Introduction: From Linear to Nonlinear Thinking 1

2 Complex Systems and the Evolution of Matter 17
2.1 Aristotle's Cosmos and the Logos of Heraclitus 18
2.2 Newton's and Einstein's Universe and the Demon of Laplace 30
2.3 Hamiltonian Systems and the Chaos of Heaven and the Quantum World 44
2.4 Conservative and Dissipative Systems and the Emergence of Order 54
2.5 Complex Systems of the Nano World and Self-Constructing Materials 71
2.6 Time Series Analysis, Fractals, and Multifractals 77

3 Complex Systems and the Evolution of Life 87
3.1 From Thales to Darwin 87
3.2 Boltzmann's Thermodynamics and the Evolution of Life 92
3.3 Complex Systems and the Evolution of Organisms 98
3.4 Complex Systems and the Ecology of Populations 112
3.5 Complex Systems and Power Laws of Life 117

4 Complex Systems and the Evolution of Mind–Brain 123
4.1 From Plato's Soul to Lamettrie's "L'Homme machine" 124
4.2 Complex Systems and Neural Networks 132
4.3 Brain and the Emergence of Consciousness 155
4.4 Intentionality and the Crocodile in the Brain 165
4.5 Complexity and the Embodied Mind 174

5 Complex Systems and the Evolution of Computability 179
5.1 Leibniz and Mathesis Universalis 179
5.2 Computability and Algorithmic Complexity 183
5.3 Information, Probability, and $1/f$-Complexity 194
5.4 Stochastic Processes, Probabilistic Attractors, and Probabilistic Complexity 200
5.5 Quantum Information, Quantum Computers, and Quantum Complexity 206
5.6 Cellular Automata, Chaos, and Randomness 217

6 Complex Systems and the Evolution of Artificial Life and Intelligence 227
6.1 Turing and Symbolic Artificial Intelligence 227
6.2 Neural Networks and Synergetic Computers 243
6.3 Cellular Neural Networks and Analogic Neural Computers 261
6.4 Universal Cellular Neural Networks and Dynamic Complexity 273
6.5 Organic Computing, Neurobionics, and Embodied Robotics 285
6.6 Embodied Artificial Intelligence and Artificial Life 300

7 Complex Systems and the Evolution of Economies 311
7.1 Smith's Economics and Market Equilibrium 311
7.2 Complex Economic Systems, Chaos, and Randomness 321
7.3 Bachelier's Financial Theory and Market Equilibrium 338
7.4 Complex Financial Markets, Turbulence, and Power Laws 345
7.5 Perspectives on Econophysics 362

8 Complex Systems and the Evolution of Human Culture and Society 367
8.1 From Aristotle's Polis to Hobbes' Leviathan 368
8.2 Complex Social and Cultural Systems 373
8.3 Complex Communication Networks, Information Retrieval, and Personalized Information Systems 390
8.4 Complex Mobile Networks, Ubiquitous Computing, and Adaptive E-Learning 405

9 Epilogue on Future, Science, and Ethics 417
9.1 Complexity, Forecasts, and the Future 417
9.2 Complexity, Science, and Technology 424
9.3 Complexity, Responsibility, and Freedom 430

References 441

Subject Index 469

Name Index 479

1 Introduction: From Linear to Nonlinear Thinking

The theory of nonlinear complex systems has become a successful problem solving approach in the natural sciences – from laser physics, quantum chaos, and metereology to molecular modeling in chemistry and computer-assisted simulations of cellular growth in biology. On the other hand, the social sciences are recognizing that the main problems of mankind are global, complex, nonlinear, and often random, too. Local changes in the ecological, economic, or political system can cause a global crisis. Linear thinking and the belief that the whole is only the sum of its parts are evidently obsolete. One of the most exciting topics of present scientific and public interest is the idea that even our mind is governed by the nonlinear dynamics of complex systems. If this thesis of computational neuroscience is correct, then indeed we have a powerful mathematical strategy to handle the interdisciplinary problems of the natural sciences, social sciences, and the humanities. But one of the main insights of this book is the following: Handling problems does not always mean computing and determining the future. In the case of randomness, we can understand the dynamical reasons, but there is no chance of forecasting. Understanding complex dynamics is often more important for our practical behavior than computing definite solutions, especially when it is impossible to do so.

What is the reason behind these successful interdisciplinary applications? The book shows that the theory of nonlinear complex systems cannot be reduced to special natural laws of physics, although its mathematical principles were discovered and at first successfully applied in physics. Thus it is no kind of traditional "physicalism" to explain the dynamics of laser, ecological populations, or our brain by similar structural laws. It is an interdisciplinary methodology to explain the emergence of certain macroscopic phenomena via the nonlinear interactions of microscopic elements in complex systems. Macroscopic phenomena may be forms of light waves, fluids, clouds, chemical waves, plants, animals, populations, markets, and cerebral cell assemblies which are characterized by order parameters. They are not reduced to the microscopic level of atoms, molecules, cells, organisms, etc., of complex systems. Actually, they represent properties of real macroscopic phenomena, such as field potentials, social or economical power, feelings or even thoughts. Who will deny that feelings and thoughts can change the world?

In history the concepts of the social sciences and humanities have often been influenced by physical theories. In the age of mechanization Thomas Hobbes described the state as a machine ("Leviathan") with its citizens as cog wheels. For

Lamettrie the human soul was reduced to the gear drive of an automaton. Adam Smith explained the mechanism of the market by an "invisible" force like Newton's gravitation. In classical mechanics causality is deterministic in the sense of the Newtonian or Hamiltonian equations of motion. A conservative system is characterized by its reversibility (i.e., symmetry or invariance) in time and the conservation of energy. Celestial mechanics and the pendulum without friction are prominent examples. Dissipative systems are irreversible, like Newton's force with a friction term, for instance.

But, in principle, nature was regarded as a huge conservative and deterministic system the causal events of which can be forecast and traced back for each point of time in the future and past if the initial state is well known ("Laplace's demon"). It was Henri Poincaré who recognized that celestial mechanics is no completely calculable clockwork even with the restrictions of conservation and determinism. The causal interactions of all planets, stars, and celestial bodies are nonlinear in the sense that their mutual effects can lead to chaotic trajectories (e.g., the 3-body problem). Nearly sixty years after Poincaré's discovery, A.N. Kolmogorov (1954), V.I. Arnold (1963), and J.K. Moser proved the so-called KAM theorem: Trajectories in the phase space of classical mechanics are neither completely regular nor completely irregular, but they depend very sensitively on the chosen initial states. Tiny fluctuations can cause chaotic developments (the "butterfly effect").

In this century quantum mechanics has become the fundamental theory of physics [1.1]. In Schrödinger's wave mechanics the quantum world is believed to be conservative and linear. In the first quantization classical systems described by a Hamiltonian function are replaced by quantum systems (for instance electrons or photons) described by a Hamiltonian operator. These systems are assumed to be conservative, i.e., non-dissipative and invariant with respect to time reversal and thus satisfy the conservation law of energy. States of a quantum system are described by vectors (wave functions) of a Hilbert space spanned by the eigenvectors of its Hamiltonian operator. The causal dynamics of quantum states is determined by a deterministic differential equation (the Schrödinger equation) which is linear in the sense of the superposition principle, i.e., solutions of this equation (wave functions or state vectors) can be superposed like in classical optics. The superposition or linearity principle of quantum mechanics delivers correlated ("entangled") states of combined systems which are highly confirmed by the EPR experiments (A. Aspect 1981). In an entangled pure quantum state of superposition an observable can only have indefinite eigenvalues. It follows that the entangled state of a quantum system and a measuring apparatus can only have indefinite eigenvalues. But in the laboratory the measuring apparatus shows definite measurement values. Thus, linear quantum dynamics cannot explain the measurement process.

In the Copenhagen interpretation of Bohr, Heisenberg, et al., the measurement process is explained by the so-called "collapse of the wave-packet", i.e., splitting up of the superposition state into two separated states of measurement apparatus and measured quantum system with definite eigenvalues. Obviously, we must distinguish the linear dynamics of quantum systems from the nonlinear act of measurement. This nonlinearity in the world is sometimes explained by the emergence of human consciousness. Eugene Wigner (1961) suggested that the linearity of Schrödinger's equation might fail for conscious observers, and be replaced by some nonlinear procedure according to which either one or the other alternative would be resolved out. But Wigner's interpretation forces us to believe that the linear quantum superpositions would be resolved into separated parts only in those corners of the universe where human or human-like consciousness emerges. In the history of science anthropic or teleological arguments often showed that there were weaknesses and failures of explanation in science. Thus, some scientists, like Roger Penrose, suppose that the linear dynamics of quantum mechanics is not appropriate to explain cosmic evolution with the emergence of consciousness. He argues that a unified theory of linear quantum mechanics and nonlinear general relativity could at least explain the separated states of macroscopic systems in the world. A measuring apparatus is a macroscopic system, and the measurement process is irreversible far from thermal equilibrium. Thus, an explanation could only succeed in a unified nonlinear theory. Even the generalization of Schrödinger's wave mechanics to quantum field theory is already nonlinear. In quantum field theory, field functions are replaced by field operators in the so-called second quantization. The quantum field equation with a two-particle potential, for instance, contains a nonlinear term corresponding to pair creation of elementary particles. In general the reactions of elementary particles in quantum field theory are essentially nonlinear phenomena. The interactions of an elementary particle cause its quantum states to have only a finite duration and thereby to violate the reversibility of time. Thus even the quantum world itself is neither conservative nor linear in general. In system theory, complexity means not only nonlinearity but a huge number of elements with many degrees of freedom [1.2]. All macroscopic systems like stones or planets, clouds or fluids, plants or animals, animal populations or human societies consist of component elements like atoms, molecules, cells or organisms. The behaviour of single elements in complex systems with huge numbers of degrees of freedom can neither be forecast nor traced back. The deterministic description of single elements must be replaced by the evolution of probabilistic distributions.

The second chapter analyzes *Complex Systems and the Evolution of Matter*. Since the presocratics it has been a fundamental problem of natural philosophy to discover how order arises from complex, irregular, and chaotic states of matter. Heraclitus believed in an ordering force of energy (*logos*) harmonizing irregular interactions and creating order states of matter. Modern thermodynamics describes the emergence of order by the mathematical concepts of statistical mechanics. We distinguish two kinds of phase transition (self-organization) for order states: conservative self-organization means the phase transition of reversible structures in thermal equilibrium. Typical examples are the growth of snow crystals or the emergence of magnetisation in a ferromagnet by annealing the system to a critical value of temperature. Conservative self-organization mainly creates order structures with low energy at low temperatures which are described by a Boltzmann distribution. An application of modern technology is pattern formation in the materials sciences. Complex systems of the nanoworld and self- constructing materials are challenges of key technologies in the future.

Dissipative self-organization is the phase transition of irreversible structures far from thermal equilibrium [1.3]. Macroscopic patterns arise from the complex nonlinear cooperation of microscopic elements when the energetic interaction of the dissipative ("open") system with its environment reaches some critical value. Philosophically speaking, the stability of the emergent structures is guaranteed by some balance of nonlinearity and dissipation. Too much nonlinear interaction or dissipation would destroy the structure. As the conditions of dissipative phase transitions are very general, there is a broad variety of interdisciplinary applications. A typical physical example is the laser. In chemistry, the concentric rings or moving spirals in the Belousov-Zhabotinski (BZ) reaction arise when specific chemicals are poured together with a critical value. The competition of the separated ring waves show the nonlinearity of these phenomena very clearly, because in the case of a superposition principle the ring waves would penetrate each other like optical waves.

The phase transitions of nonlinear dissipative complex systems are explained by synergetics. In a more qualitative way we may say that old structures become unstable and break down by changing control parameters. On the microscopic level the stable modes of the old states are dominated by unstable modes (Haken's "slaving principle") [1.4]. They determine order parameters which describe the macroscopic structure and patterns of systems. There are different final patterns of phase transitions corresponding to different attractors. Different attractors may be pictured as a stream, the velocity of which is accelerated step by step. At the first level a homogeneous state of equilibrium is shown ("fixed point"). At a higher level of velocity the bifurcation of two or more vortices can be observed corresponding to periodic and quasi-periodic attractors. Finally the order decays into deterministic chaos as a fractal attractor of complex systems. Philosophically, I want to underline that in synergetics the microscopic description of matter is distinguished from the macroscopic order states. Thus the synergetic concept of order reminds me of Heraclitus" "logos" or Aristotle's "form" which produces the order states of nature in a transformative process of matter. But, of course, in antiquity a mathematical description was excluded.

In a more mathematical way, the microscopic view of a complex system is described by the evolution equation of a state vector where each component depends on space and time and where the components may denote the velocity components of a fluid, its temperature field, or in the case of chemical reactions, concentrations of chemicals. The slaving principle of synergetics allows us to eliminate the degrees of freedom which refer to the stable modes. In the leading approximation the evolution equation can be transformed into a specific form for the nonlinearity which applies to those systems where a competition between patterns occurs. The amplitudes of the leading terms of unstable modes are called order parameters. Their evolution equation describes the emergence of macroscopic patterns. The final patterns ("attractors") are reached by a transition which can be understood as a kind of symmetry breaking [1.5]. Philosophically speaking, the evolution of matter is caused by symmetry breaking, which was earlier mentioned by Heraclitus.

Understanding complex systems and nonlinear dynamics in nature seems to yield appropriate models for the evolution of matter. But how can we find correct models in practice? Physicists, chemists, biologists, or physicians start with data mining in an unknown field of research. They only get a finite series of measured data corresponding to time dependent events of an unknown dynamical system. From the time series of these data, they must reconstruct the behavior of the system in order to guess the type of its dynamical equation. Therefore, time series analysis is a challenge to modern research in chaos theory and nonlinear dynamics.

The third chapter analyzes *Complex Systems and the Evolution of Life*. In the history of science and philosophy, people believed in a sharp difference between "dead" and "living" matter. Aristotle interpreted life as a power of self-organization (entelechy) driving the growth of plants and animals to their final form. A living system is able to move by itself, while a dead system can only be moved from outside. Life was explained by teleology, i.e., by non-causal ("vital") forces aiming at some goals in nature. In the 18th century Kant showed that self-organization of living organisms cannot be explained by a mechanical system of Newtonian physics. In a famous quotation he said that the Newton for explaining a blade of grass is still lacking. In the 19th century the second law of thermodynamics describes the irreversible movement of closed systems toward a state of maximal entropy or disorder. But how can one explain the emergence of order in Darwin's evolution of life? Boltzmann stressed that living organisms are open dissipative systems in exchange with their environment which do not violate the second law of closed systems. But nevertheless in the statistical interpretation from Boltzmann to Monod the emergence of life can only be a contingent event, a local cosmic fluctuation "at the boundary of universe".

In the framework of complex systems the emergence of life is not contingent, but necessary and lawful in the sense of dissipative self-organization. Only the conditions for the emergence of life (for instance on the planet Earth) may be contingent in the universe. In general, biology distinguishes ontogenesis (the growth of organisms) from phylogenesis (the evolution of species). In any case we have complex dissipative systems the development of which can be explained by the evolution of (macroscopic) order parameters caused by nonlinear (microscopic) interactions of molecules, cells, etc., in phase transitions far from thermal equilibrium. Forms of biological systems (plants, animals, etc.) are described by order parameters. Aristotle's teleology of goals in nature is interpreted in terms of attractors in phase transitions. But no special "vital" or "teleological" forces are necessary. Philosophically, the emergence of life can be explained in the framework of nonlinear causality and dissipative self-organization, although it may be described in a teleological language for heuristic reasons.

I remind the reader that the prebiological evolution of biomolecules was analyzed and simulated by Manfred Eigen et al. Spencer's idea that the evolution of life is characterized by increasing complexity can be made precise in the context of dissipative self-organization. It is well known that Turing analyzed a mathematical model of organisms represented as complex cellular systems. Gerisch, Meinhardt, et al. described the growth of an organism (e.g., a slime mould) by evolution equations for the aggregation of cells. The nonlinear interactions of amebas cause the emergence of a macroscopic organism like a slime mould when some critical value of cellular nutrition in the environment is reached. The evolution of the order parameter corresponds to the aggregation forms during the phase transition of the macroscopic organism. The mature multicellular body can be interpreted as the "goal" or (better) "attractor" of organic growth. Multicellular bodies, like genetic systems, nervous systems, immune systems, and ecosystems, are examples of complex dynamical systems, which are composed of a network of many interacting elements. S. Kauffman suggested studying random Boolean networks that could be programmed in a computer. In computer experiments, he found a hierarchy of dynamical behavior with fixed points and cycles of increasing complexity, which can be observed in real cells.

Even the ecological growth of biological populations may be simulated using the concepts of synergetics. Ecological systems are complex dissipative systems of plants or animals with mutual nonlinear metabolic interactions with each other and with their environment. The symbiosis of two populations with their source of nutrition can be described by three coupled differential equations which were already used by Edward Lorenz to describe the development of weather in meteorology. In the 19th century the Italian mathematicians Lotka und Volterra described the development of two populations in ecological competition. The nonlinear interactions of the two complex populations are determined by two coupled differential equations of prey and predator species. The evolution of the coupled systems have stationary points of equilibrium. The attractors of evolution are periodic oscillations (limit cycles).

The theory of complex systems allows us to analyze the nonlinear causality of ecological systems in nature. Since the industrial revolution human society has become more and more involved in the ecological cycles of nature. But the complex balance of natural equilibria is highly endangered by the linear mode of traditional industrial production. People assumed that nature contains endless sources of energy, water, air, etc., which can be used without disturbing the natural balance. Industry produces an endless flood of goods without considering their synergetic effects like the ozone hole or waste utilization. The evolution of life is transformed into the evolution of human society.

From a methodological point of view, the applicability of power laws to biological systems indicates that such systems are highly complex. Some examples of biological power-law equations include those that relate physiological variables associated with the metabolic rate or life expectancy of an organism to its body mass. Although not all power-law relationships are result from fractals, the existence of such a relationship should inspire us to test the self-similarities of cellular subsystems at different scales. Many cellular organs, such as the lungs, with their branching trees of vessels, are at least statistically self-similar. The concept of fractals and self-similarity has not only been used to describe biomedical phenomena, but it has also prompted a new approach to health by clinicians. The new idea that has emerged from nonlinear dynamics, scaling and power laws in biology is that health is homeodynamic; in other words, there are a constellation of states that determine health. A healthy physiological system has a certain amount of intrinsic variability and no fixed state.

Perhaps the most speculative interdisciplinary application of complex systems is discussed in the fourth chapter, *Complex Systems and the Evolution of Mind-Brain*. In the history of philosophy and science there have been many different suggestions for solutions to the mind-body problem. Materialistic philosophers like Democritus, Lamettrie, et al., proposed to reduce mind to atomic interactions. Idealists like Plato, Penrose, et al. emphasized that mind is completely independent of matter and brain. For Descartes, Eccles, et al. mind and matter are separate substances interacting with each other. Leibniz believed in a metaphysical parallelism of mind and matter because they cannot interact physically. According to Leibniz mind and matter are supposed to exist in "pre-established harmony" like two synchronized clocks. Modern philosophers of mind like Searle defend a kind of evolutionary naturalism. Searle argues that mind is characterized by intentional mental states which are intrinsic features of the human brain's biochemistry and which therefore cannot be simulated by computers.

But the theory of complex systems cannot be reduced to these more or less one-sided positions. The complex system approach is an interdisciplinary methodology to deal with nonlinear complex systems like the cellular organ known as the brain. The emergence of mental states (for instance pattern recognition, feelings, thoughts) is explained by the evolution of (macroscopic) order parameters of cerebral assemblies which are caused by nonlinear (microscopic) interactions of neural cells in learning strategies far from thermal equilibrium. Cell assemblies with mental states are interpreted as attractors (fixed points, periodic, quasi-periodic, or chaotic) of phase transitions.

If the brain is regarded as a complex system of neural cells, then its dynamics is assumed to be described by the nonlinear mathematics of neural networks. Pattern recognition, for instance, is interpreted as a kind of phase transition by analogy with the evolution equations which are used for pattern emergence in physics, chemistry, and biology. Philosophically, we get an interdisciplinary research program that should allow us to explain neurocomputational self-organization as a natural consequence of physical, chemical, and neurobiological evolution by common principles. As in the case of pattern formation, a specific pattern of recognition (for instance a prototype face) is described by order parameters to which a specific set of features belongs. Once some of the features which belong to the order parameter are given (for instance a part of a face), the order parameter will complement these with the other features so that the whole system acts as an associative memory (for instance the reconstruction of a stored prototype face from an initially given part of that face). According to Haken's slaving principle the features of a recognized pattern correspond to the enslaved subsystems during pattern formation.

But what about the emergence of consciousness, self-consciousness, and intentionality? In synergetics we have to distinguish between external and internal states of the brain. In external states of perception and recognition, order parameters correspond to neural cell assemblies representing patterns of the external world. Internal states of the brain are nothing other than self-referential states, i.e., mental states referring to mental states and not to external states of the world. In the traditional language of philosophy we say that humans are able to reflect on themselves (self-reflection) and to refer external situations of the world to their own internal state of feeling and intentions (intentionality). In recent neurobiological inquiries, scientists speculate that the emergence of consciousness and self-consciousness depends on a critical value of the production rate for "meta-cell-assemblies", i.e., cell-assemblies representing cell-assemblies which again represent cell-assemblies, etc., as neural realization of self-reflection. But this hypothesis (if successful) could only explain the structure of emergent features like consciousness. Of course, mathematical evolution equations of cell assemblies do not enable us to feel like our neighbour. In this sense – this is the negative message – science is blind. But otherwise – this is the positive message – personal subjectivity is saved: Calculations and computer-assisted simulations of nonlinear dynamics are limited in principle.

Anyway, the complex system approach solves an old metaphysical puzzle which was described by Leibniz in the following picture: If we imagine the brain as a big machine which we may enter like the internal machinery of a mill, we shall only find its single parts like the cog wheels of the mill and never the mind, not to mention the human soul. Of course, on the microscopic level we can only describe the development of neurons as cerebral parts of the brain. But, on the macroscopic level, the nonlinear interactions in the complex neural system cause the emergence of cell assemblies referring to order parameters which cannot be identified with the states of single cerebral cells. The whole is not the sum of its parts.

Today, we can distinguish several degrees of complexity in the CNS. Scales at the levels of molecules, membranes, synapses, neurons, nuclei, circuits, networks, layers, maps, sensory systems, and the entire nervous system are considered. Each stratum can be characterized by some order parameters that determine its particular structures, which are caused by complex interactions of subelements with respect to the particular level of hierarchy. The brain of an organism observes, maps, and monitors not only the external world, but also the internal states of the organism, especially its emotional states. To "feel" is to have an awareness of one's own emotional states. In neuromedicine, the "Theory of Mind" (ToM) even analyzes the neural correlates of social feeling, which are situated in special areas of the neocortex. From a neuropsychological point of view, the old philosophical problem of "qualia" is also solvable. Qualia are properties that are consciously experienced by a person. We can explain the dynamics of subjective feelings and experiences, but, of course, the actual feeling is an individual experience.

It is obvious that the complex system approach delivers solutions of the mind-body problem which are beyond the traditional philosophical answers of idealism, materialism, physicalism, dualism, interactionism, etc. Concerning the distinction between so-called natural and artificial intelligence, it is important to see that the principles of nonlinear complex systems do not depend on the biochemistry of the human brain.The human brain is a "natural" model of these principles in the sense that the cerebral complex system is a product of physical and biological evolution. But other ("artificial") models produced by human technology are possible, although there will be technical and ethical limits to their realization.

In Chap. 5 we discuss *Complex Systems and the Evolution of Computability.* Universal Turing machines and algorithmic complexity are the traditional concepts of computability. Computational systems are also described as information processing machines in information theory. The degree of complexity of information entropy is classified by performing a Fourier analysis of a time series in signal theory. $1/f$ spectra for signals with a frequency f are typical of processes that organize themselves to a critical state at which many small interactions can trigger the emergence of a new, unpredicted phenomenon. Earthquakes, atmospheric turbulence, stock market fluctuations, and physiological processes of organisms are typical examples. Self-organization, emergence, chaos, fractality, and self-similarity are features of complex systems that exhibit nonlinear dynamics. The fact that $1/f$ spectra are measures of stochastic noise again emphasizes the deep relationship between information theory and systems theory: any complex system can be considered to be an information processing system.

In complex systems, the behavior of single elements is often completely unknown. In this case, the degrees of complexity of stochastic processes must be distinguished. The outcomes of a stochastic process (e.g., coin tossing) are distributed with different probabilities which are characterized by different probability distribution functions. A well-known example is the bell-shaped Gaussian curve of the normal distribution. This is, however, only one example of a probabilistic attractor in the functional space of probability density functions. The set of probability density functions that fulfill the requirements of the central limit theorem with independence and finite variance of random variables constitutes the basin of attraction for the Gaussian distribution. Probabilistic attractors classify the functional space of probability density functions into regions with different complexities. Distribution curves with fluctuating tails are typical of power laws, indicating highly complex stochastic behavior. Power-law distributions are used to describe open systems. They have become increasingly important when describing, for example, complex economic and physiological systems. Turbulence in complex financial markets is also characterized by power-law distributions with wildly fluctuating tails.

Although the computability of a deterministic system is limited by the degree of algorithmic complexity (see Sect. 5.2), and the computability of a stochastic system is limited by probabilistic measures (Sects. 5.3–5.4), dynamical systems can still be considered to be computational systems that sometimes cannot deliver results in a reasonable time. The old vision of Leibniz, that the world is a gigantic computer, still holds true. From a modern physical point of view, quantum systems are the fundamental dynamical systems of nature. Quantum mechanics delivers a framework for new computational systems – quantum computers. Quantum computers open up new avenues of information processing, computation, and communication. An essential feature of the quantum world is the superposition of quantum states and the possibility of entangled states. However, quantum computing does not only imply an exponential growth in computational capacity and a reduction in computational complexity. All matter stores quantum information. Therefore, any elementary particle is a processor of quantum information. The universe is an expanding quantum computer producing quantum information.

Are there limitations to the analogies of computers with human mind and brain by Gödel's and Turing's results of incompleteness and undecidability? How can the human brain be understood as both an information processing machine and a knowledge-based system? John von Neumann's concept of celluar automata refined the idea of self-organizing cellular systems. Recent computer experiments by Stephen Wolfram have shown that all kinds of nonlinear dynamics, from fixed point attractors and oscillating behavior to chaos, can be simulated by cellular automata. Even randomness can be generated by appropriate cellular automata, though their local rules of cellular interaction may be very simple and well-known. Cellular automata deliver digital approximations of complex dynamical systems that are determined by continuous differential equations.

Computational dynamics opens up new avenues for *Complex Systems and the Evolution of Artificial Life and Intelligence* (Chap. 6). The natural evolution of life and intelligence has become an important paradigm for computational models. They are no longer restricted to symbolic knowledge representation and artificial intelligence (AI). Natural life and intelligence depends decisively on the evolution of organisms and brains. Therefore, embodied life and mind lead to embodied artificial intelligence and embodied artificial life of embodied robotics. Artificial life and neural networks have their roots in the universal methods of cellular automata. Self-organization and learning are the main features of neural networks that model intelligent systems. In synergetic computers, order parameter equations allow a new kind of (non-Hebbian) learning: a strategy to minimize the number of synapses. In contrast to neurocomputers of the spin-glass type (for instance Hopfield systems), the neurons in such systems are not threshold elements but instead perform simple algebraic manipulations like multiplication and addition. As well as deterministic homogeneous Hopfield networks, there are so-called Boltzmann machines, with have a stochastic network architecture of nondeterministic processor elements and a distributed knowledge representation which is described mathematically by an energy function. While Hopfield systems use a Hebbian learning strategy, Boltzmann machines favor a backpropagation strategy (Widrow–Hoff rule) with hidden neurons in a multilayered network.

In general it is the aim of a learning algorithm to diminish the information-theoretic measure of the discrepancy between the brain's internal model of the world and the real environment via self-organization. The recent revival of interest in the field of neural networks is mainly inspired by the successful technical applications of statistical mechanics and nonlinear dynamics to solid state physics, spin glass physics, chemical parallel computers, optical parallel computers, and – in the case of synergetic computers – to laser systems. Other reasons are the recent development of computing resources and the level of technology which make a computational treatment of nonlinear systems more and more feasible. Philosophically, traditional topics of epistemology like perception, imagination, and recognition may be discussed in the interdisciplinary framework of complex systems.

In electrical engineering, information theory, and computer science, the concept of cellular neural networks (CNN) is becoming an influential paradigm of complexity research, which has been realized in information and chip technology [1.6]. Analogic Cellular Computers are the technical response to the sensor revolution, mimicking the anatomy and physiology of sensory and processing organs. A CNN is a nonlinear analog circuit that processes signals in real time. Its architecture dates back to J. von Neumann's earlier paradigm of Cellular Automata (CA). Unlike conventional cellular automata, CNN host processors accept and generate analog signals in continuous time with real numbers as interaction values. The CNN universal chip is a technical realization of the CNN Universal Machine (CNN-UM), analogous to the Universal Turing machine of digital computers. It is a milestone in information technology because it is the first fully programmable, industrial-sized, brain-like, stored-program dynamic array computer. Further on, appropriate CNNs can simulate all kinds of pattern formation and pattern recognition, which have been analyzed in synergetics in the theory of nonlinear dynamics. Two great advantages of CNNs are their rigorous mathematical analysis and their technical realization. The dynamic complexity of cellular automata and their corresponding nonlinear dynamic systems can be characterized by a precise complexity index. An immense increase of computing speed, combined with significantly less electrical power in the first CNN chips, has led to the current intensive research activities on CNN since Chua and Yang's proposal in 1988.

An important application of the complex system approach is neurobionics and medicine. The human brain is not only a cerebral computer as a product of natural evolution, but a central organ of our body which needs medical treatment, healing, and curing. Neurosurgery, for instance, is a medical discipline responsible for maintaining the health of the biological medium of the human mind. The future well-being of the brain-mind entity is an essential aim of neurobionics. In recent years new diagnostic procedures and technical facilities for transplantation have been introduced which are based on new insights into the brain from complex dynamical systems. In this context a change of clinical treatment is unavoidable. Neural and psychic forms of illness can be interpreted as complex states in a nonlinear system of high sensitivity. Even medical treatments must consider the high sensitivity of this complex organ. Another more speculative aspect of the new technology is cyberspace. Perception, feeling, intuition, and fantasy as products of artificial neural networks? Virtual reality has become a keyword in modern philosophy of culture.

After the evolution of matter, life, mind-brain, and artificial intelligence, we consider the emergence of economic order in human societies. The seventh chapter is titled *Complex Systems and the Evolution of Economies*. From a qualitative point of view, Adam Smith's free market model can already be explained by self-organization. Smith underlined that good or bad intentions of individuals are not essential. In contrast to a centralized economical system, the equilibrium of supply and demand is not directed by a program-controlled central processor, but is the effect of an "invisible hand" (Smith), which is simply the nonlinear interaction of consumers and producers. It should be noted that Adam Smith's liberal ideas were conceived against a historical background of Newtonian physics. Like many physicists, economists believed in the exact computability of their (linear) models, and they suppressed the possibility of a "butterfly effect" that could lead to chaos and excluded long-range economic forecasts.

However, in order to describe the dynamics of an economy, it is necessary to have evolution equations for many economic quantities from perhaps thousands of sectors and millions of agents. Therefore, stochastic models are often preferred when modeling global trends. Phase transitions and bifurcations at critical points are crucial concepts for understanding the nonlinear dynamics of economies. One challenge of globalization is to model the dramatic dynamics of financial markets. Modern mathematical finance theory is still based on Louis Bachelier's assumption (1900) of an efficient market with a normal ("Gaussian") distribution of price changes and mild randomness. However, complex and global markets are actually turbulent, like the weather, following typical power laws of distribution. Stochastic processes with probabilistic attractors (see Sect. 5.4) lead to abrupt and discontinuous events (the "Noah effect") or long-term trends (the "Joseph effect"). Fractals and multifractals have been put forward as explanations for these stochastic processes. Recently, econophysics has become a fruitful research field in economic and financial sociodynamics.

After moving through matter, life, mind-brain, artificial intelligence, and economics, the book finishes in a Hegelian grand synthesis with the eighth chapter, *Complex Systems and the Evolution of Human Culture and Society*. In social sciences one usually distinguishes strictly between biological evolution and the history of human society. The reason is that the development of nations, markets, and cultures is assumed to be guided by the intentional behavior of humans, i.e., human decisions based on intentions, values, etc. From a microscopic viewpoint we may, of course, observe single individuals with their intentions, beliefs, etc. But from a macroscopic view the development of nations, markets, and cultures is not only the sum of its parts. Mono-causality in politics and history is, as we all know, a false and dangerous way of linear thinking. Synergetics seems to be a successful strategy to deal with complex systems even in the humanities. Obviously it is not necessary to reduce cultural history to biological evolution in order to apply synergetics interdisciplinarily. Contrary to any reductionistic kind of naturalism and physicalism we recognize the characteristic intentional features of human societies. Thus the complex system approach may be a method of bridging the gap between the natural sciences and the humanities that was criticized in Snow's famous "two cultures".

In the framework of complex systems the behaviour of human populations is explained by the evolution of (macroscopic) order parameters which is caused by nonlinear (microscopic) interactions of humans or human subgroups (states, institutions, etc.). Social or economic order is interpreted by attractors of phase transitions. Allen et al. analyze the growth of urban regions. From a microscopic point of view the evolution of populations in single urban regions is mathematically described by coupled differential equations with terms and functions referring to the capacity, economic production, etc., of each region. The macroscopic development of the whole system is illustrated by computer-assisted graphics with changing centers of industrialization, recreation, etc., caused by nonlinear interactions of individual urban regions (for instance advantages and disadvantages of far and near connections of transport, communication, etc.). An essential result of the synergetic model is that urban development cannot be explained by the free will of single persons. Although people of local regions are acting with their individual intentions, plans, etc., the tendency of the global development is the result of nonlinear interactions.

Another example of the interdisciplinary application of synergetics is Weidlich's model of migration. He distinguishes the micro-level of individual decisions and the macro-level of dynamical collective processes in a society. The probabilistic macro-processes with stochastic fluctuations are described by the master equation of human socioconfigurations. Each component of a socioconfiguration refers to a subpopulation with a characteristic vector of behavior. The macroscopic development of migration in a society could be illustrated by computer-assisted graphics with changing centers of mixtures, ghettos, wandering, and chaos which are caused by nonlinear interactions of social subpopulations. The differences between human and non-human complex systems are obvious in this model. On the microscopic level human migration is intentional (i.e., guided by considerations of utility) and nonlinear (i.e., dependent on individual and collective interactions). A main result of synergetics is again that the effects of national and international migration cannot be explained by the free will of single persons. I think migration is a very dramatic topic today, and demonstrates how dangerous linear and mono-causal thinking may be. It is not sufficient to have good intentions without considering the nonlinear effects of single decisions. Linear thinking and acting may provoke global chaos, although we locally act with the best intentions.

In a dramatic step, the complex systems approach has been expanded from neural networks to include global technical information networks like the World Wide Web. The information flow is accomplished through information packets with source and destination addresses. The dynamic of the Internet has essential analogies with CAs and CNNs. Computational and information networks have become technical superorganisms, evolving in a quasi-evolutionary process. The information flood in a more or less chaotic Internet is a challenge for intelligent information retrieval. We could use the analogies of the self-organizing and learning features of a living brain to find heuristic devices for managing the information flood of the Internet. Taking this further, we need personalized information systems that adapt automatically to electronic profiles of users. These are a challenge in all fields of practical appliances. Even tailored knowledge during e-learning can be packaged by personalized information systems according to the learning profile of the user.

But the complexity of global networking isn't confined to the Internet. Below the complexity of a PC, cheap, low power, and smart chip devices are distributed throughout the intelligent environments of our everyday world. Ubiquitous computing is a challenge of global networking by wireless media access, wide-bandwidth range, real-time capabilities for multimedia over standard networks, and data packet routing. Not only millions of PCs, but also billions of smart devices are interacting via the Internet. The overwhelming flow of data and information forces us to operate at the edge of chaos.

In general, economic information processes are very complex and demand nonlinear dissipative models. Recall the different attractors from economic cycles to financial chaos which can only be explained as synergetic effects by nonlinear interactions of consumers and producers, fiscal policy, stock market, unemployment, etc. Even in management possible complex models are discussed in order to support creativity and innovation by nonlinear cooperation at all levels of management and production. But experience shows that the rationality of human decision making is bounded. Human cognitive capabilities are overwhelmed by the complexity and randomness of the nonlinear systems they are forced to manage. The concept of bounded rationality, first introduced by Herbert Simon, was a reaction to the limitations of human knowledge, information, and time.

Evidently, there are some successful strategies to handle nonlinear complex systems. We shall discuss examples of applications in quantum physics, hydrodynamics, chemistry, and biology, as well as economics, sociology, neurology, and AI. What is the reason behind the successful applications in the natural sciences and humanities? The complex system approach is not reduced to special natural laws of physics, although its mathematical principles were discovered and at first successfully applied in physics (for instance to the laser). Thus, it is an interdisciplinary methodology to explain the emergence of certain macroscopic phenomena via the nonlinear interactions of microscopic elements in complex systems. Macroscopic phenomena may be forms of light waves, fluids, clouds, chemical waves, biomolecules, plants, animals, populations, markets, neural cell assemblies, traffic congestions in street networks or the Internet, which are characterized by order parameters (Table 1.1). Philosophically, it is important to see that order parameters are not reduced to the microscopic level of atoms, molecules, cells, organisms, etc., of complex systems. In some cases they are measurable quantities (for instance the field potential of a laser). In other cases they are qualitative properties (for instance geometrical forms of patterns). Nevertheless, order parameters are not mere theoretical concepts of mathematics without any reference to reality. Actually they represent properties of real macroscopic phenomena, such as field potentials, social or economic power, feelings or even thoughts. Who will deny that feelings and thoughts can change the world? If we can understand their nonlinear dynamics, it could even become possible to implement them in chips, such as CNNs.

Thus, the complex systems approach is not a metaphysical process ontology. Synergetic principles (among others) provide a heuristic scheme to construct models of nonlinear complex systems in the natural sciences and the humanities. If these models can be mathematized and their properties quantified, then we get empirical models which may or may not fit the data. The slaving principle shows another advantage. As it diminishes the high number of degrees of freedom in a complex system, synergetics is not only heuristic, mathematical, empirical and testable, but economical too. Namely, it satisfies the famous principle of Ockham's razor which tells us to cut away superfluous entities. Further on, nonlinear models may be implemented in nonlinear computer chips of high speed and miniaturized size. We can also prove basic principles of computational dynamics. But, because of chaos and randomness, understanding computational dynamics does not mean predicting and determining the future in all its details. The analysis of computational systems allows us to gain experience with nonlinear dynamics, as well as insights into and feelings about what is going on in the real world. But, as life is complex and random, we have to live it in order to experience it.

In this sense, our approach suggests that physical, social, and mental realities are nonlinear, complex, and computational. This essential result of a new episte-

Table 1.1. Interdisciplinary applications of nonlinear complex systems

DISCIPLINE	SYSTEM	ELEMENTS	DYNAMICS	ORDER PARAMETER
cosmology	universe	matter	cosmic dynamics	cosmic pattern formation (e.g., galactic structures)
quantum physics	quantum systems (e.g., laser)	atoms photons	quantum dynamics	quantum pattern formation (e.g., optical waves)
hydrodynamics	fluids	molecules	fluid dynamics	form of fluids
meteorology	weather climate	molecules	meteorological dynamics	pattern formation (e.g., clouds, hurricanes)
geology	lava	molecules	geological dynamics	pattern formation (e.g., segmentation)
chemistry	molecular systems	molecules	chemical reaction chemical dynamics	chemical pattern formation (e.g., dissipative structures)
materials science	smart materials nano system	macromolecules	macromolecular dynamics	macromolecular pattern formation (e.g., nano forms)
biology	genetic systems	biomolecules	genetic reaction	genetic pattern formation
	organisms	cells	organic growth	organic pattern formation
	populations	organisms	evolutionary dynamics	pattern formation of species
economics	economic systems	economic agents (e.g., consumer producer)	economic interaction (e.g., mechanisms of markets)	economic pattern formation (e.g., supply and demand)
sociology	societies	individuals, institutions etc.	social interaction historical dynamics	social pattern formation
neurology psychology	brain	neurons	neural rules learning algorithms information dynamics	neural pattern formation pattern recognition
computer science	cellular automata neural networks global networking (e.g., Internet, ubiquitous computing)	cellular processors	computational rules evolutionary algorithms learning algorithms information dynamics	pattern formation of computational networks

mology has severe consequences for our behavior. As we underlined, linear thinking may be dangerous in a nonlinear complex reality. Recall, as one example, the demand for a well-balanced complex system of ecology and economics. Our physicians and psychologists must learn to consider humans as complex nonlinear entities of mind and body. Linear thinking may fail to yield a successful diagnosis. Local, isolated, and "linear" therapies of medical treatment may cause negative synergetic effects. In politics and history, we must remember that mono-causality may lead to dogmatism, intolerance, and fanaticism. As the ecological, economic, and political problems of mankind have become global, complex, nonlinear, and random, the traditional concept of individual responsibility is questionable. We need new models of collective behavior depending on the different degrees of our individual faculties and insights. In short: The complex system approach demands new consequences in *epistemology* and *ethics*. Finally, it offers a chance to handle chaos and randomness in a nonlinear complex world and utilize the creative possibilities of synergetic effects.

2 Complex Systems and the Evolution of Matter

How can order arise from complex, irregular, and chaotic states of matter? In classical antiquity philosophers tried to take the complexity of natural phenomena back to first principles. Astronomers suggested mathematical models in order to reduce the irregular and complex planetary orbits as they are experienced to regular and simple movements of spheres. Simplicity was understood, still for Copernicus, as a feature of truth (Sect. 2.1). With Newton and Leibniz something new was added to the theory of kinetic models. The calculus allows scientists to compute the instaneous velocity of a body and to visualize it as the tangent vector of the body's trajectory. The velocity vector field has become one of the basic concepts in dynamical systems theory. The cosmic theories of Newton and Einstein have been described by dynamical models which are completely deterministic (Sect. 2.2).

But Poincaré discovered that those models may be non-computable in the long run (the many-body-problem). Even in a fully deterministic world, the assumption of a Laplacean demon which can calculate the universe in the long run was exposed as an illusionary fiction. Chaos can arise not only in heaven, but also in the quantum world (as quantum chaos) (Sect. 2.3). From a methodological point of view, nonlinearity is a necessary but not sufficient condition of chaos. It also allows the emergence of order. In the framework of modern physics, the emergence of the structural variety in the universe from elementary particles to stars and living organisms is modeled by phase transitions and symmetry breaking of equilibrium states (Sect. 2.4). In the present state of superstring theories and M-theory, we do not have a complete theory explaining the evolution of matter with its increasing complexity. The presocratic wondering that "there is something and not nothing" is not dissolved. But the theory of complex systems opens new avenues of pattern formation in the nano world with applications for self-constructing materials in materials science (Sect. 2.5). From a methodological point of view, the question arises, how can we detect attractors of pattern formation in an immense variety of measured data? Time series analysis, fractals, and multifractals are challenges in the current theory of complex systems. The chapter closes with a survey of the degrees of complexity of different attractors in nonlinear dynamics (Sect. 2.6).

2.1 Aristotle's Cosmos and the Logos of Heraclitus

Since the presocratics it has been a fundamental problem of natural philosophy to discover how order arises from complex, irregular, and chaotic states of matter [2.1]. What the presocratic philosophers did was to take the complexity of natural phenomena as it is experienced back to "first origins" (ἀρχή), "principles" or a certain order. Let us look at some examples. Thales of Miletus (625–545 B.C.), who is said to have proven the first geometric theorems, is also the first philosopher of nature to believe that only material primary causes could be the original causes of all things. Thales assumes water, or the wet, as the first cause. His argument points to the observation that nourishment and the seeds of all beings are wet and the natural substratum for wet things is water.

Anaximander (610–545 B.C.), who is characterized as Thales' student and companion, extends Thales' philosophy of nature. Why should water be the first cause of all this? It is only one of many forms of matter that exist in uninterrupted tensions and opposites: heat versus cold and wetness versus dryness ... Therefore Anaximander assumes that the "origin and first cause of the existing things" is a "boundlessly indeterminable" original matter (ἄπειρον) out of which the opposed forms of matter have arisen. Accordingly we have to imagine the "boundlessly indeterminable" as the primordial state in which matter was boundless, without opposites, and, therefore, everywhere of the same character. Consequently, it was an initial state of complete homogeneity and symmetry. The condition of symmetry is followed by symmetry breaking, from which the world arises with all its observable opposites and tensions:

> The everlasting generative matter split apart in the creation of our world and out of it a sphere of flame grew around the air surrounding the earth like the bark around a tree; then, when it tore apart and bunched up into definite circles, the sun, moon and stars took its place. [2.2]

The ensuing states of matter that Anaximander described in his cosmogeny were therefore by no means chaotic; instead they were determined by new partial orders. The fascination with Anaximander increases when one reads his early ideas of biological evolution. He assumes that the first human beings were born from sea animals whose young are quickly able to sustain themselves, as he had observed in the case of certain kinds of sharks. A century later searches were already being made for fossils of sea animals as evidence of the rise of humans from the sea. The third famous Milesian philosopher of nature is Anaximenes (†525 B.C.), who is thought to have been a companion of Anaximander. He regards change as the effect of the external forces of condensation and rarefaction. In his view, every form of matter can serve as basic. He chooses air (ἀέρα):

> And rarefied, it became fire; condensed, wind; then cloud; further, by still stronger condensation, water; then earth; then stones; but everything else originated by these. He, too, assumed eternal motion as the origin of transformation. – What contracts and condenses matter, he said is (the) cold; by contrast, what thins and slackens is (the) warm. [2.3]

Thus Anaximenes assumes external forces by which the various states of matter were produced out of a common original matter and were transformed into each other.

Heraclitus of Ephesus (ca. 500 B.C.), "the dark one", as he was often called, is of towering significance for our theme. His language is indeed esoteric, more phrophetic than soberly scientific, and full of penetrating metaphors. He took over from Anaximander the doctrine of struggle and the tension of opposites in nature. The original matter, the source of everything, is itself change and therefore is identified with fire:

> The ray of lightning (i.e., fire) guides the All. – This world order which is the same for all was created neither by one of the gods nor by one of the humans, but it was always, and is, and will be eternally living fire, glimmering and extinguishing according to measures. [2.4]

Heraclitus elaborated further on how all states of matter can be understood as distinguishable forms of the original matter, fire. In our time the physicist Werner Heisenberg declared:

> At this point we can interpose that in a certain way modern physics comes extraordinarily close to the teaching of Heraclitus. If one substitutes the word "fire", one can view Heraclitus' pronouncements almost word for word as an expression of our modern conception. Energy is indeed the material of which all the elementary particles, all atoms and therefore all things in general are made, and at the same time energy is also that which is moved ... Energy can be transformed into movement, heat, light and tension. Energy can be regarded as the cause of all changes in the world. [2.5]

To be sure, the material world consists of opposite conditions and tendencies which, nevertheless, are held in unity by hidden harmony: "What is opposite strives toward union, out of the diverse there arises the most beautiful harmony (*ἁρμονία*), and the struggle makes everything come about in this way." [2.6] The hidden harmony of opposites is thus Heraclitus' cosmic law, which he called "logos" (*λόγος*).

What happens when the struggle of opposites comes to an end? According to Heraclitus, then the world comes to a final state of absolute equilibrium. Parmenides of Elea (ca. 500 B.C.) described this state of matter, in which there are no longer changes and motions in (empty) spaces. Matter is then distributed everywhere equally (homogeneously) and without any preferred direction for possible motion (isotropically). It is noteworthy that infinity is thought to be imperfection and therefore a finite distribution of matter is assumed. In this way Parmenides arrived at the image of a world that represents a solid, finite, uniform material sphere without time, motion or change. The Eleatic philosophy of unchanging being was, indeed, intended as a critique of the Heraclitean philosophy of constant change, which is put aside as mere illusion of the senses. And the later historical impact of the Eleatic philosophy in Plato appears in his critique of the deceptive changes that take place in sensory perception in contrast to the true world of unchangeable being of the Ideas. But from the point of view of philosophy of nature, the world Parmenides described was not necessarily opposite to the teaching of Heraclitus; in his cosmogeny it can be understood entirely as a singular end state of the highest symmetry.

After water, air, and fire were designated as original elements, it was easy to conceive of them as raw materials of the world. Empedocles (492–430 B.C.) took that step and added earth as the fourth element to fire, water, and air. These elements are free to mix and bind in varying proportions, and to dissolve and separate. What, now, according to Empedocles, were the enduring principles behind the constant changes and movements of nature? First there were the four elements, which he thought arose from nature and chance (τύχη), not from any conscious intention. Changes were caused by reciprocal effects among these elements, that is, mixing and separation: "I shall proclaim to you another thing: there is no birth with any of the material things, neither there is an ending in ruinous death. There is only one thing: mixture and exchange of what is mixed" [2.7]. Two basic energies were responsible for these reciprocal effects among the elements; he called them "love" (φιλία) for attraction and "hatred" (νεῖκος) for repulsion. There is an analogy in the yin–yang dualism of Chinese philosophy. Empedocles taught a constant process of transformation, i.e., combination and separation of the elements, in which the elements were preserved. He did not imagine these transformation processes to be at all mechanical (as the later atomists did), but rather physiological, in that he carried over processes of metabolism in organisms to inanimate nature.

In his medical theories, equilibrium is understood to be a genuinely proportional relationship. Thus, health means a particular balance between the opposite components and illness arises as soon as one of them gets the upper hand. If we think of modern bacteriology with its understanding of the antibodies in the human body, then this view of Empedocles proves to be surprisingly apt.

Anaxagoras (499–426 B.C.) advocated what was in many regards a refinement of his predecessors' teaching. Like Empedocles he developed a mixing theory of matter. But he replaced Empedocles' four elements with an unlimited number of substances that were composed of seed particles (σπέρματα) or equal-sized particles (ὁμοιομερῆ (σώματα)). They were unlimited in their number and smallness, i.e., matter was assumed to be infinitely divisible. The idea of a granulated continuum comes forceably to mind. Anaxagoras also tried to explain mixtures of colors in this way, when he said that snow is, to a certain degree, black, although the whiteness predominates. Everything was contained in each thing, and there were only predominances in the mixing relationships.

More distinctly than some of his predecessors, Anaxagoras tried in his philosophy of nature to give physical explanations for the celestial appearances and motions that were described only kinematically in the mathematical astronomy of the Greeks. So in his cosmology he proceeded from a singular initial state: a homogeneous mixture of matter. An immaterial original power, which Anaxagoras called "spirit" (νοῦς), set this mixture into a whirling motion which brought about a separation of the various things depending on the speed of each of them. Earth clumped together in the middle of the vortex, while heavier pieces of stone were hurled outward and formed the stars. Their light was explained by the glow of their masses, which was attributed to their fast speed. Anaxagoras' vortex theory appears again in modern times with Descartes, and then in more refined form in the Kant–Laplace theory of the mechanical origin of the planetary system.

In modern natural sciences atomism has proved to be an extremely successful research program. In the history of philosophy the atomic theory of Democritus is often presented as a consequence of Heraclitus' philosophy of change and Parmenides' principle of unchanching being. The Democritean distinction between the "full" and the "empty", the smallest indestructable atoms (ἄτομος) and empty space, corresponded to Parmenides' distinction between "being" and "not-being". Heraclitean complexity and change was derived from distinguishable reconfigurations of the atoms. Empty space was supposed to be homogeneous and isotropic.

Atoms differ in their form (μορφή), their position (θέσις), and their diverse configurations (τάξις) in material combinations. The configuration of the atoms for the purpose of designation is compared with the sequence of letters in words, which has led to the presumption that atomistic ideas were developed only in cultures with phonetic alphabets. In fact, in China, where there was no phonetic alphabet but instead ideographic characters, the particle idea was unknown and a field-and-wave conception of the natural processes prevailed. The Democritean atoms move according to necessity (ἀνάγκη) in a constant whirl (δῖνος or δίνη). Here, by contrast with later Aristotelian notions, motion means only change of location in empty space. All phenomena, all becoming and perishing, result from combination (σύγκρισις) and separation (διάκρισις). Aggregate states of matter, such as gaseous, liquid, or solid, are explained by the atoms' differing densities and potentialities for motion. In view of today's crystallography, the Democritean idea that even atoms in solid bodies carry out oscillations in place is noteworthy.

Plato, in his dialogue *Timaeus*, introduced the first mathematical model of atomism. The changes, mixings, and separations on earth of which the pre-socratics had spoken were to be traced to unchangeable mathematical regularities. In Empedocles' four elements, namely fire, air, water and earth, a classification was at hand that was immediately accessible to experience. Theatetus made a complete determination of all the regular bodies that are possible in 3-dimensional (Euclidian) space: tetrahedra, octahedra, icosahedra, cubes and dodecahedra. Therefore what Plato proposed to do amounted to interpreting Empedocles' four elements with these geometric building blocks.

Plato consciously avoided the Democritean designation "atom" for his elements, since they can be decomposed into separate plan figures. Thus tetrahedra, octahedra and icosahedra have faces consisting of equilateral triangles which, when they are bisected, yield right-angled triangles with sidelenghts 1, 2 and $\sqrt{3}$, while the square faces of cubes, when bisected, yield right-angled triangles with side lengths 1, 1 and $\sqrt{2}$. A consequence is that "fluids" like water, air and fire can combine with each other whereas a solid made of earth building blocks, because of its different triangles, can only be converted into another solid.

Then Plato developed a kind of elementary particle physics in which the specific elements are transformed into each other and "reciprocal effects" can take place with the "elementary particles" (i.e., the corresponding component triangles) according to geometric laws. Transformation of the elements results, for example, from their being cut open along the edges. Plato made this possibility dependent on the acuteness of the angles of the solid. The more acute plane angles can cleave polyhedra which have a regular angle. Thus, in sequence, every tetrahedron, every cube, every octahedron, every icosahedron can, in each case, cleave the following polyhedron, but not the previous one or polyhedra of the same sort. The conclusion for the philosophy of nature is that fire can separate or dissolve all elements; earth, only air and water; air, only water.

Plato stated categorically that the elements are not all of the same size. For instance, in order to be able to explain that fire can cause water in solid form to change into water in liquid form, he maintained that in the liquid state the elements are smaller and more mobile, while in the solid state they are larger.

The escape from fire is called cooling and the state after eradicating fire, solidification. Fire and air can pass through the gaps in earth building blocks (cubes) without hindrance, without dissolution of the earth elements. Condensed air cannot be dissolved without destroying the element. Condensed air, namely, means an accumulation of octahedra under the best surface configurations possible. Even fire would not be able to penetrate into the necessarily remaining gaps, whose plane angles are smaller than those of all elements, without destroying the octahedra. In the case of water, only fire is capable of breaking the strongest condensation. The gaps between adjacent icosahedra form plane angles which do not admit penetration by either earth or air. Only fire (tetrahedra) can penetrate and dissolve the combination.

Indeed, Plato developed an internally consistent mathematical model by which various aggregate states and reciprocal effects of substances could be explained if one accepted his – albeit more or less arbitrary – initial conditions for interpretation of the elements. Naturally, a number of the consequences for the philosophy of nature are strange and ridiculous. And yet we have here the first attempt in the history of sciences to explain matter and its states by simple geometric laws. A high point up to now in this developement is modern elementary particle physics. Heisenberg made this observation about it: "… The elementary particles have the form Plato ascribed to them because it is the mathematically most beautiful and simplest form. Therefore the ultimate root of phenomena is not matter, but instead mathematical law, symmetry, mathematical form" [2.8]. In Antiquity and the Middle Ages Plato's mathematical atomism gained little support. The basic problem, for his successors, in his geometric theory of matter was already evident in the dialogue *Timaeus*. How are the functions of living organisms to be explained? The suggestion that certain corporeal forms are as they are in order to fulfill certain physiological purposes (e.g., the funnel shape of the gullet for assimilation of food) cannot, in any case, be derived from the theory of regular solids. In addition, the idea of explaining the changing and pulsating processes of life on the basis of the "rigid" and "dead" figures of geometry must have seemed thoroughly unnatural, speculative, and farfetched to the contemporaries of that time. Contemporaries of our time still have difficulties understanding the detour that today's scientific explanations take through complicated and abstract mathematical theories. This is where the Aristotelian philosophy of nature begins.

Aristotle formulated his concept of a balance or "equilibrium" in nature chiefly on the basis of the ways in which living organisms such as plants and animals function. The process and courses of life are known to us from everyday experience.

What is more obvious than to compare and explain the rest of the world, which is unknown and strange, with the familiar? According to Aristotle, the task of science is to explain the principles and functions of nature's complexity and changes. This was a criticism of those philosophers of nature who identified their principles with individual substances. The individual plant or the individual animal was not simply the sum of its material building blocks. Aristotle called the general, which made the individual being what it was, form (εἶδος). What was shaped by form was called matter (ὕλη). Yet form and matter did not exist in themselves, but were instead principles of nature derived by abstraction. Therefore Matter was also characterized as the potential (δύναμις) for being formed. Not until matter is formed does reality (ἐνέργεια) come into being.

The real living creatures that we observe undergo constant change. Here Heraclitus was right and Parmenides, for whom changes were illusory, was wrong. Changes are real. Yet according to Aristotle, Heraclitus was wrong in identifying changes with a particular substance (fire). Aristotle explained those changes by a third principle along with matter and form, namely, the lack of form (στέρησις), which was to be nullified by an adequate change. The young plant and the child are small, weak and immature. They grow because in accordance with their natural tendencies (form), they were meant to become big, strong, and mature. Therefore it was determined that movement (κίνησις) in general was change, transition from possibility to reality, "actualization of potential" (as people in the Middle Ages were to say). The task of physics was to investigate movement in nature in this comprehensive sense. Nature (φύσις) – in contrast to a work of art produced by man, or a technical tool – was understood to be everything that carried the principle of movement within itself. If the Aristotelian designations make us think, first of all, of the life processes of plants, animals, and people as they present themselves to us in everyday experience, these designations seem to us to be thoroughly plausible and apposite. Nature is not a stone quarry from which one can break loose individual pieces at will. Nature itself was imagined to be a rational organism whose movements were both necessary and purposeful. Aristotle distinguished two sorts of movement, namely quantitative change by increase or decrease in magnitude, qualitative change by alteration of characteristics, and spatial change by change of location. Aristotle designated four aspects of causality as the causes of changes. Why does a plant grow? It grows (1) because its material components make growth possible (*causa materialis*), (2) because its physiological functions determine growth (*causa formalis*), (3) because external circumstances (nutrients in the earth, water, sunlight, etc.) impel growth (*causa efficiens*), (4) because, in accordance with its final purpose, it is meant to ripen out into the perfect form (*causa finalis*).

Aristotle then employed these same principles, which are obviously derived from the life cycles of plants, animals, and humans, to explain matter in the narrower sense, that is, what was later called the inorganic part of nature. Here too Aristotle proceeded from immediate experience. What we meet with is not so and so many elements as isolated building blocks of nature. Instead we experience characteristics such as warmth and cold, wetness and dryness. Combination of these yield the following pairs of characteristics which determine the elements: warm–dry (fire), warm–wet (air), cold–wet (water), cold–dry (earth). Warm–cold and wet–dry are excluded as simultaneous conditions. Therefore there are only four elements. This derivation was later criticized as arbitrary, but it shows the Aristotelian method, namely to proceed not from abstract mathematical models, but instead directly from experience. Fire, air, water, and earth are contained more or less, more intensively or less intensively, in real bodies and they are involved in constant transformation. According to Aristotle, eliminating the coldness of water by means of warmth results in air, and eliminating the wetness of the air results in fire. The changes of nature are interpreted as maturational and transformational processes.

How could such a predominantly organic philosophy of nature deliver physical explanations for mathematical natural science, insofar as it was extant at that time? There were only two elementary spatial motions – those that proceeded in a straight line and those that proceeded in a circle. Therefore there had to be certain elements to which these elementary motions come naturally. The motions of the other bodies were determined by these elements and their natural motions, depending on which motion predominated with each of them. The most perfect motion was circular motion. It alone could go on without end, which was why it had to be assigned to the imperishable element. This was the fifth element (quintessence), which made up the unchangeable celestial spheres and the stars. The continual changes within the earthly (sublunar) world were to be set off from the unchangeable regularity of the celestial (superlunar) world. These transformational processes were associated with the four elements to which straight-line motion is peculiar, and specifically the straight-line motion toward the center of the world, toward which the heavy elements earth and water strive as their natural locus, and the straight-line motion toward the periphery of the lunar sphere, toward which the light elements strive upwards as their natural locus.

Among the natural motions there was also free fall. But Aristotle did not start out from isolated motions in idealized experimental situations as Galilei did. A falling body is observed in its complex environment without abstraction of frictional ("dissipating") forces. During its free fall a body is sinking in the medium of air like a stone in water. Thus, Aristotle imagines free fall as a hydrodynamical process and not as an acceleration in vacuum. He assumes a constant speed of falling v, which was directly proportional to the weight p of the body and inversely to the density d of the medium (e.g., air), thus in modern notation $v \sim p/d$. This equation of proportionality at the same time provided Aristotle with an argument against the void of atomists. In a vacuum with the density $d = 0$, all bodies would have to fall infinitely fast, which obviously did not happen.

A typical example for a (humanly) forced motion is throwing, which, again, is regarded in its complex environment of "dissipative" forces. According to Aristotle a nonliving body can move only as a result of a constant external cause of motion. Think of a cart on a bumpy road in Greece, which comes to a stop when the donkey (or the slave) stops pulling or pushing. But why does a stone keep moving when the hand throwing it lets go? According to Aristotle, there could be no action at a distance in empty space. Therefore, said Aristotle, the thrower imparts a movement to the continuous medium of the stone's surroundings, and this propels the stone farther. For the velocity v of a pulling or pushing motion, Aristotle asserted the proportionality $v \sim K/p$ with the applied force K. Of course, these are not mathematical equations relating measured quantities, but instead proportionalities of qualitative determinants, which first emerged in this algebraic notation in the peripatetic physics of the Middle Ages. Thus in Aristotelian dynamics, in contrast to Galilean–Newtonian dynamics, every (straight-line) change of position required a cause of motion (force). The medieval theory of impetus altered Aristotelian dynamics by attributing the cause of motion to an "impetus" within the thrown body, rather than to transmission by an external medium.

How did peripatetic dynamics explain the cosmic laws of heaven? The central symmetry of the cosmological model was based on the (unforced) circular motion of the spheres, which was considered natural for the "celestial" element, and on the theory of the natural locus in the centerpoint of the cosmos. Ptolemy was still to account for the position of the earth on the basis of the isotropy of the model and by a kind of syllogism of sufficient reason. Given complete equivalence of all directions, there was no reason why the earth should move in one direction or another.

It was Aristotle's teacher Plato who presented a centrally-symmetrical model with the earth in the center; around it the whole sky turns to the right around the celestial axis, which goes through the earth. Sun, Moon, and planets turn to the left on spheres that have different distances from the earth in the sequence Moon, Mercury, Venus, Sun, Mars, Jupiter, and Saturn. The outermost shell carries the sphere of the fixed stars. According to the Platonic–Pythagorean conception, the rotational periods are related to each other by whole numbers. There is a common multiple of all rotational times, at the end of which all the planets are exactly in the same place again. The motion of each one produces a sound, so that the tunes of the movements of the spheres jointly form a harmony of the spheres in the sense of a well-ordered musical scale. Geometry, arithmetic, and aesthetic symmetries of the cosmos ring through the universe in a harmonious music of the spheres. Soon this emphatically symmetrical model of the cosmos was called into question by exact observations. A difficult problem was presented by the irregular planetary orbits, especially their retrograde movements. The irregularities in the sky were disquieting, especially for philosophers in the Pythagorean tradition, who were accustomed to comprehending the heaven – in contrast to the earth – as the realm of eternal symmetry and harmony.

Plato posed a famous question in order to reduce the complexity of motions in the heaven: by means of what regular, ordered circular movements could the phenomena of the planets be "saved", i.e., kinematically explained? An exact model of the observed curves was achieved when Apollonius of Perga (ca. 210 B.C.) recommended that the common center of the spheres be given up. But the spherical form of planetary movement and the equal speed of the spheres were to be retained. According to this proposal, the planets rotate uniformly on spheres (epicycles), whose imagined centers move uniformly on great circles (deferents) around the centerpoint (the earth). By appropriately proportioning the speed and diameter of the two circular motions and by varying their directions of motion, it was possible to produce an unanticipated potential for curves, and these found partial application in astronomy from Kepler to Ptolemy also. The spherical symmetry of the individual models was therefore preserved, even if they no longer had a common center, but various different centers.

The following examples of the epicycle-deferent technique show what a multiplicity of apparent forms of motion can be created by appropriately combining uniform circular motions [2.9]. This makes the Platonic philosophy more comprehensible in its view that behind the changes in phenomena there are the eternal and unchangeable forms. In Fig. 2.1 an elliptical orbit is produced by combining a deferent motion and an epicycle motion. Figure 2.2 shows a closed cycloid. In this way, changing distances between planets and the earth can also be represented. In principle, even angular figures can be produced. When the epicycle diameter approaches the deferent diameter, an exact straight line results. Even regular triangles and rectangles can be produced by means of appropriate combinations of an epicycle motion and a deferent motion, if one changes the speed of the east-west motion of a planet that is moving on an epicycle with a west-east motion.

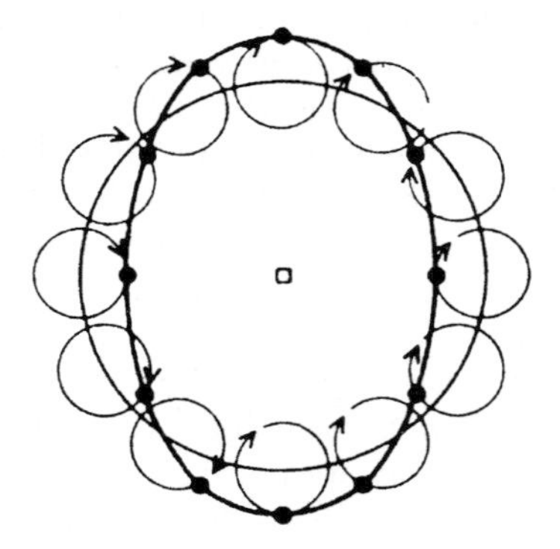

Fig. 2.1. Deferent-epicycle model of an ellipse

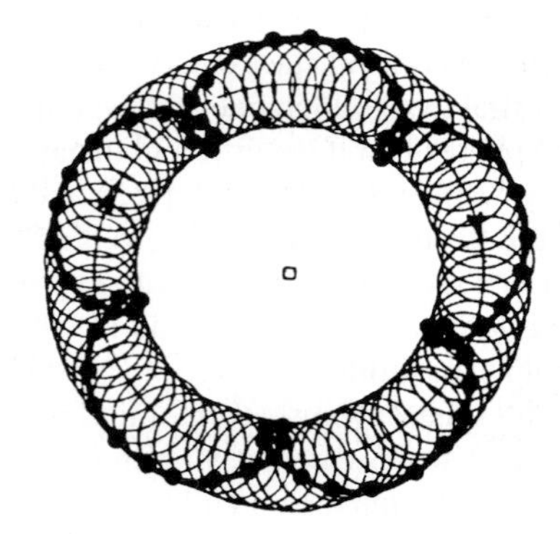

Fig. 2.2. Deferent-epicycle model of a cycloid

If one lets the celestial body circle on a second epicycle whose midpoint moves on the first epicycle, one can produce a multiplicity of elliptical orbits, reflection-symmetric curves, periodic curves, and also non-periodic and asymmetric curves. From a purely mathematical and kinetic standpoint, Plato's problem of "saving the phenomena" is completely solved. In principle, therefore, Plato's reduction of complexity in the sense of uniform circular motion (modified by Apollonius and Ptolemy) could influence the sciences right up until today. In any case, it cannot be disproved by phenomenological description of curved paths. In particular, from this standpoint not only the reversed roles of the earth and the sun in the so-called Copernican revolution, but also Kepler's change from circular to elliptical orbits, seem secondary, since both initiatives can be traced back to a combination of circular motions in accordance with the epicyle-deferent technique. This poses two questions: (1) How is the assertion mathematically substantiated? (2) If it is substantiated, why does it not play a role in modern scientific applications of curve theory? In order to answer the first question exactly and generally, it is necessary to go back to the modern structure of analytical geometry. But historically, Copernicus and Kepler also knew how the curves that they used (e.g., ellipses) could be reconstructed by the epicycle-deferent technique.

First of all, one must remember that points on the plane can be represented by complex numbers $z = x + iy = re^{i\theta}$ with the corresponding Cartesian coordinates (x, y) or polar coordinates (r, θ). The addition of complex numbers then corresponds to vector addition [2.10]. A uniform circular motion with center c, radius ϱ and period T can be represented by

$$z = c + \varrho e^{i((2\pi t/T)+\alpha)} = c + \varrho e^{(2\pi it/T)+i\alpha} \tag{2.1}$$

with time t and initial phase α for the point. Now assume a point A that is moving according to the equation $z = f(t)$. Let a point B move relative to A on a circle with radius ϱ, period T, and initial phase α. The motion of B is then described by the equation

$$z = f(t) + \varrho e^{(2\pi it/T)+i\alpha} \tag{2.2}$$

Then it is possible to describe the movement of a point B on an epicycle whose center moves around A. The addition of a new epicycle is described mathematically by the addition of a new term $\varrho e^{(2\pi it/T)+i\alpha}$ to the expression for z. Clearly, $\varrho e^{(2\pi it/T)+i\alpha} = \varrho e^{i\alpha} e^{(2\pi it/T)} = ae^{ikt}$ with a complex number $a \neq 0$ and k as a real number. In the case of a retrograde motion, T or k, respectively, is taken to be negative. A motion that results from the superposition of n epicycles is then expressed by the equation

$$z = a_1 e^{ik_1 t} + a_2 e^{ik_2 t} + \ldots + a_n e^{ik_n t} \tag{2.3}$$

Let us proceed first from a periodic motion on the plane $z = f(t)$ (e.g., with period 2π). Mathematically, we assume f continuous with limited variation. Then for f there exists a representation with a uniformly converging series

$$f(t) = \sum_{n=-\infty}^{\infty} c_n e^{int} \tag{2.4}$$

Therefore it can easily be proved mathematically that $f(t)$ can be approximated by means of sums

$$S_N(t) = \sum_{n=-N}^{N} c_n e^{int} \tag{2.5}$$

with any desired degree of exactitude for increasing N.
Function f is indeed uniformly convergent. Therefore for arbitrarily small $\varepsilon > 0$ one can choose an index N_0 so that for all $N \geq N_0$ and all t, it holds true that

$$|f(t) - S_N(t)| < \varepsilon \tag{2.6}$$

Astronomically, this result means that a constant-motion path (of limited variation) can be approximated to any desired degree of exactitude by means of finite superpositions of the epicycle motions.

Is it clear that so far we have used only superpositions with epicycle periods $\pm 2\pi$, $\pm\pi$, $\pm\frac{2}{3}\pi$, $\pm\frac{1}{2}\pi$, $\pm\frac{2}{5}\pi$, ... In particular, therefore, only commensurable superpositions were employed, which can be expressed by means of integer number ratios in accordance with the Pythagorean tradition. But in fact non-periodic curves can also be approximated by means of epicyclical superpositions if we permit incommensurable periods. This result is mathematically supported by a proposition by Harald Bohr about almost-periodic functions (1932) [2.11]. The second question, why the epicycle-deferent technique for the explanation of the paths of motion was abandoned, cannot be answered by pointing to the observation of missing curves. Mathematically, observed curves – however exotic – could still be explained in principle (under the above, very broad mathematical frame conditions) by means of the Platonic–Apollonian use of this ancient strategy for reducing the complexity of motions.

The decisive question in this case, however, is which motions the planets "really" carry out, whether they are, in fact, combined, uniform, and unforced circular motions that seem to us on earth to be elliptical paths, or whether they are in fact compelled to follow elliptical paths by forces. This determination cannot be made geometrically and kinematically, but only dynamically, i.e., by means of a corresponding theory of forces, hence by means of physics.

Besides the epicycle-deferent-technique, Ptolemy employed imaginary balance points relative to which uniform circular motions were assumed that, relative to the earth as center, appear non-uniform. This technique proved to be useful for calculation, but constituted a violation of the central symmetry and therefore had the effect of an ad hoc assumption that was not very convincing from the standpoint of philosophy of nature, a criticism later made especially by Copernicus. The reasons that Copernicus exchanged the earth for the position of the sun were predominantly kinematic. Namely, a certain kinematic simplification of the description could be achieved in that way with a greater symmetry. Thus in the heliocentric model the retrograde planetary motions could be interpreted as effects of the annual motion of the earth, which according to Copernicus moved more slowly than the outer planets Mars, Jupiter and Saturn and faster than the inner planets Mercury and Venus. But Copernicus remained thoroughly conservative as a philosopher of nature since he considered greater simplicity in the sense of "natural" circular motion to be a sign of proximity to reality.

With Johannes Kepler, the first great mathematician of modern astronomy, the belief in simplicity was likewise unbroken. In his *Mysterium cosmographicum* of 1596, Kepler began by trying once more to base distance in the planetary system on the regular solids, alternatingly inscribed and circumscribed by spheres. The planets Saturn, Jupiter, Mars, Earth, Venus, and Mercury correspond to six spheres fitted inside each other and separated in this sequence by a cube, a tetrahedron, a dodecahedron, an icosahedra, and an octahedron. Kepler's speculations could not, of course, be extended to accommodate the discovery of Uranus, Neptune, and Pluto in later centuries.

Yet Kepler was already too much of a natural scientist to lose himself for long in Platonic speculations. His *Astronomia Nova* of 1609 is a unique document for studying the step-by-step dissolution of the old Platonic concept of simplicity under the constant pressure of the results of precise measurement. In contrast to Copernicus, Kepler supplemented his kinematic investigations with original dynamic arguments. Here the sun is no longer regarded as being physically functionless at a kinematically eccentric point, as with Copernicus, but is seen as the dynamic cause for the motion of planets. The new task was to determine these forces mathematically as well. Kepler's dynamic interpretation with magnetic fields was only a (false) initial venture. Success came later, in the Newtonian theory of gravity.

The simplicity of the celestial ("superlunar") world and the complexity of the earthly ("sublunar") are also popular themes in other cultures. Let us cast a glance at the Taoist philosophy of nature of ancient China. To be sure, it is edged with myth and less logically argued than the Greek philosophy of nature, and it also invokes more intuition and empathy; nevertheless, there are parallels between the two. Taoism describes nature as a great organism governed by cyclical motions and rhythms, such as the life cycles of the generations, dynasties, and individuals from birth to death; the food chains consisting of plant, animal, and human; the alternation of the seasons; day and night; the rising and setting of the stars; etc. Everything is related to everything else. Rhythms follow upon each other like waves in the water. What forces are the ultimate cause of this pattern in nature? As with Empedocles, in Taoism two opposite forces are distinguished, namely *yin* and *yang*, whose rhythmic increase and decrease govern the world. In the book *Kuei Ku Tzu* (4th century B.C.) it says: "Yang returns cyclically to its origin. Yin reaches its maximum and makes way for yang." [2.12] While according to Aristotle all individuals carry their natural purposes and movements in themselves, the Tao of yin and yang determines the internal rhythms of individuals, and those energies always return to their origins. The cyclical rotational model of the Tao provides explanations for making calendars in astronomy, for water cycles in meteorology, for the food chain, and for the circulatory system in physiology. It draws its great persuasiveness from the rhythms of life in nature, which people experience every day and can apply in orienting themselves to life. Nature appears as a goal-directed organism.

It is noteworthy that the Chinese philosophy of nature had no notions of atomistic particles and therefore did not develop mathematical mechanics in the sense of the occidental Renaissance. Instead, at its center there was a harmonious model of nature with rhythmic waves and fields that cause everything to be connected to everything. The preference for questions of acoustics and the early preoccupation with magnetic and electrostatic effects is understandable given this philosophy of nature. The view of the Taoists bear more resemblance to the philosophy of nature of the Stoics than to Aristotle. Here too the discussion centers on effects that spread out in a great continuum like waves on water. This continuum is the Stoics' *pneuma*, whose tensions and vibrations are said to determine the various states of nature. The multifarious forms of nature are only transitory patterns that are formed by varied tensions of the pneuma. Modern thinking leaps, of course, to the patterns of standing water waves or sound waves or the patterns of magnetic fields. Nevertheless, neither the Stoic nor the Taoist heuristic background led to the development of a physical theory of acoustic or magnetic fields comparable to Galilean mechanics with its background of an atomistic philosophy of nature. The emergence of order from complex, irregular, and chaotic states of matter was only qualitatively described, using different models for earth and for heaven.

2.2 Newton's and Einstein's Universe and the Demon of Laplace

Since antiquity, astronomers and philosophers have believed that the celestial motions are governed by simple geometric laws. Simplicity was not only understood as the demand for an economical methodology, but, still for Copernicus, as a feature of truth. Thus, the astronomical doctrine from Plato to Copernicus proclaimed: reduce the apparent complexity of the celestial system to the simple scheme of some true motions! The simple building blocks were given by the basic concepts of Euclidean geometry: circle (compass) and straight line (ruler). In contrast to the simplicity of the superlunar world, the sublunar earthly world seemed to be really complex. Thus its dynamics could not be mathematized at least in the framework of Euclidean geometry. That was the reason why Plato's mathematical atomism was soon forgotten, and Aristotle's belief in a complex qualitative dynamics of nature which cannot be mathematized in principle overshadowed scientific research until the Renaissance.

Early physicists like Galileo overcame the boundary of a superlunar ("simple") and sublunar ("complex") world. They were convinced that the dynamics of nature is governed by the same simple mathematical laws in heaven and on earth. Technically, Galileo simplified the dynamics of, e.g., free fall by selecting some observable and measurable quantities and neglecting other constraints. In short, he made a simplified mathematical model of an idealized experimental situation. Of course, even the astronomical models of antiquity only considered a few parameters, such as angular velocity and position of the planets, and neglected the complex diversity of other constraints (e.g., density, mass, friction of the celestial spheres). From a modern point of view, even the presocratic philosophers suggested qualitative "models" of a complex dynamics in nature by selecting some dominant "parameters" (e.g., water, fire, air, and earth).

In general, a system which may be physical, biological, or social, is observed in different states. The strategies for making models of observed phenomena may have changed since ancient times, but the target of the modeling activity is in some sense the same: the dynamics of the changing states in the observed systems. Obviously, the real states cannot be described by only a few observable parameters, but it is assumed that they can. In the case of early astronomy and mechanics, this was the first step of mathematical idealization and led to a geometric model for the set of idealized states which is nowadays called the state space of the model. The presocratic "models" of nature differ from modern ones not only because of their mathematization and measurability, but also because the relationship between the actual states of a real system and the points of the geometric model was believed to be ontologically necessary, while in modern systems it is a fiction maintained for the sake of theory, prediction, and so on.

The simplest scheme is the one-parameter model. Early medical experience with mammals shows that the state of health or illness can be correlated with the parameter of temperature. Many animals correlate observable features with the emotional states of other animals: the ear attitude of a dog corresponds to its state of fear, while its fang exposure is a qualitative "parameter" for its degree of rage. A combination of both attitudes represents a more adequate characterization of the dog's emotional state. The state of a planet in medieval cosmology can be defined by its angular velocity and localization. States of other systems may need more than two features (e.g., temperature, blood pressure, and pulse rate for the healthy state of a mammal).

In any case, if these parameters are numerical, then the corresponding state spaces can be represented by geometric spaces. Thus the values of two numerical parameters may be represented by a single point in a two-dimensional state space visualized by the plane of Euclidean geometry. Changes in the actual state of the system are observed and may be represented as a curve in the state space. If each point of this curve carries a label recording the time of observation, then we get a trajectory of the model. Sometimes it is useful to introduce another coordinate of time and to represent the changing parameters of states by its time series. This kind of representation is called the graph of a trajectory.

The dynamical concepts of the Middle Ages included both kinds of representation. In the 1350s, the Parisian scholastic Nicole Oresme introduced the concept of graphical representations or geometrical configurations of intensities of qualities. Oresme mainly discussed the case of a linear quality whose extension is measured by an interval or line segment of space or time ("longitude of the quality"). He proposed to measure the intensity of the quality at each point of the interval by a perpendicular ordinate ("latitude of the quality") at that point. The quantity of a linear quality is visualized by the configuration of both parameters. In the case of a uniformly accelerated motion during a time interval corresponding to the longitude AB in Fig. 2.3, the latitude at each point P of AB is an ordinate PQ whose length is the velocity at the corresponding instant [2.13]. The straight line DC of the configuration is the graph of a trajectory representing the state of velocity. The so-called Merton Rule is immediately derived with a geometrical verification of Fig. 2.3: namely, it follows from the formula for the area of the trapezoid in Fig. 2.3 that the total distance traveled is $s = \frac{1}{2}(v_0 + v_f)t$.

Perhaps this interpretation was found on the basis of regarding this area as made up of very many vertical segments ("indivisibles"), each representing a velocity con-

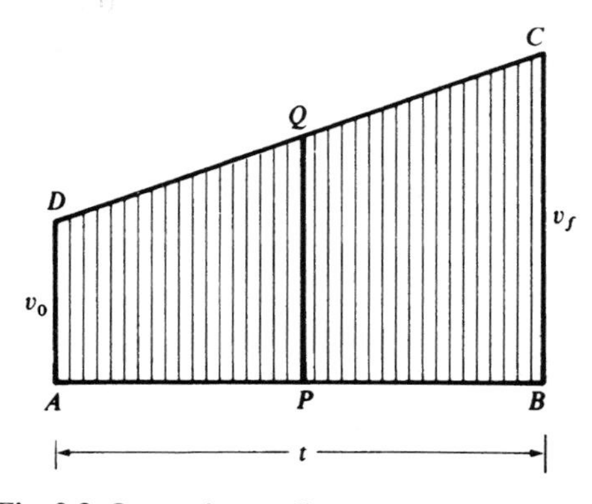

Fig. 2.3. Oresme's coordinates of a linear quality

tinued for a very short ("infinitesimal") time. The Merton Rule shows that even in the very early beginning of state space approaches a good geometric representation is not only a useful visualization, but gives new insight into the concepts of dynamics. Of course, Oresme and the Merton scholars at first only wanted to mathematize an Aristotelean-like physics of qualities. But their work was widely disseminated in Europe and led to the work of Galileo. In his famous *Discorsi* (1638), he introduced the basic concepts of modern mechanics and proceeded to the well-known distance formula $s = \frac{1}{2}gt^2$ for uniformly accelerated motion from rest (free fall) with a proof and an accompanying geometric diagram that are similar to Oresme's ideas.

With Newton and Leibniz, something new was added to the theory of dynamical systems. The calculus allows one to compute the instantaneous velocity as the derivative of a velocity function and to visualize it as the tangent vector of the corresponding curve (Fig. 2.4a). The velocity vector field has become one of the basic concepts in dynamical systems theory (Fig. 2.4b). Trajectories determine velocity

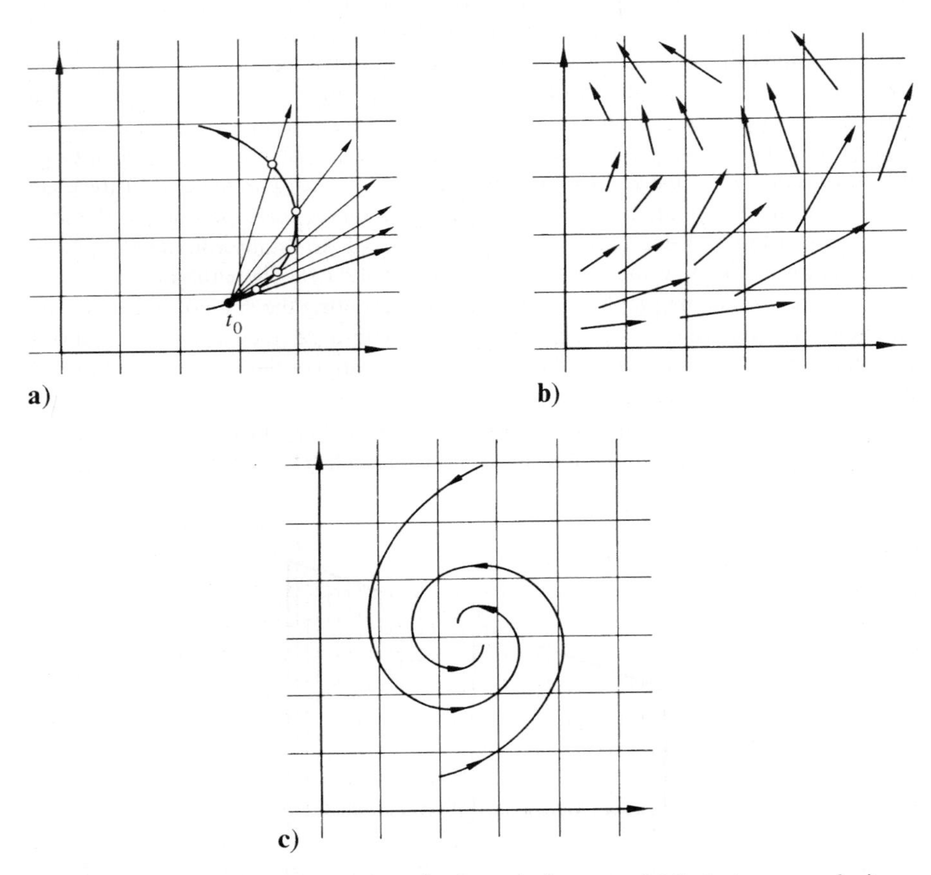

Fig. 2.4a–c. Geometric representation of a dynamical system: (**a**) Instantaneous velocity as tangent vector, (**b**) velocity vector field, (**c**) phase portrait

vectors by the differentiation procedure of the calculus. Conversely, the integration procedure of the calculus allows one to determine trajectories from velocity vectors.

The strategy of modeling a dynamical system begins with the choice of a state space in which observations can be represented by several parameters. Continuing the observations leads to many trajectories within the state space. In the sense of Newton's and Leibniz' calculus, a velocity vector may be derived at any point of these curves, in order to describe their inherent dynamical tendency at any point. A velocity vector field is defined by prescribing a velocity vector at each point in the state space. The state space filled with trajectories is called the "phase portrait" of the dynamical system (Fig. 2.4c). This basic concept of dynamical system theory was originally introduced by Henri Poincaré. The velocity vector field was derived from the phase portrait by differentiation [2.14].

Of course, the velocity vector field visualizes the dynamics of the particular system being modeled. Actually, extensive observations over a long period of time are necessary to reveal the dynamical tendencies of the system which is represented by the corresponding velocity vector field. The modeling procedure is only adequate if we assume that (a) the velocity vector of an observed trajectory is at each point exactly the same as the vector specified by the dynamical system and (b) the vector field of the model is smooth. The word "smooth" means intuitively that there are no jumps and no sharp corners. In the case of a one-dimensional state space, the vector field is specified by a graph in the plane. Thus, the graph is smooth if it is continuous and its derivative is continuous as well. Historically, condition (b) corresponds to Leibniz' famous principle of continuity playing a dominating role in the framework of classical physics.

In general, we summarize the modeling process as follows. A dynamical model is prepared for some experimental situation. We may imagine the laboratory devices of physicists like Galileo and Newton or biologists observing some organisms or even sociologists working on some social groups. The dynamical model consists of the state space and a vector field. The state space is a geometrical space (e.g., the Euclidean plane or in general a topological manifold) of the experimental situation. The vector field represents the habitual tendencies of the changing states and is called the dynamics of the model. How can we find the trajectories, thus the behaviour of the system? Technically, this problem is solved by creating the phase portrait of the system. That means we have to construct the trajectories of the dynamical system. Given a state space and a ("smooth") vector field, a curve in the state space is a trajectory of the dynamical system if its velocity vector agrees with the vector field in the sense of tangent vectors (Fig. 2.5). The point corresponding to time zero is called the initial state of the trajectory. These trajectories are supposed to describe the behaviour of systems as observed over an interval of time. Moreover, physicists have been ambitious enough to aim at making predictions indefinitily far into the future and to calculate the course of nature as if it were a huge clock.

Let us have a short glance at Newton's cosmos, which seemed to be a successful application of dynamical system theory evolving by the mathematical tools of Newton, Leibniz, Euler, etc. Newton gave three laws governing the behavior of material bodies. The first law ("lex inertiae") says that a body continues to move uniformly

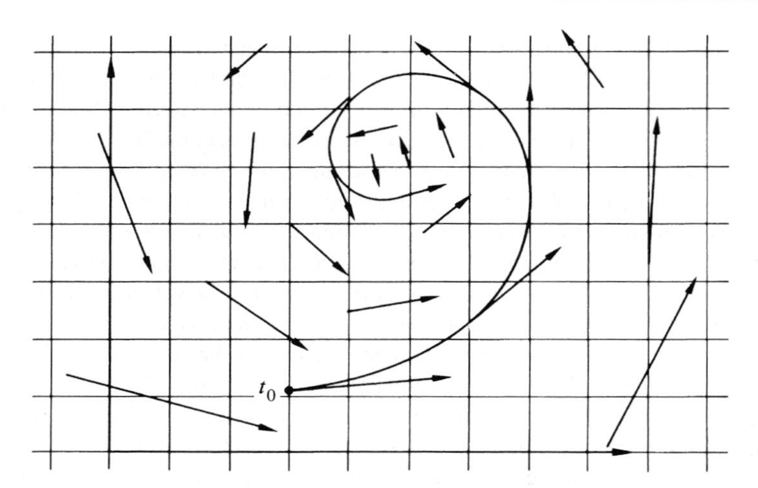

Fig. 2.5. Trajectory of a dynamical system in a vector field

in a straight line if no force acts on it. If a force does act on it, then its mass times its acceleration is equal to that force (second law). The basic framework is completed by a third law: to every action there is always opposed an equal reaction. The Newtonian cosmos consists of particles moving around in a space which obeys the laws of Euclidean geometry. The accelerations of these particles are determined by the forces acting upon them. The force on each particle is obtained by adding together all the forces of other particles in the sense of the vector addition law. If the force is a gravitational one, then it acts attractively between two bodies and its strength is proportional to the product of the two masses and the inverse square of the distance between them. But, of course, there may be other types of forces.

Actually, Newton's second law was understood as a universal scheme for all forces of nature in the macrocosmos and microcosmos. With a specific law of force the Newtonian scheme translates into a precise system of dynamical equations. If the positions, velocities, and masses of the various particles are known at one time, then their positions and velocities are mathematically determined for all later times. In short, the state of a body in Newton's cosmos is specified by the two parameters of position and velocity. The Newtonian trajectories are determined by the dynamical equations of motion. If the initial states were known, then the behavior of Newton's cosmos seemed to be determined completely. This form of determinism had a great influence on the philosophy of the 18th and 19th centuries. Newton's dynamics was understood as basic science for modeling nature. But, of course, the mechanistic models are valid only in the limiting case of vanishing friction and are never fully achieved experimentally. Nature is so complex that physicists preferred to observe unnatural ("artificial") limiting cases. Later on we shall see that the physicists' belief in simple laws completely neglected the complexity of initial conditions and constraints and, thus, created an illusory model of a deterministic as well as computable nature.

According to Newton, there is only one real world of matter in one absolute framework of space-time, in which we may choose relative reference systems. This means that for any two events it is regarded as objectively decidable whether they are simultaneous and also whether they occur at the same place. Mathematically, Newton's absolute space is represented by a 3-dimensional Euclidean space the metric of which is measurable by means of rulers, while time is taken to be a 1-dimensional Euclidean space with coordinate t which is measured by standard clocks.

Because of its absolute simultaneity the Newtonian 4-dimensional space-time is stratified by maximal subsets of simultaneous events. Each stratum is a possible 3-dimensional hyperplane $t = t(e)$ of an event e which separates its causal future, with strata $t > t(e)$, from its causal past, with strata $t < t(e)$. In Fig. 2.6a the third spatial coordinate is neglected, in order to visualize each stratum as 2-dimensional plane. This causal structure includes the Newtonian assumption that there are arbitrarily fast signals by means of instantaneous action at a distance [2.15].

Newton's relative spaces are made precise by Lange as inertial systems designating reference systems for a force-free body moving in a straight line with a steady velocity. It is not stipulated which of the many possible inertial systems is used. Particular transformations (Galilean transformations) from one inertial system to another give the corresponding coordinates. Mechanical laws are preserved (invariant) with respect to these transformations. As every Galilean transformation has ten continuous parameters (one parameter for time and three times three parameters for rotation, steady velocity and translation), we can derive ten laws of conservation. Thus, e.g., the Galilean invariance of the time coordinate implicates the law of conservation of energy. Reference systems which are not inertial systems have typical effects. A disk rotating relative to the fixed stars has radial forces which cannot be eliminated by Galilean transformations. In short, in Newtonian space-time, uniform motions are considered as absolutely preferred over accelerated motions. Its structure is defined by the group of Galilean transformations.

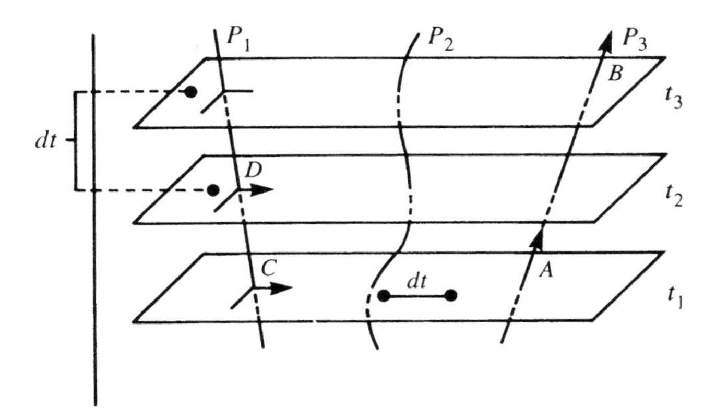

Fig. 2.6a. Newtonian space-time model with spatial strata of simultaneous events and trajectories of uniform inertial movements (*straight lines*) and accelerations (*curve*)

At the beginning of this century, Einstein proved that Newton's model of space-time is restricted to mechanical motions with slow speed relative to the speed c of light. The constancy of c independently of any moving reference system is a fact of Maxwell's electrodynamics. Thus, Newton's addition law of velocities and Galilean invariance cannot hold true in electrodynamics. In his theory of special relativity (1905), Einstein assumed the constancy of the speed of light and the invariance of physical laws with respect to all inertial systems ("principle of special relativity") and derived a common framework of space-time for electrodynamics and mechanics. Einstein's special relativistic space-time was modeled by Minkowski's four-dimensional geometry. The four-dimensionality should not surprise us, because Newton's space-time has three (Cartesian) space and one time coordinate, too.

For the sake of simplicity, the units are chosen in a way that the speed of light is equal to one, and, thus, the units of length and time can be exchanged. Each point in this space-time represents an event, which means a point in space at a single moment. As a particle persists in time, it is not represented by a point, but by a line which is called the world-line of the particle. In order to visualize the Minkowskian model, we depict a space-time system with a standard time coordinate, measured in the vertical direction, and two space coordinates, measured in the horizontal direction (Fig. 2.6b) [2.16].

Uniformly moving particles are represented by straight lines, accelerated particles by curved lines. As particles of light (photons) uniformly travel with the fundamental speed c, their world-lines are straight lines at an angle of 45° to the vertical. They form a light cone centred at the common origin 0. The system of light cones

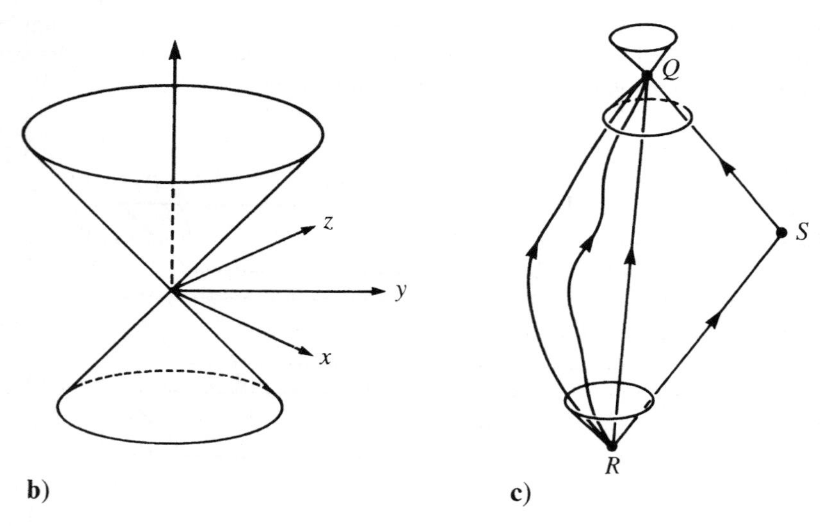

Fig. 2.6b,c. Minkowskian space-time cone in special relativity (**b**), and the twin brother paradox of special relativity: the Minkowskian distance RQ is greater than the length of RS and SQ together (**c**)

at all space-time points is regarded as the Minkowskian model of relativistic space-time.

Whereas the world-line of a photon is always along the light cone at each point, the world-line of any accelerated or uniformly moved material particle with a speed slower than c must always be inside the light cone at each point. As material particles or photons cannot travel faster than light, only the world-lines along and inside the light cone are determined physically. An event is called later than 0, if it is in the future cone above 0; it is called earlier than 0, if it is in the past cone below 0. Thus, the light cones determine the causal structure of relativistic space-time.

An essential difference between the Minkowskian model and ordinary Euclidian representations is the fact that the length of world-lines is interpreted as the time measured by physical clocks. Thus, time measurement becomes path-dependent, contrary to Newton's assumption of an absolute time. The so-called "twin paradox" visualizes this effect dramatically. In Fig. 2.6c, one twin brother remains on the earth R moving uniformly and very slowly, while the other makes a journey to a nearby star S at great speed, nearly that of light. Minkowskian geometry forecasts that the travelling brother is still young upon his return at Q, while the stay-at-home brother is an old man. This is not science fiction, of course, but a consequence of the time-measuring length of Minkowskian world-lines: the Minkowskian distance RQ is greater than the length of the distance RS and SQ together, contrary to the usual Euclidean interpretation. Today, these effects are experimentally well confirmed for elementary particles at high speeds near c.

In the framework of Minkowskian space-time, the invariance of physical laws with respect to particular inertial systems is realized by the Lorentz transformation. Newtonian space-time with Galilean invariance remains a limiting case for reference systems like celestial motions of planets or earthy motions of billiard balls with slow speed relative to the constant c. In this sense, Einstein's space-time is the culmination of classical physics rather than a revolutionary break with Newton.

An important concept which was first introduced into classical physics by Leibniz is energy, consisting of the kinetic energy T and the potential energy U of a system. The mechanical work done on a point mass which is displaced from a position 1 to a position 2 corresponds to the difference between the kinetic energy at position 1 and that of position 2. If this mechanical work is independent of the path followed from 1 to 2, then the corresponding force field is called conservative. Frictional forces are not conservative. In one dimension all forces must be conservative, since there is a unique path from one point to another point in a straight line, ignoring friction. The total energy $T + U$ is constant in a conservative field of force.

An important application of Newton's mechanics is the harmonic oscillator, such as the small amplitude pendulum, or the weight oscillating up and down on a spring. The harmonic oscillator appears as a model through all parts of physics and even chemistry and biology. For example, remember electromagnetic light waves, where the electric and magnetic field energies oscillate. Harmonic oscillations are also well known in technology, for example as oscillating electrical currents in a coil and a condenser, with friction corresponding to the electrical resistance. In the philosophy of the 18th and 19th centuries the pendulum was a symbol of the mechanistic universe which seemed to be completely determined and calculable by the Newtonian equations of motion.

Thus, the pendulum may be considered as a classical example of the dynamical modeling procedure. This model assumes that the rod is very light, but rigid. The hinge at the top is perfectly frictionless. The weight at the lower end is heavy, but very small. The force of gravity always pulls it straight down. In Fig. 2.7a, the pendulum is drawn in a two-dimensional Euclidean plane with the angle α of elevation, the force F of gravity, the pull $F \cos \alpha$ along the rod, and the force $F \sin \alpha$ turning it. In order to visualize the dynamical behavior of the pendulum we have to develop a dynamical model with a state space and a phase portrait. The state of the pendulum is fully determined by the angular variable α (with $\alpha = 0$ and $\alpha = 2\pi$ denoting the same angle) and the angular velocity v. Thus, we get a two-dimensional state space which can be visualized by the circular cylinder in Fig. 2.7b. The vertical circle in the center of this cylinder denotes the states of zero angular velocity $v = 0$. The straight line from front to back, at the bottom of the cylinder, is the axis of zero inclination with $\alpha = 0$, where the pendulum is lowest. At the origin with $(\alpha, v) = (0, 0)$, the pendulum is at rest at its lowest position [2.17].

As there is no friction and no air in the way, moving the pendulum a little to the left causes it to swing back and forth indefinitely. The full trajectory in the state space, corresponding to this oscillating motion, is a cycle, or closed loop. In the next case, the pendulum is balanced at the top, in unstable equilibrium. A tiny touch on the left causes it to fall to the right and pick up speed. The angular velocity reaches its maximum when the pendulum passes the bottom of the swing. On the way back up to the top again, the pendulum slows down. Then the pendulum balances at the top again. But when the pendulum at the beginning of its rotation shoved hard to the right, then its rate of angular velocity is rather large. Moving back up again, it

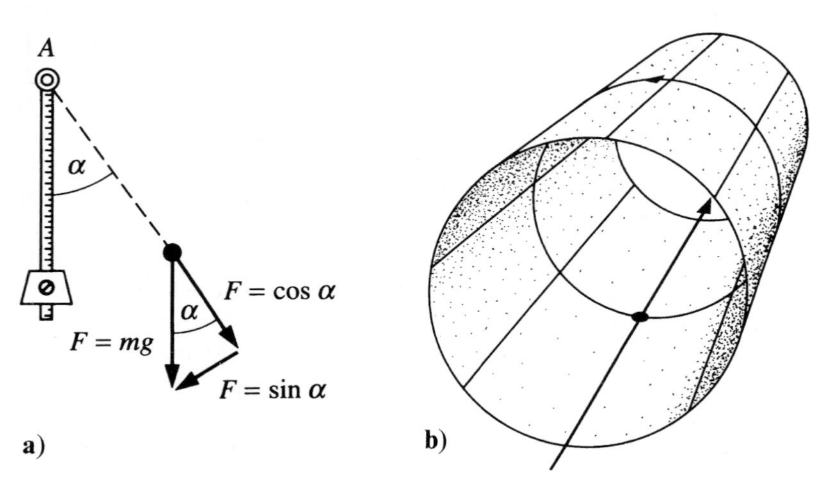

Fig. 2.7a,b. Dynamical system (pendulum) (**a**) with 2-dimensional state space (circular cylinder) (**b**) [2.17]

slows down, but not enough to come to rest at the top. Thus, the pendulum rotates clockwise indefinitely. The corresponding trajectory in the cylindrical state space is a cycle. Unlike the slow oscillation, the fast cycle goes around the cylinder. Performing many experiments would reveal the phase portrait of this dynamical model (Fig. 2.8a). There are two equilibrium points. At the top, there is a saddle point. At the origin, there is a vortex point which is not a limit point of the nearby trajectories. The phase portrait is easier to see when the cylinder is cut open along the straight line from front to back through the saddle point at the top (Fig. 2.8b).

If the system is not closed and the effects of friction are included as in physical reality, then the equilibrium point at the origin is no longer a vortex point (Fig. 2.8c). It has become a spiraling type of point attractor. As any motion of the pendulum will come to rest because of friction, any trajectory representing a slow motion of the pendulum near the bottom approaches this limit point asymptotically.

In two dimensions or more, other types of trajectories and limit sets may occur. For example, a cycle may be the asymptotic limit set for a trajectory (Fig. 2.9), or in a three-dimensional system a torus or even other more or less strange limit sets may occur.

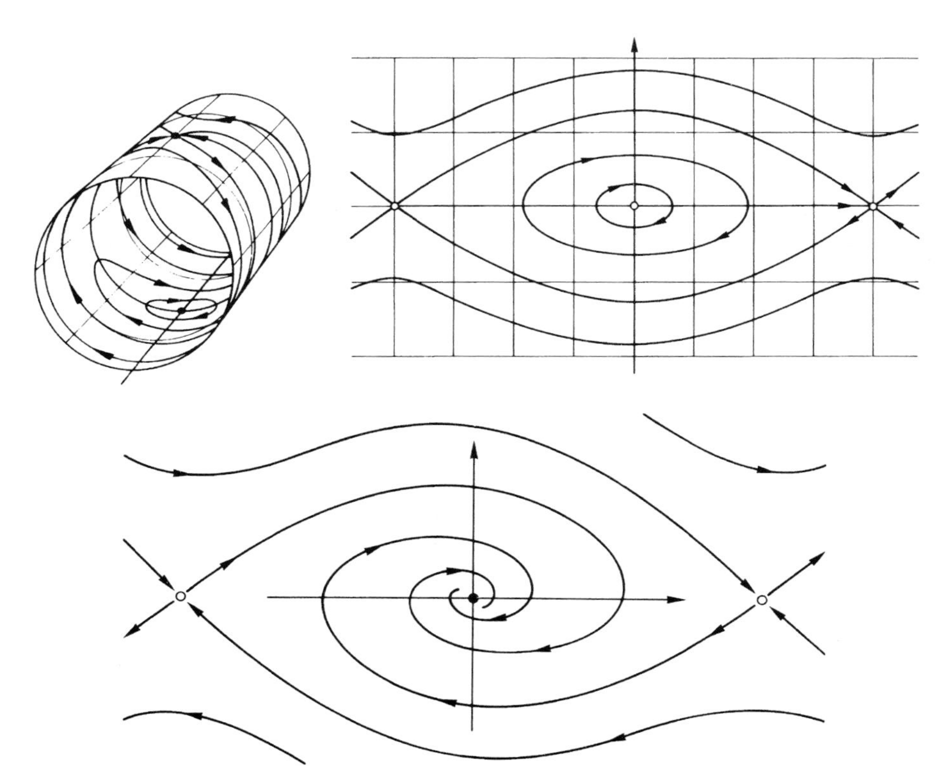

Fig. 2.8a,b. Phase portrait of the pendulum on the cylindrical state space (**a**) and cut open into a plane (**b**). (**c**) Phase portrait of the pendulum with friction

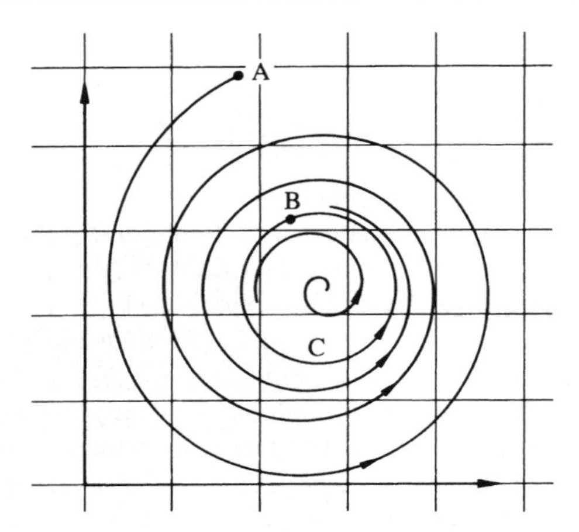

Fig. 2.9. Cycle as asymptotic limit set for a trajectory

Limit sets enable us to model a system's evolution to its equilibrium states. The key concepts are limit sets called "attractors" [2.18]. Mathematically, a limit set (limit point, cycle, torus, etc.) is called an attractor if the set of all trajectories approaching this limit set asymptotically is open. Roughly speaking, attractors receive most of the trajectories in the neighborhood of the limit set. Of all limit sets which represent possible dynamical equilibria of the system, the attractors are the most prominent. In the case of a limit point, an attractor represents a static equilibrium, while a limit cycle as attractor designates the periodic equilibrium of an oscillation. Vibrations on a pendulum, spring, or musical instrument are only a few of the mechanical applications. As we will see later on, periodic equilibria of oscillating dynamical systems play an important role in physics, chemistry, biology, and social sciences.

In a typical phase portrait, there will be more than one attractor. The phase portrait will be devided into their different regions of approaching trajectories. The dividing boundaries or regions are called separatrices. In Fig. 2.10, there are two point attractors with their two open sets of approaching trajectories and their separatrix.

In reality, a dynamical system cannot be considered as isolated from other dynamical systems. In order to get more adequate models, we will study two coupled systems. A simple example is provided by coupling two clocks. Historically, this particular system was observed by Christian Huygens in the 17th century. He noticed that two clocks hanging on the same wall tend to synchronize. This phenomenon is caused by nonlinear coupling through the elasticity of the wall. Indeed, any two dynamical systems can be combined into a single system by constructing the Cartesian product of the two corresponding state spaces. A small perturbation of this combined system is called a coupling of the two systems. The geometric model for the states of this combined system is formed as follows [2.19].

Each clock A and B is a kind of oscillator. For the sake of visualizing the asymptotic behaviour of both oscillators, the transient behavior is ignored and the two-dimensional state model of the Euclidean plane with a limit cycle around the origin for the two parameters of displacement and velocity is replaced by the limit cycle alone. A state of oscillator A is specified by an angle α corresponding to its phase (Fig. 2.11a), a state of oscillator B by an angle β (Fig. 2.11b).

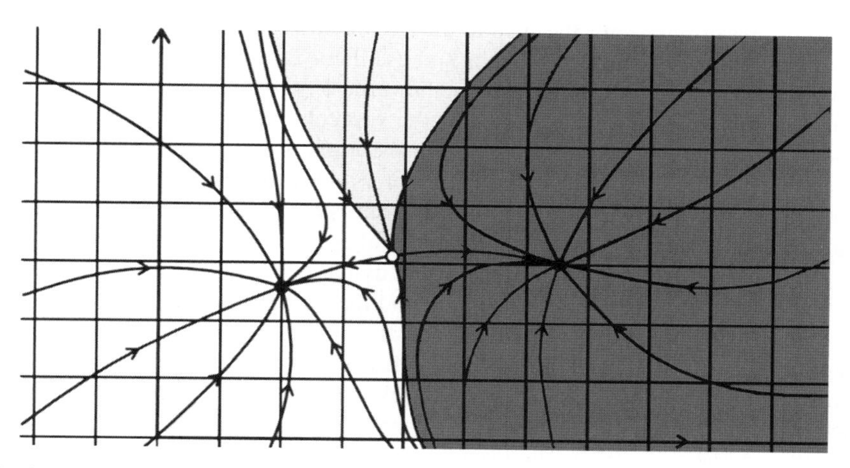

Fig. 2.10. Phase portrait with two point attractors, two open sets of approaching trajectories, and a separatrix

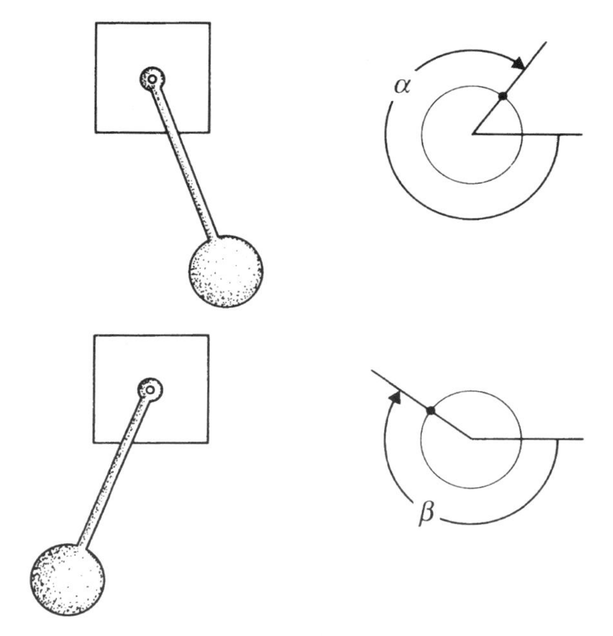

Fig. 2.11a,b. Two clocks as oscillators with two cycles as their corresponding state spaces

In order to construct the state space for the combined system of both oscillators, we consider the limit cycle of clock A in a horizontal plane. Each point of this horizontal cycle represents a phase state of A. We consider such a point as the center of the limit cycle of clock B erected perpendicular to the horizontal plane of clock A (Fig. 2.11c). Each point of this vertical cycle represents a phase state of B. The pair (α, β) of phases represents the state of the combined system [2.20].

If oscillator A is stuck at phase α and oscillator B moves through a full cycle, then the combined phase point traverses the vertical cycle in Fig. 2.11c. If oscillator A also moves through a full cycle, then the vertical cycle in Fig. 2.11c is pushed around the horizontal cycle, sweeping out the torus in Fig. 2.11d. Thus, the state space for the combined system of two oscillators is the torus, which is the Cartesian product of the two cycles. Of course, the actual state model for two oscillators is four-dimensional and not only two-dimensional as in our reduced figures.

In order to get the phase portrait of the dynamical behavior for the combined system, we have to study the vector field and the trajectories on the state space of the torus. Let us first assume that each clock is totally indifferent to the state of the other. In this case, the clocks are uncoupled. The trajectory of a point on the torus corresponding to the time phase of each clock winds around the torus. If the rate of each clock is constant, then on the flat rectangular model of the torus, the trajectory is a straight line (Fig. 2.12). The slope of this line is the ratio of the rate of clock B to the rate of clock A. If the two clocks run at the same rates, the ratio is one. Telling the same time means that both clocks have identical phases. Then the trajectory on the flat torus is the diagonal line in Fig. 2.12a.

A slight change in the system results in a slight change in the ratio of the rates or frequencies of the oscillators. Then the trajectory on the torus changes from a periodic trajectory to an almost periodic trajectory or to a periodic trajectory winding many times around, instead of just once (Fig. 2.12b). If two oscillators are coupled (for instance by Huygens' shared wall for the two clocks), then a small vector field must be added to the dynamical model representing the uncoupled system. It is a noteworthy theorem of geometric analysis that the braid of trajectories on the torus is structurally stable in the sense that a small perturbation does not result in a significant change in the phase portrait. Experimentally, this result was already confirmed by Huygens' observation of the synchronizing phenomenon of two clocks on the same wall.

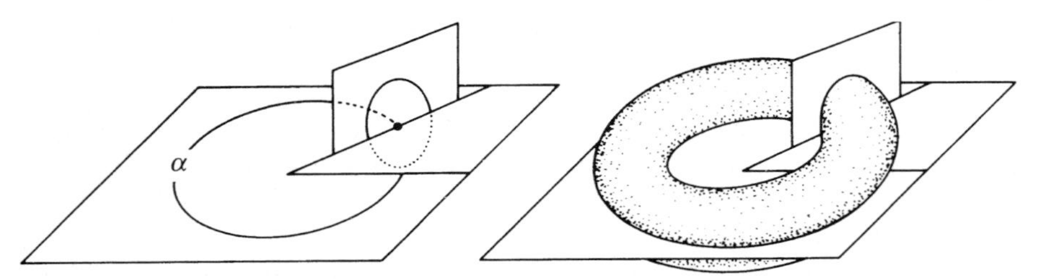

Fig. 2.11c,d. State space for the combined system of two oscillators (torus as Cartesian product of two cycles)

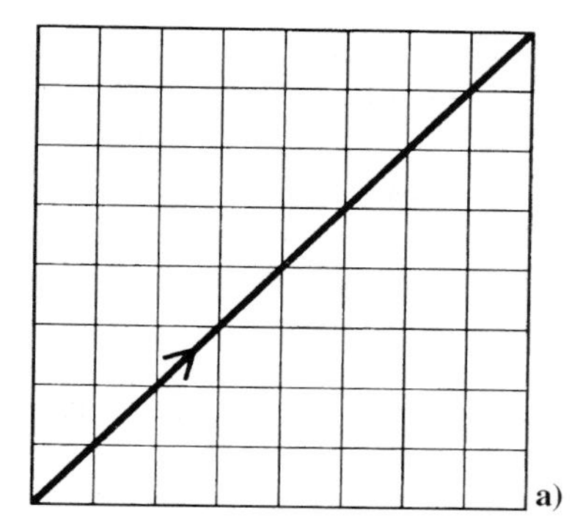

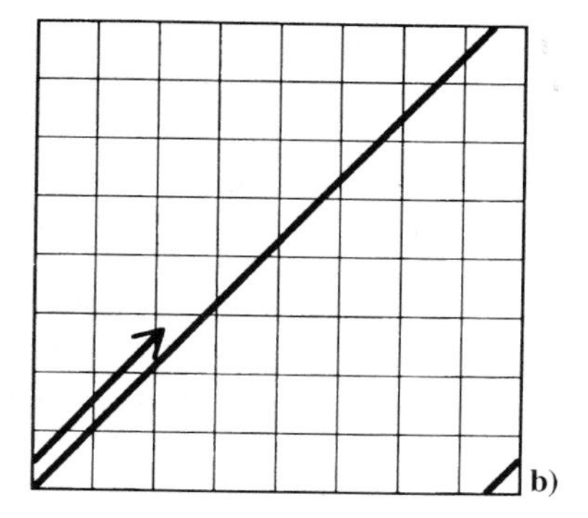

Fig. 2.12a,b. Phase portraits for the combined system of two oscillators with identical phase (**a**) and a slight change (**b**)

Oscillators are a central dynamical paradigm for the modeling procedure of nature. They are by no means restricted to mechanical applications. In the 19th century, Hermann von Helmholtz invented an electrical vibrator and Lord Rayleigh studied coupled systems of vacuum tube oscillators used in the first radio transmitters. In this century, it was van der Pol who used the further development of radio frequency electronics in understanding coupled oscillators.

In the Newtonian universe, coupled oscillators provide examples of many-body problems. What can in general be said about the mechanics of point mass systems with several moving point masses, exerting forces upon each other? Systems with two point masses have simple and exact solutions. In a two-body problem with two point masses with isotropic central forces, the (twelve) unknowns are determined by the (ten) laws of conserved quantities and Newton's laws of motion for the two particles. The problem of two-point masses can be successfully reduced to the already solved problem of a single point mass by considering Newton's law of motion for the difference vector r and reduced mass $\mu = m_1 m_2/(m_1 + m_2)$ of both point masses m_1 and m_2. Historically, Galileo assumed that the earth moves around the sun, which is at rest. Thus he reduced the celestial motions to the simple case of a two-body problem. As we all know, the sun is actually moving around the combined centre of mass of the sun–earth system, which lies inside the surface of the sun. But this assumption is still inaccurate, of course, since many planets are simultaneously moving around the sun and all of them are exerting forces upon each other.

Another example of such a many-body-problem is given by a triple collision of three billiard balls. Provided that the balls collide only in pairs, and no triple or higher-order collisions occur, then the situation is reduced to two-body problems. The outcome depends in a continuous way on the initial state. Sufficiently tiny changes in the initial state lead only to small changes in the outcome. If three balls come together at once, the resulting behavior depends critically upon which balls come together first. Thus, the outcome depends discontinuously on the input, contrary to Leibniz' principle of continuity, which he used basically to criticize

Descartes' inquiries into percussion. In the Newtonian universe, many-body problems of billiard balls and planets can be described in deterministic models in the sense that physical behavior is mathematically completely determined for all times in the future and past by the positions and velocities of the balls or planets. But the models may be non-computable in practice and in the long run. In the case of planetary theory, numerical simulations on computers for many millions of years can produce very large errors, because the initial positions and velocities are not known exactly. A very tiny change in the initial data may rapidly give rise to an enormous change in the outcome. Such instabilities in behavior are typical for many-body problems. Even in a fully deterministic world, the assumption of a Laplacean demon which can calculate the Newtonian universe in the long run will eventually be exposed as an illusory fiction.

2.3 Hamiltonian Systems and the Chaos of Heaven and the Quantum World

In the 18th and 19th centuries, Newtonian mechanics seemed to reveal an eternal order of nature. From a modern point of view, Newtonian systems are only a useful kind of dynamical system for modeling reality. In order to specify the initial state of a Newtonian system, the positions and the velocities of all its particles must be known. Around the middle of the 19th century, a very elegant and efficient formalism was introduced by the mathematician William Hamilton [2.21]. His fruitful idea was to characterize a conservative system by a so-called Hamiltonian function H which is the expression for the total energy (= sum of kinetic and potential energy) of the system in terms of all the position and momentum variables. While the velocity of a particle is simply the rate of change of its position with respect to time, its momentum is its velocity multiplied by its mass. Newtonian systems are described with Newton's second law of motion in terms of accelerations, which are rates of change of rates of change of position. Thus, mathematically, they are defined by second-order equations. In the Hamiltonian formulation, there are two sets of equations. One set of equations describes how the momenta of particles are changing with time, and the other describes how the positions are changing with time. Obviously, Hamiltonian equations describe the rates of change of quantities (i.e., position or momentum). Thus, we get a reduction of mathematical description with first-order equations which are, of course, deterministic. For dynamical systems of n unconstrained particles with three independent directions of space, there are $3n$ position coordinates and $3n$ momentum coordinates.

With suitable choices of the Hamiltonian function H, Hamiltonian equations can be used to characterize any classical dynamical system, not just Newtonian systems. Even in Maxwell's electrodynamics, Hamiltonian-like equations deliver the rate of change with time of the electric and magnetic fields in terms of what their values are at any given time. The only difference is that Maxwell's equations are field equations rather than particle equations, needing an infinite number of parameters to describe the state of the system, with field vectors at every single point in space, rather than the finite number of parameters with three coordinates of position and three of momentum for each particle. Hamiltonian equations also hold true for special relativity and (in somewhat modified form) for general relativity. Even, the crucial step from classical mechanics to quantum mechanics is made by Bohr's correspondence principle in the framework of the Hamiltonian formalism. These applications will be explained later on. Just now, it is sufficient to recall that Hamiltonian equations deliver a universal formalism for modeling dynamical systems in physics.

The corresponding state spaces allow us to visualize the evolution of the dynamical systems in each "phase". Thus they are called phase spaces. For systems with n particles, phase spaces have $3n + 3n = 6n$ dimensions. A single point of a phase space represents the entire state of a perhaps complex system with n particles. The Hamiltonian equations determine the trajectory of a phase point in a phase space. Globally, they describe the rates of change at every phase point, and therefore define a vector field on the phase space, determining the whole dynamics of the corresponding system.

It is a well-known fact from empirical applications that states of dynamical models cannot be measured with arbitrary exactness. The measured values of a quantity may differ by tiny intervals which are caused by the measuring apparatus, constraints of the environment, and so on. The corresponding phase points are concentrated in some small regions of a neighborhood. Now, the crucial question arises if trajectories starting with neighboring initial states are locally stable in the sense that they have neighboring final states. In Fig. 2.13a, a phase state region R_0 of initial states of time zero is dragged along by the dynamics of the vector field to a region R_t at later time t (of course, the actual large number of coordinates must be neglected in such a visualization of a phase space) [2.22].

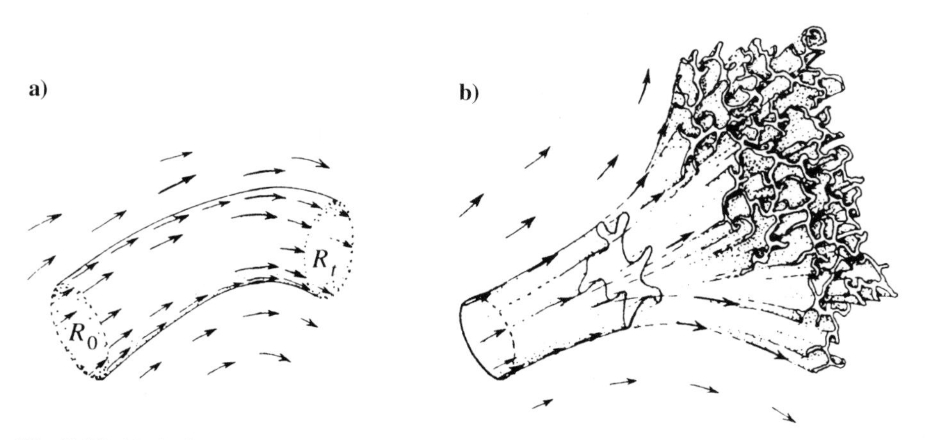

Fig. 2.13. (**a**) A phase state region R_0 at time 0 is dragged along by a Hamiltonian dynamics to a region R_t at later time t [2.22]. (**b**) According to Liouville's theorem, the volume of an initial phase state region is conserved under a Hamiltonian dynamics, although its shape may be distorted, stretched, and spread outwards [2.22]

In this case, similiar initial states lead to similiar final states. This assumption is nothing else than a classical principle of causality in the language of Hamiltonian dynamics: similar causes lead to similiar effects. Historically, philosophers and physicists from Leibniz to Maxwell believed in this causal principle, which seemed to secure the stability of the measuring process and the possibility of forecasts despite an appreciable interval of inaccuracy.

It is noteworthy that the representation in the Hamiltonian formalism allows a general statement about the causality of classical dynamical systems. Due to a famous theorem of the mathematician Liouville, the volume of any region of the phase space must remain constant under any Hamiltonian dynamics, and thus for any conservative dynamical system. Consequently, the size of the initial region R_0 in Fig. 2.13a cannot grow by any Hamiltonian dynamics if we understand "size" in the right manner as phase-space volume. But its conservation does not exclude that the shape of the initial region is distorted and stretched out to great distances in the phase space (Fig. 2.13b) [2.22].

We may imagine a drop of ink spreading through a large volume of water in a container. That possible spreading effect in phase spaces means that the local stability of trajectories is by no means secured by Liouville's theorem. A very tiny change in the initial data may still give rise to a large change in the outcome. Many-body problems of celestial mechanics and billiard balls remain non-computable in the long run, although their dynamics are deterministic. Nevertheless, Liouville's theorem implies some general consequences concerning the final regions which can be displayed by Hamiltonian dynamics, and thus by conservative dynamical systems. Remember the phase portrait Fig. 2.8c of a pendulum with friction (which is not a conservative system) with a different equilibrium point at the origin. While the non-conservative system has a spiraling type of point attractor (Fig. 2.14a), the conservative system has a vortex point (Fig. 2.14b) which is not an attractor [2.23].

In Fig. 2.14a, trajectories are attracted to a field point, and the volume of an intial area shrinks. In Fig. 2.14b, the trajectories rotate around a vortex point, and the volume of an initial area is conserved. Thus, due to Liouville's theorem, we can generally conclude that in any conservative system attracting points must be excluded.

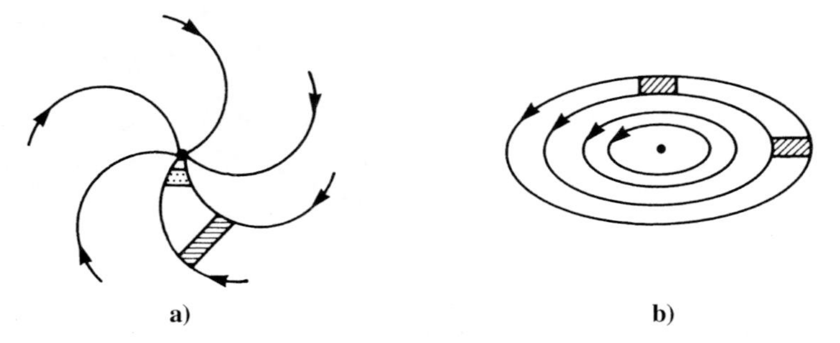

Fig. 2.14a,b. Point attractor of a non-conservative system without conservation (**a**), vortex point of a conservative system with conservation (**b**)

The effect of shrinking initial areas can easily be visualized for the trajectories of limit cycles, too. So, limit cycles as attractors are also not possible in conservative systems for the same mathematical (a priori) reasons.

These results are derived a priori by a far-reaching mathematical theorem of Hamiltonian systems. We must be aware that conservative physical systems like planetary systems, pendula, free fall, etc., are only some of the empirical applications of Hamiltonian systems. Hamiltonian systems are defined by a particular kind of mathematical equation (Hamiltonian equations). Features of Hamiltonian systems are derived from the mathematical theory of the corresponding equations. Consequently, modeling reality by Hamiltonian systems means that we can epistemically forecast some a priori features, e.g., that no static equilibrium of a limit point attractor and no periodic equilibrium of a limit cycle attractor can be expected.

Philosophically, this point of view obviously conforms to Kant's epistemology in some modified sense. If we assume the mathematical framework of some dynamical systems, then, of course, we can assert some a priori statements about our empirical models, independently of their empirical applications in several sciences. But Kant's epistemology and the dynamical system approach differ in the following sense: not only is there one categorial framework (e.g., Newtonian systems), but there are many kinds of systems modeling reality with more or less success. So, it will not be physicalist or reductionist to apply conservative systems even in cognitive and economical science, later on.

A further a priori result of Hamiltonian (conservative) systems says that there are irregular and chaotic trajectories. In the 18th and 19th centuries, physicists and philosophers were convinced that nature is determined by Newtonian- or Hamiltonian-like equations of motion, and thus future and past states of the universe can be calculated at least in principle if the initial states of present events are well known. Philosophically, this belief was visualized by Laplace's demon, which like a huge computer without physical limitations can store and calculate all necessary states. Mathematically, the belief in Laplace's demon must presume that systems in classical mechanics are integrable, and, thus are solvable. In 1892, Poincaré was already aware that the non-integrable three-body problem of classical mechanics can lead to completely chaotic trajectories [2.24]. About sixty years later, Kolmogorov (1954), Arnold (1963) and Moser (1967) proved with their famous KAM theorem that motion in the phase space of classical mechanics is neither completely regular nor completely irregular, but that the type of trajectory depends sensitively on the chosen initial conditions [2.25].

As celestial mechanics is an empirically well confirmed dynamical model of a Hamiltonian system, the KAM theorem refutes some traditional opinions about the "superlunar" world. Heaven is not a world of eternal regularity, either in the sense of Aristotle's cosmos or in the sense of Laplace's demon. Obviously, it is not the seat of the Gods. Nevertheless, it is not completely chaotic. Heaven, as far as recognized by Hamiltonian systems, is more or less regular and irregular. It seems to have more similarity with our human everyday life than our forefathers believed. This may be a motivation for writers to be curious about Hamiltonian systems. But now let us see some mathematical facts.

One of the simplest examples of an integrable system is a harmonic oscillator. Practically, the equations of motion of any integrable system with n degrees of freedom are the same as those of a set of n uncoupled harmonic oscillators. The corresponding phase space has $2n$ dimensions with n position coordinates and n momentum coordinates. For a harmonic oscillator with $n = 1$ we get a circle, and for two harmonic oscillators with $n = 2$ a torus (compare Fig. 2.11d). Thus, the existence of n integrals of motion confines the trajectories in the $2n$-dimensional phase space of an integrable system to an n-dimensional manifold which has the topology of an n-torus. For an integrable system with two degrees of freedom and a four-dimensional phase space, the trajectories can be visualized on a torus. Closed orbits of trajectories occur only if the frequency ratios of the two corresponding oscillators are rational (Fig. 2.15). For irrational frequency ratios, the orbit of a trajectory never repeats itself, but approaches every point on the torus infinitesimally closely [2.26].

A nonintegrable system of celestial mechanics was studied by Hénon and Heiles in 1964. The dynamical model consists of an integrable pair of harmonic oscillators coupled by nonintegrable cubic terms of coordinates. If the initial state of the model with two position coordinates q_1, q_2 and two momentum coordinates p_1, p_2 is known, then its total energy E is determined by the corresponding Hamiltonian function H depending on these position and momentum coordinates. The trajectories of the system move in the four-dimensional phase space on a three-dimensional hyperplane which is defined by $H(q_1, q_2, p_1, p_2) = E$.

The values of E can be used to study the coexistence of regular and irregular motions which was forecast by the KAM theorem. For small values of E, the dynamical system has regular behavior, while for large values it becomes chaotic. In order to visualize this changing behavior, we consider the intersections of the trajectories with the two-dimensional plane of coordinates q_1 and q_2 (Poincaré maps). For $E = \frac{1}{24}$ (Fig. 2.16a) and $E = \frac{1}{12}$ (Fig. 2.16b), the Poincaré maps show the intersections of somewhat deformed tori which signal regular motion. Above a critical value of $E = \frac{1}{9}$, most, but not all, tori are destroyed, and spots of irregular points appear to be random. For $E = \frac{1}{8}$ (Fig. 2.16c), the Poincaré map illustrates a state of transition with the coexistence of regular and irregular motions. For $E = \frac{1}{6}$ (Fig. 2.16d), the motion seems to be almost completely irregular and chaotic [2.27].

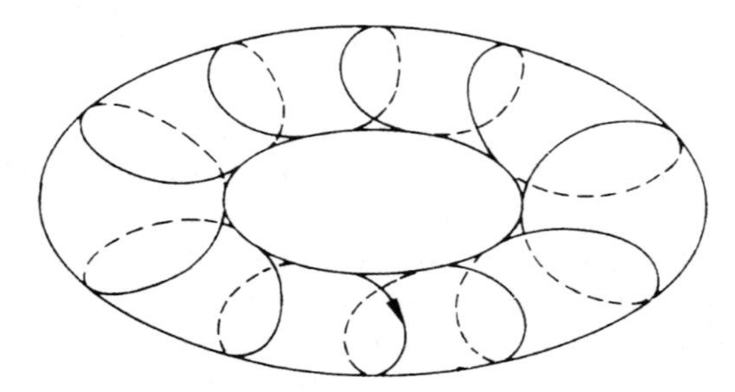

Fig. 2.15. Integrable system with two degrees of freedom on a torus and a closed orbit of a trajectory

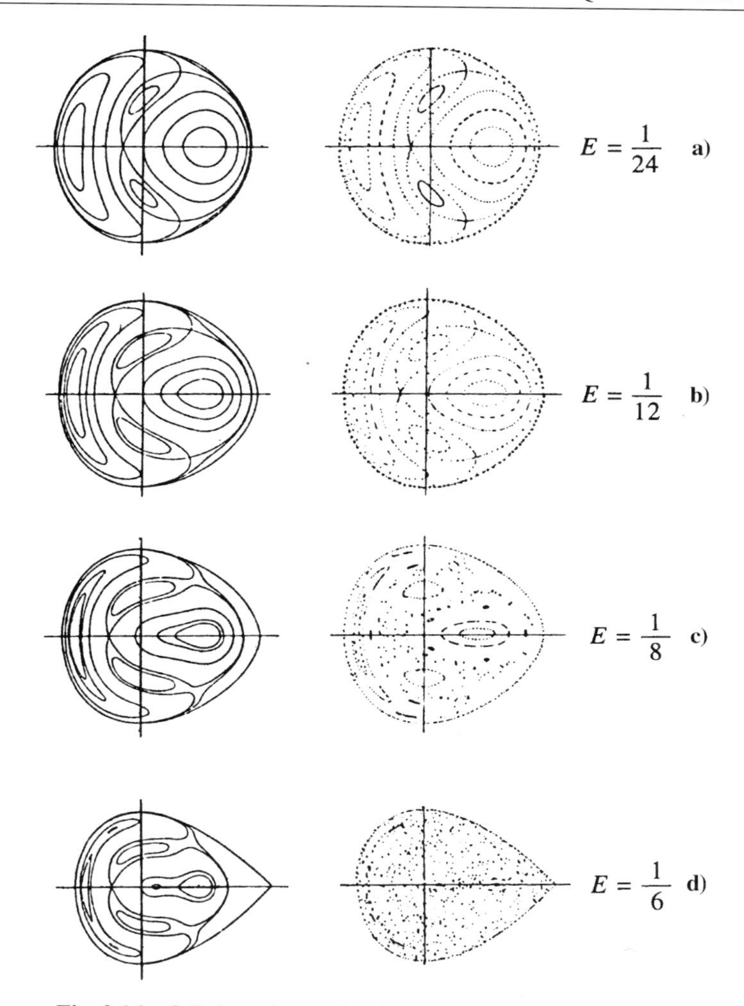

Fig. 2.16a–d. Poincaré maps for the Hénon-Heiles system [2.27]

An empirical application is given in the following three-body problem of celestial mechanics, which is nonintegrable. Consider the motion of Jupiter perturbing the motion of an asteroid around the sun (Fig. 2.17).

Jupiter and the asteroid are interpreted as two oscillators with certain frequencies. According to the KAM theorem, stable and unstable motions of the asteroid can be distinguished, depending on the frequency ratio.

In general, we must be aware that stable as well as unstable trajectories are mathematically well defined. Consequently, even nonintegrable many-body problems describe deterministic models of the world. Metaphorically, we may say that the God of Leibniz and Newton would have no difficulty in forecasting regular and irregular trajectories *sub specie aeternitatis* and does not need to calculate their de-

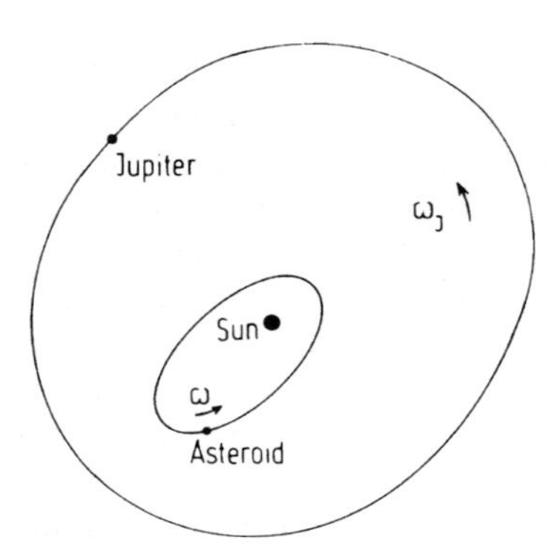

Fig. 2.17. Perturbation of an asteroid's motion by Jupiter

velopment step by step. The observed chaotic behavior is neither due to a large number of degrees of freedom (a celestial three-body problem has rather few degrees of freedom) nor to the uncertainty of human knowledge. The irregularity is caused by the nonlinearity of Hamiltonian equations which let initially close trajectories separate exponentially fast in a bounded region of phase. As their initial conditions can only be measured with finite accuracy, and errors increase exponentially fast, the long-term behavior of these systems cannot be predicted. Mathematically, initial conditions are characterized by real values which may be irrational numbers with infinite sequences of digits. Thus, computer-assisted calculations will drive the errors faster and faster with improved measurement of more and more digits.

The macrocosmos of celestial mechanics, the world of asteroids, planets, stars, and galaxies, is determined by the coexistence of regular and irregular behavior. Deterministic chaos in the heavens is not everywhere, but locally possible, and thus may cause cosmic catastrophes which cannot be excluded in principle. What about the microcosmos of quantum mechanics, the quantum world of photons, electrons, atoms, and molecules? Is there chaos in the quantum world? In order to answer this question, we first must remind the reader of some basic concepts of Hamiltonian systems and phase spaces corresponding to objects in the quantum world [2.28].

In 1900, Max Planck proposed that electromagnetic oscillations occur only in quanta, whose energy E bears the definite relation $E = h\nu$ to the frequency ν depending on the constant h ("Planck's quantum"). Besides Einstein's huge constant c of light's speed, Planck's tiny constant of quanta is the second fundamental constant of nature, according to 20th century physics. Planck's relation was experimentally supported by, e.g., the radiation of black bodies. In 1923, Louis de Broglie proposed that even the particles of matter should sometimes behave as waves. De Broglie's wave-frequency ν for a particle of mass m satisfies the Planck relation. Combined with Einstein's famous theorem $E = mc^2$ of Special Relativity ("mass is a particular state of energy and can therefore be transformed into energy by radiation"), we get a relation telling us that ν is related to m by $h\nu = mc^2$. It follows that in the quantum world, fields oscillating with some frequency can occur only in discrete units of mass, depending on Planck's and Einstein's constants. Obviously, in the quantum world, phenomena can be considered as waves as well as particles. This so-called particle-wave duality was well confirmed by many experiments which reveal features of waves or particles for quantum systems like photons or electrons, depending on the preparation of experimental conditions.

In 1913, Niels Bohr introduced his "planetary" model for the atom which could explain the observed and measured discrete stable energy levels and spectral frequencies with suprising accuracy. Bohr's rules required that the angular momentum of electrons in orbit about the nucleus can occur only in integer multiples of $\hbar = h/2\pi$. His successful, albeit somewhat ad hoc rules only delivered an approximate geometric model which must be derived from a dynamical theory of the quantum world, corresponding to Newtonian or Hamiltonian classical mechanics which can explain Kepler's planetary laws. The dynamics of the quantum world was founded by Heisenberg's and Schrödinger's quantum mechanics, which became the fundamental theory of matter in 20th century physics.

The main concepts of quantum mechanics can be introduced heuristically by analogy with corresponding concepts of Hamiltonian mechanics if some necessary modifications depending on Planck's constant are taken into account. This procedure is called Bohr's correspondence principle (Fig. 2.18). So, in quantum mechanics, classical vectors like position or momentum must be replaced by some operators satisfying a non-commutative (non-classical) relation depending on Planck's constant. If h disappears ($h \to 0$), then we get the well known classical commutative relations of, e.g., position and momentum which allow us to measure both vectors together with arbitrary accuracy. An immediate consequence of non-commutative relations in quantum mechanics is Heisenberg's uncertainty principle $\Delta p \Delta q \geq \hbar/2$. If one measures the position q with precision Δq, then one disturbs the momentum p by Δp. Thus, it is obvious that there are no trajectories or orbits in the quantum world which demand precise values of both the position and the momentum of a particle. Bohr's popular electronic orbits are only very rough geometric visualizations [2.29].

According to Bohr's correspondence principle, classical systems described by Hamiltonian functions must be replaced by quantum systems (e.g., electrons or pho-

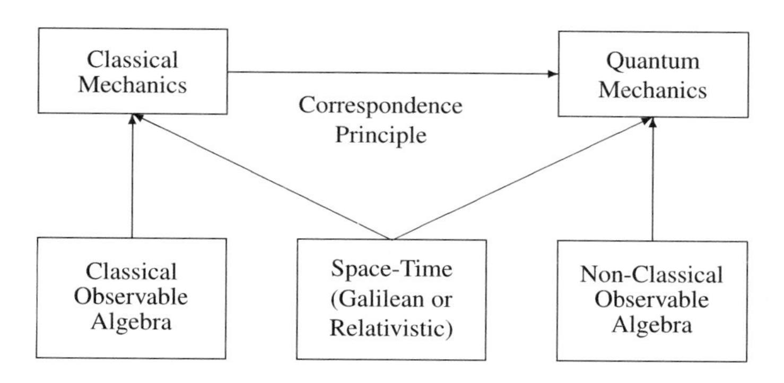

Fig. 2.18. Bohr's correspondence principle

tons) described by a Hamiltonian operator depending on operators (for position and momentum) instead of vectors. In classical physics, the states of Hamiltonian systems are determined by the points of a phase space. In quantum mechanics, the appropriate analogous concept is that of a Hilbert space. States of a quantum system are described by vectors of a Hilbert space spanned by the eigenvectors of its Hamiltonian operator.

In order to illustrate this mathematical remark a little bit more, let us imagine a single quantum particle. Classically, a particle is determined by its position in space and by its momentum. In quantum mechanics, every single position which the particle might have is a single alternative combined in a collection of all possible positions with complex-number weightings. Thus, we get a complex function of position, the so-called wave function $\psi(x)$. For each position x, the value of $\psi(x)$ denotes the amplitude for the particle to be at x. The probability of finding the particle in some small fixed-sized interval about this position is obtained by forming the squared modulus of the amplitude $|\psi(x)|^2$. The various amplitudes for the different possible momenta are also determined by the wave function. Thus, the Hilbert space is a complex state space of a quantum system.

The causal dynamics of quantum states is determined by a partial differential equation called the Schrödinger equation. While classical observables are commutative with always definite values, non-classical observables of quantum systems are non-commutative with generally no common eigenvector and consequently no definite eigenvalues. For observables in a quantum state only statistical expectation values can be calculated.

An essential property of Schrödinger's quantum formalism is the superposition principle demonstrating its linearity. For example, consider two quantum systems which once interacted (e.g., a pair of photons leaving a common source in opposite directions). Even when their physical interaction ceases at a large distance, they remain in a common superposition of states which cannot be separated or located. In such an entangled (pure) quantum state of superposition an observable of the two quantum systems can only have indefinite eigenvalues. The superposition or linearity principle of quantum mechanics delivers correlated (entangled) states of combined systems which are highly confirmed by the EPR experiments. Philosophically, the (quantum) whole is more than the sum of its parts. Non-locality is a fundamental property of the quantum world which differs from classical Hamiltonian systems [2.30]. We shall return to this question in discussing the emergence of mind–brain and artificial intelligence (Chaps. 4–6).

Bohr's correspondence principle lets the question arise of whether the existence of chaotic motion in classical Hamiltonian systems leads to irregularities in the corresponding quantum systems [2.31]. Our summary of basic quantum mechanical concepts gives some hints of changes which must be expected in passing from a classically chaotic system to its corresponding quantum mechanical version. In contrast to classical mechanics, quantum mechanics only allows statistical predictions. Although the Schrödinger equation is linear in the sense of the superposition principle and can be solved exactly, e.g., for a harmonic oscillator, and although the wave function is strictly determined by the Schrödinger equation, this does not mean that the properties of a quantum state can be calculated exactly. We can only calculate the density of probability to find a photon or electron at a space-time point.

Because of Heisenberg's uncertainty principle, there are no trajectories in the quantum world. Therefore, the determination of chaos with the exponentially fast separation of close trajectories is not possible for quantum systems. Another aspect of the uncertainty principle concerning chaos is noteworthy: remember a classical phase space with chaotical regions like in Fig. 2.16. The uncertainty principle implies that points in a $2n$-dimensional phase space within a volume h^n cannot be distinguished. The reason is that chaotic behavior smaller than h^n cannot be represented in quantum mechanics. Only the regular behavior outside these chaotic regions could be expected. In this sense, the tiny, but finite value of Planck's constant could suppress chaos.

In quantum mechanics, one distinguishes between time-independent stationary and time-dependent Hamiltonian systems. For systems with stationary Hamiltonians the Schrödinger equation can be reduced to a so-called linear eigenvalue problem which allows one to calculate the energy levels of, e.g., a hydrogen atom. As long as the levels are discrete, the wave function behaves regularly, and there is no chaos. The question arises of whether there are differences between the energy spectra of a quantum system with a regular classical limit and a quantum system whose classical version displays chaos. Time-dependent Hamiltonians are used to described the time-evolution of, e.g., elementary particles and molecules.

According to Bohr's correspondence principle, quantum chaos can be detected by starting with the investigation of some classical Hamiltonian systems. They may be integrable, almost integrable, or chaotic. Thus, the trajectories in the hyperplane of energy may be regular, almost regular, or almost chaotic. Quantizing the Hamiltonian function by replacing the vectors of position and momentum with the corresponding operators, we get the Hamiltonian operator of the corresponding quantum system. In the next step, the Schrödinger equation and eigenvalue equation can be derived. Now, we may ask if the properties of the classical system with its integrable, almost integrable, or chaotic behavior can be transferred to the corresponding quantum system. What about the spectrum, eigenfunctions, etc.? These questions are summarized under the title "quantum chaos". For instance, there are calculations which show that the energy spectrum of a free quantum particle in a stadium, for which the classical motion is chaotic, differs drastically from that of a free quantum particle in a circle, for which the classical motion is regular.

In Fig. 2.19, the distribution of distances between neighboring levels is illustrated with two examples [2.32]. In Fig. 2.19a,b, a system consisting of two coupled oscillators is shown for two different values of the coupling coefficient. While the corresponding classical dynamics of Fig. 2.19a is regular, the classical dynamics of Fig. 2.19b is almost chaotic.

Figure 2.19c,d shows the example of a hydrogen atom in a uniform magnetic field. While the corresponding classical dynamics of Fig. 2.19c is regular, the classical dynamics of Fig. 2.19d is almost chaotic. The regular and chaotic cases can be distinguished by different distributions of energy levels (Poisson and Wigner distributions) which are calculated by solving the corresponding Schrödinger equation.

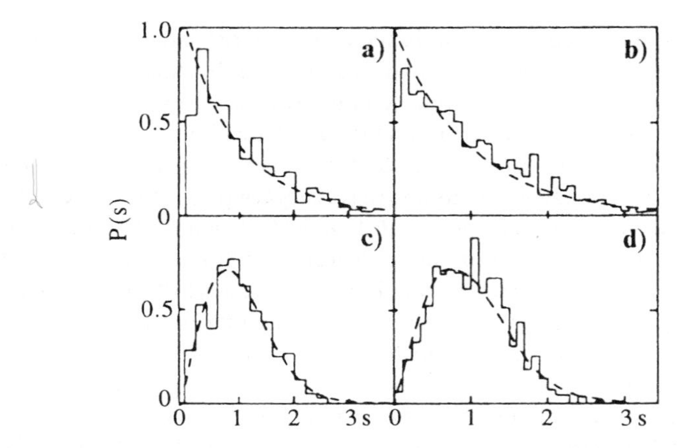

Fig. 2.19a–d. Two coupled oscillators with regular (**a**) and almost chaotic classical dynamics, (**b**) Hydrogen atom in a uniform magnetic field with corresponding regular (**c**) and almost chaotic classical dynamics (**d**) [2.32]

They are confirmed by several numerical models as well as by measurements in laboratories with laser spectroscopy. In this sense, quantum chaos is no illusion, but a complex structural property of the quantum world. Hamiltonian systems are a key to discovering chaos in the macro- and microcosmos. But, of course, we must not confuse the complex mathematical structure of deterministic chaos with the popular idea of disorder.

2.4 Conservative and Dissipative Systems and the Emergence of Order

Since Poincaré's celestial mechanics (1892), it was mathematically known that some mechanical systems whose time evolution is governed by nonlinear Hamiltonian equations could display chaotic motion. But as long as scientists did not have suitable tools to deal with nonintegrable systems, deterministic chaos was considered as a mere curiosity. During the first decades of the 20th century, many numerical procedures were developed to deal with the mathematical complexity of nonlinear differential equations at least approximately. The calculating power of modern high speed computers and refined experimental techniques have supported the recent successes of the nonlinear complex system approach in natural and social sciences. The visualizations of nonlinear models by computer-assisted techniques promote interdisciplinary applications with far-reaching consequences in many branches of science. In this scientific scenario (1963), the meteorologist Edward Lorenz, a student of the famous mathematician Birkhoff [2.33], observed that a dynamical system with three coupled first-order nonlinear differential equations can lead to completely chaotic trajectories. Mathematically, nonlinearity is a necessary, but not sufficient condition of chaos. It is necessary condition, because linear differential equations can be solved by well-known mathematical procedures (Fourier transformations) and do not lead to chaos. The system Lorenz used to model the dynamics of weather differs from Hamiltonian systems à la Poincaré mainly by its dissipativity. Roughly speaking, a dissipative system is not conservative but "open", with an external control parameter that can be tuned to critical values causing the transitions to chaos.

More precisely, conservative as well as dissipative systems are characterized by nonlinear differential equations $\dot{\boldsymbol{x}} = F(\boldsymbol{x}, \lambda)$ with a nonlinear function F of the vector $\boldsymbol{x} = (x_1, \ldots, x_d)$ depending on an external control parameter λ. While for conservative systems, according to Liouville's theorem, the volume elements in the corresponding phase space change their shape but retain their volume in the course of time, the volume elements of dissipative systems shrink as time increases (compare Figs. 2.13, 2.14) [2.34].

Lorenz's discovery of a deterministic model of turbulence occurred during simulation of global weather patterns. The earth, warmed by the sun, heats the atmosphere from below. Outer space, which is always cold, absorbs heat from the outer shell of the atmosphere. The lower layer of air tries to rise, while the upper layer tries to drop. This traffic of layers was modeled in several experiments by Bénard. The air currents in the atmosphere can be visualized as cross-sections of the layers. The traffic of the competing warm and cold air masses is represented by circulation vortices, called Bénard cells. In three dimensions, a vortex may have warm air rising in a ring, and cold air descending in the center. Thus, the atmosphere consists of a sea of three-dimensional Bénard-cells, closely packed as a hexagonal lattice. A footprint of such a sea of atmospheric vortices can be observed in the regular patterns of hills and valleys in deserts, snowfields, or icebergs.

In a typical Bénard experiment, a fluid layer is heated from below in a gravitational field (Fig. 2.20a). The heated fluid at the bottom tries to rise, while the cold liquid at the top tries to fall. These motions are opposed by viscous forces. For small temperature differences ΔT, viscosity wins, the liquid remains at rest, and heat is transported by uniform heat conduction. The external control parameter of the system is the so-called Rayleigh number Ra of velocity, which is proportional to ΔT. At a critical value of Ra, the state of the fluid becomes unstable, and a pattern of stationary convection rolls develops (Fig. 2.20b) [2.35].

Beyond a greater critical value of Ra, a transition to chaotic motion is observed. The complicated differential equations describing the Bénard experiment were simplified by Lorenz to obtain the three nonlinear differential equations of his famous model. Each differential equation describes the rate of change for a vari-

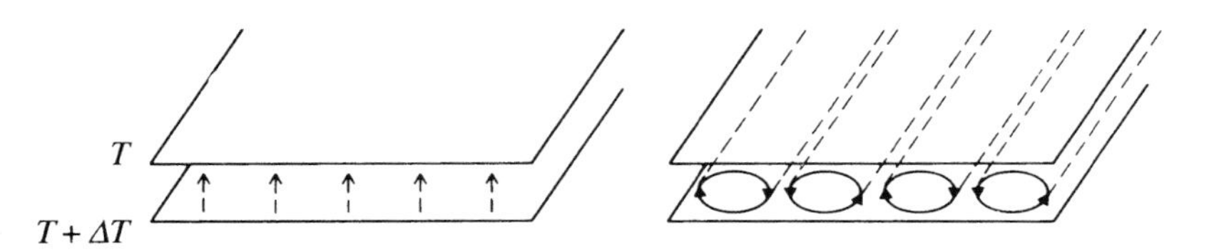

Fig. 2.20a,b. Bénard experiment: a heated fluid layer

able X proportional to the circulatory fluid flow velocity, a variable Y characterizing the temperature difference between ascending and descending fluid elements, and a variable Z proportional to the deviation of the vertical temperature profile from its equilibrium value. From these equations, it can be derived that an arbitrary volume element of some surface in the corresponding phase space contracts exponentially in time. Thus, the Lorenz model is dissipative.

This can be visualized by computer-assisted calculations of the trajectories generated by the three equations of the Lorenz model. Under certain conditions, a particular region in the three-dimensional phase space is attracted by the trajectories, making one loop to the right, then a few loops to the left, then to the right again, etc. (Fig. 2.21) [2.36].

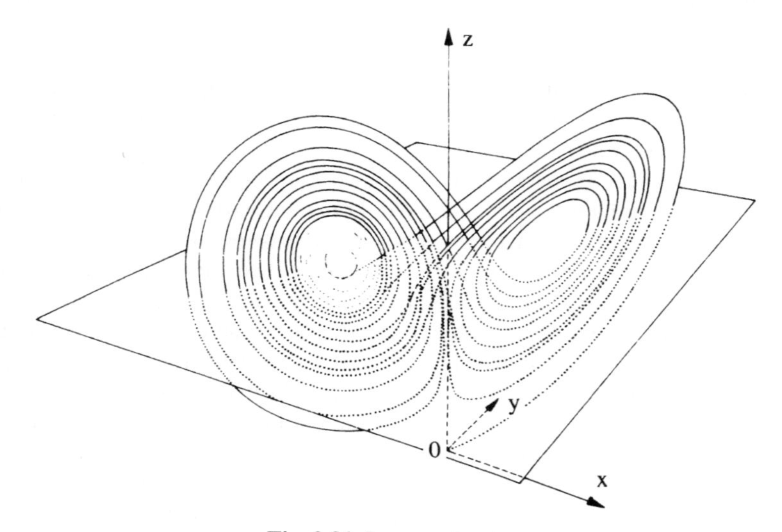

Fig. 2.21. Lorenz attractor

The paths of these trajectories depend very sensitively on the initial conditions. Tiny deviations of their values may lead to paths which soon deviate from the old one with different numbers of loops. Because of its strange image, which looks like the two eyes of an owl, the attracting region of the Lorenz phase was called a "strange attractor". Obviously, the strange attractor is chaotic. But which topological structure do the trajectories achieve by winding more and more densely without intersecting each other? An example illustrates the definition of so-called fractal dimensions [2.37]:

Let M be the subset of the attractor in the n-dimensional phase space. Now, the phase space is covered by cubes with edge length ε. Let $N(\varepsilon)$ be the number of cubes which contain a piece of the attractor M. If ε contracts to zero ($\varepsilon \to 0$), then the negative limit of the ratio of the logarithm of $N(\varepsilon)$ and the logarithm of ε, i.e., $D = -\lim \ln N(\varepsilon)/\ln \varepsilon$, is called the fractal dimension.

If the attractor is a point (Fig. 2.14a), the fractal dimension is zero. For a stable limit circle (Fig. 2.9) the fractal dimension is one. But for chaotic systems the fractal dimension is not an integer. In general, the fractal dimension can be calculated only numerically. For the Lorenz model, the strange attractor has the fractal dimension $D \approx 2.06 \pm 0.01$.

Another dissipative system in which chaotic motion has been studied experimentally is the Belousov–Zhabotinsky reaction. In this chemical process an organic molecule is oxidized by bromate ions, the oxidation being catalyzed by a redox system. The rates of change for the concentrations of the reactants in a system of chemical reactions are again described by a system of nonlinear differential equations with a nonlinear function. The variable which signals chaotic behavior in the Belousov–Zhabotinsky reaction is the concentration of the ions in the redox system. Experimentally, irregular oscillations of these concentrations are observed with a suitable combination of the reactants. The oscillations are indicated by separated colored rings. This separation is a fine visualization of nonlinearity. Linear evolutions would satisfy the superposition principle. In this case the oscillating rings would penetrate each other in superposition.

The corresponding differential equations are autonomous, i.e., they do not depend on time explicitly. For computer-assisted visualization it is often convenient to study the flow in a dynamical system described by differential equations of motion via discrete equations which construct the intersecting points of the trajectories with the $(d-1)$-dimensional Poincaré map in the corresponding d-dimensional phase space (compare Fig. 2.16). The constructed points are denoted by $\boldsymbol{x}(1), \boldsymbol{x}(2), \ldots, \boldsymbol{x}(n), \boldsymbol{x}(n+1), \ldots$ with increasing time points n. The corresponding equation has the form $\boldsymbol{x}(n+1) = G(\boldsymbol{x}(n), \lambda)$ for the successor point $\boldsymbol{x}(n+1)$ of $\boldsymbol{x}(n) = (x_1(n), \ldots, x_{d-1}(n))$. The classification of conservative and dissipative systems can be generalized from flows to Poincaré maps. A discrete map equation is called dissipative if it leads to a contraction of volume in phase space.

A famous example of a discrete map is the so-called logistic map with many applications in the natural sciences as well as the social sciences. The basic concepts of complex dynamical systems from nonlinearity to chaos can be illustrated by this map with rather simple computer-assisted methods. Thus, let us have a short glance at this example. Mathematically, the logistic map is defined by a quadratic (nonlinear) recursive map $x_{n+1} = \alpha x_n(1 - x_n)$ of the interval $0 \leq x \leq 1$ onto itself with control parameter α varying between $0 \leq \alpha \leq 4$. The function values of the sequence $x_1, x_2, x_3, \ldots$ can be calculated by a simple pocket computer. For $\alpha < 3$ the sequence converges towards a fixed point (Fig. 2.22a). If α is increased beyond a critical value α_1, then the values of the sequence jump periodically between two values after a certain time of transition (Fig. 2.22b). If α is increased further beyond a critical value α_2, the period length doubles. If α is increased further and further, then the period doubles each time with a sequence of critical values $\alpha_1, \alpha_2, \ldots$. But beyond a critical value α_c, the development becomes more and more irregular and chaotic (Fig. 2.22c) [2.38].

The sequence of period doubling bifurcations which is illustrated in Fig. 2.23a is governed by a law of constancy which was found by Grossmann and Thomae for

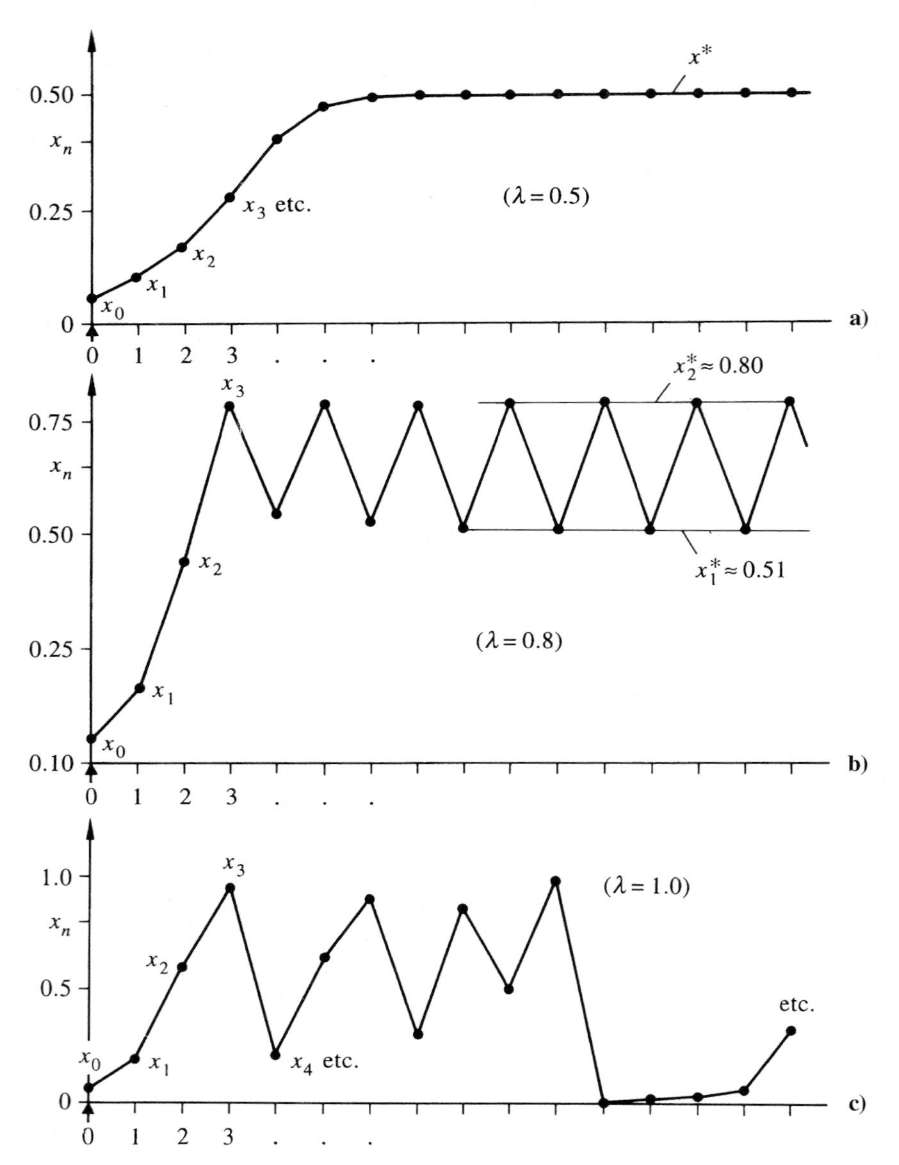

Fig. 2.22a–c. Logistic curve as nonlinear recursive map with control parameter $\alpha = 4\lambda = 2$ (**a**), $\alpha = 4\lambda = 3, 2$ (**b**), and $\alpha = 4\lambda = 4$ (**c**)

the logistic map and recognized by Feigenbaum as a universal property for a whole class of functions (the Feigenbaum-constant) [2.39]. The chaotic regime beyond α_c is shown in Fig. 2.23b.

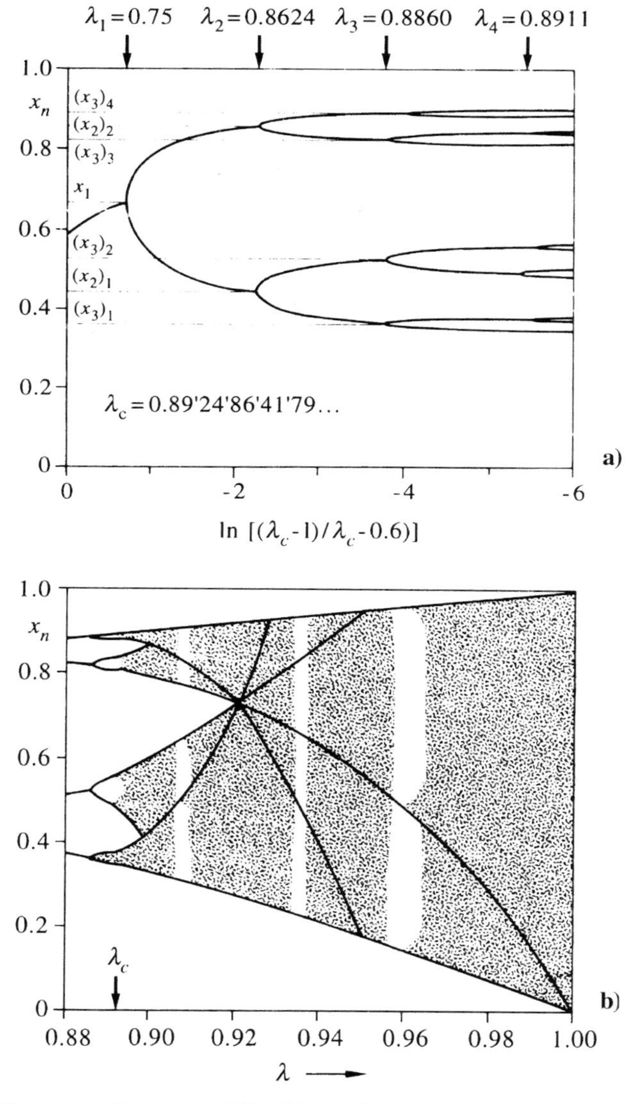

Fig. 2.23a,b. Sequence of period doubling bifurcations (**a**) and chaotic regime of the logistic map beyond $\alpha_c = 4\lambda_c$ (**b**)

In Fig. 2.24a–c the mappings of x_n onto x_{n+1} are illustrated for different control parameters, in order to construct the corresponding attractors of a fixed point, periodic oscillation between two points, and complete irregularity without any point attractor or periodicity.

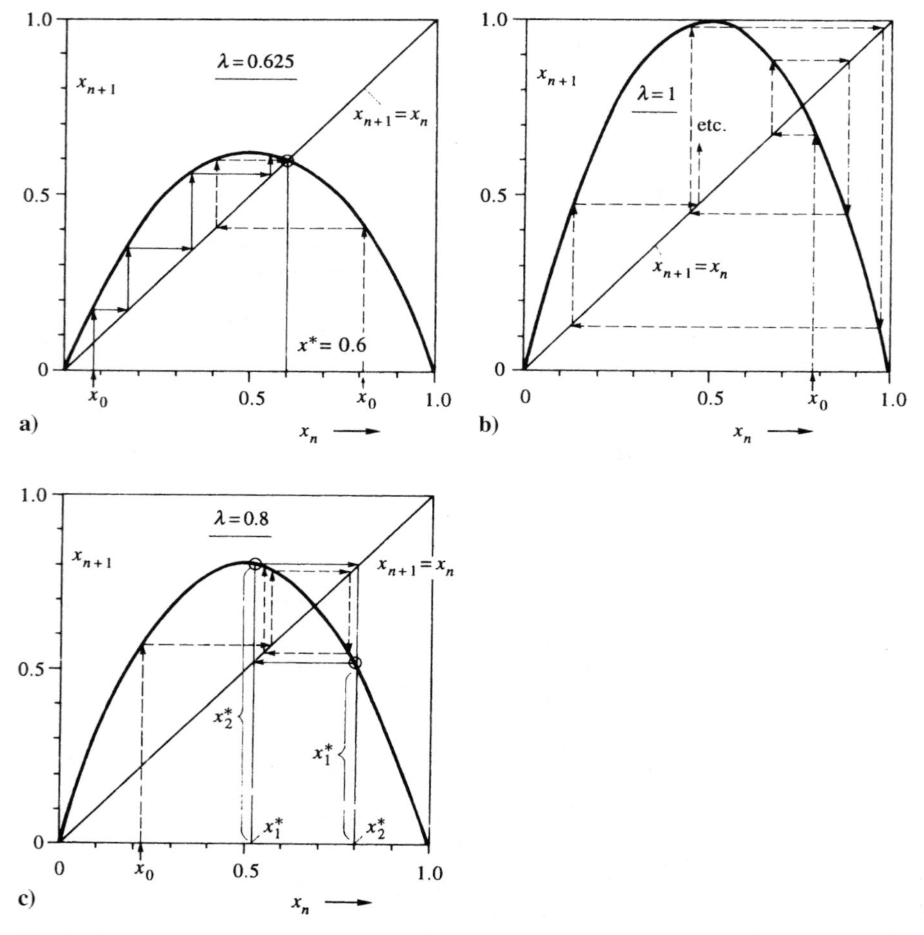

Fig. 2.24a–c. Attractors of logistic map with differing control parameter: fixed point attractor (**a**), periodic oscillation (**c**), and chaos (**b**)

It is rather astonishing that a simple mathematical law like the logistic map produces a complexity of bifurcations and chaos for possible developments as shown in Fig. 2.23a,b. A necessary, but not sufficient reason is the nonlinearity of the equation. In this context, the degrees of increasing complexity are defined by the increasing bifurcations which lead to chaos as the most complex and fractal scenario. Each bifurcation illustrates a possible branch of solution for the nonlinear equation. Physically, they denote phase transitions from a state of equilibrium to new possible states of equilibria. If equilibrium is understood as a state of symmetry, then phase transition means symmetry breaking which is caused by fluctuational forces.

Mathematically, symmetry is defined by the invariance of certain laws with respect to several transformations between the corresponding reference systems of an observer. In this sense the symmetry of Kepler's laws is defined by its Galilean transformations (compare Fig. 2.6a). The hydrodynamical laws describing a fluid layer heated from below (Fig. 2.20a) are invariant with respect to all horizontal translations. The equations of chemical reactions (in an infinitely extended medium) are invariant with respect to all translations, rotations, and reflections of a reference system used by an observer [2.40].

Nevertheless, these highly symmetric laws allow phase transitions to states with less symmetry. For example, in the case of a Bénard experiment, the heated fluid layer becomes unstable, and the state of stationary convection rolls develops (Fig. 2.20b). This phase transition means symmetry breaking, because tiny fluctuations cause the rolls to prefer one of two possible directions. Our examples show that phase transition and symmetry breaking is caused by a change of external parameters and leads eventually to a new macroscopic spatio-temporal pattern of the system and emergence of order.

Obviously, thermal fluctuations bear in themselves an uncertainty, or more precisely speaking, probabilities. A particle which is randomly pushed back or forth (Brownian motion) can be described by a stochastic equation governing the change of the probability distribution as a function of time. One of the most important means to determine the probability distribution of a process is the so-called master equation. To visualize the process we may think of a particle moving in three dimensions on a lattice.

The probability of finding the system at point x at time t increases due to transitions from other points x' to the point under consideration ("rate in"). It decreases due to transitions leaving this point ("rate out"). As the "rate in" consists of all transitions from initial points x' to x, it is composed of the sum over these initial points. Each term of the sum is given by the probability of finding the particle at point x', multiplied by the transition probability (per unit time) for passing from x' to x. In an analogous way the "rate out" can be found for the outgoing transitions. Thus, the rate of change for the probability distribution of a process is determined by a stochastic differential equation which is defined by the difference between "rate in" and "rate out".

Fluctuations are caused by a huge number of randomly moving particles. An example is a fluid with its molecules. So a bifurcation of a stochastic process can only be determined by the change of probabilistic distribution. In Fig. 2.25 the probabilistic function changes from a sharp centration at a single attractor (Fig. 2.25a) to

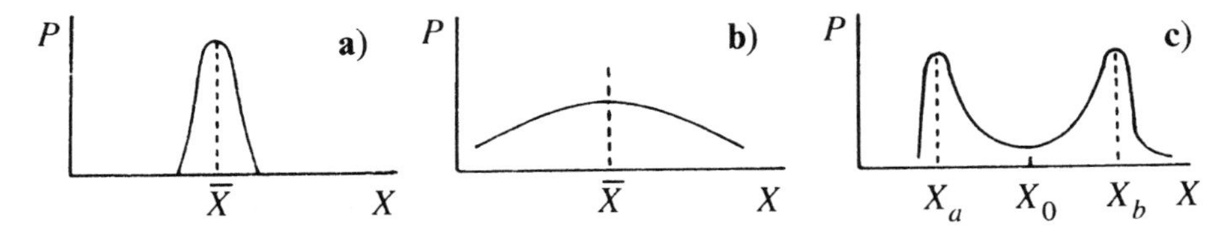

Fig. 2.25a–c. Probabilistic function with single attractor (**a**), flat distribution (**b**), and two attractors as stochastic symmetry breaking (**c**)

a flat distribution (Fig. 2.25b) and finally to a distribution with two maxima at two attractors (Fig. 2.25c), when the control parameter increases beyond corresponding critical values. Figure 2.25c illustrates stochastic symmetry breaking [2.41].

In this context, complexity means that a system has a huge number of degrees of freedom. When we manipulate a system from the outside we can change its degrees of freedom. For example, at elevated temperature the molecules of water vapor move freely without mutual correlation. When the temperature is lowered, a liquid drop is formed. This macroscopic phenomenon is formed when the molecules keep a mean distance between each other with correlated motion. At the freezing point water is transformed into ice crystals with a fixed molecular order. Since the early days of mankind people have been familiar with these phase transitions. The different aggregate states may have been a reason for philosophical ideas that water is a basic element of matter (compare Sect. 2.1).

Another example is taken from the material sciences. When a ferromagnet is heated, it loses its magnetization beyond a critical value. But the magnet regains its magnetization when the temperature is lowered. Magnetization is a macroscopic feature which can be explained by changing the degrees of freedom at the microscopic level. The ferromagnet consists of many atomic magnets. At elevated temperature, the elementary magnets point in random directions. If the corresponding magnetic moments are added up, they cancel each other. Then, on the macroscopic level, no magnetization can be observed. Below a critical temperature, the atomic magnets are lined up in a macroscopic order, giving rise to the macroscopic feature of magnetization (Fig. 4.9a). In both examples, the emergence of macroscopic order was caused by lowering the temperature. The structure is formed without loss of energy at low temperature. Thus, it is a kind of conservative (reversible) self-organization. Physically, it can be explained by Boltzmann's law of distribution demanding that structures with less energy are mainly realized at low temperatures.

On the other hand, there are systems whose order and functioning are not achieved by lowering temperature, but by maintaining a flux of energy and matter through them. Familiar examples are living systems like plants and animals which are fed by biochemical energy. The processing of this energy may result in the formation of macroscopic patterns like the growth of plants, locomotion of animals, and so on. But this emergence of order is by no means reserved to living systems (compare Chap. 3). It is a kind of dissipative (irreversible) self-organization far from thermal equilibrium which can be found in physics and chemistry as well as in biology.

As is well-known from the second law of thermodynamics, closed systems without any exchange of energy and matter with their environment develop to disordered states near thermal equilibrium. The degree of disorder is measured by a quantity called "entropy". The second law says that in closed systems the entropy always increases to its maximal value. For instance, when a cold body is brought into contact with a hot body, then heat is exchanged so that both bodies acquire the same temperature, i.e., a disordered and homogeneous order of molecules. When a drop of milk is put into coffee, the milk spreads out to a finally disordered and homogeneous mixture of milky coffee. The reverse processes are never observed. In this sense, processes according to the second law of thermodynamics are irreversible with a unique direction [2.42].

An example from hydrodynamics is the Bénard instability, which was already described in the beginning of Sect. 2.4. When the heated fluid layer (Fig. 2.20a) reaches a critical value, it starts a macroscopic motion (Fig. 2.20b). Thus a dynamic well-ordered spatial pattern emerges out of a disordered and homogeneous state as long as a certain flux of energy is maintained through the system.

Another example from fluid dynamics is the flow of fluid round a cylinder. The external control parameter is the Reynolds number *Re* of fluid velocity. At low speed the flow happens in a homogeneous manner (Fig. 2.26a). At higher speeds, a new macroscopic pattern with two vortices appears (Fig. 2.26b). With yet higher speeds

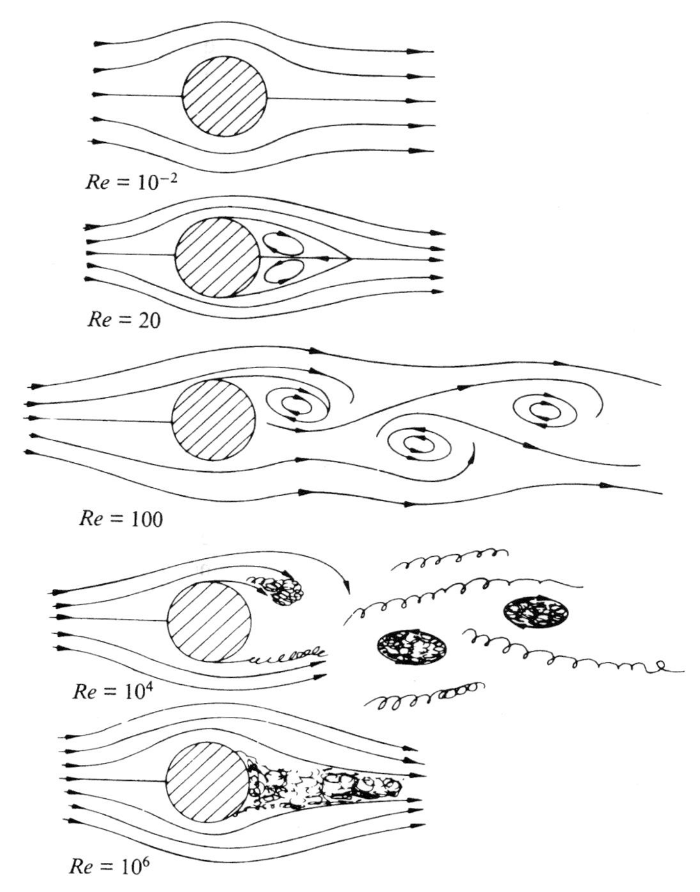

Fig. 2.26a–e. Macroscopic patterns of fluid dynamics with homogeneous state (**a**), two vortices (**b**), oscillations (**c**), quasi-oscillations (**d**), and chaos (**e**) behind a cylinder depending on increasing fluid velocity as control parameter

the vortices start to oscillate (Fig. 2.26c–d). At a certain critical value, the irregular and chaotic pattern of a turbulent flow arises behind the cylinder (Fig. 2.26e). Figure 2.26a–e presents a survey of possible attractors with one and more fixed points, bifurcations, oscillating, and quasi-oscillating attractors, and finally fractal chaos [2.43].

A famous example from modern physics and technology is the laser. A solid state laser consists of a rod of material in which specific atoms are embedded. Each atom may be excited by energy from outside leading it to the emission of light pulses. Mirrors at the end faces of the rod serve to select these pulses. If the pulses run in the axial direction, then they are reflected several times and stay longer in the laser, while pulses in different directions leave it. At small pump power the laser operates like a lamp, because the atoms emit independently of each other light pulses (Fig. 2.27a). At a certain pump power, the atoms oscillate in phase, and a single ordered pulse of gigantic length emerges (Fig. 2.27b) [2.44].

The laser beam is an exampel of macroscopic order emerging by a dissipative (irreversible) self-organization far from thermal equilibrium. With its exchange and processing of energy, the laser is obviously a dissipative system far from thermal equilibrium.

In former days of history, scientists would have postulated certain demons or mystic forces leading the elements of these systems to new patterns of order. But, as in the case of conservative self-organization, we can explain dissipative self-organization by a general scheme which is made precise by well-known mathematical procedures. We start with an old structure, for instance a homogeneous fluid or randomly emitting laser. The instability of the old structure is caused by a change of external parameters, leading eventually to a new macroscopic spatio-temporal structure. Close to the instability point we may distinguish between stable and unstable collective motions or waves (modes). The unstable modes start to influence and determine the stable modes which therefore can be eliminated. Hermann Haken calls this process very suggestively a "slaving principle". Actually, the stable modes are "enslaved" by the unstable modes at a certain threshold.

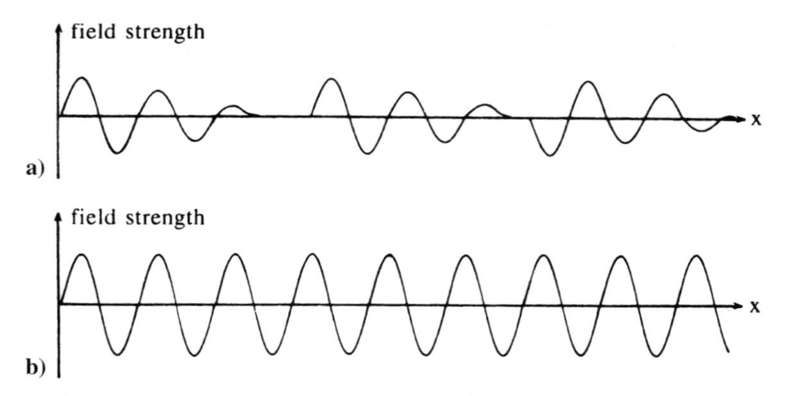

Fig. 2.27a,b. Wave patterns emitted from a lamp (**a**) and from a laser (**b**)

Mathematically, this procedure is well known as the so-called "adiabatic elimination" of fast relaxing variables, for instance, from the master equation describing the change of probabilistic distribution in the corresponding system. Obviously, this elimination procedure enables an enormous reduction of the degrees of freedom. The emergence of a new structure results from the fact that the remaining unstable modes serve as order parameters determining the macroscopic behavior of the system. The evolution of the macroscopic parameters is described by differential equations. In contrast to properties of the elements of a system at the microscopic level (for instance, atoms, molecules, etc.), the order parameters denote macroscopic features of the whole system. In the case of the laser, some slowly varying ("undamped") amplitudes of modes may serve as the order parameters, because they start to enslave the atomic system. In the language of biology, the order parameter equations describe a process of "competition" and "selection" between modes. But, of course, these are only metaphoric formulations which can be made precise by the mathematical procedure mentioned above [2.45].

In general, to summarize, a dissipative structure may become unstable at a certain threshold and break down, enabling the emergence of a new structure. As the introduction of corresponding order parameters results from the elimination of a huge number of degrees of freedom, the emergence of dissipative order is combined with a drastic reduction of complexity. Dissipative structures are a fundamental concept of complex systems which are used in this book to model processes in natural and social sciences. The irreversibility of dissipative structures may remind us of Heraclitus' famous quotation that nobody can enter a stream in the same state. Obviously, irreversibility violates the time-invariance symmetry which characterizes the classical (Hamiltonian) world of Newton and Einstein. But the classical view will turn out to be a special case in a steadily changing world. On the other hand, Heraclitus believed in an ordering law harmonizing irregular interactions and creating new order states of matter. We have to see wether the mathematical scheme of a dissipative system will satisfy the universal features of such a law.

A general framework for the evolution of matter would be based on a unified theory of all physical forces (Fig. 2.28). The standard models of cosmic evolution which are derived from Einstein's general theory of relativity must be explained by the principles of quantum theory. Until today there are only several more or less satisfying mathematical models of cosmic evolution which can only partially be tested and confirmed by experiments. Nevertheless, it is the general idea of these models that the emergence of structures with increasing complexity (elementary particles, atoms, molecules, planets, stars, galaxies, etc.) can be explained by cosmic phase transitions or symmetry breaking [2.46].

In cosmic evolution an initial state is assumed to be nearly homogeneous and symmetric in the sense that in general no elementary particles can be distinguished, but they can be transformed into one another. During cosmic evolution, critical values have been realized step by step at which symmetries break down by deviations and fluctuations and new particles and forces emerge: "C'est la dissymétrie, qui crée le phénomène," said Pierre Curie [2.47]. But we must be aware that the cosmic pro-

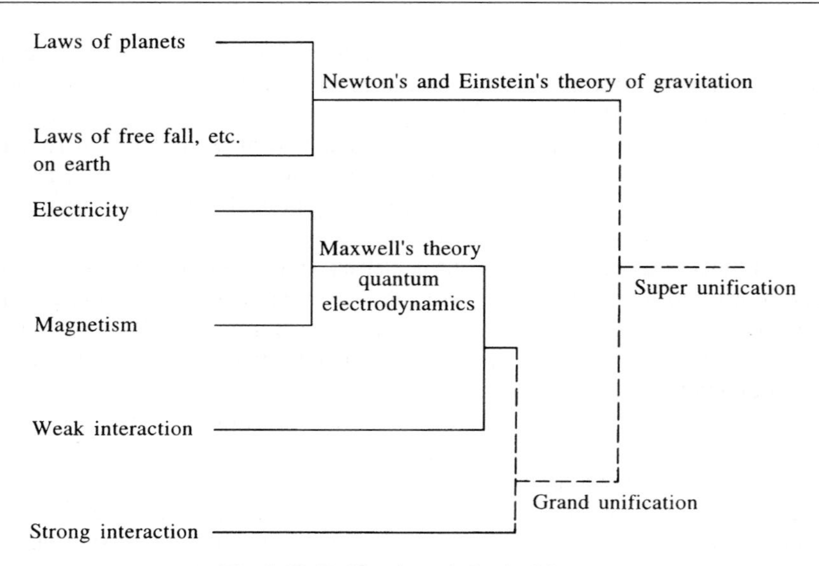

Fig. 2.28. Unification of physical forces

cesses of symmetry breaking and phase transitions are mathematical extrapolations from experiments and theories in high energy physics.

Nowadays, physics distinguishes four fundamental forces, the electromagnetic, strong, weak, and gravitational forces which are mathematically described by so-called gauge fields. Elementary particle physics aims to unify the four physical forces in one fundamental force corresponding to the initial state of the universe. Electromagnetic and weak forces have already been unified at very high energies in an accelerator ring at CERN (Fig. 2.28). Unification means that at a state of very high energy the particles that "feel" the weak force (electrons, neutrinos, etc.) and those that "feel" the electromagnetic force cannot be distinguished. They can be described by the same symmetry group ($U(1) \times SU(2)$), i.e., they are invariant with respect to transformations of this group. At a particular critical value of lower energy the symmetry breaks down into partial symmetries ($U(1)$ and $SU(2)$) corresponding to the electromagnetic and weak forces.

Physically, this kind of symmetry breaking means a phase transition which is connected with the emergence of two new physical forces and their elementary particles. The process of spontaneous symmetry breaking is well known. For instance, our breakfast egg is not stable in its symmetric position on its top. Any tiny fluctuation causes it to fall spontaneously down to an asymmetric, but energetically stable position. The phase transition of a ferromagnet from a non-magnetic to a magnetic state is caused by cooling down the temperature to a critical point. The elementary dipoles spontaneously take one of the two possible magnetic orientations, break the spin-rotation symmetry, and cause the emergence of a new macroscopic property (magnetization).

The complex variety of baryons (protons, neutrons, etc.) and mesons interacting via the strong force are constructed from the so-called quarks with three degrees of freedom, i.e., the so-called "colors" red, green, and blue. A baryon, for instance, is built up from three quarks which are distinguishable by three different colors. These three colors are complementary in the sense that a hadron is neutral (without color) to its environment. The mathematical symmetry group (SU(3)) characterizing the color transformation of quarks is well known.

After the successful unification of the electromagnetic and weak interactions physicists try to realize the "grand unification" of electroweak and strong forces, and in a last step the "superunification" of all four forces (Fig. 2.28). There are several research programs for superunification, such as supergravity and superstring theory. Mathematically, they are described by extensions to more general structures of symmetry ("gauge groups") including the partial symmetries of the four basic forces. Technically, the unification steps should be realized with growing values of very high energy. But the "grand unification" demands states of energy which cannot be realized in laboratories. Thus, the high energy physics of grand unification could only be confirmed by certain consequences which could be tested in a laboratory or observed in the universe (e.g., the decay of protons). The superunification of all forces would demand infinitely increasing states of energy whose physical principles are still unknown.

The theory of the so-called "inflationary universe" assumes an early state of the universe with small size, but very high energy ("quantum vacuum") which expands very rapidly to macroscopic dimensions driven by a repulsive force of the quantum vacuum state ("anti-gravity"). This cosmic phase transition allows one to explain some well-known properties of the observed universe such as the relatively homogeneous distribution of stars and matter. During the inflationary period, some tiny deviations from symmetry and uniformity would have been amplified until they were big enough to account for the observed structures of the universe. In the expanding universe the density of matter varied slightly from place to place. Thus, gravity would have caused the denser regions to slow down their expansion and start contracting. These local events led to the formation of stars and galaxies [2.48].

In general, the emergence of the structural variety in the universe from the elementary particles to stars and living organisms is explained by phase transitions, corresponding to symmetry breaking of equilibrium states (Figs. 2.29, 2.30). In this sense the cosmic evolution of matter is understood as a self-organizing process with the emergence of conservative and dissipative structures. But we must be aware that cosmic self-organization is today only a "regulative idea of research", as Kant had said: we have more or less plausible dynamical models which are more or less empirically confirmed. The very beginning of cosmic evolution is still unknown.

If we only assume the classical principles of Einstein's general relativity, then, as Roger Penrose and Stephen Hawking have mathematically proved, the standard models of cosmic evolution have an initial singularity which may be interpreted as the Big Bang, i.e., the emergence of the universe from a mathematical point. But if we assume a unification of the general theory of relativity (i.e., Einstein's relativistic theory of gravitation) and quantum mechanics with imaginary (instead of real)

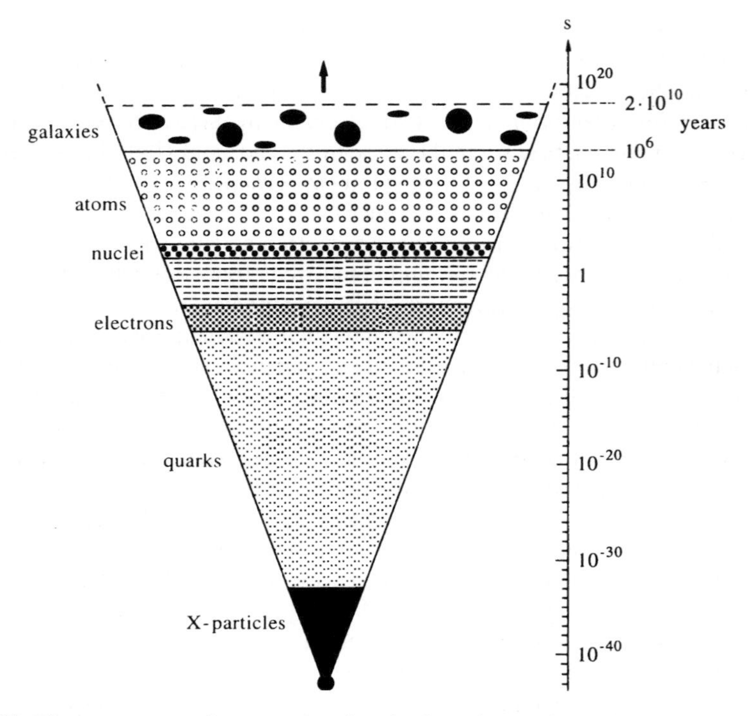

Fig. 2.29. The emergence of structural variety in the universe from elementary particles to galaxies [2.48]

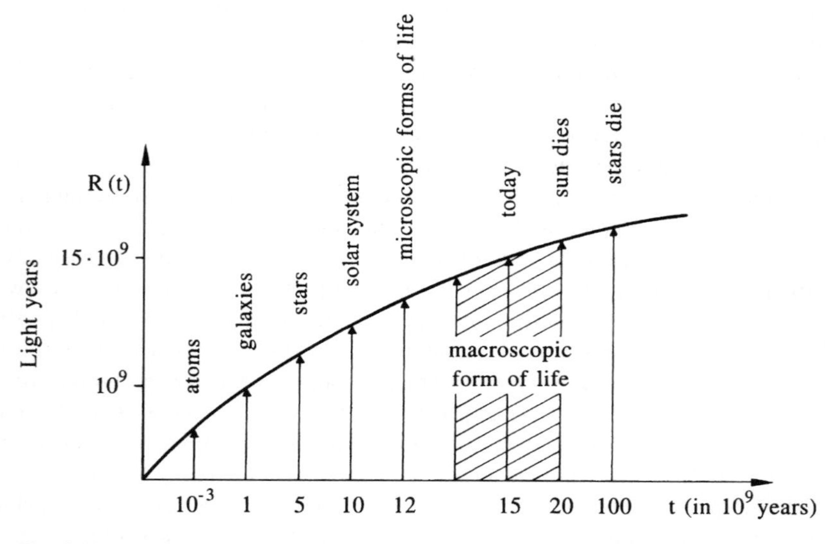

Fig. 2.30. The evolution of matter with increasing and decreasing complexity [2.51]

time, then, as Hawking has mathematically proved, a "smooth" cosmic model is possible without any beginning, which simply exists, according to the mathematical principles of a unified relativistic quantum physics [2.49].

The singularity theorems of Penrose and Hawking started with predictions of small regions of space where space-time is so warped that gravity becomes infinitely great. The existence of such singularities, in the form of black holes, for example, suffer from a methodological disadvantage: Classical and relativistic laws of physics are not applicable in regions with infinite curvature, so it is not possible to predict events in time. That consequence is, of course, quite more dramatic than the exponentially increasing difficulties in predicting the long-term future of chaotic systems. That is why James B. Hartle and Stephen Hawking have suggested a singularity-free model of the universe, in which quantum theory and general relativity theory are unified, and the real time axis is replaced by an imaginary one (in the sense of real and imaginary numbers) [2.50]. In Hawking's model, in contrast to Einstein's relativistic theory, the three spatial axes, together with a complex time axis, lead to a closed early quantum universe that lacks boundaries and edges. This space-time not only would always have existed, but every physical event could be explained according to its laws. In this model, the traditional concepts of everything having somehow "begun" or been "created" are methodologically inappropriate and are revealed to be human imaginings stemming from our having adapted to the limited space-time facets of our everyday experience.

Hawking's theory is not only mathematically consistent, but is also, at least in principle, experimentally testable. It is, therefore, a scientific theory and not mere speculation. Among the testable consequences of this singularity-free model is the prediction of black holes in which not all world lines of photons ("light beams") disappear entirely, but are reemitted as measurable amounts of radiation. As in the explanation of the initial singularity of the universe, the reason lies in the possibility of quantum fluctuations rooted in the uncertainty relation. But radiating black holes lose energy and mass. In time, they will disintegrate and, with them, the history of their stars will be lost. In their place, memory gaps will appear in the universe. With the collapse of its galactic structures, a featureless universe expanding into a void is heading for a "cosmic Alzheimer's disease".

Philosophically, Hawking's early quantum universe without a beginning reminds us of Parmenides world of unchangeable being. But the uncertainty principle of quantum mechanics implies that the early universe cannot have been completely uniform because there must have been some uncertainties or fluctuations in the positions and velocities of the particles. Thus, the universe would have undergone a period of rapid expansion which is described by the inflationary model, leading to our complex universe in the long run. The equilibrium of the Parmenidean world broke down and changed to the evolutionary and complex world of Heraclitus, caused by a basic principle of quantum physics under the hypothesis of a "smooth" time without singularities.

A cosmological model of an "eternal" universe without beginning and without end was already introduced by Hermann Bondi, Thomas Gold, and Fred Hoyle in 1948. These authors did not only assume spatial homogeneity and isotropy of the universe at every time ("Cosmological Principle" of the standard models with Big Bang), but also temporal homogeneity and isotropy: Their "perfect Cosmological Principle" suggests that the universe globally looks the same not only at all points and in all directions, but at all times, leading to a steady state model. According to Hubble, there is a correlation between the red shift and the increasing distances of expanding galaxies. So if the average number of galaxies per unit proper volume is to remain constant, new galaxies must appear to fill up the holes in the widening comoving coordinate mesh. An ad-hoc hypothesis of steady state cosmology was the necessity of continuous creation of matter.

In recent quasi-steady state cosmologies, the strange assumption of a contingent and nonlocal creation of matter is explained by the local birth of new galaxies everywhere and at every time in the universe. The conditions of local big bangs are assumed to be realizable in the supermassive centers of old galaxies. The red shifts seems also to indicate the age of a galaxy. The uniform evolution with the sequential emergence of elementary particles, atoms, molecules, galaxies, stars, etc. after the global Big Bang (Fig. 2.29) is replaced by an autocatalytically self-reproducing universe without global beginning and without end, but with local births, growths, and deaths of galaxies. In this case, old dying galaxies create the matter of new galaxies like plants and organisms bearing the seed of new life. The universal dynamics would be a gigantic never ending nonlinear recycling process of matter [2.49].

But, perhaps, the laws of quantum mechanics open loopholes ("wormholes") of escape from the fate of our universe. According to general relativity theory, time travel cannot be faster than the speed of light. As light is curved by gravitational fields, time travelers must pass curved paths in space-time with high speed, limited by the speed of light. Therefore, in order to overcome disruption of space-time by gravitational fields, space-time regions would have to be explored using vast curved detours. According to Heisenberg's principle of uncertainty, quantum fluctuations could open short-lived wormholes in space-time. So, the laws of quantum mechanics make it at least conceivable that wormholes can be employed as fleeting shortcuts between folded regions. However, if our universe is not alone but is instead interwined with a fractal multiverse, along with many other bifurcating universes, as was suggested in Andrei Linde's inflationary theory, then wormholes could also be used as escape routes for fleeing a universe that is aging with cosmic Alzheimer's disease and growing hostile to life as it loses energy.

From a theological point of view, these models do not need any creator, because their worlds simply have been and will be self-contained and self-organizing without beginning and without end. From a mathematical point of view, these models may be very elegant. But from a methodological point of view, we must conclude that we do not yet have a complete and consistent theory combing quantum mechanics and relativistic gravity which could explain the evolution of matter with its increasing complexity. Thus we are only certain of some of the properties by gravitational fields such a unified theory could have. Today, different approaches of string theories exist for achieving this unification on a sublevel of elementary particles. If all kinds of elementary particles of gravitational, strong, weak, and electromagnetic interactions are generated by oscillating strings, then there is even a chance to avoid the ultimate loss of information in the black holes of an aging universe: The information could be stored by the vibrating membranes of more-dimensional strings on the substructure of matter [2.52].

2.5 Complex Systems of the Nano World and Self-Constructing Materials

In the evolution of matter, self-organizing processes can be observed from the level of elementary particles to the cosmic structures of galaxies. They are not only interesting from an epistemic point of view, but for applications in materials and life science, too. At the boundary between materials science and life science, supramolecular systems play a tremendous role. In this case molecular self-organization means the spontaneous association of molecules under equilibrium conditions into stable and structurally well-defined aggregates with dimensions of $1–10^2$ nanometers ($1\,\text{nm} = 10^{-9}\,\text{m} = 10\,\text{Å}$).

Nanostructures may be considered as small, familiar, or large, depending on the view point of the disciplines concerned. To chemists, nanostructures are molecular assemblies of atoms numbering from 10^3 to 10^9 and molecular weights of 10^4 to 10^{10} daltons. Thus, they are chemically large supramolecules. To molecular biologists, nanostructures have the size of familiar objects from proteins to viruses and cellular organelles. But to materials scientists and electrical engineers, nanostructures are at the current limit of microfabrication and thus they are rather small [2.53].

In the beginning of nanoscience there was the vision of an ingenious physicist. In an article entitled "There's Plenty of Room at the Bottom", Richard Feynman declared:

> The principles of physics, as far as I can see, do not speak against the possibility of maneuvering things atom by atom. It would be, in principle, possible ... for a physicist to synthesize any chemical substance that the chemist writes down ... How? Put the atoms down where the chemist says, and so you make the substance. The problems of chemistry and biology can be greatly helped if our ability to see what we are doing, and to do things on an atomic level, is ultimately developed – a development which I think cannot be avoided. [2.54]

Feynman proclaimed his physical ideas of the nanoworld in the late 1950s. The belief in a new world needs new instruments of observation and measurement for confirmation. Since the start of the 1980s, the nanoworld could actually be explored using the scanning tunnel microscope. At the end of the 1980s, Eric Drexler described a revolutionary vision of technological applications:

> Nature shows that molecules can serve as machines because living things work by means of such machinery. Enzymes are molecular machines that make, break, and rearrange the bonds holding other molecules together. Muscles are driven by molecular machines that haul fibers past one another. DNA serves as a data-storage system, transmitting digital instructions to molecular machines, the ribosomes, that manufacture protein molecules. And these protein molecules, in turn, make up most of the molecular machinery just described. [2.55]

With nanotechnology, atoms will be specifically placed and connected in a fashion similar to processes found in living organisms. Complex organisms, such as plants and animals, make use of molecular machinary to manufacture and undertake repairs at the cellular and subcellular levels. A cell can be considered a factory of nanomachines consisting of molecular prototypes such as protein, nucleic acid, lipid, and polysaccharide. They are used for energy production, information processing, self-replication, self-repairing, and moving. A ribosome, for example, is a cellular nanomachine that reads information off a RNA strand in order to construct the amino acids of a protein. It reminds us of the assembly-like production of cars by robots in the motor industry. Biological micro-organisms have been understood as cellular systems driven and controlled by nanomachines. For example, bacteria such as Escherichia coli use whip-like tails for moving around in fluids. The tails like a propeller fueled by biochemical nanomachines. These nanomachines consist of proteins in membranes generating the rotation of whip-like tails. They use motor shafts and armatures like electric motors. But the similarity of nanomachines and electric motors is only illustrative. A biochemical nanomachine does not use electric current to generate a magnetic field; it changes the shape of molecules by biochemical procedures, such as decomposing ATP, in order to rotate the shaft [2.56].

Genetic engineering and computer programming have begun to inspire the development of new materials. Using special bacterium-sized assembler devices, nanotechnology should permit the exact control and fast manipulation of molecular structures. A fast enzyme can process almost a million molecules per second, even without conveyors and power-driven mechanisms to slap a new molecule into place as soon as an old one is released. Drexler assumed that an assembler arm would be about fifty million times shorter than a human arm and, accordingly, would be able to move back and forth about fifty million times more rapidly. According to Feynman's vision, such machines would seize individual atoms using selectively sticky manipulator arms, then plug those atoms together like Lego blocks until chemical bonding took place. Following the line of computer programming, one would expect general-purpose chemical synthesizers acting like a general-purpose computer using nanotechnology. The desired molecules would be modeled on a computer screen and an appropriate assembler would allow the mass-production of the desired substances. Perhaps someday, specially designed nanodevices the size of bacteria will be programmed to destroy arterial plaque or cancer cells, or to repair cellular damage caused by aging. They could be injected into the body with an induction to self-destruct or integrate themselves into the body's cells. Finally, it still seems to be science fiction that smart nanodevices distributed throughout the brain might permit the copying of thought patterns and mind uploading, so that a copy of a person's personality and memories could be placed in storage, or even run as a form of naturally created artificial intelligence.

Nanostructures are complex systems which evidently lie at the interface between solid-state physics, supramolecular chemistry, and molecular biology. It follows that the exploration of nanostructures may deliver hints about the emergence of life and about the fabrication of new materials. But engineering of nanostructures cannot be mastered in the traditional way of mechanical construction. There are no man-made tools or machines for putting together their building blocks like the elements of a clock, motor, or computer chip. We must thus understand the principles of self-organization which are used by nanostructures in nature. Then, we only need to arrange the appropriate constraints under which the atomic elements of nanostructures associate themselves in a spontaneous self-construction: The elements adjust their own positions to reach a thermodynamic minimum without any manipulation by a human engineer.

Historically, the idea of supramolecular interactions dates back to a famous metaphor of Emil Fischer (1894), who described a selective interaction of molecules as the lock and key principle. Today, supramolecular chemistry has far surpassed its original focus. Molecular self-assemblies combine several features of covalent and noncovalent synthesis to make large and structurally well-defined assemblies of atoms. The strengths of individual van der Waals interactions and hydrogen bonds are weak relative to typical covalent bonds and comparable to thermal energies. Therefore, many of these weak noncovalent interactions are necessary in order to achieve molecular stability in self-assembled aggregates. In biology, there are many complex systems of nanoscale structures such as proteins and viruses which are formed by self-assembly. Living systems sum up many weak interactions between chemical entities to make large ones. How can one make structures of the size and complexity of biological structures, but without using biological catalysts or the informational devices coded in genes?

Many nonbiological systems also display self-organizing behavior and furthermore provide examples of useful interactions. Molecular crystals are self-organizing structures. Liquid crystals are self-organized phases intermediate in order between crystals and lipids. Micelles, emulsions, and lipids display a broad variety of self-organizing behavior. An example is the generation of cascade polymers yielding molecular bifurcational superstructures of fractal order [2.57]. Their synthesis is based on the architectural design of trees. Thus, these supramolecules are called dendrimers (from the Greek word *dendron* for tree and polymer). The generation of dendrimers has followed two basic procedures for monomer addition. A divergent construction begins at the core and builds outward via an increasing number of repeating bifurcations. A convergent construction begins at the periphery and builds inward via a constant number of transformations. The divergent construction transforms the chemical reaction centers from the center into the periphery, generating a network of bifurcating branches around the center. The bifurcations increase exponentially up to a critical state of maximal size. They yield fractal structures such as molecular sponges which can contain smaller molecules, which can then be dispersed in a controlled way for medical applications.

Examples of cave-like supramolecules are the Buckminsterfullerenes, forming great balls of carbon [2.58]. The stability of these complex clusters is supported by their high geometric symmetry. The Buckminsterfullerenes are named after the geodesic networks of ball-like halls which were constructed by the American architect Richard Buckminster Fuller (1895–1983). The cluster C_{60} of 60 carbon atoms has a highly Platonic symmetry of atomic pentagons forming a completely closed spheroid.

Cave-like supramolecules can be arranged using chemical templates and matrices to produce complex molecular structures. Several giant clusters comparable in size to small proteins have been obtained by self-assembly. Figure 2.31 shows a ball-and-stick model of the largest discrete cluster (700 heavy atoms) ever characterized by X-ray structure analysis. This cluster containing 154 molybdenum, 532 oxygen, and 14 nitrogen atoms has a relative molecular mass of about 24 000. The highly symmetric "big wheel" was synthesized by Achim Müller and coworkers [2.59]. Giant clusters may have exceptional novel structural and electronic properties: There are planes of different magnetization which are typical for special solid-state structures and of great significance for materials science. A remarkable structural property is the presence of a nanometer-sized cavity inside the giant cluster. The use here of templates and the selection of appropriate molecular arrangements may well remind us of Fischer's lock and key principle.

Molecular cavities can be used as containers for other chemicals or even for medicaments which need to be transported within the human organism. An iron-storage protein that occurs in many higher organisms is ferritin. It is an unusual host-guest system consisting of an organic host (an aprotein) and a variable inorganic guest (an iron core). Depending on the external demand, iron can either be removed from this system or incorporated into it. Complex chemical aggregates

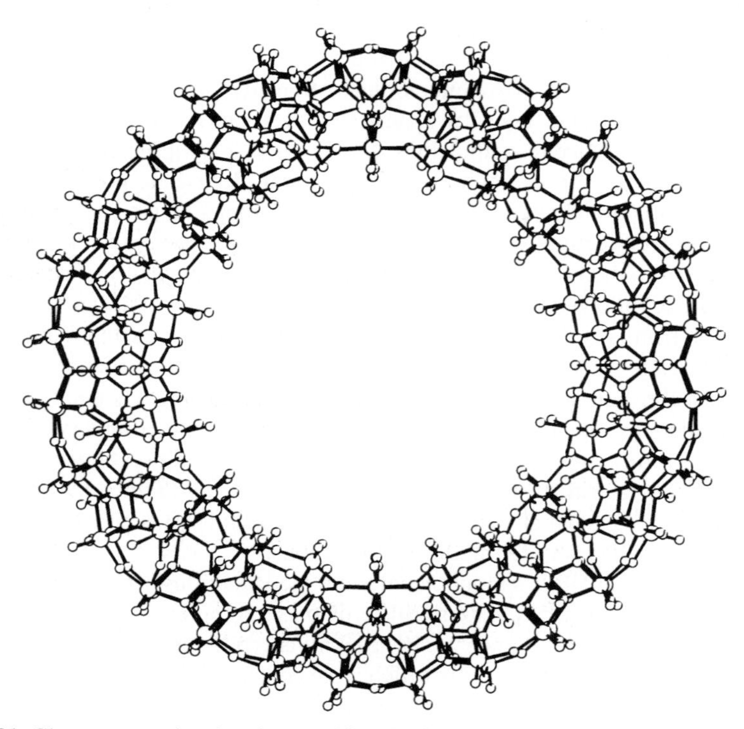

Fig. 2.31. Giant supramolecular cluster ("big wheel") in a ball-and-stick representation: An example of a complex near-equilibrium system [2.59]

like polyoxometalates are frequently discovered to be based upon regular convex polyhedra, such as Platonic solids. But their collective electronic and/or magnetic properties cannot be deduced from the known properties of these building blocks. According to the catchphrase "from molecules to materials" supramolecular chemistry applies the "blue-prints" of conservative self-organization to build up complex materials on the nanometer scale with novel catalytic, electronic, electrochemical, optical, magnetic, and photochemical properties. Multi-property materials are extremely interesting.

The exploration of the nanoworld and applications in nanotechnology depend on better instruments of observation and measurement. The scanning force microscope is a further development of the scanning tunnel microscope and can be used like a fountain pen to write down molecular structures of nano size. A thin film of thiolmolecules is used as "nano ink". In a tiny drop of water the thiolmolecules organize themselves as mono layer. Nanocrystals of a few hundred atoms can organize themselves with cadmium ions, selen ions, and organic molecules in to a ball-like structure (Fig. 2.32). In ultraviolet light they fluoresce with a certain color. Thus, they could be used as markers ("quantum dotes") of molecules, cells, and substances in medicine, for example. Complex systems of carbon molecules can organize themselves as tiny tubes of 1nm diameter according to certain catalysts and templates. Their symmetric order of bonding results in great hardness and toughness. Carbon nanotubes might be used as conductors for miniaturized chips beyond the limits of silicon technology.

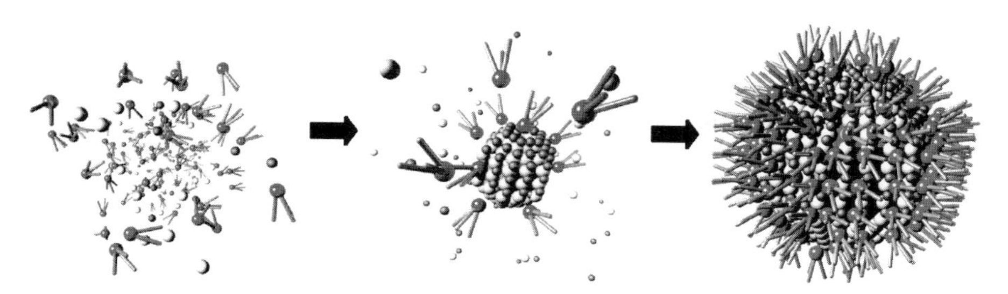

Fig. 2.32. Self-organizing nanocrystals ("quantum dots") [2.60]

Supramolecular transistors are an example that may stimulate a revolutionary new step in the development of chemical computers. Actually, there is a strong trend towards nanostructures in electronic systems which may realize small, fast devices and high-density information storage. But one can also imagine nonelectronic applications of nanostructures. They could be used as components in microsensors or as catalysts and recognition elements in analogy to enzymes and receptors in living systems. In natural evolution very large complex molecular systems are also produced by stepwise gene-directed processes. The conservative self-organization processes of nanomolecular chemistry are non-gene-controlled reactions. Only a clever combination of conservative and non-conservative self-organization could have initiated prebiotic evolution before genes emerged. But even during the evolution of complex organisms, conservative self-organization must have occurred. Open ("dissipative") physical and chemical systems lose their structure when the input of energy and matter is stopped or changed (e.g., laser, BZ-reaction). Organismic systems (like cells) are able to conserve much of their structure at least for a relatively long time. On the other hand, they need energy and matter within a certain interval of time to keep their structure more or less far from thermal equilibrium. In the technical evolution of mankind, the principles of conservative and dissipative self-organization have once more been discovered and open new avenues of technical applications.

The complex systems approach enables engineers to endow materials with more and more of the attributes of living organisms. Self-regulation and self-adaption to a changing environment are well-known capabilities of living systems. They can be considered as specific forms of self-organizing open systems in a changing environment. Analogously, engineers aim to create complex materials systems that can sense their own state, the state of their environment and respond to it. Dramatic examples are materials for bridges that could detect and counter corrosion before a pylon gives way, buildings that could brace themselves against seismic waves, or skins of aeroplanes that could spontaneously react against dangerous material fatigue.

Actuators are materials which can change their features according to changing states of the system [2.61]. Examples are piezoelectric ceramics and polymers acting either as pressure sensors or as mechanical actuators. The electrical polarity of their crystal or molecular structures allows a transformation of mechanical forces exerted on them into electrical current or, conversely, a transformation of electrical stimuli into vibrations. Piezoelectric polymers could be embedded in the skin of a robotic hand in order to get a high degree of sensitivity (e.g., to decipher braille).

Other examples are alloys with a so-called shape memory that can be used as actuators. Below a certain control value of transition temperature, a shape-memory wire will take any shape it is bent into. When the wire is heated beyond the transition, it returns to its original shape. Engineers propose the incorporation of a shape-memory metal into a material system in its low- temperature shape. It exerts a force whenever it is heated. The force-generating transition takes place as the atoms in the alloy's crystal grains toggle between different geometric arrangements. Damage-resisting bridges or airplane wings would be possible applications of these control structures.

There are even actuator materials that can reversibly transform their mechanical properties from a liquid to a solid state. They consist of fine polarizable particles of ceramic or polymer suspended in a liquid such as silicone oil. When subjected to strong electric fields, such fluids organize themselves into filaments and networks which stiffen the material into a gel-like solid. When the electric field is removed, the organization dissipates, and the material becomes fluid again. Other applications are optical fibers acting as sensor materials. The properties of these hair-thin fibers are affected by changes in temperature, pressure or other physical or chemical conditions within the materials. They can be considered as "glass nerves" providing optical signals of the material's internal "health".

Sometimes, modern materials scientists call their systems smart or even "intelligent" materials. The goal of their research is sometimes described as "animation of the inanimate world" [2.62]. From a philosophical point of view, this slogan seems to hark back to alchemistic traditions. Some philosophers of science may perhaps criticize the vocabulary of materials scientists as non-scientific animism. But, from the view point of complex systems, there is a hard scientific core. Properties of self-organization are not necessarily combined with conscious behavior on the basis of nervous systems. They even do not necessarily depend on biological catalysts or the informational devices coded in genes. Thus, there is no break between the so-called inanimate and animate world. In the evolution of matter, we observe systems with more or less high degrees of organization. It is clear that we have only made the very first steps in understanding their full potential.

Concerning the future of technology, the question arises, how realistic is the vision of self-replicating nanorobots? They would be the equivalent of a new parasitic life form. Pathogenic bacteria and cancer cells are dangerous examples of self-replicating biological systems. Computer viruses with self-replicating strings of bits are the first, at least virtual, examples of artificial self-organizing systems. Bill Joy, the chief scientist of Sun Microsystems, has already raised concerns about the societal implications of proliferating nanobots [2.63]. In an artificial evolution, Joy says, hostile agents could evolve into populations of embodied biochemical agents of nano size. As autonomous, self-interested beings, they could attack the foundations of human life. Richard E. Smalley, who received the Nobel Prize in chemistry for the discovery of fullerenes, dismisses the notion of out-of-control nanorobots [2.64]. Following Feynman's slogan, "There's plenty of room at the bottom", Smalley argues that not much room is needed to manipulate atoms one by one with nano-sized atomic instruments. He calls these constraints the fat and sticky fingers problem: The nanobot's manipulating "fingers" are not only too large ("fat") but also too sticky, because their atoms will adhere to the atom that is being moved. Smalley's picture of fingers underlines the fact there are no counterparts of our today technology at nanometer sizes. In living systems, evolution has developed examples of biochemical nanomachines, and there is no reason to believe that there cannot be others on different material grounds. But the technological strategy should follow the natural idea of self-organization under appropriate constraints, not the old- fashioned mechanical idea of picking and placing atoms with nanoscale pincers. We should not look for assemblers, but self-assemblers. From the point of view of computer science, the idea of a universal fabricator of any kind of structure, including itself, is not strange. A universal Turing machine (compare Sect. 5.2) is already embodied by our general-purpose computers, which process all kinds of programs. Why not on the nanoscale?

2.6 Time Series Analysis, Fractals, and Multifractals

Understanding complex systems and nonlinear dynamics in nature seems to yield appropriate models for the evolution of matter. But how can we be sure that our models are correct? The mathematical theory of nonlinear dynamics distinguishes different types of time-dependend equations, generating different types of behavior, such as fixed points, limit cycles, and chaos. For application, they are related to natural systems in the micro, nano, and macroworld. We use our understanding of the special mechanisms to write an appropriate dynamical equation. For example, Lorenz's understanding of the dynamics of weather led to his famous nonlinear equations, which were also applied to biological and economic systems by people familiar with those fields.

From a methodological point of view, this is the top-down approach to model building: We start with an assumed mathematical model of a natural system and deduce its behavior by solving the corresponding dynamical equations under certain initial and secondary conditions. The solutions can be represented geometrically as trajectories in the phase space of the dynamical system and classified by different types of attractors. They forecast the types of behavior that we are likely to observe in a specialized field of research. Especially chaotic dynamics can be derived from the given equations if certain criteria are satisfied. But, in practice, we often must take the opposite, bottom-up approach. Physicists, chemists, biologists, or physicians start with data mining in an unknown field of research. They only get a finite series of measured data corresponding to time- dependend events of a dynamical system. From these data they must reconstruct the behavior of the system in order to guess its type of dynamical equation. Therefore, the bottom-up approach is called time series analysis [2.65]. In many cases, we have knowledge of the system from which the data came. Time series analysis then aims to construct a black box, which takes the measured data as input and provides as output a mathematical model describing the data. In practice, the realistic strategy of research is a combination of the top-down approach with model building and the bottom-up approach with time series analysis.

The bottom-up approach starts with data as results of measurements, not with the idealized variables of a model. The measurements approximate the variables of a dynamical model. Their difference is called measurement error, which can be caused by several factors of noise. Noise of measurements refers to fluctuations of data that differ from a well-defined average behavior and arise from chance. While measurement noise is caused by the intrinsic behavior of the real system, the outside influence of the system also affects a kind of noise. Many variables of outside influence must be excluded in order to reduce the complexity of model building. The outside influence on the actual behavior of a system is considered random noise affecting the measured variables of the model.

In classical measurement theory, measurement error is analyzed by statistical methods, such as correlation coefficient and autocorrelation function. But these standard procedures are not able to distinguish between data from linear and nonlinear models. In nonlinear data analysis, the measured data are used in a first step to reconstruct the dynamics of the system in a reconstructed phase space. A simple example is the finite difference equation of the logistic map, which we studied in Sect. 2.4: The nonlinear equation $x_{t+1} = f(x_t)$ describes a relationship between x_{t+1} and x_t. In Fig. 2.24, the coordinates are plotted x_{t+1} versus x_t. If there is no measurement noise, we can identify the measurement data D_t and the variable x_t at time t. It is no surprise then, that a scatter plot of the measured data D_{t+1} versus D_t delivers the same relationship as the model.

If data are collected from a continuous-time dynamical system with differential equations, rather than finite-difference equations, the corresponding phase plane or phase space must be reconstructed from the measured data of the continuous system. The heuristic idea is that the measured data in the reconstructed phase space show the same dynamical behavior as the trajectories in the phase space of the dynamical model. Consider, for example, the data generated by a harmonic oscillator with the 2nd order differential equation $d^2x/dt^2 = -bx$. The corresponding phase plane is given by the variables x and y, which are determined by the two 1st order differential equations $dx/dt = y$ and $dy/dt = -bx$. We suppose that a time series $D(t) = x(t)$ is measured without measurement noise. In order to reconstruct the phase plane from the measured data, we remember that the state of the system at any instant t is represented by the position (x, y) on the phase plane. The time series of measurements yields us only one coordinate $D = x$ at every instant. But we can calculate the other coordinate $y = dD/dt$ from the 1st order differential equation of the phase space. A plot dD/dt versus D generates a continuous phase plane. In the reconstructed discrete phase plane of the measured data D_{t+1} versus D_t the trajectory shows the same cyclic behavior as in the continuous phase plane of the model.

In general, the dynamics on a phase plane are given by a pair of coupled differential equations $dx/dt = f(x, y)$ and $dy/dt = g(x, y)$. Sometimes we can only measure x. But then we can calculate dx/dt and get the value $f(x, y)$, which also contains some information about y. This information is often sufficient to reconstruct the dynamics of trajectories in the (x, y) phase plane. The 1st derivative of x at time t is calculated using the well-known formula

$$dx(t)l\,dt = \lim_{h\to 0}[x(t+h) - x(t)]/h.$$

A time series of measurements $D(t) = x(t)$ without noise consists of measurement data $D_0, D_1, D_2, \ldots$ at discrete times $t = 0, 1, 2, \ldots$. The derivative of x at time t can be approximated by differences of corresponding measurement data

$$dD_t/dt = [D_{t+h} - D_t]/h \text{ with } h = 1, 2, \ldots.$$

The smallest useful value of h is 1. But sometimes it is appropriate to select a larger time-lag h. By plotting D_{t+h} versus D_t the phase plane dynamics of a system can often be reconstructed from measurements D_t without the direct measurement of the variable y of the model. In this case, the dynamics in the reconstructed (D_t, D_{t+h}) phase plane are similar to the original (x, y) phase plane of the dynamical system.

Nonlinear dynamical systems generating chaos must be determined by at least three equations. As an example, the Lorenz attractor (Fig. 2.21) is generated in a phase space with three coordinates $x(t)$, $y(t)$, and $z(t)$, which are determined by three nonlinear differential equations. Figure 2.33a shows a time series of measured data D_t from the Lorenz system. If only one variable $D(t) = x(t)$ can be measured, a Lorenz attractor in a (D_t, D_{t-h}, D_{t-2h}) phase space (Fig. 2.33c) can be reconstructed with great similarity to the original Lorenz attractor of the (x, y, z) phase space (Fig. 2.33b). In general, a time series can be embedded in a p-dimensional space with p-coordinates $\mathbf{D}_t = (D_t, D_{t-h}, D_{t-2h}, \ldots, D_{t-(p-1)h})$ and time-lag h. According to Takens' embedding theorem [2.66], the reconstructed dynamics are geometrically similar to the original for both continuous-time and discrete-time systems. The sequence of points created by embedding a time series is called the trajectory of the time series.

In practice, decisions about chaotic dynamics are rather difficult. How can we decide that a time series of measured data is not generated by noisy irregularity but by highly structured chaotic attractors? A chaotic attractor is determined by a trajectory in a bounded region of a phase space with aperiodic behavior and sensitive dependence on initial conditions. These criteria – determinism, boundedness, aperiodicity, and sensitivity – can be checked by several techniques of time series analysis. A system is called deterministic when future events are causally set by past events. For example, a finite-difference equation like $x_{t+1} = f(x_t)$ is deterministic if $f(x_t)$ has only one value for each value of x_t and the future value x_{t+1} can be calculated from the past value x_t by function f.

How can we decide that measured data of past events D_t determine the future events D_{t+1}? We suppose that measurements are made up to time T, and that a prediction of the value at time $T + 1$ should be made. Again, we use the afore mentioned procedure to embed

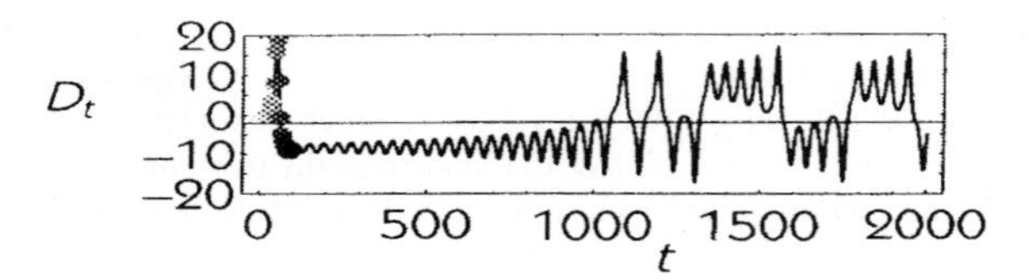

Fig. 2.33a. Measured time series of Lorenz system [2.67]

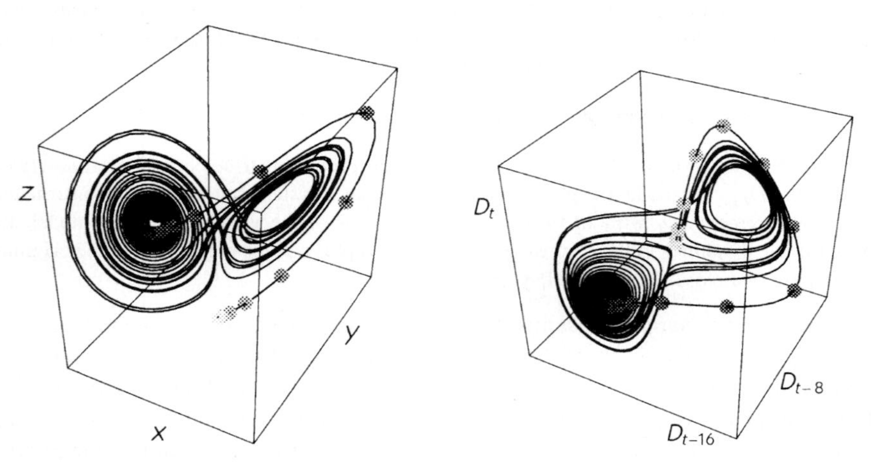

Fig. 2.33b,c. Trajectory in (x, y, z) phase space (**b**) and reconstructed trajectory in (D_t, D_{t-h}, D_{t-2h}) phase space with time-lag h of Lorenz attractor (**c**) [2.68]

the time series in a p-dimensional space with time-lag h. The embedding point at time T, representing measurements of past events, is $\mathbf{D}_T = (D_T, D_{T-h}, \ldots, D_{T-(p-1)h})$. We look through the finite rest of the embedded time series for the closest point to $\mathbf{D}_T$, which is called $\mathbf{D}_c$ at time c. $\mathbf{D}_c$ represents the past events to the measurement D_{c+1}. As $\mathbf{D}_T$ is close to $\mathbf{D}_c$, the measured value D_{c+1} is expected to be close to D_{T+1} in deterministic dynamics. Thus, the prediction of D_{T+1} is identified with the measured value D_{c+1}. The difference between the prediction and D_{T+1} is the prediction error which indicates the quality of the prediction. A more meaningful indication of determinism uses the average of many prediction errors.

Dynamics are bounded if they stay in a finite range of the phase space and do not approach $+\infty$ and $-\infty$ when time increases. In the case of noise, the trajectories spread unbounded all over the phase space. A chaotic attractor is always bounded in a certain region of the phase space. But practically measured data are, of course, always in a finite range, because the physical universe is finite. Thus, boundedness of measured data is related to the concept of stationarity. A time series is stationary if the mean and standard deviation remain the same throughout the time series. Aperiodicity means that the states of a dynamical system never return to their previous values. But values of states may return more or less to previous values. Thus, aperiodicity is a question of degree. How can we determine the degree of aperiodicity in measured data?

Again, we embed the time series of measurements in a p-dimensional space with time-lag h. Each point $\mathbf{D}_t = (D_t, D_{t-h}, \ldots, D_{t-(p-1)h})$ represents the state of the dynamical system at time t. The distance of two states is measured by the distance between two points at times i and j by $\delta_{ij} = |\mathbf{D}_i - \mathbf{D}_j|$ (Fig. 2.34). If the time series is periodic with time T, the values of states are repeated after T values for several times. In this case, the distance δ_{ij} of the points representing times t and j is zero for $|i - j| = nT$ with $n = 0, 1, 2, \ldots$. The degrees of periodicity and aperiodicity can be studied in recurrence plots of points (i, j) if the distance of $\mathbf{D}_i$ and $\mathbf{D}_j$ is smaller than a given distance r.

Such plots depict how the reconstructed trajectory recurs or repeats itself. The number of dots in a recurrence plot shows how many times the trajectory came

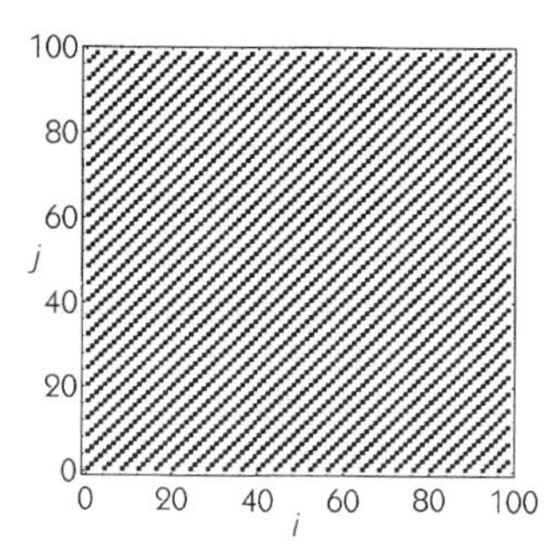

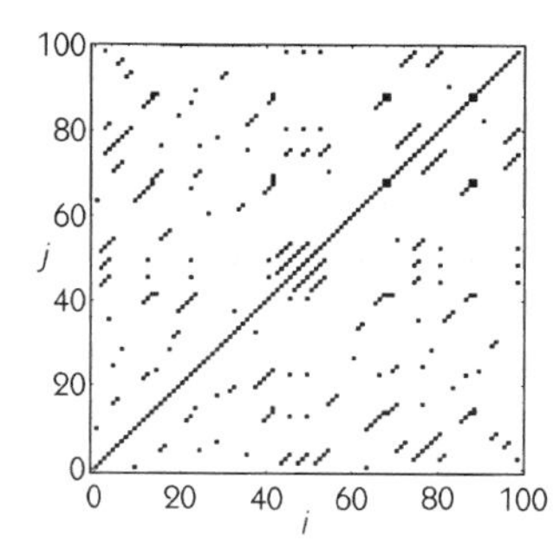

Fig. 2.34a,b. Recurrence plots with periodicity for quadratic map $x_{t+1} = 3.52x_t(1 - x_t)$ $(p = 2, r = 0.001)$ (**a**) and aperiodicity for chaotic map $x_{t+1} = 4x_t(1 - x_t)$ $(p = 2, r = 0.001)$ (**b**) [2.69]

within a distance r of a previous value. The correlation integral $C(r)$ defines the density of points (i, j) in a recurrence plot where the measured time series $\mathbf{D}_i$ and $\mathbf{D}_j$ are closer than r for $i \neq j$ (Fig. 2.34). The correlation integral is an effective concept of chaotic time series analysis [2.70]. If the distance r increases, more dots appear in the recurrence plots with an increasing density $C(r)$. The characteristic curves of $C(r)$ are flat for a periodic system, with a gentle slope for a chaotic system, and with a steeper slope for a random system.

There is an important relationship between the correlation integral and the concept of fractal dimension (compare Sect. 2.4). Consider the scattered points in an area within a distance r to a reference point on a 2-dimensional surface (e.g., a circle with radius r and area πr^2) or in a 3-dimensional space (e.g., a sphere with radius r and volume $4/3\pi r^3$). In general, for points scattered throughout an object in a ν-dimensional space, the number of points closer than distance r to a reference point is proportinal to r^ν. The correlation integral was introduced as a measure for the density of scattered points within a distance r to a reference point of a recurrence plot. Thus, the correlation integral of a scattering of points throughout a ν-dimensional object is proportional to r^ν i.e., $C(r) = qr^\nu$ with a constant q of proportionality. The correlation dimension ν of the ν-dimensional object can be calculated by the logarithm of this equation, i.e. $\log C(r) = \nu \log r + \log q$. In order to find the correlation dimension ν, we can plot $\log C(r)$ versus $\log r$ and determine the slope of the resulting line. This procedure can also be used to find the fractal dimension of an object.

In time series analysis the correlation dimension is sometimes used to find attractors. It is well known that chaotic attractors are often self-similar with fractal dimension. If a time series is generated by a chaotic system, the trajectory of the time series, which is reconstructed from the measurement data by embedding, has the same topological properties as the original attractor of the system, as long as the embedding dimension is large enough. Takens proved a method for finding an appropriate embedding dimension for the reconstruction of an attractor: If the original attractor has the dimension ν, then a dimension $p = 2\nu + 1$ is adequate for the embedding space of the reconstructed attractor. But this method yields no procedure for finding a chaotic attractor, because its existence has been assumed in order to determine its dimension from the measurement data.

Another way to characterize chaotic dynamics is to measure the strength of their sensitive dependence on initial data. Consider two trajectories starting from nearly the same initial data. In chaotic dynamics only a tiny difference in the initial conditions can result in the two trajectories diverging exponentially quickly in the phase space after a short period of time (Fig. 2.35). In this case, it is difficult to calculate long-term forecasts, because the initial data can only be determined with a finite degree of precision. Tiny deviations in digits behind the decimal point of measurement data may lead to completely different forecasts. This is the reason why attempts to forecast weather fail in an unstable and chaotic situation. In principle, the wing of a butterfly may cause a global change of development. This "butterfly effect" can be measured by the so-called Lyapunov exponent. A trajectory $\mathbf{x}(t)$ starts with an initial state $\mathbf{x}(0)$. If it develops exponentially fast, then it is approximately

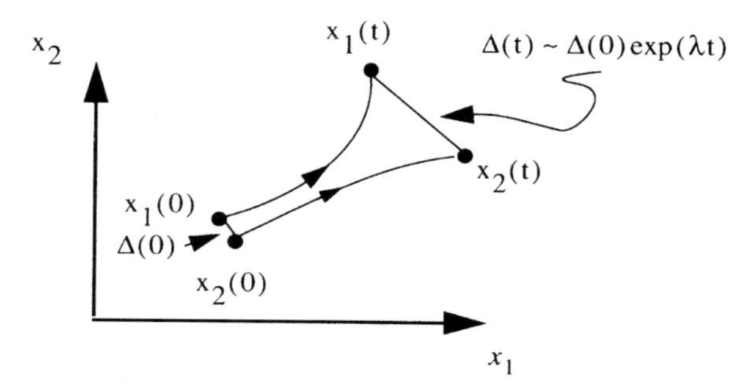

Fig. 2.35. Exponential dependence on initial conditions measured by Lyapunov exponent Λ [2.71]

given by $|\mathbf{x}(t)| \sim |\mathbf{x}(0)|e^{\Lambda t}$. The exponent Λ is smaller than zero if the trajectory is attracted by attractors, such as stable points or orbits. It is larger than zero if it is divergent and sensitive to very small perturbations of the initial state.

Let us consider a finite-difference equation $x_{t+1} = f(x_t)$ with two nearby initial positions x_0 and y_0 in the phase space. By iterated application of the function f we get $x_t = f(x_{t-1}) = f^t(x_0)$ and $y_t = f(x_{t-1}) = f^t(y_0)$ with $t = 0, 1, 2, \ldots$. If the positions x_t and y_t are separated exponentially fast by iterations, then their distances are $|y_t - x_t| = |y_0 - x_0|e^{\lambda t}$ with $\Lambda > 0$. For increasing $t \to \infty$ it follows $(1/t)|y_t - x_t|/|y_0 - x_0| \to \Lambda$. If the path of the trajectory is within a bounded region, the exponential separation only occurs when the initial positions are very close to each other. In this case, we decrease the difference $|y_0 - x_0|$ before we determine the limit for increasing $t \to \infty$. The Lyapunov exponent of the trajectory $x_t = f^t(x_0)$ can then be defined by the constant

$$
\begin{aligned}
\Lambda &= \lim_{t\to\infty} 1/t \lim_{|y_0 - x_0|\to 0} \ln |y_t - x_t|/|y_0 - x_0| \\
&= \lim_{t\to\infty} 1/t \lim_{|y_0 - x_0|\to 0} \ln |f^t(y_0) - f^t(x_0)|/|y_0 - x_0| \\
&= \lim_{t\to\infty} 1/t \ln |df^t(x_0)/x_0| = \lim_{t\to\infty} 1/t \sum_{i=0}^{t-1} \ln |df(x_i)/x_i|.
\end{aligned}
$$

For continuous dynamical systems with differential equations, the trajectory is a vector $\mathbf{x}(t)$ with a Lyapunov exponent $\Lambda = \limsup 1/t \ln |\mathbf{x}(t)|$. The Lyapunov exponent provides a measure for the mean convergence and divergence rate of neighboring trajectories of a dynamical system. For an n-dimensional system, the n Lyapunov exponents $\Lambda_1 \geq \Lambda_2 \geq \ldots \geq \Lambda_n$ describe different types of attractors. For non-chaotic attractors we can distinguish asymptotically stable equilibrium with $\Lambda_i < 0$ $(i = 1, \ldots, n)$, asymptotically stable limit cycle with $\Lambda_1 = 0$ and $\Lambda_i < 0$ $(i = 2, \ldots, n)$, asymptotically stable two-torus with $\Lambda_1 = \Lambda_2 = 0$ and $\Lambda_i < 0$ $(i = 3, \ldots, n)$, and asymptotically stable m-torus with $\Lambda_1 = \ldots = \Lambda_m = 0$ and $\Lambda_i < 0$ $(i = m+1, \ldots, n)$. A chaotic system must have at least one positive Lyapunov exponent. In the 3-dimensional case, the only possibilty for chaos is $\Lambda_1 > 0$, $\Lambda_2 = 0$, $\Lambda_3 < 0$ with $\Lambda_3 < -\Lambda_1$.

Dynamical systems can be classified by attractors with increasing complexity from fixed points, periodic and quasi-periodic up to chaotic behavior. This classification of attractors can be characterized by different methods, such as typical patterns of time series, their power spectrum, phase portraits in a phase space, Lyapunov exponents, (fractal) dimension, and a measure of their information flow (Kolmogorov-Sinai-Entropy), which will be discussed in Sect. 5.3 in more detail. Table 2.1 yields an overview of these degrees of dynamic complexity, which form the framework for the complex dynamical approach of this book.

One of the most significant concepts is that of the fractal dimension, a measure of the roughness of an object. Fractality seems to be a natural feature of reality. Rocky coastlines consist of cliffs and crannies. Rocks with rough surfaces erode. In organic growth, such as that for the airways of the lungs, a fractal process of iterative division is the natural outcome of the genetic rules for animal development. In Euclidean geometry, we are familiar with the single dimension of a straight line, or the two dimensions of a plane. An example of a fractal dimension is that of Koch's curve (Fig. 2.36). In order to measure its length, one starts with a ruler that is one-third of the breadth of the object (the curve). This ruler corresponds to each line inside the curve in the top panel. The line fits inside the curve four times. The ruler is then shortened to a third of its original length, as shown in the bottom diagram. Because this shorter ruler can fit into more "crannies" of the curve, the length of curve obtained using this ruler is greater than that given by the original ruler, (by four-thirds). For each change in state, the length measured is multiplied by the same fraction, four-thirds. The fractal dimension is then defined as the ratio of the logarithm of 4 to the logarithm of 3, or 1. 2618... The intuitive sense of this "fractal" number is obvious: the curve is crinkly, so it fills more space than a one-dimensional straight line does. However, it does not completely fill the two-dimensional plane.

phase portrait	time-history response	power spectrum	auto-correlation	Lyapunov-exponents	(fractal) dimension	KS-entropy
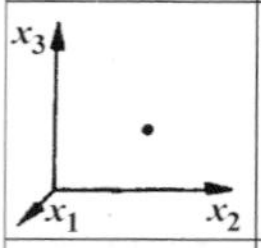 x_3, x_1, x_2	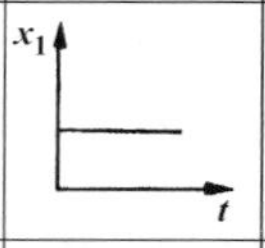 x_1, t	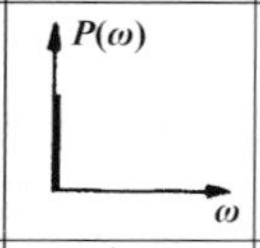 $P(\omega)$, ω	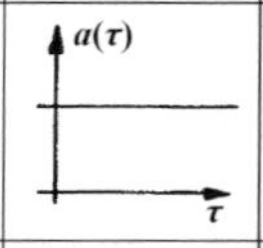 $a(\tau)$, τ	*(−, −, −)*	*0*	*0*
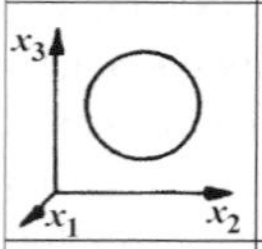 x_3, x_1, x_2	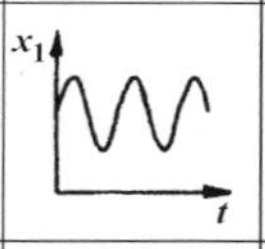 x_1, t	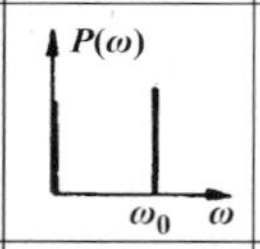 $P(\omega)$, ω_0, ω	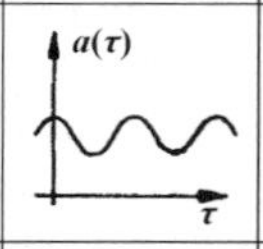 $a(\tau)$, τ	*(0, −, −)*	*1*	*0*
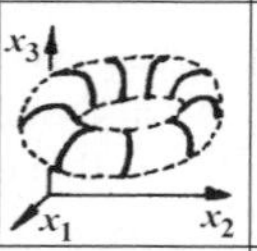 x_3, x_1, x_2	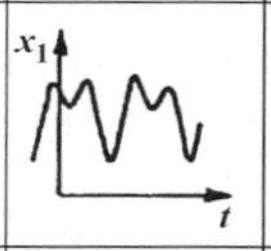 x_1, t	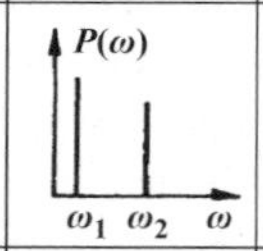 $P(\omega)$, ω_1, ω_2, ω	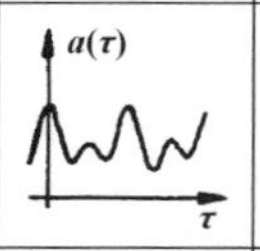 $a(\tau)$, τ	*(0, 0, −)*	2	*0*
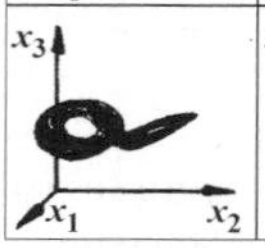 x_3, x_1, x_2	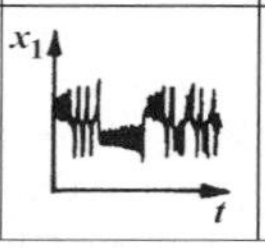 x_1, t	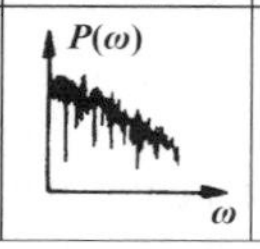 $P(\omega)$, ω	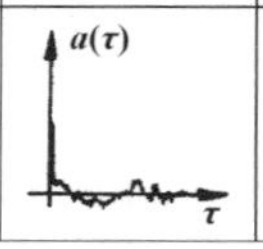 $a(\tau)$, τ	*(+, 0, −)*	$2 < h < 3$	$e_{KS} > 0$

Table 2.1. Dynamic complexity of attractors for 3-dimensional systems [2.72]

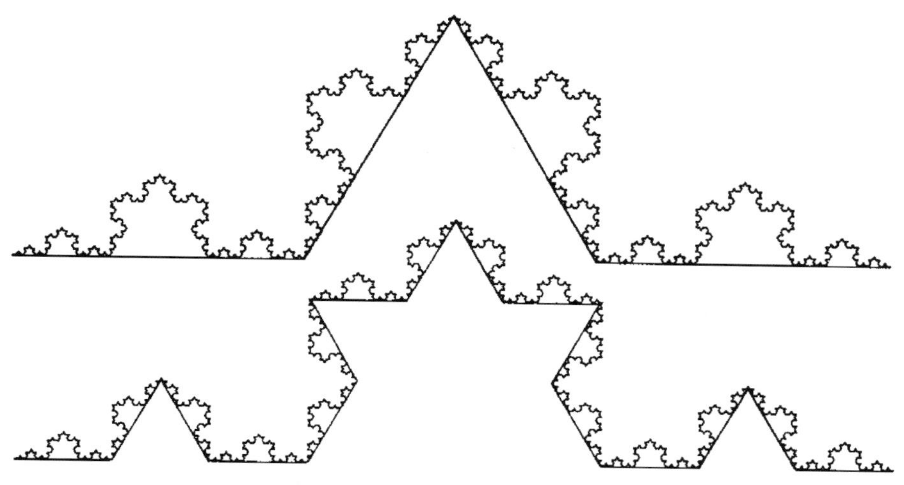

Fig. 2.36. Fractal dimension of Koch's curve

Using analog recursive procedures, we can construct Hilbert and Sierpinski curves that fill a plane with self-similar patterns in iterated steps of increasing density. These curves seem to be more than lines, with one dimension, but less than planes, with two dimensions. Their dimensions are "fractions" between the integers one and two. The fractal dimension can be illustrated by the geometrical dimension D of similarity. For a Euclidean object of dimension D, the length, area or volume of an object with edge length ε is proportional to ε^D. For example, a square with edge length ε has an area of ε^2, while a cube has a volume of ε^3. For self-similar objects, one way to measure the length, area or volume of an object is to count the number of self-similar copies. If there are N copies each with an edge length ε, then the length, area or volume of the object is related to its dimension: N is proportional to ε^D. Thus, one obtains $D \approx \log N / \log \varepsilon$. For Koch's curve, the number of self-similar copies is $N = 4$ and the edge length is $\varepsilon = 3$. If phase portraits of chaos attractors have a fractal dimension (Table 2.1), they are termed "strange." Time series are sometimes characterized by statistical self-similarity on different scales (e.g., Fig. 8.16). Thus, a fractal dimension could hint at chaos, but its presence alone does not indicate chaos.

Self-similar mathematical objects consist exclusively of smaller self-similar copies of themselves. Our procedure for calculating the dimension of a fractal object is only useful if we know the number N of self-similar copies and the size ε of the original relative to each copy. For practical applications (e.g., a map or picture of a fractal object or real objects in the three-dimensional world), we need a better procedure for estimating the fractal dimension. The following procedure comes directly from the definition of the fractal dimension. In a first step, all points in the object are covered with $N(\varepsilon_0)$ or cubes of edge length ε_0. This step is repeated with squares or cubes of edge length $\varepsilon_1 = \varepsilon_0/2$, then with $\varepsilon_2 = \varepsilon_1/2$, and so on. By doing this, we obtain a function $N(\varepsilon)$ sampled at the values $\varepsilon = \varepsilon_0, \varepsilon_1, \ldots$ In theory, the

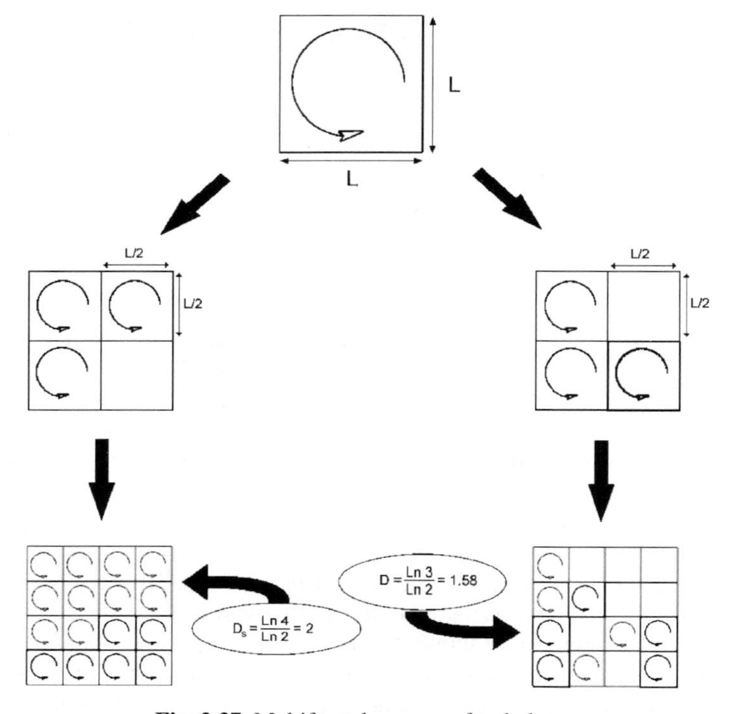

Fig. 2.37. Multifractal process of turbulance

dimension D is defined by $\lim N(\varepsilon) = k \cdot \varepsilon^{-D}$, with a constant k. In practice, D can be estimated as $D \approx (\log(N(\varepsilon_{i+1})/N(\varepsilon_i)))/(\log(\varepsilon_i/\varepsilon_{i+1}))$. However, the squares or cubes should not be made smaller than the cells or particles that are considered to be the building blocks of the object [2.73].

Intuitively, a fractal is a pattern or object whose parts echo the whole, only scaled down. By contrast, a multifractal has more than one scaling ratio in the same object. Some parts of the object shrink quickly, others slowly. Multifractals resemble the way in which many aspects of nature really work more closely than fractals. Different clusters are formed on the surface of the Earth in multifractal processes; they are not always distributed and scaled in the same way. On a stormy day, the wind velocities form clusters of high gusts interspersed with gentler breezes. One can think of a multifractal as being composed of an infinite hierarchy of different fractal sets. An example is given in Fig. 2.37, which shows a hierarchy for the vertical cross-section of stratified turbulence. The generic multifractal process of turbulence is a cascade of cells with different distributions of whirls. Mathematically, multifractals are defined by two groups. One determines the statistics (more precisely, they vary as a function of scale), while the second defines the notion of scale itself [2.74].

3 Complex Systems and the Evolution of Life

How can one explain the emergence of order in the Darwinian evolution of life? In the history of philosophy and biology, life was explained teleologically by non-causal ("vital") forces aiming at some goals in nature. In a famous quotation Kant said that the "Newton for explaining a blade of grass" could never be found (Sect. 3.1). Boltzmann could show that living organisms are open dissipative systems which do not violate the second law of thermodynamics: Maxwell's demons are not necessary to explain the arising order of life in spite of the increasing entropy and disorder in closed systems according to the second law. Nevertheless, in the statistical interpretation from Boltzmann to Monod the emergence of life is only a contingent event, a local cosmic fluctuation at the boundary of the universe (Sect. 3.2). In the framework of complex systems the emergence of life is not contingent, but necessary and lawful in the sense of dissipative self-organization. The growth of organisms and species is modeled as the emergence of macroscopic patterns caused by nonlinear (microscopic) interactions of molecules, cells, etc., in phase transitions far from thermal equilibrium (Sect. 3.3). Even ecological populations are understood as complex dissipative systems of plants and animals with mutual nonlinear interactions and metabolism with their environment (Sect. 3.4). In the life sciences, relations between physiological, morphological or ecological variables often lead to power laws with an underlying fractal process. Complex organs like the lungs or heart are obviously structured fractally. Power laws are important criteria for complexity (Sect. 3.5). Therefore, Spencer's idea that life is determined by a structural evolution with increasing complexity can be mathematized through complex dynamical systems. Has the "Newton of life" been found? The theory of complex dynamical systems does not explain what life is, but it can model how forms of life can arise under certain conditions. Thus, our existence remains a source of wonder to us just as it did for our ancestors, even if we will eventually be able to model the complex dynamics of life.

3.1 From Thales to Darwin

Before we discuss complex systems and the evolution of life, let us have a glance at early philosophies of life [3.1]. It is a surprising fact that many aspects of modern ecology remind us of early ideas of self-organization. In mystic interpretations, life was understood as a cyclic movement of growth and decay, birth and death. Animals and humans only survived in adapting to the great cycles of nature such as high tide and low tide, the change of seasons, changing constellations of stars, fertile and sterile periods of nature, and so on. Nature itself seemed to be a great organism, and humans were considered as partly involved in its natural development. Mythologies of natural religions and their rituals were used to conjure the forces of nature and to live in harmony with the natural order.

Mythology was given up in favour of natural philosophy, when people asked for the basic principles of life, and when they no longer accepted demons and gods as personified forces of nature. In the 6th century B.C. the presocratic philosopher Thales of Miletus declared water to be the fundamental source of life. Anaximander seems to have some early ideas of evolution:

> It is said that in the wet element the first living beings come to be, in a husk of prickly rinds; with increasing age they climbed onto the dry element, the rind tore of on all sides and so, for a short time they took on a different live form. [3.2]

Regarding the derivation of humans, Anaximander expressed an utterly modern conception. Observing the long period that human children need for care and protection, he drew the conclusion that if they had always required that, humans would have not been able to survive. So, earlier they must have been different. Empedocles explained the processes of life with certain mixtures and transformations of the familiar elements water, air, fire, and earth.

While these organic explanations of life seemed to be intuitively convincing for former contemporaries, Democritus' atomism with his reduction of life to the interaction of invisible atoms was considered as rather abstract. Even consciousness and the soul of man was explained by microscopic interactions of tiny material elements. So, Democritus and his school were not only attacked as materialists, but as atheists too. Plato tried to model the first elements of matter and their combination by geometric figures and constructions.

From a modern scientific point of view, Democritus' atomism and Plato's mathematical models were early reductionist programs for life. They tried to reduce physiological and biological processes to the interactions of physical elements. But the idea of explaining the changing and pulsating processes of life on the basis of the rigid and dead figures of geometry or material atoms must have seemed thoroughly unnatural, speculative, and far-fetched to the contemporaries of that time. In short, "real" life seemed to be hopelessly "complex", and Euclid's mathematics too "simple". So, Euclid's mathematics was reserved for the "superlunar" world of stars, but not applied to the "sublunar" world of earthly life.

This is where the Aristotelian philosophy of life begins. While Plato, in the Pythagorean tradition, drew his concept from geometry, Aristotle formulated his concept of processes in nature mainly on the basis of the ways in which living organisms such as plants and animals function. The processes and courses of life are known to us from everyday experience. What is more obvious than to compare and explain the rest of the world, which is unknown and strange, with the familiar? According to Aristotle, the task of physics is to explain the principles and functions of nature's complexity and changes. In modern terms, Aristotle rejected atomic reductionism as well as the mathematization of life as speculative and unrealistic.

Life was defined by the feature of self-movement, in contrast to a dead stone which must be pushed from outside in order to move. In this Aristotelean sense, life meant "having a soul", which was understood as an organizing force (entelechy) of matter (vitalism). In modern terms, the self-organisation of life was interpreted by Aristotle as a functionally governed process aiming at certain "attractors" of purposes (teleology). For instance, a tree grows out of a seed with the purpose of reaching its final form. In modern terms, the change of forms characterizing the growth of an organism is something like the (qualitative) evolution of an order parameter which Aristotle called the "potentiality" of that organism. But, of course, the main difference compared to modern concepts of order parameters is the fact that Aristotle criticized any reduction of macroscopic forms to atomic or microscopic interactions.

It is noteworthy that Aristotle proposed a continuous scale of more or less animated states of nature (*scala naturae*) and denied an absolute contrast of "alive" and "dead". He was always seeking for the intermediate or connecting links between organisms with different complexity. For instance, for a Greek like Aristotle, living by the Mediterranean Sea with its plentiful flora and fauna, it was easy to observe organisms like water-lilies "which may be doubted to be animals or plants, because they grow to the floor like plants, but eat fishes like animals" [3.3]. On the background of continuity, Aristotle suggested a kind of biogenetic law: "In the beginning the fetus of an animal seems to have a kind of life like a plant; during its later development, we may speek of a sensitive and thinking soul" [3.4].

Aristotle was not only a theorist, but one of the first observing botanists, zoologists, and physiologists. He designed a taxonomy of plants and animals according to different features, and tried to describe the physiological processes of life. His leading paradigm of life was the idea of a self-organizing organism, rejecting any atomic, molecular, or anorganic reductionism. Aristotle's philosophy of life has overshadowed the development of biology until today.

In the Roman period, even medicine was influenced by Aristotelean tradition. Galen, the physician of the Roman emperor Marcus Aurelius, taught that organs had to be adapted completely to their functions in our body. Following Aristotelean teleology, he described the digestive organs selecting the "purposeful" parts of food for the life processes and separating the "useless" ones. In the Middle Ages, Albertus Magnus combined Aristotelean philosophy of life with Christianity. On the background of Aristotle's teleology, Albertus developed an early ecology demanding that humans have to live in harmony with their natural environment. Organisms and their environments are connected with each other by numerous exchanges of air, food, excreta, etc. which are in a natural balance ("equilibrium") governed by divine ordinances. Albertus thought that even the health of the human soul depends on a healthy environment with healthy air, climate, plants, and animals. Soul and body are not separated, but an organic whole.

The decisive condition of modern physics was the connection of mathematics, observation, experiment, and engineering which was realized by Galilei in the Renaissance. Newton founded a new mathematical and experimental philosophy of nature which he called *Philosophiae naturalis principia mathematica* (1687). Geometry and mechanics became the new paradigm of natural sciences. In the history of science this period is called the mechanization of nature, which was imagined to be nothing else than a huge mechanical clock. The mathematician and philosopher René Descartes and the physicist Christian Huygens taught that every system in nature consists of separated elements like the cog wheels of a clock. Every effect of nature was believed to be reducible to linear causal chains like sequences of cog wheels of a clock. Obviously, Cartesian mechanism is contrary to Aristotelian holism.

Even the physiology of life processes should be explained mechanically. The heart, for instance, was considered as a pumping machine. In general, Descartes believed that the motions of an animal and human body can be derived from the mechanism of organs "and that with the same necessity as the mechanism of a clock from the position and form of its weights and wheels" [3.5]. The anatomy of human bodies which was practised by dissection since the Renaissance was an application of the analytical method of Descartes. According to Descartes, each system can be separated into its basic building blocks, in order to explain its functions by the laws of geometry and mechanics.

The Italian physicist and physiologist Borelli (1608–1679) founded the so-called iatrophysics as an early kind of biophysics. He transferred a famous quotation of Galileo from physics to biology, and declared in his book *De motu animalium* (About motions of animals) emphatically:

> As the scientific recognition of all these things is founded on geometry, it will be correct that God applied geometry by creating animal organisms, and that we need geometry for understanding them; therefore it is the only and suitable science, if one wants to read and understand the divine script of the animal world. [3.6]

While Descartes still believed in an immortal soul of man, Lamettrie reduced man to an automaton without soul, according to his motto *L'homme machine* (1747). Human and animal bodies were only distinguished by their level of complexity and organization. After physics, teleology in the tradition of Aristotle should be eliminated in physiology and medicine, too. During the Enlightenment, the mechanism of life was understood as materialistic and atheistic philosophy. The following story by Voltaire about Lamettrie is rather amusing: when Lamettrie suddenly fell ill during a plentiful banquet, and died because of indigestion a few days later, the God-fearing contemporaries were said to be thankful about the fact that a materialist had to die because of his own insatiability.

Nevertheless, some Aristotelean concepts were discussed during the age of mechanization. For instance, Leibniz assumed a hierarchical order of nature with a continuous scale of animation from the smallest building blocks ("monads") to the complex organisms. Leibniz tried to combine Aristotelean ideas with physical mechanics, and became one of the early pioneers for a theory of complex dynamical systems. Concerning the status of man in nature, Leibniz declared:

Thus, every organic body of a living being is a kind of divine machine or natural automaton surpassing all artificial automata infinitely. [3.7]

Inspired by Leibniz, the zoologist Bonnet (1720–1793) proposed a hierarchy of nature ("Echelles des êtres naturelles") with a measure of complexity which seems to be rather modern. Bonnet underlined "organization" as the most important feature of matter. An organization realizing the most effects with a given number of different parts is defined as the most perfect one [3.8].

At the end of the 18th century, Immanuel Kant criticized the application of Newtonian mechanics to biology: "The Newton explaining a blade of grass cannot be found." The main reason for Kant's critique is that in the 18th century the concept of a machine was only made precise in the framework of Newtonian mechanics. Thus, in his famous *Kritik der Urteilskraft*, Kant wrote that an organism "cannot only be a machine, because a machine has only moving force; but an organism has an organizing force ... which cannot be explained by mechanical motion alone" [3.9]. Kant also criticized Aristotelean teleology and the assumption of "aims" and "purposes" in nature as a metaphorical anthropomorphism. An organism must be described by the model of a "self-organizing being".

Like Kant, Goethe rejected the materialistic-mechanical explanation of life which was defended by, e.g., the French encyclopedist Holbach in his *Système de la Nature*. For Goethe, the mechanist model of nature is "grey, ... like death ... like a ghost and without sun" [3.10]. He believed that life develops organically and harmonically like the metamorphosis of a plant or the mental maturity of man.

On the background of Goethe's age and Kant's critique of mechanistic rationalism, a romantic philosophy of nature arose in Germany in the beginning of the 19th century. It was a renaissance of the organic paradigm against mechanism. Friedrich Schelling (1775–1854) designed a "science of the living" assuming that organization and reproduction are main features of the living [3.11]. Oken (1779–1851), physician and philosopher of nature, described a "planetary process", in which living organisms were explained by a synthesis of magnetism, chemism, and galvanism. From a modern point of view, "self-organization" and "self-reproduction" were far-reaching concepts of the romantic philosophy of nature. But in those days they were only speculations or inspired intuitions, because the experimental and mathematical base was still missing.

The peaceful picture of an organic and harmonic metamorphosis was soon pushed aside by biology. Charles Darwin's theory of evolution does not need teleological forces to explain life. The "survival of the fittest" (Herbert Spencer) depends on the greater advantage of selection with respect to certain conditions of the environment (for instance nourishment, climate) [3.12]. Darwin was inspired by some ideas of Lamarck (1744–1829), for instance the heredity of acquired properties. Darwin's evolution is governed by (genetic) variability of species ("mutation") and natural selection driving development in one direction. Spencer taught that life is driving to more complexity, controlled by selection. Many contemporaries considered Darwinism not only as a theory of natural sciences. Darwin's theory seemed to present a scenario of life with a strong analogy to the society of the 19th century. The "selection of the fittest" became a slogan of the "social Darwinism" as a political attitude.

In the second half of the 19th century, Haeckel generalized the evolution of life from monocellular organisms to humans. But in those days the theory of evolution could not be compared with the highly confirmed physical and chemical theories. Darwin could only deliver some comparative studies of morphology. He described the variability of species and natural selection, but he could not explain it by mathematized and testable laws like physics. Mendel's laws of heredity (1865) were still unknown to Darwin as well as to many contemporaries. Nevertheless, one of the great physicists of the 19th century, Ludwig Boltzmann, declared, casting a retrospective glance at his century:

> When you ask me for my deepest conviction if our century will sometimes be called the iron century or the century of steam or the century of electricity, then I answer without hesitation, it will be called the century of Darwin. [3.13]

3.2 Boltzmann's Thermodynamics and the Evolution of Life

In the 19th century, the dominant topics of natural science, social science, and philosophy became "evolution" and "history". While the biological sources of these ideas date back to Darwin's theory of evolution, physical examples of irreversible processes were at first discussed in thermodynamics. The initial principles of thermodynamics were developed by Carnot (1824). His principles were discovered in analyzing mechanical forces produced by steam engines. Roughly speaking, the first law of thermodynamics says that energy cannot be created or destroyed. Despite mechanical work, electrical energy, or chemical transformations that energy is constantly undergoing in nature, the total energy within a closed system remains unchanged. In accordance with Einstein's equivalence of mass and energy (compare Sect. 2.2), the first law has been enlarged to a conservation principle of mass and energy in this century.

The basic importance of the second law in the context of physical evolution was recognized by Clausius (1865) who borrowed the term "entropy" from the Greek word for evolution or transformation [3.14]. Mathematically, the entropy change of a system is defined by the reversible heat addition to the system divided by its absolute temperature. According to Ilya Prigogine, one must refer to the fact that every system has surroundings [3.15]. Thus, the variation of entropy during a time moment is more generally the sum of the rate at which entropy is supplied to the system by its surroundings and the rate at which entropy is produced inside the system. The second law of thermodynamics demands that the rate at which entropy is produced inside the system is greater than or equal to zero. For closed and isolated systems without an entropy supply from (or sink to) the surroundings we get the classical statement of Clausius that entropy increases or remains constant when thermodynamic equilibrium has been reached. In other words, there is no process in nature involving physical, chemical, biological, or (as we shall see) informational transformations occurring spontanously without some energetic cost in terms of entropy.

Entropy is a macroscopic property of systems like volume and size. Therefore, thermodynamics was at first only a phenomenological theory describing possible heat distributions of macroscopic systems. Boltzmann was not happy with this positivistic attitude and tried to deliver a statistical-mechanical explanation reducing such macroscopic states of systems as, e.g., heat to the mechanics of microscopic molecules. Inspired by the microstate-macrostate distinction which has become crucical for the theory of evolution, Boltzmann gave thermodynamics its first statistical interpretation [3.16]. Irreversibility in statistical thermodynamics is based on this distinction.

In general, statistical mechanics explains a macrostate like density, temperature, etc., by microstates. In this sense, an observable macrostate is said to be realized by a large number W of microstates. In order to define the number W, a large number of independent mechanisms of the same kind like atoms, molecules, crystals, etc., are considered. They develop their microstates according to their equations of motion with different initial phase states. If a macrostate is realized by W microstates of this kind, then Boltzmann's entropy quantity H of the corresponding macrostate is assumed to be proportional to the logarithm of W, i.e., $H = \mathrm{k} \ln W$ with the Boltzmann constant k. In a continous phase space, the Boltzmann expression can be generalized by an integral of a velocity distribution function. For Boltzmann, H is a measure for the probability of molecular arrangements corresponding to the observable macrostates of the system.

Boltzmann's reductionism met historically with violent objections from physicists, mathematicians, and philosophers. Positivistic physicists and philosophers like Ernst Mach criticized Boltmann's hypothesis of molecules and atoms which were not empirically confirmed in those days. But after their successful discovery this critique is only of historical interest.

One of the most important objections is Loschmidt's reversibility paradox. Since the laws of mechanics are invariant (symmetric) with respect to the inversion of time, to each process there belongs a corresponding time-reversed process. This seems to be in contradiction with the existence of irreversible processes. Boltzmann answered that the second law of thermodynamics in the form of his so-called H-theorem cannot be derived only from the (reversible) mechanical laws, but requires the additional assumption of extremely improbable initial conditions, too. The second law is assumed to hold true with very high probability, but not with security. Irreversible processes are only frequent or probabilistic, reversible ones seldom and improbable. Thus, the second law allows local deviations or fluctuations (for instance Brownian motion) [3.17].

Another objection, by Henri Poincaré and Ernst Zermelo, underlined that each state of a mechanical system with finitely many degrees of freedom must recur at least approximately after a certain time [3.18]. Thus, an arrow of time connected with an increase of entropy cannot exist. Boltzmann answered that the times of return become extremely long with increasing number of degrees. Cosmologically, there are two possible points of view in the sense of Boltzmann: (1) the universe started with extremely improbable initial conditions, or (2) when the universe is large enough, there may be deviations from the equal distribution in some places. Figure 3.1 illustrates Boltzmann's hypothesis of fluctuations. He assumed that the whole universe is in thermal equilibrium, i.e., in maximum disorder. Boltzmann believed that both directions of time are completely symmetric. So, the curve of local entropy increases similarly in both directions of time, becoming flat with maximum entropy [3.19].

Life as a developing system of order is only possible in regions with strongly changing entropy, i.e., on the two slopes of the entropy curve in Fig. 3.1. The two arrows denote Boltzmann's local worlds in which life may occur. So, in the sense of Boltzmann, there cannot be an objectively unique arrow of time, but only one of the two possible directions of increasing entropy which people subjectively experience, living in one of the two possible local worlds on the slopes in Fig. 3.1. Before we criticize Boltzmann's view in detail, let us have a glance at his theory of life against the background of his thermodynamics of thermal equilibrium, which overshadowed science in this century until recently.

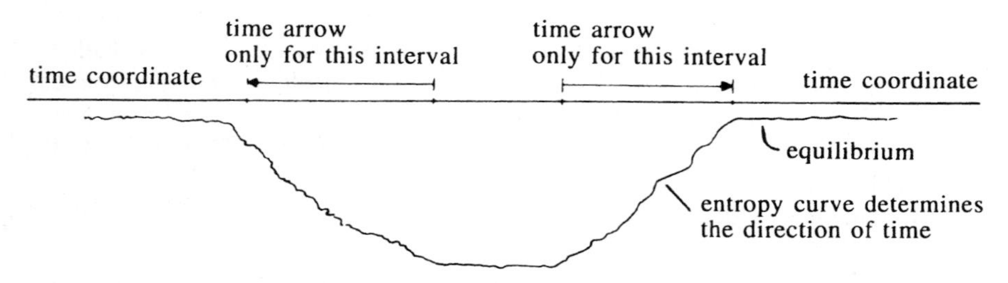

Fig. 3.1. Entropy curve in Boltzmann's universe in thermal equilibrium with symmetric directions of time

Ludwig Boltzmann (1844–1906) was the first scientist who tried to reduce the biological theory of evolution to the thermodynamics and chemistry of the 19th century. For scientists at the end of the last century, it was a great problem that the second law of thermodynamics seemed to forecast the final disorder, death, and decay of nature, while Darwinian evolution seemed to develop living systems of order with increasing complexity. Of course, the second law is reserved to closed systems, and living systems must be open, in a permanent exchange of energy, matter, and information with their environment. Nevertheless, how is the local increase of complexity possible in a sea of disorder and thermal equilibrium?

Boltzmann suggested some explanations which already remind us of modern biochemical concepts of molecular autocatalysis and metabolism. The origin of first primitive living beings like cells was reduced to a selection of inanimate molecular building blocks which Boltzmann imagined as a process like Brownian motion. Plants as cellular aggregates are complex systems of order. Thus, in the sense of the second law of thermodynamics they are improbable structures which must fight against the spontaneous tendency of increasing entropy in their body with sunlight.

Because of the high temparature of the sun, the earth gets energy with relatively low entropy which can be used to compensate the spontaneous increase of entropy in the plants. This process is realized by photosynthesis, which was physically explained by Boltzmann in 1886:

> Thus, the general struggle of life is neither a fight for basic material . . . nor for energy . . . but for entropy becoming available by the transition from the hot sun to the cold earth. [3.20]

Boltzmann extended his physically founded theory of evolution to the history of the nervous system and the emergence of memory and consciousness. He argued that the sensitivity of primitive organisms to outer impressions has led to the development of special nerves and organs of seeing, hearing, feeling, moving, etc.:

> The brain is considered as the apparatus or the organ to develop models of the world. Because of the great utility of these models for the survival of the race, the human brain has been developed according to Darwin's theory with the same perfection as the giraffe's neck or the stork's bill. [3.21]

Even the ability to develop concepts and theories was explained by evolution. Boltzmann tried to justify human categories of space, time, and causality as tools developed by the brain for the racial survival of the fittest. He did not hesitate to extend biological evolution even to the socio-cultural development and history of mankind. In 1894, the Viennese physician S. Exner wrote about the subject "Morality as weapon for the struggle of life" in the sense of Boltzmann. In 1905, Boltzmann himself gave a lecture with the amazing title "Explanation of the entropy law and love by the principles of the probability calculus". Obviously, Boltzmann's Darwinism had reached its limits.

At the beginning of this century, life still could not be explained by physical and chemical foundations. Classical mechanics, the foundation of natural sciences in the 17th and 18th centuries, assumed deterministic and time-reversible laws of nature delivering no explanation of irreversible processes of life. A frictionless pendulum clock moves time-reversibly as an oscillating mechanical system, in principle without limit. Humans are born, grow, and die – why? The thermodynamics of the 19th century deals with irreversible processes of closed systems being driven to a state of maximum entropy or disorder. But how can the development of complex living systems be explained? In the sense of Boltzmann's statistical interpretation, the emergence of order and biological complexity can only be an improbable event, a local cosmic fluctuation "at the edge of the universe" (as Jacques Monod said later), which will disappear without significance for the whole universe in thermal equilibrium [3.22]. Following Monod, we have only the philosophical choice of an existentialism à la Camus, to perish in a finally senseless biological and cultural evolution with human dignity. The tragic death of the genius Ludwig Boltzmann, who committed suicide in 1909, seems to be a symbol of this attitude. But Boltzmann's thermodynamics did not definitely explain the origin of life. He only proved that his statistical interpretation of the second law is not contrary to Darwinian evolution.

After classical mechanics in the 17th and 18th centuries and thermodynamics in the 19th century, quantum mechanics has become the fundamental theory of physics. In spite of Heisenberg's principle of uncertainty, the laws of quantum mechanics and classical mechanics are characterized by the reversibility of time. Concerning the reductionist program for treating complexity, it was a great success that the quantum chemistry of molecules could be explained by the laws of quantum mechanics. In 1927, Heitler and London succeeded in modifying Schrödinger's equation of atomic and subatomic systems for molecules. There are no particular chemical forces in chemistry besides the well-known physical forces. After physics, teleology seemed to be eliminated in chemistry.

But is chemistry completely reducible to physics [3.23]? Definitely only in a restricted sense! The structural models of molecular orbits can only be introduced by abstraction from quantum mechanical correlations. While, for instance, the electrons of an atom cannot be distinguished in the sense of the Pauli principle, they are used by chemists as quasi-classical objects moving around the atomic nucleus in well-distinguished orbits. There are well-known chemical procedures of abstraction (Born–Oppenheimer and Hartree–Fock procedure) for introducing electronic orbits in approximately quasi-classical models of the non-classical quantum world. Further on, we have to consider practical limitations of computability with Schrödinger's equations for complex molecules in spite of all the fantastic successes of numerical quantum chemistry. This weak reduction of chemistry to physics seemed to prove that scientists should continue on the path of reductionism, in order to reduce elementary particles, atoms, molecules, cells, and finally organisms to physics and chemistry.

In the 1920s and 1930s, the struggle between physical reductionism and neovitalism could not actually be decided. For example, the physicist Heitler, the biologist Driesch, and the philosophers Bergson and Whitehead supported explicit neovitalistic opinions in the Aristotelean tradition [3.24]. They argued that particular biological laws may sometimes invalidate the laws of physics and chemistry. From Aristotle to Goethe and Schelling, teleological self-organization and the spontaneity of life from living cells to consciously acting humans have been mentioned to demonstrate that physical reductionism is impossible. Wholeness is a primary feature of an organism which cannot be reduced to the sum of its building blocks. Inspired by Niels Bohr's so-called Copenhagen interpretation of quantum mechanics, some physicsts tried to mediate between physicalism and vitalism with Bohr's concept of complementarity. Bohr used complementarity to justify such excluding concepts of quantum mechanics as, e.g., particle–wave dualism. Thus, complementarity is assumed for the two classes of physical-chemical and biological laws, which are believed to be incommensurable. We must remember that complementarity is not a physical law, but a philosophical interpretation of quantum mechanics which was not supported by Erwin Schrödinger. He knew that in the 1930s and 1940s the struggle of physicalism and vitalism could not be decided, and complementarity was only a concept to describe the status quo. In his book *What is Life?* Schrödinger wrote:

> After all we have heard about the structure of living matter we must be ready for the fact that it works in a way which cannot be reduced to the usual physical laws. The reason is not that a "new force" or something similiar governs the behavior of the single atoms in

a living organism, but because its structure differs from all that we have ever studied in our laboratories. [3.25]

Schrödinger recalls the image of an engineer who is familiar with a steam engine and wants to explore an electromotor. As the two motors work in quite different ways, he will hit upon the idea that the electromotor is driven by a ghost. In the tradition of Leibniz, Schrödinger expects to understand a living organism as "the finest master piece which was ever achieved according to the leading principles of God's quantum mechanics" [3.26].

The problem for Schrödinger was that he as well as Monod tried to describe the emergence of order and life in the framework of Boltzmann's thermodynamics. He was right with his critique of teleological forces or ordering demons, which were even postulated by physicists at the end of the last century. The fiction of a demon which can reverse the increase of entropy in a closed system according to the second law without outer influence, and therefore cause it to act as a *perpetuum mobile* of the second kind, dates back to James Clerk Maxwell. In 1879 William Thompson (the later Lord Kelvin) introduced "Maxwell's sorting demon", which was able to separate the gas molecules in a closed container in static equilibrium and with a homogeneous distribution of molecular velocities spontaneously into two parts with faster and slower molecules [3.27].

Obviously, "sorting demons" are an ad hoc hypothesis which cannot be explained in the physical framework of the 19th century. Boltzmann's thermodynamics as well as Newton's mechanics are insufficient to model the emergence of complex order, and thus the origin and growth of living systems. The first and second laws of thermodynamics are subjected to an important condition that is generally not true of all nature. These laws assume that all energy exchanges take place in a closed and isolated system. As energy and material fluxes through most regions of the universe, natural systems are rarely closed. As solar energy bombards the earth, it cannot be considered as a closed and isolated system.

So the first and second laws of thermodynamics are not false, but they are empirically restricted to approximately isolated microscopic subsystems, cosmic systems, or prepared conditions in laboratories. The situation can be compared with Newton's classical mechanics. After Einstein's special theory of relativity, it has not become false, but it is no longer the universal scheme of physics, and is restricted to motions that are slow relative to the speed of light. Most of nature must be modeled by dynamical systems which do not live in Boltzmann's general condition of equilibrium, because they are subjected to energy and material fluxes.

Historically, fundamental contributions such as those of Maxwell or Gibbs deal uniquely with situations corresponding to equilibrium or to situations infinitesimally near to equilibrium. Pioneering work on non-equilibrium thermodynamics was started by, e.g., Pierre Duhem at the beginning of this century. But his work remained unnoticed until Onsager (1931), and later the Prigogine school, the Haken school, and others began to study the behavior of complex systems far from thermal equilibrium. From a historical point of view, the situation can be compared with the development of chaos theory and complex Hamiltonian systems (compare

Sect. 2.3). Chaotic phenomena were already discovered and well known to Poincaré, Maxwell, and others. But the mathematical problems connected with nonlinear complex systems deterred most scientists from searching for corresponding models.

3.3 Complex Systems and the Evolution of Organisms

Open systems not only have internal sources of entropy production but also an external source of entropy production associated with energy or mass transformations to or from their surroundings. These systems maintain their structure by dissipation and consumption of energy and are called "dissipative structures" by Ilya Prigogine. We have already become acquainted with nonliving dissipative systems like fluids, lasers, and clouds which are dependent on outside energy fluxes to maintain their structure and organization. Nonequilibrium systems exchange energy and matter with their environment, maintaining themselves for some period of time in a state far from thermal equilibrium and at a locally reduced entropy state. Small instabilities and fluctuations lead to irreversible bifurcations and thus to an increasing complexity of possible behavior.

A mathematical theory of dissipative structures with nonlinear evolution equations seems to offer the framework for modeling Aristotle's "sublunar" world of growing and dying nature. It is amazing to recognize that Aristotle's idea of a cyclic nature corresponds to periodic attractors or limit cycle solutions of corresponding differential equations. The cyclic nature of these systems allows them not only to develop stability but also to develop a hierarchy of complex structures within themselves. A cycle of living systems as it was already described in antiquity becomes autocatalytic by virtue of an evolutionary feedback.

The main idea was already expressed by Spencer and Boltzmann when they assumed that a pre-biological system may evolve through a whole succession of transitions leading to a hierarchy of more and more complex states. But, contrary to Boltzmann, these transitions can only arise in nonlinear systems far from thermal equilibrium. Beyond a critical threshold the steady-state regime becomes unstable and the system evolves to a new configuration. Evolving through successive instabilities, a living system must develop a procedure to increase the nonlinearity and the distance from equilibrium. In other words, each transition must enable the system to increase its entropy production. The evolutionary feedback of Ilya Prigogine, Manfred Eigen, and others means that changing the control parameter of the system beyond a certain threshold leads to an instability through fluctuations, which causes increased dissipation and thus influences the threshold again.

It follows that life did not originate in a single extraordinarily unlikely event and that the evolution of life did not proceed against the laws of physics. As we have learnt, Boltzmann's and Monod's idea of a gigantic fluctuation which would unfold over the time of biological evolution stems from equilibrium thermodynamics. While the probability of a dissipative structure (for instance, a periodic temporal process like a Bénard problem) is tiny in equilibrium statistical mechanics, it occurs with probability one in conditions far from equilibrium. Thus, Prigogine can argue:

Far from being the work of some army of Maxwell's demons, life appears as following the laws of physics appropriate to specific kinetic schemes and too far from equilibrium conditions. [3.28]

In the mathematical framework of nonlinear complex systems, many models have already been suggested to simulate the molecular origin of life. Complexity on the molecular scale is characterized by a large potential number of states which could be populated given realistic limits of time and space.

For instance, a typical small protein molecule contains a polypeptide chain which is made of about 10^2 amino-acid residues. Given the 20 classes of natural amino acids, there are 20^{100} or 10^{130} alternative sequences of this length. The DNA molecule that comprises the total genome of a single bacterial cell represents one or few choices out of more than $10^{1000000}$ alternative sequences. Obviously, only a minute fraction of all such alternatives could have been tested by nature. Mathematically, a sequence containing ν residues of λ classes allows for $\binom{\nu}{k}(\lambda-1)^k$ alternative copies having substitutions at k positions. Figure 3.2 shows a gene which codes for a sequence of 129 amino acids [3.29].

Certain microstates may strongly influence macroscopic behavior. Such fluctuations may amplify and cause a breakdown of formerly stable states. Nonlinearity comes in through processes far from the thermal equilibrium.

Classical and only necessary conditions for life demand: (1) self-reproduction (in order to preserve a species, despite steady destruction), (2) variability and selection (in order to enlarge and perfect the possibility of a species, biased by certain value criteria), and (3) metabolism (in order to compensate for the steady production of entropy) [3.30].

Manfred Eigen has suggested a realization of these criteria by a mathematical optimization process. In this model, the nucleation of a self-reproducing and further evolving system occurs with a finite expectation value among any distribution of

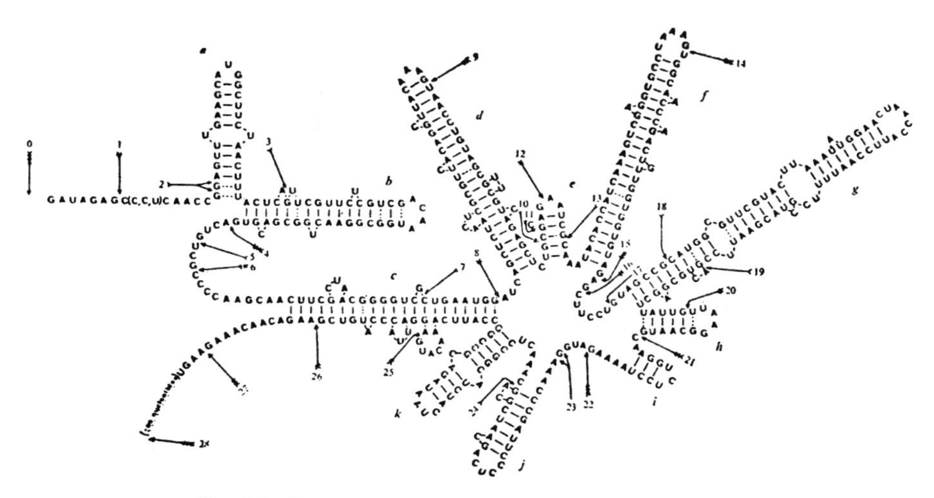

Fig. 3.2. Gene coding for a sequence of 129 amino acids

random sequences of macromolecules such as proteins and nucleic acids. The initial copy choice for self-reproduction is accidental, but the subsequent evolutionary optimization to a level of unique efficiency is guided by physical principles. In this model, life should be found wherever the physical and chemical conditions are favorable, although some molecular structures should show only slight similarity with the systems known to us.

The final outcome will be a unique structure, e.g., an optimized molecular sequence. Darwin's principle of the survival of the fittest is mathematized by an optimization principle for possible microstates of molecular sequences. It is assumed that in simple cases biomolecules multiply by autocatalysis. For instance, two kinds of biomolecules A and B from ground substances GS are multiplied by autocatalysis, but in addition the multiplication of one kind is assisted by that of the other kind and vice versa (Fig. 3.3a). In more complicated cases with more kinds of biomolecules, the latter are assumed to multiply by cyclic catalysis (Eigen's "hypercycles") (Fig. 3.3b). This mechanism combined with mutations is able to realize an evolutionary process.

Eigen suggests the following simplified model of an evolutionary optimization [3.31]: the machinery of a biological cell is codified in a sequence of four chemical substances A, T, G, C which consititute the genes. Each gene represents a functional unit, which is optimally adapted to the special purpose of its environment. The length of a gene in nature is seldom more than 1000 sequential positions. Thus for 4 symbols, there are 4^{1000} alternative genes ("mutations") of length 1000. In scientific notation that means about 10^{600} possibilities. In order to get an impression of this huge number, we should recall that the content of matter in the whole universe corresponds to 10^{74} genes and that the age of the universe is less than 10^{18} seconds.

Thus if all the matter of the universe since its very beginning (the "Big Bang") were used to alter and to produce new genes of length 1000 in each second, then by today only 10^{93} mutations could have been tested. Eigen concludes that genes which represent optimal functional unities cannot be produced by random processes, but must be developed by a self-optimizing process.

Mathematically, the process of adaptation can be imagined as a successive replacement of positions which aims at a final ("optimal") sequence. This is a typ-

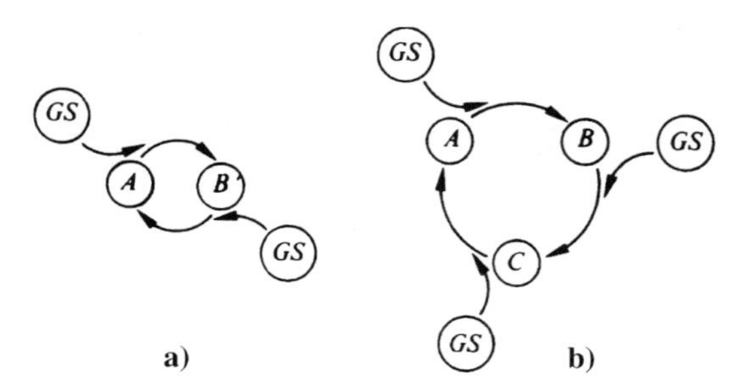

Fig. 3.3a,b. Autocatalysis with two kinds of biomolecules (**a**) and cyclic catalysis (hypercycles) with more kinds (**b**)

ical interpretation of the problem-solving approach of computer science. In order to solve a problem successfully, we must find an adequate spatial representation of the self-optimizing strategies. Because of the huge numbers, a 3-dimensional space is obviously not suitable. The length of strategies, i.e., the distance from a gene to its optimal variant, is too long. One would go astray. Further, sequences with a high degree of similarity cannot be represented adequately by a neighborhood in a 3-dimensional space. Thus it is proposed to alter the dimensions in the following way.

A sequence with n positions is defined as a point in n-dimensional space. For two symbols 0 and 1 there are 2^n alternative sequences, which are the points of the space. Each point has n nearest neighbors, which represent the mutations differing only in one position ("1-mistake mutation"). Between the two extremal points with only 0 or only 1, there are $n!$ possible connections. In Fig. 3.4a–d there are some examples of n-dimensional sequential spaces for the dual case. The great advantages of these spaces are the very short distances and the dense network of possible connections. As an example, the longest distance in a 1000-dimensional space is only 1000 units, in a 23-dimensional space with 10^{14} points only 23 units.

The 23-dimensional space suffices to represent all the points on the earth's surface in units of one meter. In this space optimal strategies can be given to find the highest mountain in some region of the earth. For this purpose we introduce a valuation function which associates each point with a numerical "height". Imagine a tour in the Alps. You have no fixed aim (for instance a special mountain), but a rough idea of orientation: you want to wander "uphill" without losing too much height. Mathematically, the gradient of your path is known and determines your decision as to direction. In the real Alps you encounter 1-dimensional edges and passes in the mountains, and your chances of reaching optimal points are restricted. In a 23-dimensional space you can go in 23 directions and distinguish pathways with different gradients, i.e., k in direction "uphill" and $23 - k$ "downhill" ($k \leq 23$). The chances of reaching optimal points in your neighborhood are very high.

In the n-dimensional sequential space of genes, the valuation of points is given by "selection values". Mutations do not arise completely irregularly or chaotically, but depend on which predecessors occur in the distribution most frequently. The question of which predecessors occur in the distribution most frequently depends again on their selection value relative to the optimal variant in the distribution. The selection values are not distributed irregularly, but in connected regions. For instance, on the earth a high mountain like Mount Everest is not situated in flat countryside, but among the Himalayas.

For the replication and self-reproduction of life, Eigen presupposes a self-optimizing machinery of high efficiency. Freeman Dyson proposed a mathematical model, according to which primitive life systems first occur without the correct machinery for replication and selection and fulfill only metabolic functions in relations to their environment [3.32]. The essential characteristic of such molecular systems is their homeostasis, i.e., the ability to maintain a stable and more or less constant equilibrium in a changing environment. According to Dyson, the configuration of self-replication mechanisms occurs only in a second step. The nucleic acids

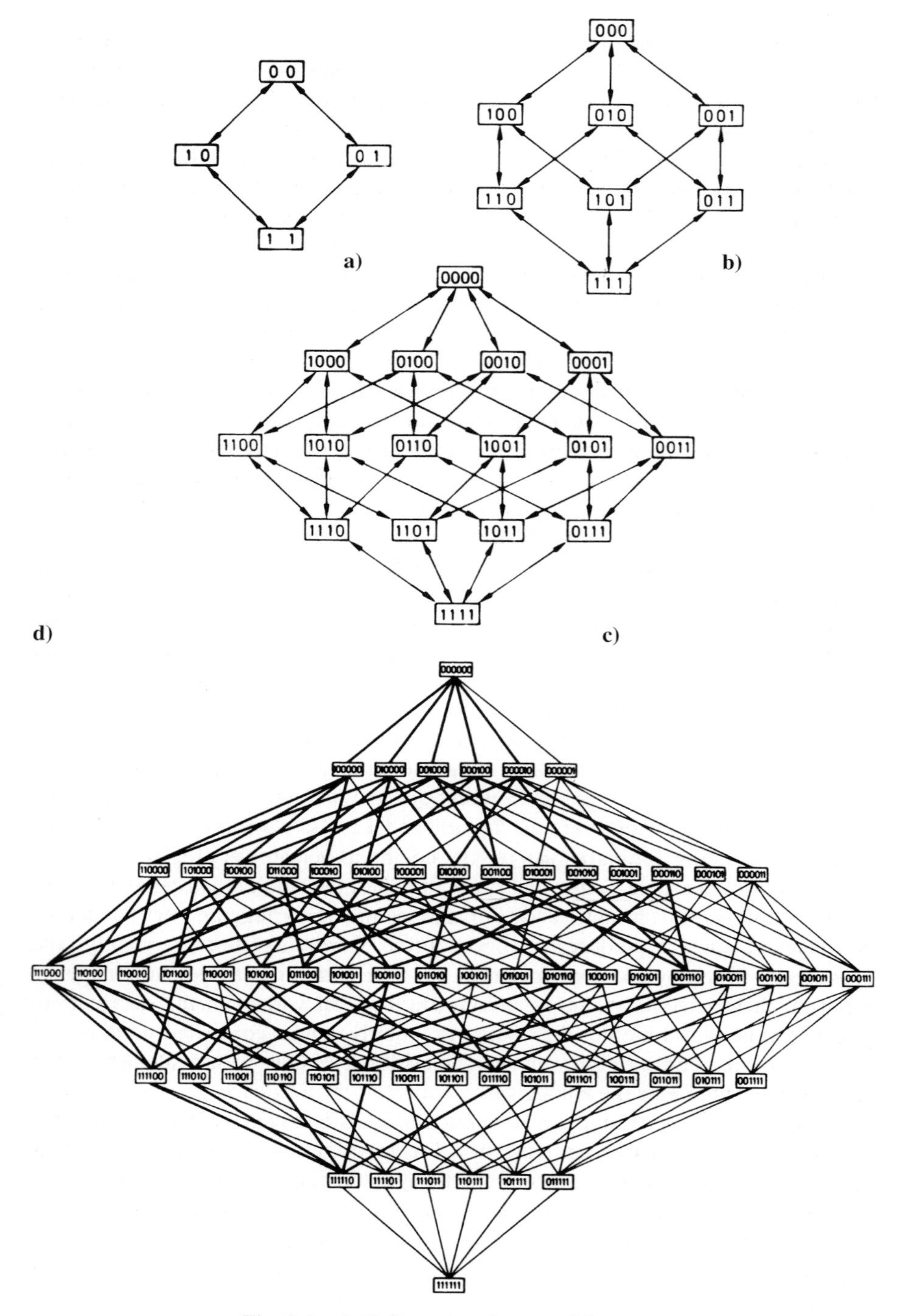

Fig. 3.4a–d. N-dimensional sequential spaces

required for that purpose are explained by Dyson as "non-assimilated" byproducts of the early metabolic life process, which first assumed a parasitic status in the whole system, and then ultimately developed through symbiotic intermediate states to fully-integrated functional mechanisms for reproduction and evolution.

The dual function of metabolism and replication in Dyson's model has much in common with the complex system of prebiotic evolution that was suggested by Stuart Kauffman [3.33]. He rejects the idea that life sprung from an RNA world. His hypercycle-like system is a complex autocatalytic network of reactions self-organizing themselves in such a way that metabolism, as a macroscopic order state of the system, becomes possible. Metabolism draws materials and energy from the environment around to increase and maintain internal order. Thus, it is an open dissipative system.

Examples of complex biological systems are genetic systems, nervous systems, immune systems, and ecosystems, all of which are composed of a network of multiple interacting elements as agents. The nonlinear dynamics of these complex networks can only be modeled by some simplifications. So it is assumed that time is discrete, and that the behavior of a network at one time depends on the state of the network at a preceding time. Further on, the elements of a network have only a limited number of different states, e.g., a gene is turned on or off, or a neuron is firing or not firing. A network is a collection of connected elements as agents that can be visualized by a set of nodes and a set of edges connecting pairs of nodes. Each element is characterized by a single output and several inputs from elements of the network. There is also a rule for each element telling what the output should be given the inputs. In the case of Boolean elements, there are only two values 1 ("on") or 0 ("off"). A rule determining a Boolean output by Boolean inputs is called a Boolean function. The state of a Boolean network specifies whether each element is "on" or "off". For a network of n elements, there are 2^n possible states. Boolean networks of elements with a single input have only geometries of strings, simple loops, and loops with strings. Thus, their dynamics are restricted to fixed points, cycles, and multi-stability. When elements have more than one input, there can be multiple, connected loops with much more complicated Boolean functions. Examples of Boolean functions in biochemistry are control mechanisms, in which activities of proteins and genes are regulated by circulating molecules. The regulatory gene networks in living organisms can be understood by the complex dynamics of Boolean networks.

In order to manage the high complexity of gene networks in living organisms, S. Kauffman suggested studying random Boolean networks. They are ordinary Boolean networks, where the choice of connections and Boolean functions is made randomly when the network is designed. For a network of n nodes with k inputs of each node, there are 2^{2^k} Boolean functions. A random-number generator selects the inputs to each node. Kauffman programmed a computer to iterate the dynamics of random Boolean networks. In his experiments, he found a hierarchy of dynamical behavior with fixed points and cycles of increasing complexity, which can be observed in real cells.

In general, an evolutionary process is expected to produce new kinds of species [3.34]. A species may be considered as a population of biomolecules, bacteria, plants, or animals. These populations are characterized by genes which undergo mutations producing new features. Although mutations occur at random, they may be influenced by external factors in the environment, such as changing temperature or chemical agents. At a certain critical mutation pressure new kinds of individuals of a population come into existence. The rate of change of these individuals is described by an evolution equation. As these individuals have new features, their growth and death factors differ. A change (mutation) is only possible when fluctuations occur in the population and the environment. Thus, the evolution equation determines the rate of change as the sum of fluctuations and the difference of growth and death factors.

A selection pressure can be modeled when different subspecies compete for the same living conditions (e.g., the same food supply). If the mutation rate for a special mutant is small, only that mutant survives which has the highest gain factor and the smallest loss factor and is thus the fittest. The competition procedure can be simulated by the slaving principle: unstable mutants begin to determine the stable ones. It is noteworthy that the occurrence of a new species due to mutation and selection can be compared with a nonequilibrium phase transition of a laser [3.35].

A living cell is an open system with a flow of energy passing through it. As already shown by Erwin Schrödinger, the energy flow creates conditions that allow strong deviations from thermodynamic equilibrium. According to Prigogine et al., this results in models of dissipative self-organization and pattern formation, the parameters of which are set up by genetic as well as epigenetic constraints. However, it would be misleading to expect that the process of self-organization in living cells simply represents a reduced copy of the pattern-formation phenomena in macroscopic reaction–diffusion systems. The laws of physics, when applied at a different scale typical for intracellular processes, can influence the mechanisms involved and produce a wealth of new properties. This also makes spatial pattern formation based on such reactions and diffusion impossible at very small length scales. The temporal self-organization of chemical processes, expressed in the generation of different periodicities and interactions between them, plays a fundamental role in living cells [3.36]. Thus, from a methodological point of view, it is not sufficient to know the general scheme of dissipative self-organization. But we must inquire experimentally into its cellular application under particular temporal, spatial, and chemical constraints.

Nevertheless, the link between physical-chemical systems and biological structures can be modeled by dissipative structures which may be involved in living systems. An important example is provided by the immune system, the disturbance of which causes many very dangerous illnesses, such as AIDS. Concerning the antibody-antigen kinetics, new types of antibodies can be generated successively, where some antibodies act as antigens. This process leads to a very complex dynamics of the total system [3.37].

As we have learnt, among the most striking features of dissipative systems are oscillatory phenomena. At the subcellular level, there are series of oscillating enzymatic reactions. Glycolysis is a process of great importance in living cells. The regulatory enzyme gives rise to oscillations with periods ranging from two to ninety minutes. Experimental oscillations can be identified with limit cycle type oscillations arising when the uniform state is unstable.

Another metabolic oscillation is that of the periodic synthesis of cyclic AMP in cellular slime molds. This species exhibits a transition between two different states of organization. First, the amebas are independent and separate cells. The transition to an aggregation and finally to a multicellular fruiting body takes place after starvation. Single cells, deprived of nutrient, aggregate in concentric waves around centers as a response to cyclic AMP being emitted from the centers. The synthesis of cyclic AMP realizes the limit cycle type. The aggregation process itself represents a self-organization occurring beyond instability.

Modeling this process in the framework of complex systems, we first consider a population of separated and homogeneous cells. A control parameter denotes the supply of nutrition which can be tuned to a critical value of starvation. Then the cyclic AMP is emitted and overcomes the random diffusive motion of amebas, and the uniform state becomes unstable. On the macroscopic level, the cells start to differentiate into several functions and to cooperate. On the macroscopic level, intermediate states of aggregation can be observed which finally lead to the new form of the mature multicellar body. Producing isolated spores, the life cycle of the slime mold is repeated in the described states of phase transition (Fig. 3.5) [3.38].

The spontaneous emergence of organic forms has seemed to be a miracle of life. Thus, in the history of science morphogenesis was a prominent counterexample against physical reductionism in biology. Today, morphogenesis is a prominent ex-

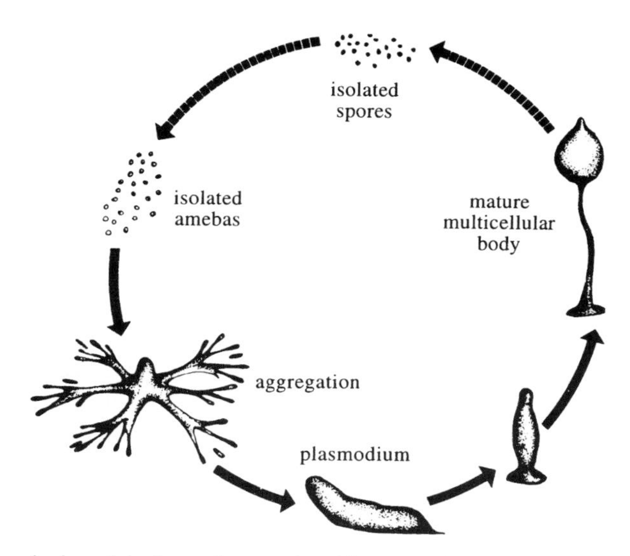

Fig. 3.5. Dynamical model of morphogenesis with states of cell formation (life cycle of the slime mold)

ample for modeling biological growth by complex dynamical systems. What would Goethe have said about a mathematical model of his beloved metamorphosis? In this context, pattern formation is understood as a complex process wherein identical cells become differentiated and give rise to a well-defined spatial structure. The first dynamic models of morphogenesis were suggested by Rashevsky, Turing, and others. Let us regard Rashevsky's model for the morphogenesis of plant growth ("phyllotaxis") [3.39].

Figure 3.6a shows an idealized vine stalk sprouting one branchlet at a time with a symmetrically rotating direction for the three branchlets. At the tip of the growing stalk is a growth bud containing a mass of undifferentiated and totipotent cells. The problem of phyllotaxis was the emergence of the growth pattern with differentiated cells as the leaf bud cells, the branch cells, and others leading to the leaf buds and branchlets. The Rashevsky model is based on a ring of growth cells around the circumference of the stalk, near the growth bud at the top.

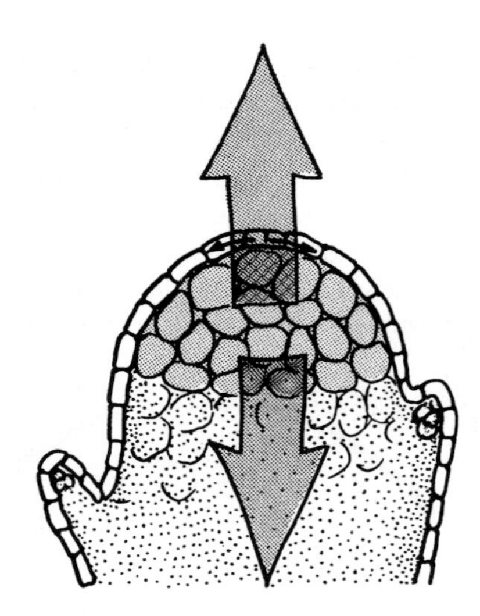

Fig. 3.6a. Dynamical model of morphogenesis with states of cell differentiation (Rashevsky model of phyllotaxis) [3.39]

A cell is considered as a bag of fluid with homogeneous chemical composition. One of the chemical constituents is a growth hormone called morphogen. The concentration x of this morphogen is the observed parameter of the model. As the parameter varies between 0 and 1, the state space of the model is a line segment (Fig. 3.6b). If the concentration of this morphogen exceeds a certain critical value, the growth function of the cell is turned on, the cell devides, and a branchlet comes into existence.

In the next step, two cells are considered as an open system in which one morphogen can be exchanged between the two-celled system and its environment. If the concentration of the morphogen in the second cell is denoted by y, then the

Fig. 3.6b. A cell with concentration x of morphogen and the corresponding state space with states x on a line segment

state of the whole system corresponds to a point (x, y) in the unit square which is interpreted as the state space of the system. In Fig. 3.6c, the state space is divided into four regions corresponding to the situations "cell 1 off and cell 2 growing" (A), "both cells' growth turned off" (B), "cell 1 growing, cell 2 off" (C), "both cells growing" (D).

In the final step three cells form a ring with a uniform concentration of the morphogen in each. The point (x, y, z) in the unit cube represents a state of the system. In a three-dimensional space, the state space for the closed system of three cells with one morphogen is realized by a triangle with $x + y + z = 1$, i.e., the sum of the concentrations is constant (Fig. 3.6d).

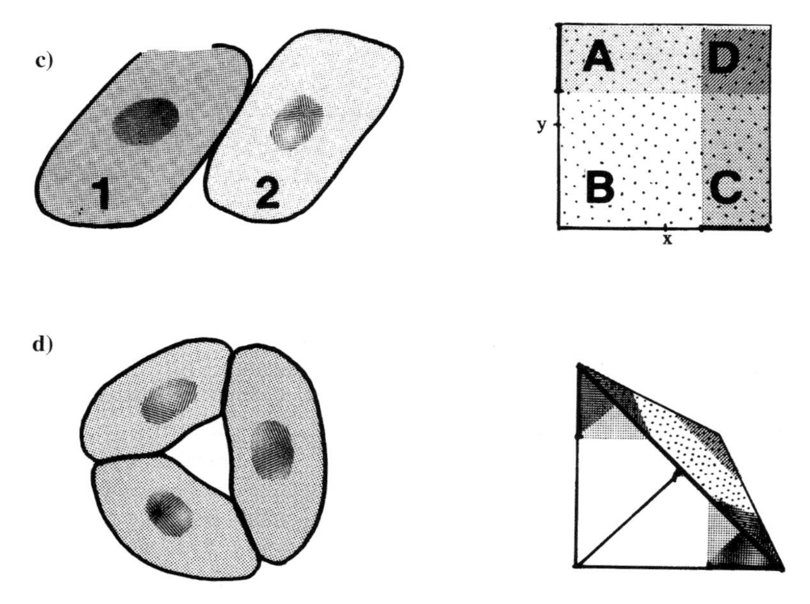

Fig. 3.6c,d. A two-celled system with morphogens x and y and the corresponding state space with states (x, y) in a unit square (**c**) and a three-celled system with a uniform concentration of morphogens x, y, and z and the corresponding state space with states (x, y, z) in the unit cube (**d**) [3.39]

In Fig. 3.6e, a dynamical system with a periodic attractor is added to the state space. One after another of the three cells is turned on, then off, periodically. In Fig. 3.6f, the stalk is modeled as a stack of rings of cells, each ring represented as an identical copy of the triangular model of Fig. 3.6d. Growth of the stalk upwards in time is represented by associating time with the upward direction. The periodic attractor of Fig. 3.6e is transformed to a periodic time series spiraling upwards in time.

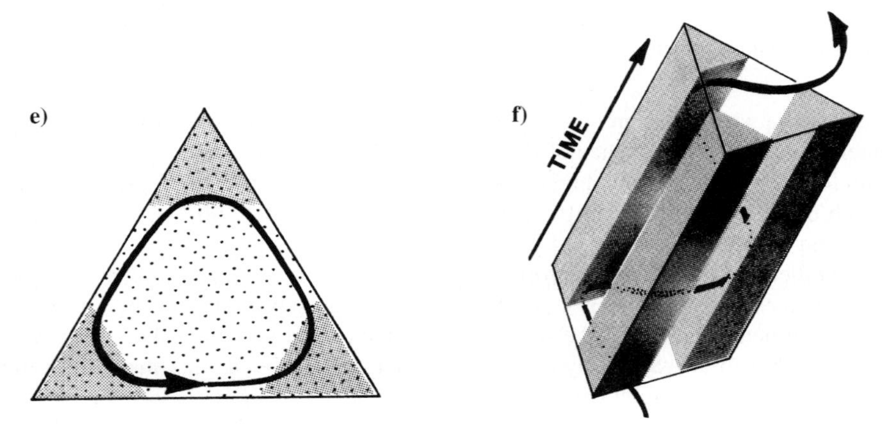

Fig. 3.6e,f. A three-celled system with a periodic attractor (**e**) and the growth of a stalk modeled as a stack of three-celled systems (Fig. 3.6d) with periodic attractor (Fig. 3.6e) transformed to a spiraling time series (**f**) [3.39]

In this simplified dynamical model of morphogenesis, a central problem remains open. How do the originally undifferentiated cells know where and in which way to differentiate? Experiments indicate that this information is not originally given individually to the cells but that a cell within a cellular system receives information on its position from its surroundings. A famous example is the hydra, which is a tiny animal, consisting of about 100 000 cells of about 15 different types. Along its length it is subdivided into different regions, e.g., its head at one end. If a part of a hydra is transplanted to a region close to its old head, a new head grows by an activation of the cells. There is some experimental evidence that activator and inhibitor molecules actually exist.

In a mathematical model due to Gierer and Meinhardt, two evolution equations were suggested, describing the rate of change of activator and inhibitor concentrations, which depend on the space-time coordinates. The change of rates is due to a production rate, decay rate, and diffusion term. Obviously, inhibitor and activator must be able to diffuse in some regions, in order to influence the neighboring cells of some transplant. Further on, the effect of hindering autocatalysis by the inhibitor must be modeled. In Fig. 3.7 the interplay between activator and inhibitor leads to a growing periodic structure which is calculated and plotted by computer-assisted methods [3.40].

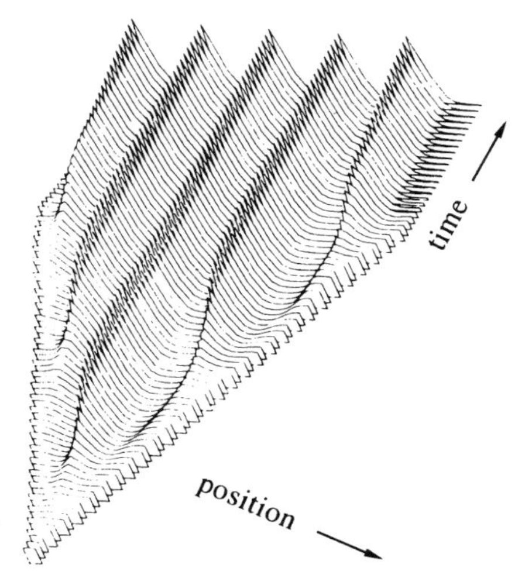

Fig. 3.7. Computer-assisted model of morphogenesis leading to a periodic structure [3.40]

To derive such patterns, it is essential that the inhibitor diffuses more easily than the activator. Long range inhibition and short range activation are required for a non-oscillating pattern. By methods of mathematical analysis, the evolving patterns described by the evolution equations of Gierer and Meinhardt can be determined. A control parameter allows one to distinguish stable and unstable configurations ("modes").

At a critical value, the unstable modes start to influence and dominate the stable ones according to the slaving principle. Mathematically, the stable modes can be eliminated, and the unstable ones deliver order parameters determining the actual pattern. Thus, actual patterns come into existence by competition and selection of some unstable solution. Selection according to the slaving principle means reduction of the complexity which stems from the huge number of degrees of freedom in a complex system.

Biochemically, this kind of modeling of morphogenesis is based on the idea that a morphogenetic field is formed by diffusion and reaction of certain chemicals. This field switches genes on to cause cell differentiations. Independently of the particular biochemical mechanism, morphogenesis seemes to be governed by a general scheme of pattern formation in physics and biology. We start with a population of totipotent cells corresponding to a system with full symmetry. Then, cell differentation is effected by changing a control parameter which corresponds to symmetry breaking. The consequence is an irreversible phase transition far from thermal equilibrium. In Fig. 3.8, the phase transition of activator and inhibitor concentration is illustrated in a computer simulation.

Independently of the common scheme of symmetry breaking, there is an important difference between physico-chemical and biological pattern formation. Physical and chemical systems lose their structure when the input of energy and materials is

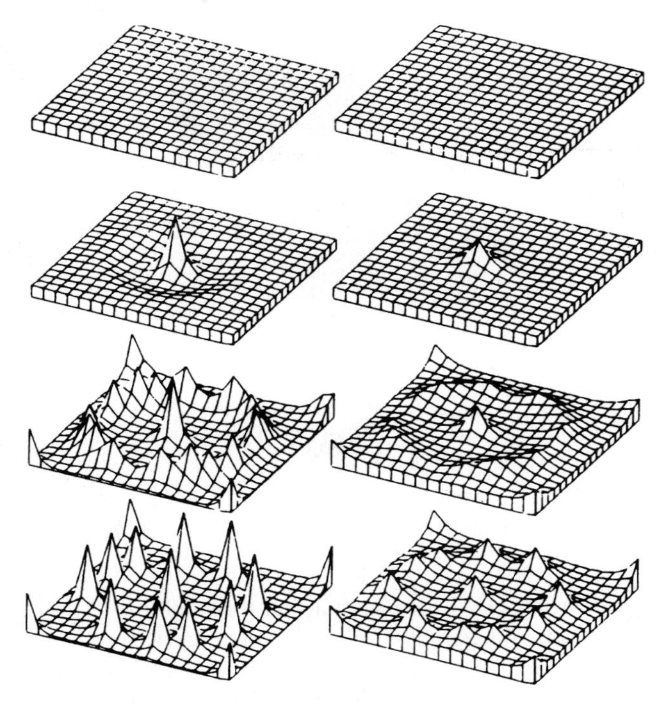

Fig. 3.8. Two computer-assisted models of morphogenesis simulating the phase transitions of activator and inhibitor concentrations

stopped (compare, e.g., the laser or the Zhabotinsky reaction). Biological systems are able to conserve much of their structure at least for a relatively long time. Thus they seem to combine approximately conservative and dissipative structures.

Since antiquity, living systems were assumed to serve certain purposes and tasks. Organs of animals and humans are typical examples of functional structures which are explored by physiology and anatomy. How can the functional structures of medicine be understood in the framework of complex systems [3.41]?

The complex bifurcations of vessel networks are examples of fractal structures. The form of trees, ferns, corals, and other growing systems are well described by fractals. In Chap. 5, we shall discuss recursive and computer assisted procedures to simulate the fractal growth of trees. The vascular tree in the heart reminds us of a complex network of branches and roots. This is quite natural when one appreciates that vascular growth occurs by the budding of capillaries into regions of cell division and differentiation.

Trees branching into open space have room to expand. But hearts, lungs, and other organs occupy a limited space. The networks of nerves or vessels that penetrate them are servants to the principal occupants of the space. The structure of the microvascular network is virtually completely defined by the cells of the organ. In skeletal and cardiac muscles the capillaries are arrayed parallel to the muscle cells, with some cross-branches. The system is guided in its growth by the need for the nerve or the vascular system to follow the lines of least resistance.

This leads to medically quite interesting questions of whether the fractal growth and form of the vascular network can give rise to observed heterogeneities of flow in the heart. A simple algorithm for a branching network as shown in Fig. 3.9 leads to an appropriate probability density function of regional flow. The fractal system of an organ has become a functional structure [3.42].

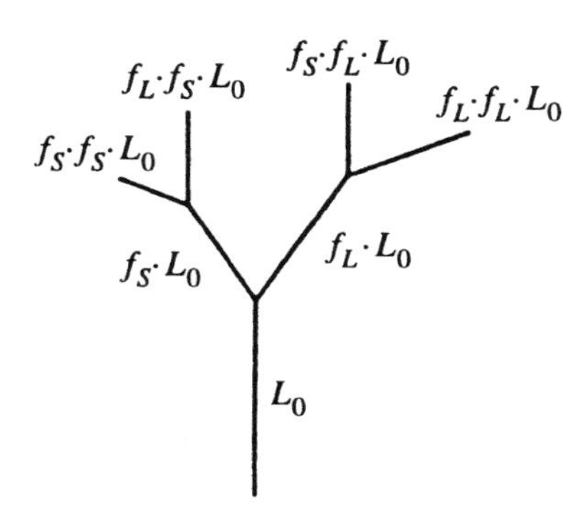

Fig. 3.9. A branching vascular network of the heart with fractal recursions for branch lengths starting with length L_0 of a main stem vessel and decreasing lengths of following daughter vessels by factors f_L and f_S (subscripts L and S indicating longer and shorter vessels)

Fractal illustrations of a bronchial network are an inspiration for physicians to apply these approaches to the lung. Physical systems, from galactic clusters to diffusing molecules, often show fractal behavior. Obviously, living systems might often be well described by fractal algorithms. The vascular network and the processes of diffusion and transmembrane transport might be fractal features of the heart. These fractal features provide a basis which enables physicians to understand more global behavior such as atrial or ventricular fibrillation and perfusion heterogeneity.

As we have seen in Sect. 2.4, nonlinear dynamics allows us to describe the emergence of turbulence, which is a great medical problem for blood flow in arteries. Turbulence can be the basis of limit cycling, as can be shown with water flowing through a cylindrical pipe. A variety of control systems produce oscillations. It might also be expected that some oscillating control systems show chaotic behavior.

Atrial and ventricular fibrillation are the classic phenomena that appear chaotic. The clinical statement on the heart rate in atrial fibrillation is that it is irregularly irregular. The observations are that the surface of the atrium is pulsing in an apparently chaotic fashion. However, the studies of reentry phenomena and of ventricular fibrillation show that there are patterns of excitation, again illustrating that this is organized ("mathematical") chaos. Fractal and chaotic algorithms for this have been described. The two curves of Fig. 3.10 show a regular and chaotic heart rate [3.43].

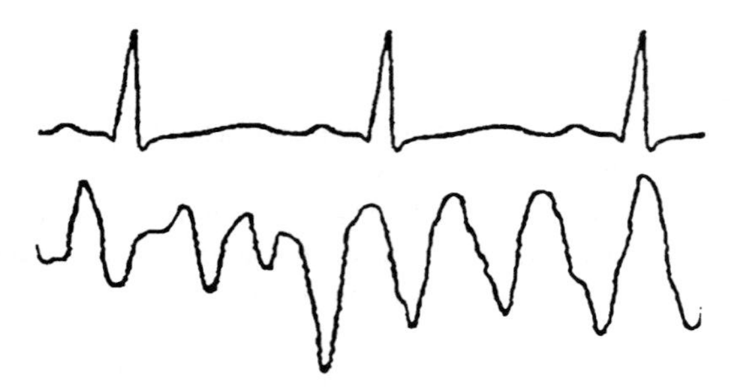

Fig. 3.10. Regular and chaotic heart rate

Nevertheless, chaotic states cannot generally be identified with illness, while regular states do not always represent health. There are limited chaotic oscillations protecting the organism from a dangerous inflexibility. Organs must be able to react in flexible ways, when circumstances change rapidly and unexpectedly. The rates of heart beat and respiration are by no means fixed like the mechanical model of an idealized pendulum.

Single organs and the whole organism of the human body must each be understood as a system of nonlinear complex dynamical systems of high sensibility. Tuning their control parameters to critical values may cause phase transitions of irreversible developments representing more or less dangerous scenarios of human health. Dissipative complex structures are open systems which cannot be separated from their surrounding environment. Thus, on the background of the complex dynamical system approach, the classical "mechanical" view of medicine separating the human body into particular parts for highly specialized experts must be heavily criticized. The whole body is more than the sum of its parts. It is amazing to recognize that from the modern view of complex dynamics an old demand of traditional physicians since antiquity is supported again, namely that medicine is not only an analytical science, but an art of healing which has to consider the wholeness of health and illness.

3.4 Complex Systems and the Ecology of Populations

Ecosystems are the results of physical, chemical, and biotic components of nature acting together in a structurally and functionally organized system. Ecology is the science of how these living and nonliving components function together in nature. Obviously, in the framework of the complex system approach, ecology has to deal with dissipative and conservative structures of very high complexity depending on the complexity of the individual physical, chemical, and biotic systems involved in them, and the complexity of their interactions [3.44].

In 1860, one of the first empirical case studies on ecology was provided by Henri Thoreau in a lecture about "the succession of forest trees". He observed that nature displayed a process of plant development resulting in a sequential change of species that seemed to be observable and predictable. If the ecosystem is left undisturbed, the progression from bare field to grassland to grass-shrub to pine forest and finally to an oak-hickory forest is a predictable, 150-year process, at least in Massachusetts in the 19th century [3.45].

In nearly the same year, Charles Darwin published his famous theory of evolution based on the mechanisms of variation and selection. Darwin saw progressive changes by organisms resulting from competition and adaption to fit optimally into ecological niches. It is the flux of energy from the sun and chemical reactions that sets the process of life in motion and maintains it. Boltzmann already recognized that the biosphere extracts a high energy-entropy cost for the organization of living things. These processes are not only based on the biotic components of an ecosystem but affect the nonbiotic components as well.

James Lovelock has proposed that living systems drive the major geochemical cycles of the earth. He proposed that the global atmospheric composition was not only developed by living systems but also controlled by the global ecosystem. The "balance of nature" has become the popular title for a complex network of equilibria characterizing the human ecosystem on earth [3.46].

The mathematical theory of complex systems allows one to model some simplified ecological case studies. The phenomena to be explained are, mainly, the abundance and distribution of species. They may show typical features of dissipative structures like temporal oscillations. At the beginning of the 20th century, fishermen in the Adriatic Sea observed a periodic change of numbers in fish populations. These oscillations are caused by the interaction between predator and prey fish. If the predators eat too many prey fish, the number of prey fish and then the number of predators decreases. The result is that the number of prey fish increases, which then leads to an increase in the number of predators. Thus, a cyclic change of both populations occurs.

In 1925, Lotka and Volterra suggested a nonlinear dynamical model. Each state of the model is determined by the number of prey fish and the number of predator fish. So the state space of the model is represented by a two-dimensional Euclidean plane with a coordinate for prey fish and a coordinate for predator fish. The observations, over time, of the two populations describe a dotted line in the plane. Births and deaths change the coordinates by integers, a few at a time. To apply continuous dynamics, the dotted lines must be idealized into continuous curves.

The vector field on the state space can be roughly described in terms of four regions (Fig. 3.11a). In region A, both populations are relatively low. When both populations are low, predator fish decrease for lack of prey fish while prey fish increase because of less predation. The interpretation of this habitual tendency as a bound velocity vector is drawn as an arrow. In region B, there are many prey fish, but relatively few predators. But when there are many prey fish and few predator fish, both populations increase. This is interpreted by the vector in region B. In region C, both populations are relatively large. The predator fish are well fed and multiply, while the prey fish population declines. This tendency is shown by the vector in region C. In region D, there are few prey fish but many predator fish. Both populations decline. This tendency is shown by the vector in region D. The phase portrait

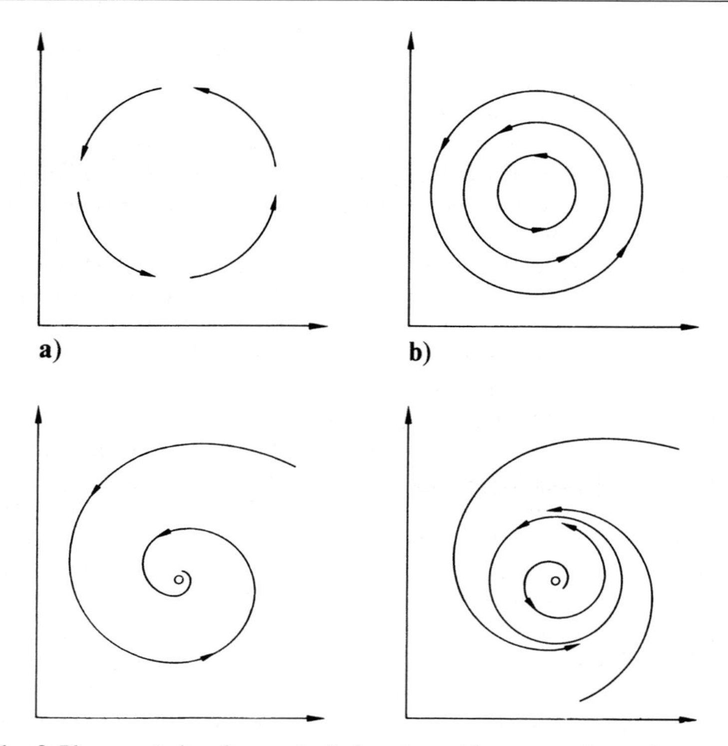

Fig. 3.11a–d. Phase portraits of an ecological system with a prey and a predator population (Lotka-Volterra): (**a**) a closed trajectory, (**b**) a nest of closed trajectories, (**c**) a point attractor, (**d**) a periodic trajectory [3.47]

of this system can be visualized by a closed trajectory, because the flow tends to circulate.

In Fig. 3.11b, the phase portrait is a nest of closed trajectories, around a central equilibrium point. As dynamical systems theory tells what to expect in the long run, the phase portrait enables the ecologist to know what happens to the two populations in the long run. Each initial population of predator fish and prey fish will recur periodically [3.47].

If some kind of ecological friction were added to the model, the center would become a point attractor. This would be a model for an ecological system in static equilibrium (Fig. 3.11c). A different but perhaps more realistic modification of the model results in a phase portrait like Fig. 3.11d, with only one periodic trajectory.

From an analytical point of view, the evolution of a population is governed by an equation for the rate of change of its size [3.48]. Obviously, the number of individuals in the population changes according to its growth rate minus its death rate. A further parameter which has to be considered refers to the limited food supply or depletion of the food resources. There are several living conditions of populations which must be modeled. If different species live on different kinds of food and do not interact with each other, they can coexist.

If different species live in similiar conditions, then the overlapping food supplies must be considered in the evolution equations of the populations. An enormous reduction of complexity is realized if the temporal change of the food supply is neglected. The resulting evolution equations allows several scenarios of coexistence, when stable configurations are realized.

Biologically, stable states correspond to ecological niches which are important for the survival of a species. The predator-prey relation of two populations is realized by the Lotka–Volterra equations characterizing the phase portraits in Fig. 3.11. A particular form of cooperation in nature is the symbiosis of two species. Modeling a symbiosis by evolution equations, it must be considered that the multiplication rate of one species depends on the presence of the other.

Animal populations can be characterized on a scale of greater or lesser complexity of social behavior. There are populations of insects with a complex social structure which is rather interesting for sociobiology. Nicolis and others have tried to model the social organization of termites by a complex dynamical system. The interactions between individuals are physically realized by sound, vision, touch, and the transmission of chemical signals.

The complex order of the system is determined by functional structures like the regulation of the castes, nest construction, formation of paths, the transport of materials or prey, etc. Ants synthesize chemical substances which regulate their behavior. They have a tendency to follow the same direction at the place where the density of the chemical molecules reaches a maximum. Collective and macroscopic movements of the animals are regulated by these chemical concentrations.

In order to model the collective movements, two equations are suggested, considering the rate of change for the concentrations of insects and chemical substances. There is a critical value of an order parameter ("chemicotactic coefficient") for which a stationary homogeneous solution becomes unstable. The system then evolves to an inhomogeneous stationary state. Accordingly, different branching structures will appear, as observed in different ant societies. Figure 3.12 shows the collective movement of ants with two types of structure characteristic of two different species [3.49].

The social complexity of insects can also be characterized by such coordinated behavior as nest construction. This activity has been well observed and explored by experimental studies. A typical observation is that the existence of a deposit of building material at a specific point stimulates the insects to accumulate more building material there. This is an autocatalytic reaction which, together with the random displacement of insects, can be modeled by three differential equations. These equations refer to the observation that the termites, in manipulating the construction material, give it the scent of particular chemical substance, which diffuses in the atmosphere and attracts the insects to the points of highest density, where deposits of building material have already been made.

Thus, the first equation describes the rate of change of the concentration of building material, which is proportional to the concentration of insects. A second

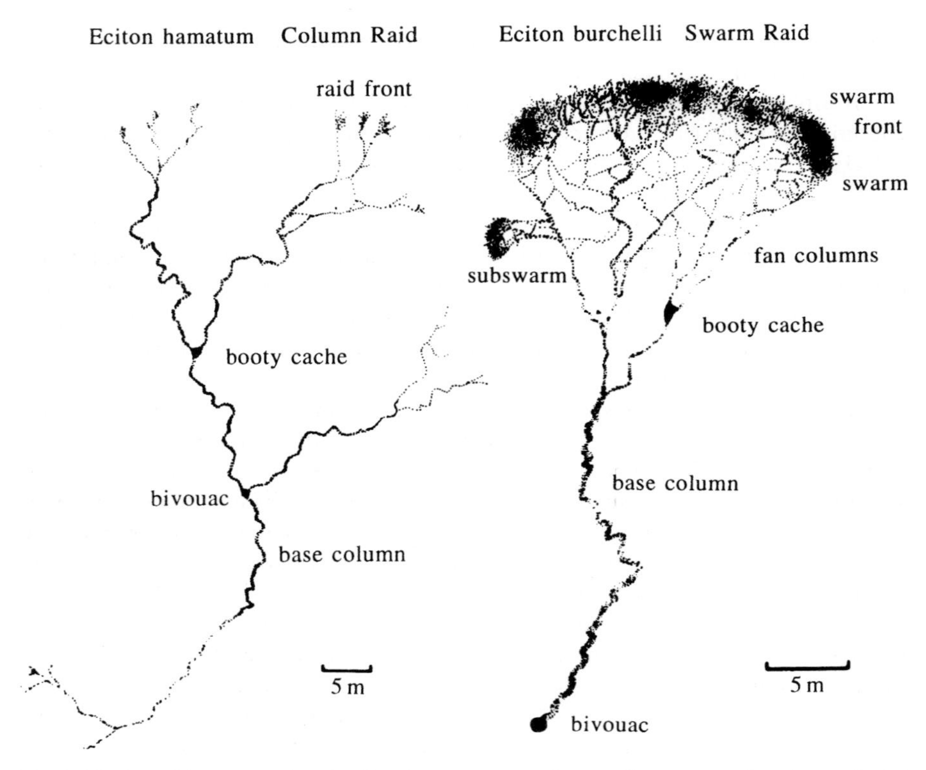

Fig. 3.12. Branching networks of collective movements in two different species of ants [3.49]

evolution equation refers to the rate of change of the scent with a certain diffusion coefficient. A third evolution equation describes the rate of change of the concentration of insects including the flow of insects, diffusion, and motion directed toward the sources of the scent.

The complex social activity of nest construction correponds to the solutions of these equations. Thus, an uncoordinated phase of activity in the beginning corresponds to the homogeneous solution of these equations. If a sufficiently large fluctuation with a larger deposit of building material is realized somewhere, then a pillar or wall can appear. The emergence of macroscopic order, visualized in the insect's architecture of nests, has been caused by fluctuations of microscopic interactions.

Models of the above types are now often used in ecology. It must be mentioned that they are still on a rather simplified level. In reality, one has to take into account many additional effects, such as time-lag, seasons, different death rates, and different reaction behavior. In general, there is not only the interaction of one or two complex populations with their (simplified) environments, but a huge number of different interacting populations. The phase portraits of their nonlinear dynamics at least allow global forecasts in the long run [3.50].

In the traditional Darwinian view, there are two important forces at work in biological evolution which must be modeled in complex dynamical systems: mutation pressure and selection. In biological populations in which the behavior of individuals is governed uniquely by their genes, the amplification of a new type individual corresponds to the Darwinian evolution by natural selection of mutants which appear spontaneously in the system. In the case of higher animals, there is the possibility of behavioral change ("innovation") and its adaption by information. In ecological evolution, new ecological niches have arisen which are occupied by specialized species. Obviously, there is no simple scheme of evolution, but a complex hierarchy of changing and stabilizing strategies which have been layered from prebiological evolution to ecological and finally cultural evolution with human learning strategies (compare Chap. 7).

The complex system approach shows a great variety of possible evolutions with unexpected directions, caused by stochastic fluctuations. There is no global optimizer, no global utility function, no global selection function, no other simplified strategy of evolution but successive instabilities near bifurcation points. In short, Darwin's view is only a particular aspect of evolution. For many contemporaries, he seemed to replace a personal deity called "God" by an impersonal deity called "evolution" governing the world by simple laws. This secularized religious attitude of the 19th century was later on continued by political thinkers like Karl Marx, who believed in an impersonal deity called "history" governing human fate by simplified laws of society.

In the 18th century, Kant had already complained that the term "nature" seems to denote an impersonal deity. But "nature" is, as Kant argued, nothing more than a "regulative idea" of man. From a modern point of view we can actually only recognize dynamical models with more or less high degrees of complexity which may fit observational data with more or less accuracy. The abandonment of some mighty supervisor in nature and human history may leave us feeling alone with perhaps dangerously chaotic fluctuations. But, on the other hand, these fluctuations may enable real innovations, real choices, and real freedoms.

3.5 Complex Systems and Power Laws of Life

The goal of modeling in biology, as elsewhere in science, is to obtain appropriate models that capture the essential features of the structure or process being investigated. Simple linear laws and Euclidean forms and shapes are usually not applicable to biology. Even fractal objects with geometrical self-similarity, where the parts that make up the object are smaller, exact duplicates of the object itself (see Sect. 2.6), cannot be found in real life. Parts of biological objects are rarely exact reduced copies of the whole object. Rather than being geometrically self-similar, they are statistically self-similar. In this case, the statistical properties of the pieces are proportional to the statistical properties of the whole. An example is the average rate at which new vessels branch off from their parent vessels in physiological structures. This rate is the same for large and small vessels. Arteries in the lung or branching vascular networks in the heart (Fig. 3.9) satisfy the conditions for statistical self-similarity. Further examples of branching patterns that are similar at different spatial scales can be found in the dendrites of neurons, the ducts in the liver, the blood vessels in the circulatory system, and flow distributions inside them. Statistical self-similarity can also occur in a hierarchical structure. Processes where local interactions between neighboring pieces produce a global statistical self-similar pattern are termed "self-organizing." Such patterns can be generated at the molecular level, as in the binding of ligands to enzymes, at the cellular level, as in the differentiation of the embryo, and at the organism level, as in slime mold aggregation (Fig. 3.5).

Mathematically, statistical self-similarity means that a property measured at high resolution for a part of an object is proportional to the same property measured for the entire object at coarser resolution. Therefore, the value of a property $L(r)$ when it is measured at resolution r is compared to the value $L(ar)$ when it is measured at a finer resolution ar, where $a < 1$. Statistical self-similarity means that $L(r)$ is proportional to $L(ar)$, or $L(ar) = k\,L(r)$, where k is a constant of proportionality, which depends on a. The size of individual features depends on the measurement resolution. In fractal objects, there is no true value for a measurement. The relationship between the value measured and the measurement resolution is called the scaling relationship. Self-similarity determines the scaling relationship. The self-similarity relationship mentioned above implies that there is a scaling relationship that describes how the measured value of a property $L(r)$ depends on the scale r at which it is measured. The simplest scaling relationship determined by self-similarity takes the form of the power law $L(r) = A\,r^{\alpha}$, where A and α are constant for any particular fractal object or process. Taking the logarithms of both sides of this equation yields the linear equation $\log L(r) = \alpha \log(r) + b$ with $b = \log A$. Thus, power law scalings are found to be straight lines when the logarithm of the measurement is plotted against the logarithm of the scale at which it is measured [3.51]. Although not all power-law relationships are due to fractals, the existence of such a relationship should prompt us to test for self-similarity.

Examples of power-law scaling include the diameters of the bronchial passages for the successive generations of branching in the lung. Another example is the length of the transport pathway through the junctions between pulmonary endothelial cells. The time courses of chemical reactions have also been studied in order to determine whether the time delays resulting from the diffusion of the substrate are long compared to the time required for enzymatic reactions. Scaling in mammalian physiology, from small animals like mice to big ones like elephants, has been examined in relation to metabolism and structure. It is well known that small and light animals usually move rapidly, whereas large and heavy ones move slowly. This is also true for the frequency of the heartbeat, which is higher for a mouse than for an elephant. Elephants also live longer than mice. Therefore, a relationship between the lifetime or activity of a living being and its body mass is assumed.

The activity of an animal can be determined by the velocity of the metabolism. The so-called metabolic rate X_{MB} indicates the velocity of energetic exchange between an organism and its environment. Is there a relationship between the metabolic rate and the mass of an organism? Obviously, the mass M of a body is proportional to the volume L^3 of an organism with a typical length L. Metabolism, the exchange of energy with the environment, takes place on the surface of an organism and so it scales as L^2. From $X_{MB} \sim L^2$ and $M \sim L^3$, it follows that $X_{MB} \sim M^{2/3}$. This expected relation is generally expressed as the so-called allometric equation $X = X_o \cdot M^{\gamma}$. This power-law equation relates a biological variable X (for example, the metabolic rate or life expectation of an organism) to the body mass M. X_o is a reference value that gauges the scale, and γ is the scaling exponent. While X_o varies with the individual and typical properties of an organism, γ only takes a few values. At first it was assumed that γ are multiples of one-third, because the mass of a body depends on its three-dimensional volume. But experiments demonstrate that the metabolic rate actually scales with $M^{3/4}$. Examples of this are the heartbeat, which scales as $M^{-1/4}$, and both life expectation and blood circulation, which scale as $M^{1/4}$. G.B. West, J.H. Brown, and B.J. Enquist [3.52] suggested that these power laws could be explained by the fractality of organisms. Their hypothesis is based on three principles:

1. The natural selection pressure present in nature causes metabolic capacity of an organism to be optimized by maximizing its surface area a and minimizing the transport distance l and time t inside it.
2. Internal supply networks of an organism can be fractal.
3. There is a smallest typical unit of length l_m in biological systems that does not scale with the size of an organism, but remains constant.

It is remarkable that applying only two of these three principles leads to an (empirically false) exponent of 1/3 and multiples of 1/3, but applying all three principles leads to (empirically confirmed) multiples of 1/4. Where does the fourth dimension come in? First, the whole area a of an organism changes if all characteristic lengths l_i of an organism are stretched by a factor Γ as $l_i \rightarrow \Gamma \cdot l_i$. In this case we get a new area $a'(l_1, l_2, \ldots) = \Gamma^{\alpha} \cdot a(l_1, l_2, \ldots)$. If all lengths can be scaled in an organism (i.e., the third principle is not true), then it follows that $\alpha = 2$. This is the expected scaling of a normal area, which can easily be illustrated using an example: a rectangle with side lengths $l_1 = 3\,\text{m}$ and $l_2 = 2\,\text{m}$ has an area $a(l_1, l_2) = 6\,\text{m}^2$. If each side is made three times longer, then the area of the rectangle increases ninefold, to $a' = 36\,\text{m}^2$. For $\Gamma = 3$, it follows that $\Gamma^2 = 9$ and that the new area $a' = \Gamma^2 \cdot a$ is nine times larger than the old area a. In general, $\alpha = 2$ means the scaling of an area, independent of the fractality of a. A similar argument holds for the typical unit of length of an organism $l'(\Gamma \cdot l_1, \Gamma \cdot l_2, \ldots) = \Gamma^{\lambda} \cdot l(l_1, l_2, \ldots)$ with $\lambda = 1$.

Now let us assume that the third principle holds for a normal area $a(l_1, l_2)$, which means that l_1 cannot be stretched. In this case, the area after stretching is $a' = l_1 \cdot \Gamma \cdot l_2$, which is only threefold larger the old area. The scaling law $a' = \Gamma^{\alpha} \cdot a$ does not have the usual exponent (2) for an area; instead, $\alpha = 1$, corresponding to a length. With the scaling exponents α and λ for area and length, the scaling of a volume is given by $\nu' = \Gamma^{\alpha+\lambda} \cdot \nu$. In the case of a uniform density of tissue, the mass M of an organism is proportional to its volume ν . The dependence of the internal area a on the mass M of an organism is then determined by $a \sim M^{\alpha/(\alpha+\lambda)}$.

For normal values $\alpha = 2$ and $\lambda = 1$ in our three-dimensional world, the (empirically false) law $a \sim M^{2/3}$ follows. This also holds if there is no scale-resistant unit in an organism. However, according to the second principle, biological systems have a smallest length l_m, which cannot be scaled (for example capillary diameter). In this case, a possible fractal structure in a and l (as postulated by the second principle) can be considered: the exponents α and λ do not necessarily take the values they do in the three-dimensional world. According to the laws of fractals, they can vary between 1 and 2 (for λ) and between 2 and 3 (for α). For $\alpha = 3$, we get a fractal that fills a volume, while $\alpha = 2$ gives a fractal that fills an area. In order to determine the values of the equation $a \sim M^{\alpha/(\alpha+\lambda)}$, the first principle must be taken into account. The area a becomes maximal if α takes the maximal value 3 and λ takes the minimal value 1. Using these values, we obtain the observable law $a \sim M^{3/4}$ and, in general, exponents with multiples of 1/4. This structure satisfies the first principle: the maximal internal area is a fractal that fills a volume. The shortest transport connections are normally geometric lines that are not enlarged by fractal structures, so $\lambda = 1$. In Fig. 3.13, geometric Euclidean and biological fractal dimensions are compared. The relationships with biomass M assume a constant density of tissue. Living beings act in three-dimensional space, but their internal physiology seems to suggest a four-dimensional structure [3.53]. These results depend on empirical observations and measurements. Therefore, future research could change and improve these results. With more precise statistics, more precise deviations in the 1/4 exponents could be identified.

In any case, the presence of power laws indicates the high complexity of all physiological systems. At a static level, the bronchial system of the mammalian lung serves as a useful example of anatomic complexity. The treelike network involves a complicated hierarchy of airways, beginning with the trachea and branching down through increasingly smaller scales to the level of the smallest branchioles. The human lung has two dominant features, irregularity and richness of structure, along with organization. The essential concept underlying this kind of constrained randomness is that of scaling. The corresponding power law scales are similar to the scaling principles of allometry.

The fractal concept arises in the three distinct but related guises of geometry, statistics, and dynamics. The first context in which we find fractals is complex geometric forms. A fractal structure is not smooth and homogeneous but instead reveals

variable	**Euclidean scaling**	**biological scaling**
length	$L \sim A^{1/2} \sim V^{1/3} \sim M^{1/3}$	$l \sim a^{1/3} \sim v^{1/4} \sim M^{1/4}$
area	$A \sim L^2 \sim V^{2/3} \sim M^{2/3}$	$a \sim l^3 \sim v^{3/4} \sim M^{3/4}$
volume	$V \sim L^3 \sim V^{3/2} \sim M$	$v \sim l^4 \sim a^{4/3} \sim M$

Fig. 3.13. Scaling of length, area, and volume for biological networks and normal Euclidean space

greater and greater levels of detail. The lungs, the heart, and many other anatomic structures may also process fractal structures. The second context in which we find fractals involves the statistical properties of a process. The statistics are inhomogeneous and irregular rather than smooth in this case (instead of the structure). A fractal statistical process is one in which there is a statistical rather than a geometrical sameness to the process at all levels of magnification. Thus, just as the geometrical structure satisfies a scaling relation, so too does the stochastic process. The third context in which fractals are observed involves time, and is related to dynamical processes. The chaotic dynamics that occur in nonlinear dynamical systems arise in part because the attractor on which the dynamics take place has a fractal dimension. This deep relation between chaotic time series and fractal structure was introduced in Sect. 2.6. A second way in which a dynamical quantity can be related to a fractal is when the conduit for the measured quantity has a fractal dimension. An example is the voltage measured from the cardiac pulses that emerge from the conduction system of the heart. Again, the small-scale structure is similar to the large-scale form. The apparent lack of a characteristic time scale in the time series is a consequence of the structure of the conduction system. This is one of the connections between geometric structures and dynamics.

In applying the scaling ideas to physiology, we note that irregularity should be seen as being fundamental rather than treated as a pathological deviation from some classical ideal. The concept of fractals and self-similarity has not only entered into descriptions of biomedical phenomena, but it has prompted a new health paradigm for the clinician [3.54]. The traditional notion of health is one of homeostasis, which is based on the idea that there is an ideal state in which the body is operating in a vaguely defined, maximally efficient way. In this model, illness is considered to be the deviation of the body from this state. It is the task of a physician to help the patient to regain this state. The new idea that has emerged from nonlinear dynamics, scaling, and power laws in biology is that health is homeodynamic; in other words there are a constellation of states that determine health. A healthy person occupies many of these states during the course of normal activity. Flexibility of response and tolerance of error are typical features of this new paradigm. The most important consequences of these concepts can be found in physiology and medicine, where they have changed long-believed views about order and variability in health and disease. A healthy physiological system has a certain amount of intrinsic variability, and a transition to a more ordered or less complicated state may be indicative of disease. Strange attractors may determine the dynamical maps of healthy fluctuations in the heart and brain.

4 Complex Systems and the Evolution of Mind–Brain

How can one explain the emergence of brain and mind? The chapter starts with a short history of the mind-body problem. Besides religious traditions, the concepts of mind and body held by our ancestors were often influenced by the most advanced standards in science and technology (Sect. 4.1). In the framework of complex systems the brain is modeled as a complex cellular system with nonlinear dynamics. The emergence of mental states (for instance pattern recognition, feeling, thoughts) is explained by the evolution of (macroscopic) order parameters of cerebral assemblies which are caused by nonlinear (microscopic) interactions of neural cells in learning strategies far from thermal equilibrium. Pattern recognition, for instance, is interpreted as a kind of phase transition by analogy with the evolution equations which determine pattern emergence in physics, chemistry, and biology (Sect. 4.2). In recent studies in neurobiology and cognitive psychology, scientists even speculate that the emergence of consciousness and self-consciousness depends on the production rate of "meta-cell-assemblies" as neural realizations of self-reflection. The Freudian unconscious is interpreted as a (partial) switching off of order parameters referring to certain states of attention. Even our dreams and emotions seem to be governed by nonlinear dynamics (Sect. 4.3).

Is the "Newton of the human brain and mind" found? Of course not. The complex system approach cannot explain what mind is. But we can model the dynamics of some mental states under certain conditions. Even the modeling of intentional behavior cannot be excluded in principle. Complex systems do not need a central processor like the fiction of "a little man" in the brain. Thus, Virchow's cynical observation that he did not find any soul in human bodies even after hundreds of operations is obsolete. A mental disposition is understood as a global state of a complex system which is caused by the local nonlinear interactions of its parts, but which cannot be reduced to its parts (Sect. 4.4). The wonder of our feeling, imagination, and creativity which has been celebrated by poets and artists since the beginning of human culture is not touched by the complex system approach, although we shall sometimes model some aspects of their nonlinear dynamics.

4.1 From Plato's Soul to Lamettrie's "L'Homme machine"

One of the most complex organs in nature is the human brain. Nowadays we know that it enables the emergence of human mind, consciousness, and personality, which has been considered as one of the greatest miracles of mankind since the early beginning of human thinking. The complex system approach allows us to model the emergence of human perception and thinking with respect to nonlinear interactions of complex neural networks. Thus, models of complex systems help us to understand *how* the mind-brain process may work, and *how* it may have emerged in natural evolution under certain conditions. From this point of view, it will be, in the long run, no wonder *how* consciousness and mind emerge by well-known laws in the evolution of nature. But it still remains a wonder *that* it has arisen.

Before we explore complex systems and the evolution of mind-brain, let us glance at the early philosophy of mind and the history of neural physiology. On the historical background we can decide which questions of the traditional mind-body problem have been solved by the complex system approach and which questions are still unsolved.

In the previous chapters, we have already remarked that early myths and religious beliefs were attempts to explain the human living world and to conjure the forces of nature. Obviously, human desire, fear, anger, and imagination govern the human living world like the might of nature. Consciousness or spirit or mind or soul are experienced in life, and they seem to "leave" the body of dead people. Humans have tried to model these unknown processes by familiar experiences of interacting physical things. Mental or conscious states are hypostatized as a particular substance called "soul", or something like that, which is responsible for the intentional behavior of humans. With the hypostatization of mental states, the problem of the position of the soul in the body has been raised, and is usually answered by the idea that it pervades the body, or is centered in some organs, such as the heart and the lungs. Although the effects of this miraculous "thing" are obviously real, it cannot be seen or seized like a God or ghost. Thus, it has been generally believed to have divine origin. Criticizing traditional myths and religious beliefs, the presocratic philosophers searched for natural causes and principles. Some thinkers regarded the "soul" as material stuff like "air" or "fire" because they were believed to be the finest and lightest forms of matter. For Anaxagoras mind is the principle of motion and order, and therefore the principle of life. For Heraclitus soul is like a flame of fire ruled by the law (*logos*) of the universe. The soul, like fire, is killed by water: "It is death for souls to become water" [4.1]. These approaches are nothing else than modeling the unknown by the familiar and known.

It is noteworthy that one of the early medical thinkers, the Pythagorean Alcmaeon of Croton, seems to have been the first Greek thinker to locate sensation and thought in the brain [4.2]. Like the early Greek astronomical model of a heliocentric universe, this genial idea was soon overshadowed by the authority of Aristotle who taught that the heart is the seat of consciousness, while the brain is only a mechanism for cooling by means of air. Although Aristotle was greatly influenced by the early Greek medical thinkers, he disagreed with Hippocrates' great insight that the brain "is the messenger to consciousness (*sunesis*) and tells it what is happening".

An early reductionistic philosophy of mind was defended by Democritus, who tried to reduce the mental states to interactions of the smallest atoms [4.3]. The problem with his reductionism is, of course, that "soul" is only identified with particular material (but unobservable) atoms. In contrast to material identifications or analogies, the Pythagorean philosophers explained that the human soul or mind must be an immaterial essence because it is able to think immaterial ideas like mathematical numbers and relations. In other words, the soul is modeled as a mathematical proportional system whose harmonies or disharmonies represent mental states like musical melodies.

The Pythagorean concept influenced Plato's philosophy of the human soul, which is related to his theory of forms or ideas. In his dialogue *Menon* Plato demonstrates that an untrained slave can solve mathematical problems. The reason is that in Plato's view every man has an eternal knowledge before (*a priori*) any empirical experience. Man has this kind of a priori knowledge, e.g., in mathematics, by participation in the eternal forms and ideas which are *ante rem*, which means independent of the fuzzy and transitory appearance of being [4.4].

Aristotle criticized Plato's hypothesis of an idealistic world behind reality. Ideas are human abstractions of forms which are acting in nature (*in re*). The soul is described as the form ("essence") of the living body, the "first entelechy", which is a teleological force. But it is not separated from matter. The soul is involved as potentiality in a human body. According to Aristotle, the human organism is understood as a wholeness.

Nevertheless, in the Aristotelean and Stoic tradition the anatomy of the nervous system had been discovered. Galen believed that the nerves transport a psychic pneuma to the muscles which are caused to produce movements. The psychic pneuma was not only a material stuff like breath or air, but a kind of vital spirit [4.5]. In the Middle Ages, the Aristotelean and Stoic philosophy of nature had a great influence on medical thinkers in the Islamic tradition like Avicenna, who founded a medical school with impressive activities in the fields of surgery, pharmacology, and practical healing and helping ill people [4.6]. Later on, these medical standards of the Persian and Arabic world were only realized by a few thinkers of the Christian Middle Ages like Albertus Magnus. Concerning the philosophy of mind and brain, the scientific discussions on, for example, the problem of human consciousness were always overshadowed by religious ideologies, and for a long time it was rather dangerous to dissect dead bodies in anatomy.

In contrast to holistic philosophy in the tradition of Aristotle and Avicenna, Descartes' rationalism taught a dualistic ontology strictly separating mind and matter, soul and body. The human body (*res extensa*) is a material machine constructed by the laws of mechanics and geometry. It is directed and controlled by innate ideas (*ideae innatae*) which are incorporated in the human mind (*res cogitans*). In his *Meditations*, Descartes gets to his primal intuition of the human mind by methodically doubting everything. Methodical doubt is designed to uncover what, if anything, is indubitable. Whereas Descartes can doubt the results of all sciences, of common sense, of perception, he cannot doubt that he exists as a kind of thing which is capable of engaging in cognitive processes such as doubting:

> But what then am I? A thing which thinks. What is a thing which thinks? It is a thing which doubts, understands, affirms, denies, wills, refuses, which also imagines and feels. [4.7]

The difficulty with Descartes' theory is, of course, the interaction of mind and body. He assumes that the human organism with its various organs is directed by the mind with its seat in the brain. The nerves are message cables to and from the brain. They work as causal chains between the ordering mind and the executing muscles. According to the clockwork paradigm of his mechanics, Descartes believed in tiny material particles called "animal spirits" moving and pushing each other very quickly in the cables of the nerves, in order to transport some input from the brain to the muscles.

In contrast to all mechanical effects in nature, the human mind is able to decide the direction of a movement voluntarily. Thus, the action of the mind on animal spirits is to deflect the direction of their motion. This can be done without violating Descartes' laws of physics as long as the "amount of motion" (the later so-called conservation law of momentum) was conserved. In Fig. 4.1, Descartes' mechanical model of perception is illustrated: the minute particles of light beams bomb the human eyes, exciting the brain by the transmission of particular nerves and their "animal spirits". The movements of arm and hand are coordinated with perception by the mind in the brain [4.8].

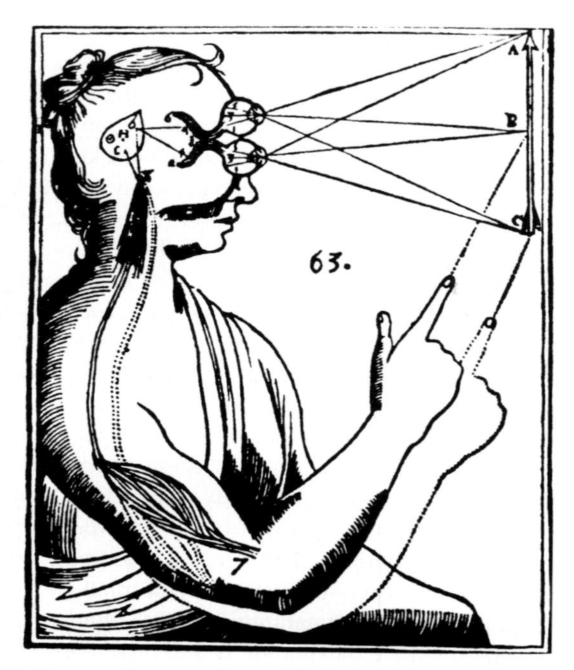

Fig. 4.1. Descartes' geometric model of perception and arm movement

In his book *Passions of the soul* Descartes even tried to analyze all emotional states like fear and love as the passive physical outcomes of the way various "animal spirits" are induced to flow by external events. If Descartes' mechanical model of animal spirits is replaced by biochemical substances and electrophysical effects like hormones and neurotransmitters, then his concept of nervous activity seems to be rather modern.

His main difficulty is the immediate interaction of an immaterial, and therefore unextended and indivisible mind (*res cogitans*) with a material, and therefore extended and divisible body (*res extensa*). Descartes locates the mind within a very small organ of the brain, the pineal gland, steering the movements of the animal spirits. But how can the unextended mind exert a push on an extended particle like an animal spirit? In the framework of mechanics, this problem of interaction was unsolvable in principle and gave rise to s several developments in the philosophy of mind.

For occasionalistic philosophers like Malebranche all causation is miraculous. God must intervene on the occasion of every particular case of causal action. Thus, the mind-body problem is explained by a theological and adhoc hypothesis. Spinoza reduced Descartes' dualism of mind and matter to a monism of one unique substance. God alone is the only substance of everything. All appearances of nature, mind, and body are only attributes ("states") of the universal substance. There are no miracles at any occasion. But God and human mind are naturalized, and nature has become divine in a universal pantheism [4.9].

According to the complex system approach, it was Leibniz who delivered a most remarkable philosophy of mind. Concerning his philosophy of nature, we recall of Leibniz' universe that, contrary to Descartes and Spinoza, consists of infinitely many substances ("monads"), corresponding to different points of view in space and with a more or less clear perspective on the whole. Thus, the monads are considered as soul-like substances endowed with perception and memory, differing in the degree of clarity of their consciousness. There are substances with a rather great perspective and rather high degree of consciousness like humans, when they are compared with animals, plants, and stones with decreasing degrees of consciousness. Even God may be embedded in Leibniz' monadology as the central monad with the highest degree of consciousness and best perspective on the whole, but still as an individual and different entity [4.10].

Obviously, Leibniz did not have the metaphysical problems of Descartes' interactionism. He actually tried to combine classical mechanics with the traditional Aristotelean teleology of nature, because he was aware of the mechanistic lack of an adequate philosophy of mind. From a modern point of view, Leibniz' notion of more or less animated soul-like substances with perception and memory seems to be rather strange. But it was no problem for him to model his monadology in the framework of automata with more or less complexity. Leibniz suggested that each substance can be modeled by an automaton with different states corresponding to the monad's perceptions. Its degree of consciousness is measured by its degree of complexity, characterizing the size of the monad's state space and its information processing capacity. The states of Leibniz' more or less complex automata are correlated with each other in networks, according to his famous quotation that the monads have "no windows" and do not interact, but reflect each other like the mirrors in a baroque palace. Leibniz' complex networks of monads will be discussed in Chap. 5 in more detail. In short, Leibniz assumed that mind is not reserved to humans, but is a feature of systems which emerges with different intensity according to the system's degree of complexity.

The English empiricist philosophers like Locke and Hume criticized the Cartesian-Platonic belief that mental states can be analyzed by introspection and pure thinking without sensory experience. For empiricist philosophers mind is nothing more than a *tabula rasa*, an empty store for receiving sensory data, in order to form concepts by association and abstraction. Images are merely less vivid copies of sense-impressions which can be imaginatively combined, like for instance the notion of an unicorn.

According to the complex system approach, Hume developed a remarkable psychological theory of association. He proclaimed that there is neither a causal mechanism in nature nor a causal law in our mind, but only an unconscious reflex of associating those sense-impressions which occur in a correlated way on several occasions like flashes of lightning and thunder. We may say that the brain has an unconscious capacity to build up patterns of sense-impressions. Notions are nothing more than terms designating more or less complex patterns of sense-impressions. Apart from mathematics, there are no sharp and definite concepts founded on perception, but only more or less fuzzy patterns allowing more or less probable assertions about events. In *A treatise of human nature*, Hume wrote:

> The table before me is alone sufficient by its view to give me the idea of extension. This idea, then, is borrow'd from, and represents some impression, which this moment appears to the senses. But my senses convey to me only the impressions of colour'd points, dispos'd in a certain manner. If the eye is sensible of any thing farther, I desire it may be pointed out to me. But if it be impossible to shew any thing farther, we may conclude with certainty, that the idea of extension is nothing but a copy of these colour'd points, and of the manner of their appearance. [4.11]

According to Descartes' rationalism, the human mind rules the body's mechanics like a monarch governing the state in his century of absolutism. For Hume, there is no separate substance of the human mind, but only a self-organizing field of permanently emerging and disappearing patterns excited by associations of more or less intense sense-impressions. Hume's spontaneously associating and separating sense-impressions can be compared with the free citizens of a democratic society who may associate in groups and parties without the orders and prohibitions of a sovereign.

Kant tried to synthesize rationalism and empiricism. According to empiricism, cognition starts with experience and sensory data. But rationalism is right, because we need mental structures, cognitive schemes, and categories in order to organize experience and cognition. Kant tried to introduce the philosophical categories which found the axioms of Newtonian mechanics. It is the main feature of his epistemology that recognition does not arise by passive impressions of the external world on the tabula rasa of our brain. Recognition in the Kantian sense is an active process producing models of the world by *a priori categories*. The spatial and temporal order of physical events is reduced to geometrical forms of intuition. Perception in the Kantian sense is active information processing regulated by a priori anticipations. The causal connections of events is made possible philosophically by an a priori category of causality.

Hume was right that a causal relation cannot be perceived. But, in order to forecast and calculate the path of a pushed billiard ball exactly, it is not sufficient to repeat several pushes of billiard balls and to associate several sense-impressions à la Hume. We must anticipate that causes and effects can be connected by some deterministic relation. That is done by Kant's general scheme of (deterministic) causality in epistemology. But the question is which particular causal function is adequate to decide and test by physical experience. Cognitive schemes are already applied in everyday life. They are even modeled by data schemes of programming languages in computer science (compare Sect. 5.2). Thus, Kant's epistemology may be interpreted as an important forerunner of modern cognitive sciences, where cognitive schemes are presumed to order the mass of experienced data. But, contrary to Kant, they may be changed in the development of history, as was shown by the change from Euclidean space to non-Euclidean spaces, e.g., in the theory of relativity [4.12].

While Spinoza suggested a spiritual monism as a way out from Descartes' dualism, Lamettrie supported a kind of materialistic monism. Descartes' assumption of a separate soul-like substance (*res cogitans*) was criticized as superfluous, because all mental states should be reduced to mechanical processes in the human body: "L'Homme machine". Lamettrie claimed there were no fundamental differences between humans and animals. Intelligent and reflex behavior should be explained by "irritation" of the nerves and not by a "ghost in the machine". But, in the mechanistic framework of the 18th century, Lamettrie's revolutionary ideas could only be an inspiring program of physiological research [4.13].

The famous mathematical physicist and physiologist Hermann von Helmholtz (1821–1894) was a post-Kantian philosopher [4.14]. He supported a kind of naturalized framework of cognitive categories which must be presumed before any particular perception of the world can be constructed. Of course, the categories had changed since Kant. Nevertheless, there are some general schemes like the concept of space, numbers, measurement, and causality characterizing the physical theories of the 19th century. Helmholtz was aware, for instance, of the mathematical possibility of non-Euclidian spaces. Thus, he thought that the correct physical geometry must be decided by physical measurement.

Concerning his theories of physiology, Helmholtz started as a student of Johannes Müller (1801–1858), who is sometimes called the father of modern physiology [4.15]. Müller defended a law of specific nerve energies demanding that each nerve has its own particular energy or quality. He found that sensations could be elicited by mechanical or chemical influences, heat, electricity, etc. A Kantian aspect of perception is now naturalized, because it became evident that the brain has to reconstruct the world from its effects on nerves. Nevertheless, Müller defended an immaterial conception of animal spirit. He believed that animal spirits cannot be measured, because their speed is too high.

Helmholtz had explored the mathematical conservation law of energy. As energy could be transformed but neither created nor destroyed, an immaterial energy of life beyond the conservation law seemed to be senseless. Helmholtz preferred the theory that so far as the question of energy was concerned, the body could be viewed as a mechanical device for transforming energy from one form to another without special forces and spirits. Chemical reactions were capable of producing all the physical activity and heat generated by the organism. Muscular activity was realized by chemical and physical changes in the muscles. Furthermore, Helmholtz measured the velocity of nerve conduction and demonstrated that it was slower even than the speed of sound [4.16]. Philosophically, these results were interpreted as a refutation of Müller's vitalism.

Emil du Bois-Reymond (1818–1896), who was another student of Müller, showed that nervous affect was actually a wave of electrical activity. In those days, histologists began to discover separate cell bodies and fibers through the microscope. According to these results, nervous activity and the brain seemed to be a complex system of nerve cells ("neurons") with a complicated network of connections. The communication structure of neurons transmitting a signal from one neuron to another one was first described at the beginning of this century. But observations of synaptic junctions were not possible before the use of the electron microscope around the middle of this century.

How can the emergence of perceptions, thoughts, and feelings be explained by these descriptions of neuroanatomy and neurophysiology? One of the first thinkers who explained mental states by cell assemblies of neural networks was the American philosopher and psychologist William James. In his brief course "Psychology" (1890), James defended the Darwinian and evolutionary position that the brain is not constructed to think abstractly, but is constructed to ensure survival in evolution. In a pragmatic way, he assumed that the brain has many of the features of a good engineering solution applied to mental operations:

> Mental facts cannot properly be studied apart from the physical environment of which they take cognizance .... Mind and world in short have evolved together, and in consequence are something of a mutual fit. [4.17]

Brain organization seems to be very poor at doing arithmetic and formal logic. But the ability to form concepts and associations, to make good guesses and to assume hypotheses is a characteristic feature of the brain. James presents a mechanistic model of association that stems from Hume's pioneering work and reminds us of the later associative neural networks. In a more qualitative way, he formulated some principles which are partially incorporated in the modern mathematical models of complex neural networks:

1. James believed that association was mechanistic and a function of the cerebral cortex.

2. James' principle of association:

When two elementary brain processes are active together or in immediate succession, one of them, on reoccurring tends to propagate its excitement into the other.

3. James' summing rule for brain activity:

The amount of activity at any given point in the brain cortex is the sum of the tendencies of all other points to discharge into it, such tendencies being proportionate (1) to the number of times the excitement each other point may have accompanied that of the point in question; (2) to the intensities of such excitements; and (3) to the absence of any rival point functionally disconnected with the first point, into which the discharges might be diverted. [4.18]

If in the second principle the term "brain process" is replaced by "neuron", then we get a description of a synapse which was later introduced by Hebb (compare Sect. 4.2). If in the third rule the term "point in the brain cortex" is replaced by "neuron", we get a linear summation rule of synaptic inputs which is very close to some network models of the Hebbian type. James also discussed the ability of networks of partial associations to reconstruct the missing pieces through some particular procedure of cell connecting. Although James was, of course, not familiar with computer-assisted modeling, he had the essential insight of the complex system approach that complex events are made up of numerous subassociations which are interconnected by elementary mechanisms like synapses.

In his chapter on "Association", James considered someone thinking of a certain dinner-party. The only thing which all the components of the dinner-party could combine to recall would be the first concrete occurrence which ensued upon it. All the details of this occurrence could in turn only combine to awaken the next following occurrence, and so on. In relation to Fig. 4.2, James described this process schematically:

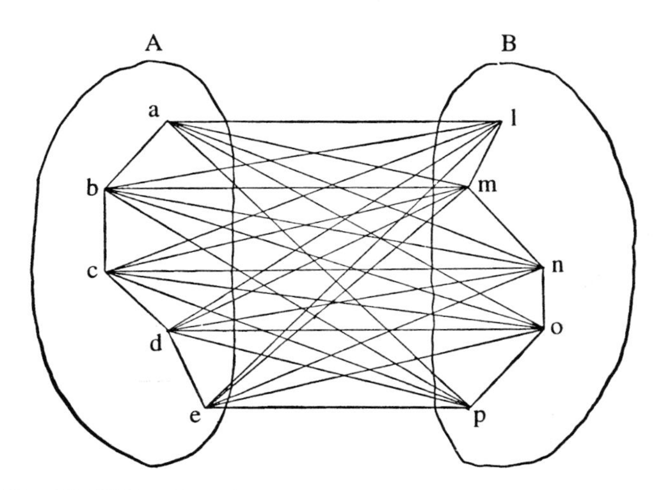

Fig. 4.2. William James' geometric model of an association network

> If a, b, c, d, e, for instance, be the elementary nervetracts excited by the last act of the dinner-party, call this act A, and l, m, n, o, p be those of walking home through the frosty night, which we may call B, because a, b, c, d, e will each and all discharge into l through the paths by which their original discharge took place. Similarly they will discharge into m, n, o, and p; and these latter tracts will also each reinforce the other's action because, in the experience B, they have already vibrated in unison. The lines in Fig. 4.2 [4.19, Fig. 57] symbolize the summation of discharges into each of the components of B, and the consequent strength of the combination of influences by which B in its totality is awakened. [4.19]

James is convinced that "the order of presentation of the mind's materials is due to the cerebral physiology alone." In the modern complex system approach, order parameters are used to describe mental states which are caused by macroscopic neural cell assemblies. In the following sections we will see that many basic insights into the operations of the mind from the presocratic philosophers to Kant and James have not been fundamentally altered even today.

4.2 Complex Systems and Neural Networks

In the 19th century, the physiologists discovered that macro-effects like perception, vision, muscular motion, etc., displayed by the nervous system depend on individual cells. These cells are able to receive and transmit signals by causing and responding to electric current. Obviously, the nervous system and the brain have turned out to be one of the most complex systems in the evolution of nature. There are at least ten billion nerve cells (neurons) in the human brain. Each neuron receives inputs from other cells, integrates the inputs, generates an output, and sends it to other neurons. The inputs are received by specialized synapses, while outputs are sent by specialized output lines called axons.

A neuron itself is a complex electrochemical device containing a continuous internal membrane potential. If the membrane potential exceeds a threshold, the neuron propagates a digital action potential to other neurons. The nerve impulses originate in the cell body, and are propagated along the axon with one or more branches. Neurologists usually distinguish excitatory and inhibitory synapses, which make it more or less likely that the neuron fires action potentials. The dendrites surrounding the neuron might receive incoming signals from tens or thousands of other neurons. The activity of a neuron is measured by its firing frequency. Biological neurons are not binary, because outputs are continuous. However, many models of neural networks are simplified and use binary computing elements [4.20].

Brains are complex systems of such cells. But while an individual neuron does not see or reason or remember, brains are able to do so. Vision, reasoning, and remembrance are understood as higher-level functions. Scientists who prefer a bottom-up strategy recommend that higher-level functions of the brain can be neither addressed nor understood until each particular property of each neuron and synapse is explored and explained.

An important insight of the complex system approach discloses that emergent effects of the whole system are system effects which cannot be reduced to the single elements. Philosophically, the whole is more than the sum of its parts. Thus, a purely bottom-up-strategy of exploring the brain functions must fail. On the other hand, the advocates of a purely top-down strategy proclaiming that cognition is completely independent of the nervous system are caught in the old Cartesian dilemma "How does the ghost drive the machine?".

Traditional positions in the philosophy of mind (compare Sect. 4.1) have more or less defended one of these strategies of research. In the 18th century, Leibniz and later on the zoologist Bonnet already suggested that there is a scale of complexity in nature with more or less highly developed levels of organization. In Fig. 4.3 the levels of organization in the nervous system are illustrated [4.21]. The hierarchy of anatomical organizations varies over different scales of magnitude, from molecular dimensions to that of the entire central nervous system (CNS).

The scales consider molecules, membranes, synapses, neurons, nuclei, circuits, networks, layers, maps, systems, and the entire nervous system. On the right side of the figure, a chemical synapse is shown at the bottom, in the middle a network model of how ganglion cells could be connected to simple cells in visual cortex, at the top a subset of visual areas in visual cortex, and on the left the entire CNS.

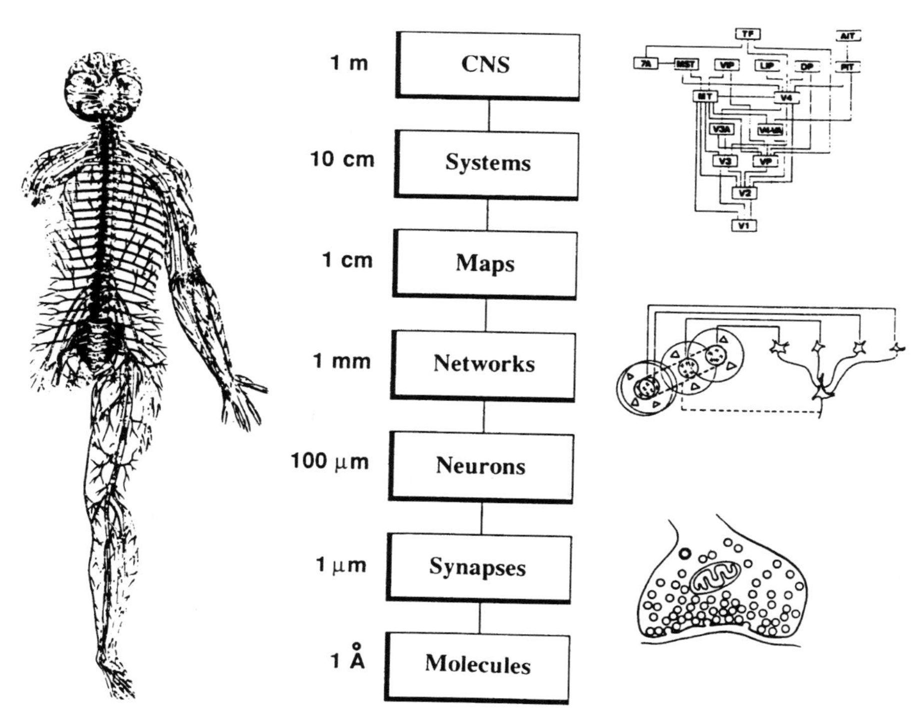

Fig. 4.3. Degrees of complex systems in the central nervous system (CNS): a chemical synapse, a network model of cellular connections in the visual cortex, and subsystems of the visual cortex [4.21]

The research perspectives on these hierarchical levels may concern questions, for example, of how signals are integrated in dendrites, how neurons interact in a network, how networks interact in a system like vision, how systems interact in the CNS, or how the CNS interact with its environment. Each stratum may be characterized by some order parameters determining its particular structure, which is caused by complex interactions of subelements with respect to the particular level of hierarchy. Beginning at the bottom we may, for instance, distinguish the orders of ion movement, channel configurations, action potentials, potential waves, locomotion, perception, behavior, feeling, and reasoning.

It is quite obvious that an important function of the nervous system is to monitor and control the living conditions of the organism relative to its environment. An example of an elementary controllable state is, for instance, the temperature of an organism. At the highest level the change of states in the environment needs anticipatory planning and social interactions, which have led to the human skills of verbal communication, creating art, solving mathematical problems, etc., during a complex cultural evolution.

From a Darwinian point of view, the evolution of the nervous system with its levels of increasing complexity seems to be driven by the fundamental purpose in nature to survive as the fittest. Some scientists of the brain even defend the strong opinion that the emergence of mental phenomena like abstract thinking is only some kind of "epiphenomenon" which was not originally intended by nature. But the belief in intentions and purposes of nature is, of course, only a human metaphor presuming some secularized divinity called "nature" governing evolution. According to the complex system approach, each level of the CNS has its own functional features which cannot be reduced to the functional features of lower levels. Thus, abstract thinking can only be regarded as an "epiphenomenon" from the perspective of a level like, say, the control system of the body's temperature.

In order to model the brain and its complex abilities, it is quite adequate to distinguish the following categories. In neuronal-level models, studies are concentrated on the dynamic and adaptive properties of each neuron, in order to describe the neuron as a unit. In network-level models, identical neurons are interconnected to exhibit emergent system functions. In nervous-system-level models, several networks are combined to demonstrate more complex functions of sensory perception, motor functions, stability control, etc. In mental-operation-level models, the basic processes of cognition, thinking, problem-solving, etc. are described. Their simulation is closely related to the framework of artificial intelligence (compare Chap. 6).

From a methodological point of view, we must be aware that models can never be complete and isomorphic mappings of reality. In physics, for instance, models of the pendulum neglected friction. In chemistry, models of molecules treated electrons in orbitals like planets in the solar system, contrary to Heisenberg's uncertainty principle. Nevertheless, these models are useful with respect to certain conditions of application. The conditions of brain models are given by the levels of brain organization. If a function of a certain level of brain organization is modeled, the model should take into account the conditions from the levels below and above. Higher-level properties are often not relevant. In general, the methodology of modeling is determined by a calculation of methodological costs and benefits. A model of the human brain which is intended to be realistic in every respect needs over-expensive analysis and construction. It would never satisfy the desired purpose, and thus it is impractical. Scientists will be more successful if they try to model each level of brain organization with simplifications concerning the levels below. On the other side, the models must be rich enough to reveal the essential complex features of brain organization.

According to the complex system approach, brain functions should be modeled by an appropriate state space and a phase portrait of its dynamical trajectories describing the brain's activities. René Descartes, the French mathematician and philosopher, already described the coordination of perception, arm moving, and brain in the framework of (Euclidian) geometry (Fig. 4.1).

Today, neural networks are geometrically characterized by vector spaces and neural matrices. The electrochemical input of neurons are connected with the outputs by weights. In a schematic section of the cerbellum (Fig. 4.4) the weights w_{ij} from a neural matrix allow the network to calculate the output vector from the input vector by matrix multiplication [4.22].

The example of Fig. 4.4 concerns a 3×4-neuron matrix. Neural physiology demands great flexibility of modeling, because the neural network may be rather complex. But the connectivity matrices can effect transformations on state spaces of high dimensionality into others with different dimensionality. Mathematically, these transformations of high dimensionalities may provoke geometrical problems which cannot be solved by the elementary formalism of analytical geometry. In this case, a generalized tensor network theory is necessary, in order to manage complex coordination tasks. From a historical point of view, it is amazing that the change from Euclidian to more general topological and metric spaces can be stated not only for the physics of the outer world in general relativity but also for the intrinsic features of the nervous system.

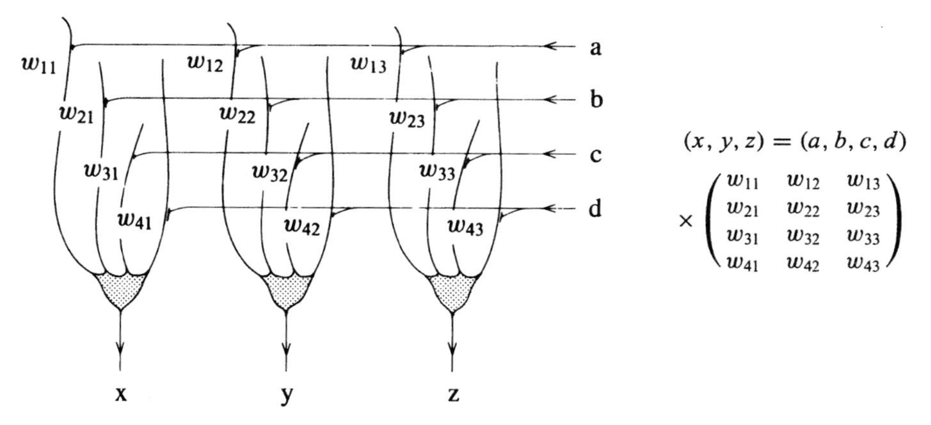

Fig. 4.4. Schematic section of the cerebellum modeled by neural matrix multiplication [4.22]

With respect to Descartes' early approach, let us regard an elementary sensorimotor coordination which is represented by vector or tensor transformations. How can an animal seize an object which it perceives by its sensory organs (Fig. 4.5a)? In a simplified model, the position of the two eyes is at first codified in a 2-dimensional space of sensory data. The state space can be visualized by a 2-dimensional topo-

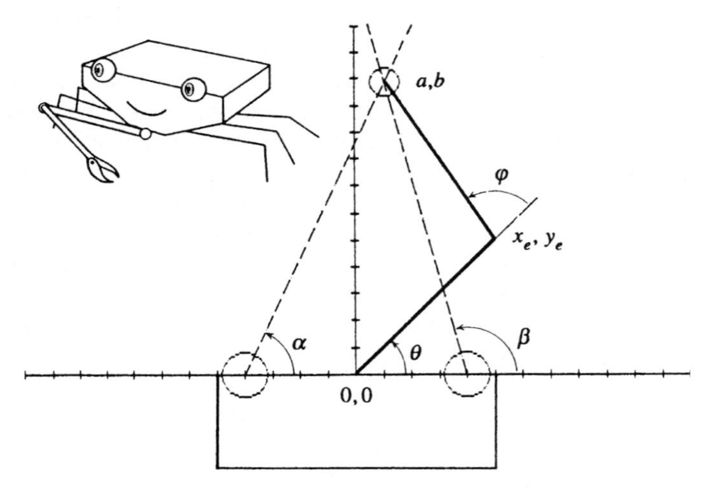

Fig. 4.5a. Sensorimotor coordination of perception and arm movement [4.23]

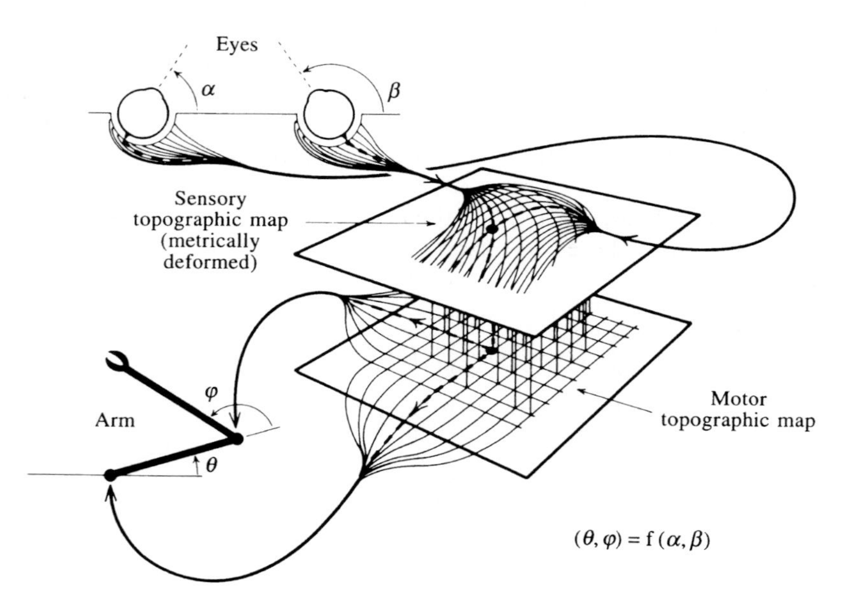

Fig. 4.5b. Geometric model of sensorimotoric coordination by sensory and motor topographic maps [4.23]

graphic map. An impulse is sent from a point of the sensory state space to a corresponding point of the motor state space which is also represented by a 2-dimensional topographic map. A point of the motor state space codifies the corresponding position of the arm (Fig. 4.5b) [4.23].

Another example of sensorimotor coordination is given by the vestibulo-ocular reflex. This is the neural arrangement whereby a creature can stabilize an image on the retina by short-latency movements of the eyes in the opposite direction to head movement. There are two neural structures involved in this neural system, which can be represented by different coordinate systems intrinsic to the CNS. First we have to analyze the semicircular canals of the vestibular apparatus in the ear, three canals on each side, which can be represented by a 3-dimensional coordinate system. Second, each eyeball has six extraocular muscles corresponding to a 6-dimensional coordinate system. Thus, the sensorimotor coordination of the vestibulo-ocular reflex is geometrically described by a tensor transformation of a 3-dimensional (covariant) vector. The mathematical scheme can be used for calculations of any eye-muscle activation emerging from a given vestibular input.

On the level of neurons and networks, nets of artificial units are used to simulate and explore the brain organization [4.24]. These units are assumed to vary between 0 and 1. Each unit receives input signals from other units via synaptic connections of various weights. The incoming and outgoing representations are ordered sets of values, and the output units are activated appropriately. Mathematically, the procedure can be interpreted as a mapping of some inputs as arguments onto corresponding outputs as function values. The function rule is determined by the arrangement of the weights, which depends on the topology of the neural network.

In the brain, neurons sometimes constitute a population as input layer (Fig. 4.6). The axons of these cells are sent to a second layer of neurons. Axons from cells in this second layer can then project to a third population of cells, and so on. The assembled set of simultaneous activation levels in all the input units is the network's representation of the input stimulus, as input vector. This input vector with its activation levels is propagated upward to the middle layer. The result is a set of activation levels determined by the input vector of the input layer and the several connection weights at the ends of the terminal branches of the input units to the neurons of the middle layer. This activation vector of the middle layer is propagated upward to the

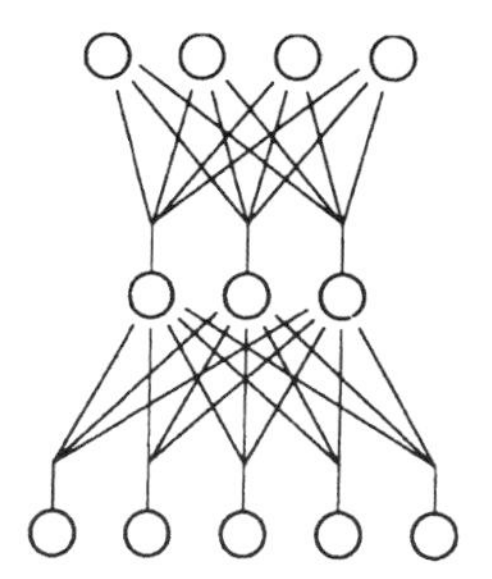

Fig. 4.6. Three-layer network with hidden units

topmost layer of units, where in the case of a three-layer network an output vector is produced. Again, this ouput vector is determined by the activation vector at the middle layer and the connection weights at the ends of the terminal branches of the middle units to the output units.

A two-layer network with only an input and an output layer corresponds to a simple stimulus-response scheme with observable and measurable inputs and outputs. In the case of a three-layer network, the units of the middle layer with their weights are sometimes not directly measurable or only hypothesized as a hidden mechanism in a black box. Thus, they are called hidden units.

Real nervous systems, of course, display many more units and layers. In humans, for instance, the structure of the cerebral cortex contains at least six distinct layers of neurons. By the way, the mapping of inputs onto outputs by a neural matrix in the cerebellum in Fig. 4.4 can equivalently be described by a two-layer network with input and output layer. A three-layered network is equivalent to a pair of neural matrices connected in series. But this kind of a many-layered network cannot be generalized for the whole brain and nervous system, because cell populations in the real brain often show extensive cell-to-cell connections within a given layer, which will be considered in some different models.

According to the complex system approach, the neurons of a particular layer can be interpreted as the axes of a state space representing the possible activity states of the layer. The development of states, their dynamics, is illustrated by trajectories which may be caused by some learning process of the particular network.

For instance, a perception can be explained by vector-processing in a neural network. At first, there is a sample of stimuli on the input neurons from the outer world (e.g., electromagnetic light signals, colors in the eye, or sound waves in the ear), which is processed in the neural network to produce an output vector which represents, e.g., a visual or auditory picture of the outer world. But the neural networks must learn to distinguish and recognize the correct forms, colors, sounds, etc., in a huge mass of input data.

The learning procedure is nothing else than an adjustment of the many weights so that the desired output vector (e.g., a perception) is achieved. The learning procedures can be simulated by mathematical algorithms which are an important topic of research in artificial intelligence (compare Sect. 6.2). They produce weight configurations at each neural layer which can be represented in terms of vectors, too. At any given time, the complete set of synaptic values defines a weight space with points on each axis specifying the size of a particular weight. In general, learning means minimizing the errors or differences between a most adequate solution (perception, idea, etc.) and a less adequate one. Thus, a learning process can be visualized by a trajectory in the weight space, starting from the initial randomly set position to the final minimal-error position (Fig. 4.7a). The key to this kind of modeling means that weights in a network can be set by an algorithmic procedure to embody a function. It is assumed that any representable world can be represented in a network, via configurations of the weights.

Figure 4.7a shows a trajectory in the synaptic weight space during a learning process. This space, simplified for three weights, represents all possible weight

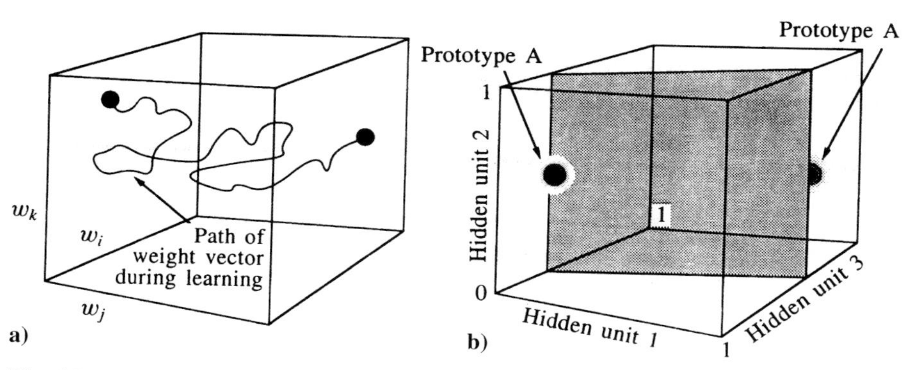

Fig. 4.7a,b. Synaptic weight space (**a**) and activation vector space (**b**) of the three-layer network in Fig. 4.6 [4.25]

combinations from the synapses in the three-layer network (Fig. 4.6). Figure 4.7b shows a corresponding activation-vector space whose axes are the hidden units of the three-layer network (Fig. 4.6) [4.25].

The weight space and the activation space are similarity spaces, because similar vectors representing similar things are reflected by proximity of position. Weight configurations cluster similar things, taking into account that weight configurations may be sensitive to very tiny differences between things. Thus, in the activation space, we can distinguish prototype vectors representing similar things with tiny differences measured by their distances to the prototype vector. On the macroscopic level of observation and behavior, these prototype vectors may represent particular categories of trees, plants, fruits, persons, etc., which are more or less similar. In the framework of complex system dynamics, prototype vectors can be interpreted as point attractors dividing the state space into several regions.

Similar motor behavior (like seizing, walking, etc.) is represented by similar trajectories in a motor state space. Learning, as we saw, means reconfiguration of weights according to some algorithmic procedure. The crucial question arises: how do thousands of cells and synapses know when they should change their states without the guiding hand of a demon?

In his famous book *The Organization of Behavior* (1943), Donald Hebb suggested that learning must be understood as a kind of self-organization in a complex brain model. As in the evolution of living organisms, the belief in organizing "demons" could be dropped and replaced by the self-organizing procedures of the complex system approach. Historically, it was the first explicit statement of the physiological learning rule for synaptic modification. Hebb used the word "connectionism" in the context of a complex brain model. He introduced the concept of a synapse, which was later called a "Hebb-synapse". The connection between two neurons should be strengthened if both neurons fired at the same time:

> When an axon of cell A is near enough to excite a cell B and repeatedly or persistently takes part in firing it, some growth process or metabolic change takes place in one or both cells such that A's efficiency, as one of the cells firing B, is increased. [4.26]

In 1949, the "Hebb-synapse" could only be a hypothetical entity. Today, its neurophysiological existence is empirically confirmed. Hebb's rule is by no means a mathematically exact statement. Later on, we shall see that many Hebb-like learning rules of connectionism are possible. A simple mathematical version of the Hebb rule demands that the change Δw_{BA} of a weight w_{BA} between a neuron A projecting to neuron B is proportional to the average firing rate v_A of A and v_B of B, i.e., $\Delta w_{BA} = \varepsilon v_B v_A$ with the constant ε.

Hebb-like rules suggest schemes tending to sharpen up a neuron's predisposition "without a teacher" from outside. In this sense, it is a self-organizating method for a neuron's firing to become better and better correlated with a cluster of stimulus patterns. Hebb was aware that the brain uses global patterns of connected neurons to represent something. He explicitly used the term "cell assemblies", which has become a key to modern neuroscience. Active cell assemblies could correspond to complex perceptions or thoughts. Philosophically, Hebb's idea of cell assemblies reminds us of Hume's concept of association, which was only psychological without the physiological basis of the brain.

How are Hebb's physiological ideas incorporated in modern complex systems of neural networks? The basic concept of an associative network demands that an input vector is "associated" with an output vector by some transformation. Mathematically, the similarity of two vectors can be measured by their inner product, which is the result of multiplying both vectors, component by component, and then adding up the products. Geometrically, the inner product is proportional to the cosine of the angle between the vectors. In the case of total congruence of the vectors, the angle is zero, which means that the similarity is complete.

Thus, the similarity of a stored prototype vector (for instance the prototype picture of a typical tree) with an input vector (for instance the perception of a particular tree) can be calculated in an associative network by their inner product. The prototype vector is assumed to be stored in the matrix of the weights connecting the input and output of the network. Figure 4.8a shows a net with horizontal input lines for the input components, vertical output lines, and weights on the connections (which are considered to be binary, with open circles for zero and closed circles for one).

If in general the input vector (x_j) is associated with the output vector (y_j) by a linear transformation $y_i = \sum_j w_{ij} x_j$ with respect to the stored weight vector w_{ij}, then we get the simple case of a linear associator. This kind of an associative network is able to classify vectors representing examples of some category which is realized by a stored prototype vector. This task is actually crucial for the survival of animals. In reality, a variety of more or less similar perceptions (for instance, of a hostile animal) must be identified and subsumed under a category.

Another kind of associative network can perform vector completion or vector correction. A so-called autoassociative network can produce an output which is as close as possible to a prestored vector given only part of the vector as input. In reality, noisy versions of an input vector (for instance, a picture of a person) must be completed according to a stored picture. A Hebb-like rule can fulfill this task by strengthening the connection weights between neurons with respect to the degree of their correlated activity.

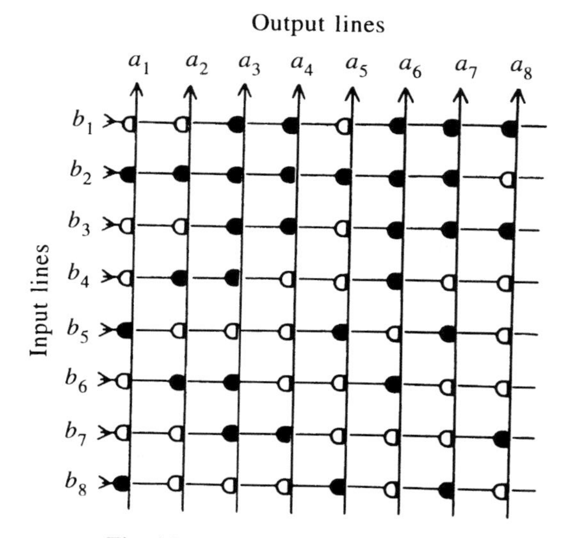

Fig. 4.8a. Linear associative network

A method for increasing the capacity of such a complex network is to introduce a nonlinear threshold for the output units. A linear associative network (for instance, Fig. 4.8a) has a feedforward topology with information flowing from input units to output units. Hebb-like learning procedures suggest local interactions of neural units converging to the correct global output by self-organization. Circulating information in the network means a feedback architecture. In Fig. 4.8b, each unit receives inputs from outside and feedback from intrinsic units of the network. The weights are represented by the intersections of horizontal with vertical lines [4.27].

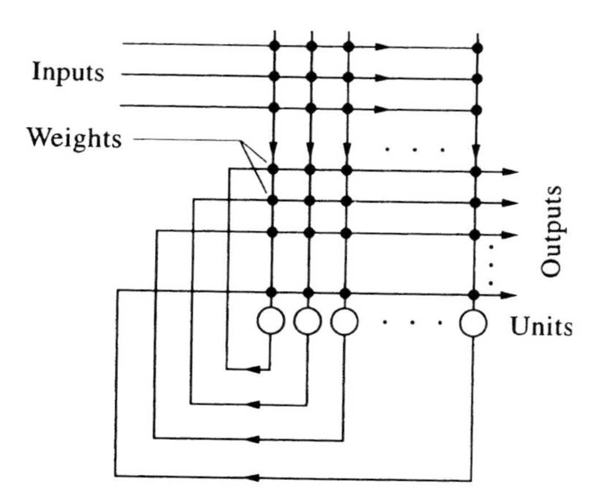

Fig. 4.8b. Nonlinear feedback network [4.27]

Obviously, the complex system of Fig. 4.8b models a nonlinear feedback network which allows a wide range of possible dynamics. A famous example was explored by John Hopfield (1982). His class of nonlinear feedback networks has a dynamics converging to a solution. They have been interesting not only for modeling brain functions, but also (as we shall see in Chap. 6 on artificial intelligence) for the development of a new network technology. Concerning our complex system approach, it is noteworthy that Hopfield is physicist who has applied mathematical equations from spin glass physics to neural networks [4.28].

The dynamics of a ferromagnet is a well-known example of conservative self-organization in thermal equilibrium. In the Ising model a ferromagnet consists of a lattice of spins, each of which can be either up (↑) or down (↓). Each spin can interact with its nearest neighbors. The state with the lowest energy has all the spins lined up in the same direction. At a high temperature the directions of the spins are random because the thermal energy which causes the fluctuations is larger than the energies of interaction. If the temperature is reduced, the spins become aligned in the same direction. Evidently, the spins behave like a magnet (compare Sect. 2.4). Dynamically, it seems to seek the nearest local energy minimum as an attractor state (Fig. 4.9a). But a single energy minimum with all spins pointing in the same direction is only provided if all interactions are attractive. In the case of mixed attractive and repulsive interactions a complex system like a spin glass may have many local energy minima [4.29].

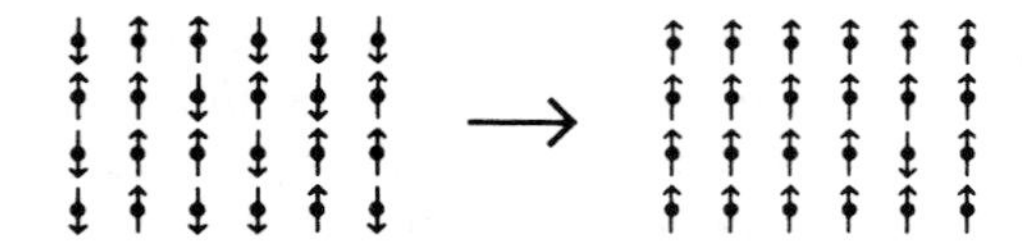

Fig. 4.9a. Phase transition in a 2-dimensional Ising model of a ferromagnet (annealing)

Hopfield assumed that the function of the nervous system is to develop a number of locally stable points in state space. Other points in state space flow toward the stable points as attractors of the system. As deviations from the stable points disappear, this dynamics is a self-correcting procedure. On the other hand, the stable point appropriately completes missing parts of an incomplete initial state vector. Thus, the dynamics can be used to complete noisy inputs.

Hopfield's model is rather simplified and involves threshold logic units, summing synaptic inputs, and comparing the sum with a threshold. If the sum is at or above threshold, they yield 1, and 0 otherwise. The network is recurrent in the sense that the neurons connect to each other with the exception of self-connection. Mathematically, the corresponding connectivity matrix has zeros along the main diagonal. Hopfield suggests a Hebb-like learning rule for constructing elements of the connectivity matrix. The complex system evolves like an Ising model of a spin glass according to a nonlinear feedback dynamics. The term isomorphic to energy decreases until it reaches a – perhaps local – minimum.

A simple application is given by the well-known problem of alphanumeric character recognition. The complex network is composed of interacting Boolean variables represented on a 2-dimensional grid. A pattern (for instance, the letter A) can be associated to the grid with a dark point for all active variables (with a value 1), and a blank point for those with a value 0. It is assumed that the letters of the alphabet are associated to attractors ("fixed points") as the desired states of the dynamical system. We may imagine that a human brain has stored the correct shapes of the letters by seeing many correct examples. If an incomplete and partly ruined letter is shown to the system, it should be able to reconstruct the correct shape that was learnt before (Fig. 4.9b) [4.30].

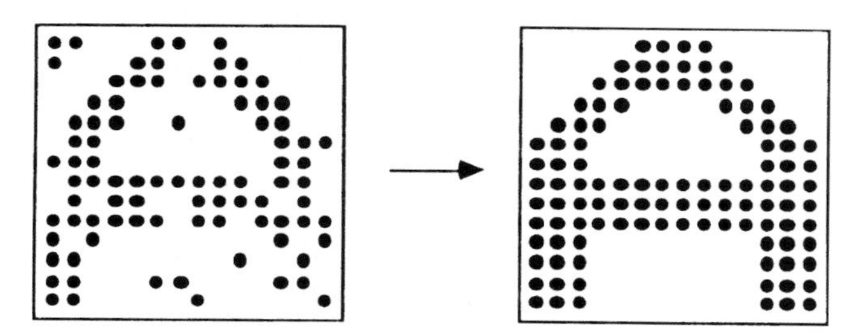

Fig. 4.9b. Phase transition in a Hopfield system for pattern recognition

Thus, pattern recognition means pattern evolution by self-organization. This process aims at some attractors as the desired states of the system. We remember that an attractor is a state towards which the system may evolve, starting from certain conditions. The basin of attraction is defined by the set of initial conditions that drive the trajectories of the system in the direction of the attractor. As we saw in earlier sections, an attractor may be a unique state referring to a fixed point or stable state, as in the examples of Hopfield networks and spin glass systems. But a periodic succession of states (a "limit cycle") or several forms of chaotic attractors (in dissipative systems) are also possible. Thus, the Hopfield networks are only a first and simplified approach to modeling neural states by attractors of complex systems.

Hopfield saw the analogy between the local energy minima in spin glasses and the prototypes in an associative brain. In the formal framework of a spin glass, attractors can be designated as prototype vectors. In Fig. 4.10a, the state space of a Hopfield system is visualized by an energy landscape by analogy with the thermodynamics of spin glasses. All possible states of the network are represented by points in a plane. The height of the surface refers to the energy of the corresponding state of the network.

The phase portrait of the system in Fig. 4.10b shows the convergence of the trajectories to stable local minima from different starting points. Each point in the plane is a state of the network. The energy landscape has basins attracting the trajectories of the Hopfield dynamics. The stable points ("attractors") are at the bottom of the basin. In the example of pattern recognition, the prototype letters are connected

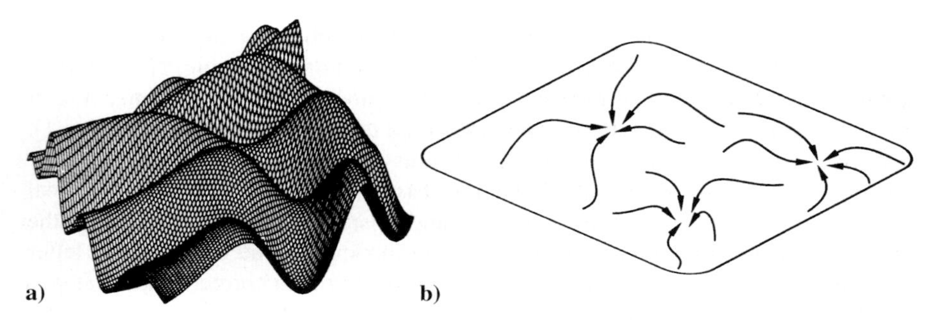

Fig. 4.10a,b. State space of a Hopfield system as energy landscape (**a**) with local minima as attractors (**b**) [4.31]

with the stable points. Thus, the process of pattern completion is a form of relaxation which can be compared formally with the annealing processes of conservative self-organization. In the physical examples, the final states are ordered structures of a spin glass, a magnetized ferromagnet, or a frozen crystal [4.31].

In general, Hopfield networks only converge to local minima in a state of lower energy. In some applications, the local minima are associated with particular stored items, and there may be no need to reach a global minimum. However, in many cases the global minimum is required. A solution of this problem was offered by making the individual units stochastic rather than deterministic.

Figure 4.11a visualizes the solution by a ball traveling along a curve of an energy landscape to probably end up in the deepest minimum. Starting from a given initial situation, the ball will move towards an energy minimum or the bottom of a well. If the energy landscape is characterized by a muliplicity of minima close together, the result depends upon the initial conditions. How can the network be prevented from getting stuck in a local minimum? The idea is to shake the energy landscape with a certain energy increment which is required to escape the valley of the local minimum B to enter the attractor of the global minimum A.

Then, mechanically, the ball is more likely to go from B to A than from A to B. On average, the ball should end up in the valley of A. In the language of thermody-

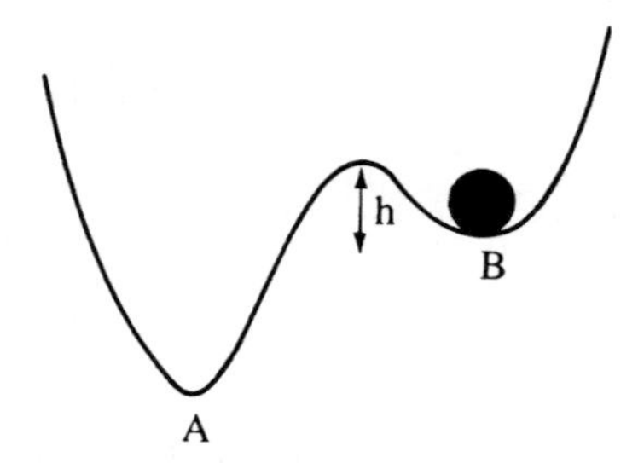

Fig. 4.11a. Phase transition from local minimum B to global minimum A in an energy landscape (simulated annealing)

namics, the kinetic energy added by shaking the landscape corresponds to increasing the temperature of the system. For fairly high temperatures, the probability of transition between the valleys is no longer negligible. At thermal equilibrium, the probability of occupying the various valleys only depends on their depth.

In practice, the method of simulated annealing is well known and used for global optimization. As we already mentioned, annealing is the process of heating a material (e.g., metal or glass) to a high temperature, and then gradually lowering the temperature. But the material will only end up at its global energy minimum if the annealing process is gradual enough. Sudden cooling of, e.g., metal will leave the material only in a local minimum with a brittle state. Simulated annealing makes the escape from local minima likely by allowing jumps to higher energy states.

In the thermodynamics of gases, the gas is described by its probability of phase transition. It was Boltzmann who derived a probability distribution for the states of a gas when it had reached a uniform distribution of temperature. Hinton, Sejnowski, and others claimed that the distribution could be used to describe neural interactions. In this case of modeling, the low temperature term added to the system is interpreted as a small noise term. It is the neural analog to random thermal motions of molecules in gases.

This formal equivalence is the reason that the network under consideration is called a "Boltzmann machine" [4.32]. But, of course, no physicalism is intended, reducing neural interactions to the molecular interactions of gases. In Boltzmann's formalism, it can be proved that a Boltzmann machine is guaranteed to find the desired global minimum as long as it is cooled slowly enough. Obviously, a neural network with the dynamics of simulated annealing is capable of searching a state space for the pattern giving the global energy minimum.

A possible learning rule according to this dynamics matches probabilities between the network and its environment. All possible states of the network are possible at thermal equilibrium, with the relative probabilities of a Boltzmann distribution. If the probabilities of the states in the network are the same as the probabilities of states of the environment, then the network has an adequate model of the environment. Thus, a learning rule must be able to adjust the weights in the Boltzmann machine so as to decrease the discrepancy between the network's model and the environment.

At first, the rule lets the system run free. The probabilities of the states taken by each unit can be estimated. Then, the input and output units are clamped or forced to take appropriate values. Again, values of the probabilities of the states of the units are estimated. The local change of weights is proportional to the difference in the probabilities of the units coupled by that weight [4.33].

Formally, the weight modification rule demands that

$$\Delta w_{ij} = \varepsilon(\langle s_i s_j \rangle \text{ clamped} - \langle s_i s_j \rangle \text{free})$$

where ε is the constant of proportionality ("rate of learning"), s_i is the binary unit of the ith unit and $s_i s_j$ is averaged over time to $\langle s_i s_j \rangle$ after the network has reached equilibrium. In the clamped condition, the input and output units are fixed to their correct values. In the free condition, none of the units is fixed. Then, the learning rule is unsupervised. If the inputs are fixed in the free condition, the learning rule is supervised.

In Fig. 4.11b, the units in the network of a Boltzmann machine have binary values and the connections between them are reciprocal [4.33]. The weights of the connections can be trained by presenting patterns to the input units in the presence and the absence of output patterns and applying the Boltzmann learning rule. During the learning process, every weight in the network is modified. The hidden units which do not receive direct information from outside enable the network to yield complex associations between input and output patterns. Thus, Boltzmann machines with hidden units in their middle layer have internal representations of the environment which are not possible for networks with only visible (input and output) units.

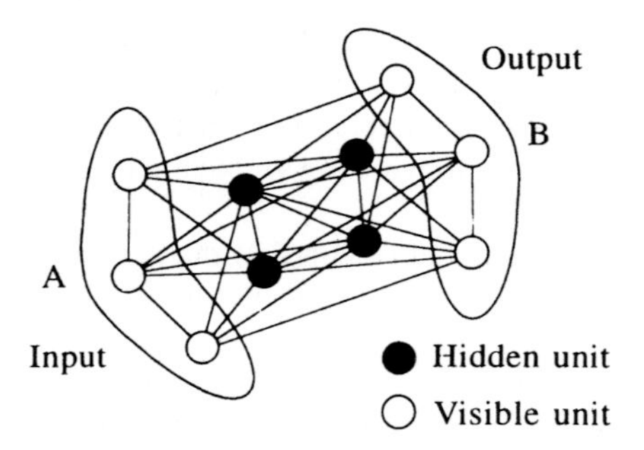

Fig. 4.11b. Network of a Boltzmann machine [4.33]

From a neurobiological point of view, supervised learning with a "teacher" seems to be rather unrealistic in nature. Feature extraction or categorization by an animal must be self-organized from an analysis of the sensory inputs. The more frequently a feature occurs in the input vectors, the more likely it is to belong to a certain category. The outputs of the network must learn to converge to the corresponding prototype vectors as attractors.

How can a network be designed to invent criteria of classification without the supervision of an external teacher? Some authors assume that this kind of self-organization depends on the nonlinear interactions and selective reinforcement of the connections in a multi-layered system. The learning procedure is organized in a Darwinian process of selection and competition.

In Fig. 4.12, the multi-layered architecture of a competitive learning system is designed to produce such eminently cognitive tasks as classification and categorization [4.34]. Active units are represented by filled dots, inactive ones by open dots. The connections from the input layer to each element in the second layer are excitatory. The second layer is subdivided into clusters within which each element inhibits all the others. Elements of the same cluster compete with each other in responding to the input pattern. According to the rules of Rumelhart and Zipser, a unit can learn only if it can win the competition with the other units within the same

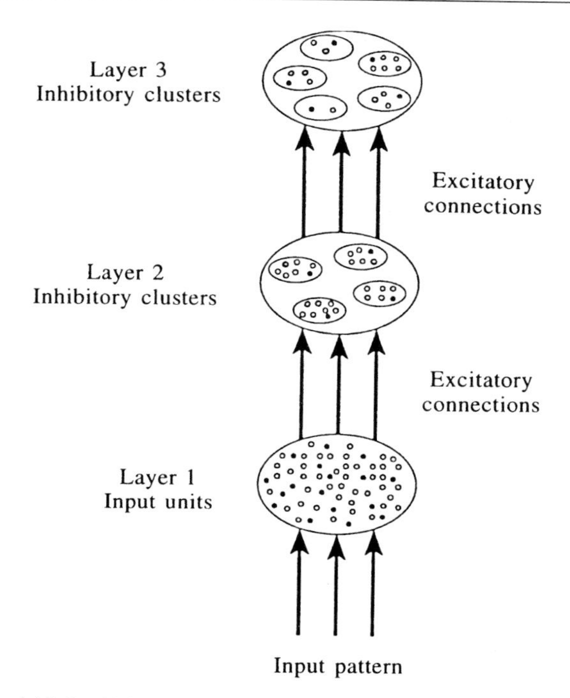

Fig. 4.12. Multi-layered network with competitive learning [4.34]

cluster. Learning means an increase in the active connections and a decrease in the inactive connections.

A simple task of classification refers to word recognition by a child. Obviously, the two-letter words AA, AB, BA and BB may be classified in several categories, for instance, the set {AA, AB} of words beginning with A or the set {BA, BB} of words beginning with B or the set {AA, BA} of words ending with A or the set {AB, BB} of words ending with B. In a computer-assisted experiment, the two-letter words were presented to a layered network with one level of competing units organized in a cluster of two units. The system was able to detect the position of the letters. One of the units spontaneously learnt to act as a detector of A as beginning letter, while the other one detected B as the beginning letter.

In further experiments, the number of letters was increased, with a modified network structure. Although these experiments seem only to illustrate limited capabilities, they demonstrate the emergence of cognitive behavior from unsupervised neural systems, at least in principle. They have started some interesting research linking neurophysiology with the cognitive sciences in the framework of complex systems which will be explored in more detail in Sect. 4.4.

Another approach to self-organizing cognitive systems through competitive learning was proposed by Teuvo Kohonen. He is a physicist who also has worked physiologically on associative memory. His mathematical modeling of neural systems has been important for engineering applications in artificial intelligence (compare Chap. 5). Kohonen's ideas of brain modeling by self-organizing feature maps stems back to anatomically and physiologically well-confirmed facts. Most neural networks in the brain are two-dimensional layers of processing units which may be cells or cellular modules. These units are interconnected through lateral feedback. For instance, in the neocortex there are 10 000 interconnections for every principal cell.

The synaptic coupling from a neuron to its neighbors is excitatory for all those neurons whose distance is smaller than a certain critical value. It is inhibitory for neurons lying at a greater distance. At some yet greater distance, the coupling is weakly excitatory again. The degree of lateral interaction is mathematically modeled by a curve with the form of a Mexican hat [4.35] (Fig. 4.13a).

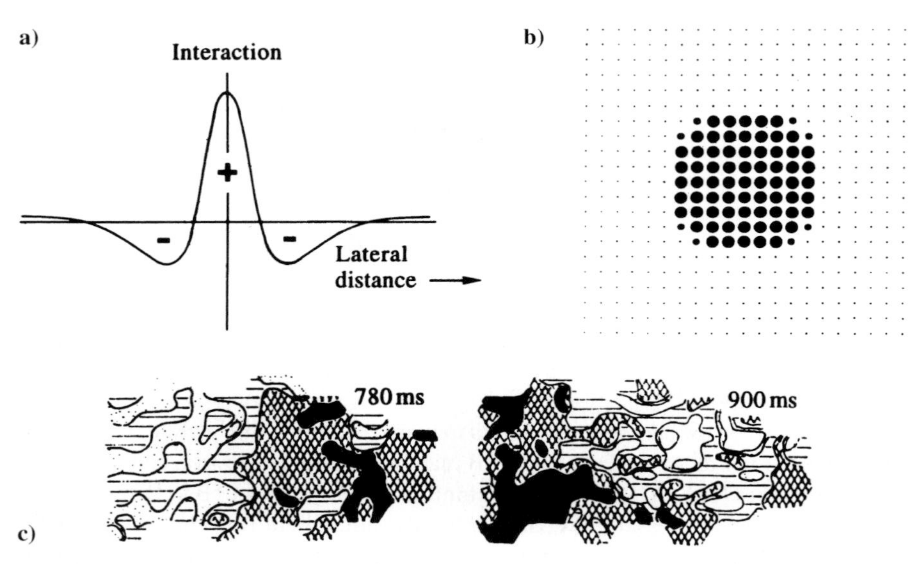

Fig. 4.13a–c. Mexican hat of neural interaction (**a**). Distribution of neural activity in a 2-dimensional model (**b**) and a raccoon's cortex (**c**) [4.36]

Obviously, the activity of lateral couplings tends to a spatially bounded cluster. Figure 4.13b shows a two-dimensional example of clustering which was simulated by a network with a 21×21 square array of processing units. The clustering phenomenon ("activity bubble") depends on the degree of positive or negative feedback, which may be influenced by chemical effects in the neural network. In neural reality, "activity bubbles" do not have the regular form of computer-assisted simulations. Figure 4.13c shows distributions of activity on a raccoon's cortex, which does not represent a regularly-shaped figure, but a rather diffuse map [4.36].

Nevertheless, the cluster phenomenon can be shown to be useful in the self-organizing processes of the brain. While initially the activities of the neural network are homogeneously distributed, a progressive specialization of neural regions can be observed, according to a self-organizing learning process. After presentation of an input pattern, the neuron with highest activation and its neighbors are chosen for learning. The neural weights are modified according to the circular neighborhood of given radius, centered around the neuron with highest activation. This learning rule can be used to detect and categorize similarities among input data of visual or speech patterns.

Formally, Kohonen considered a nonlinear projection P from the space V of input signals v onto a two-dimensional map A. Figure 4.14 illustrates the learning step: the input value v selects a center s. In the neighborhood of s, all neurons shift their weights w_s in the direction of v. The degree of shift decrease with increasing distance from the center s and is visualized by differing grey values [4.37].

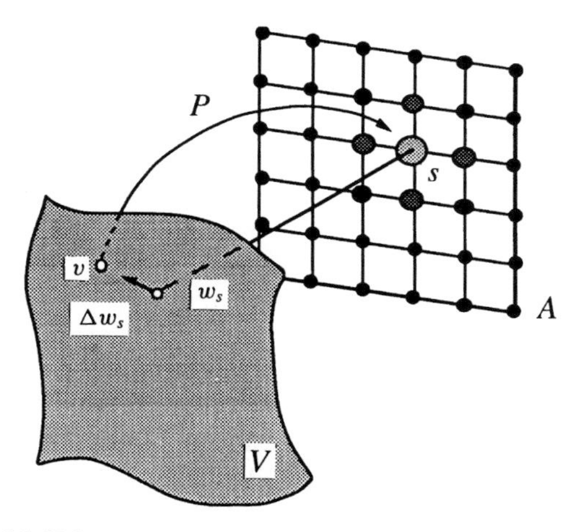

Fig. 4.14. Kohonen's model of self-organizing neural maps [4.37]

The map converges to a state of equilibrium with different regions of activity by self-organization. The projection should map the regularities of the input signals onto the neural map. Thus, P is mathematically called a topologically invariant mapping. Actually, the structure of the brain's environment, which is represented by the regularities of the sensory input signals, should be projected onto a neural map of the brain: the brain should get an adequate model of the world.

How realistic is the modeling of the brain by self-organizing maps? The magnitude of a neural field varies, depending on the importance of the perceived sensory stimuli for the survival of the species. In a neural field there are centers which can analyze and represent the stimuli with more accuracy than their environment. For instance, in the eye of a mammal the fine analysis of visual information is performed by the "fovea", which is a very small region around the optical axis of the retina with a very high density of light-sensitive receptors. Thus, the dissolution of signals is essentially higher in this center than in the surrounding region of the neural field. Similiar unproportional representations can be observed in the somatosensory system and in the motor cortex. The importance of the hand for human survival is represented by a rather great region on the somatosensory and motor cortex, relative to the representation of the body's surface.

Contrary to these results, the auditory cortex of cats, dogs, and apes does not project the frequencies of the outer world with special centers. An exception is the bat with its specialized system of orientation, which is necessary for its survival. Bats can send many different supersonic frequencies and determine the distance and magnitude of objects by the reflections of the signals. The bat's velocity relative to other objects can be determined by the Doppler effect in supersonic echo sounding. Even the movements of tiny insects can be detected by this sensitive system.

The specialization of the bat can be experimentally confirmed by a self-organizing map on its auditory cortex. Figure 4.15a shows the brain of a bat with the auditory cortex in the rectangle. Figure 4.15b is an enlargement of the rectangle with a distribution of the best frequencies on the auditive cortex. The one-dimensional frequency spectrum is represented continuously and monotonicly from the poste-

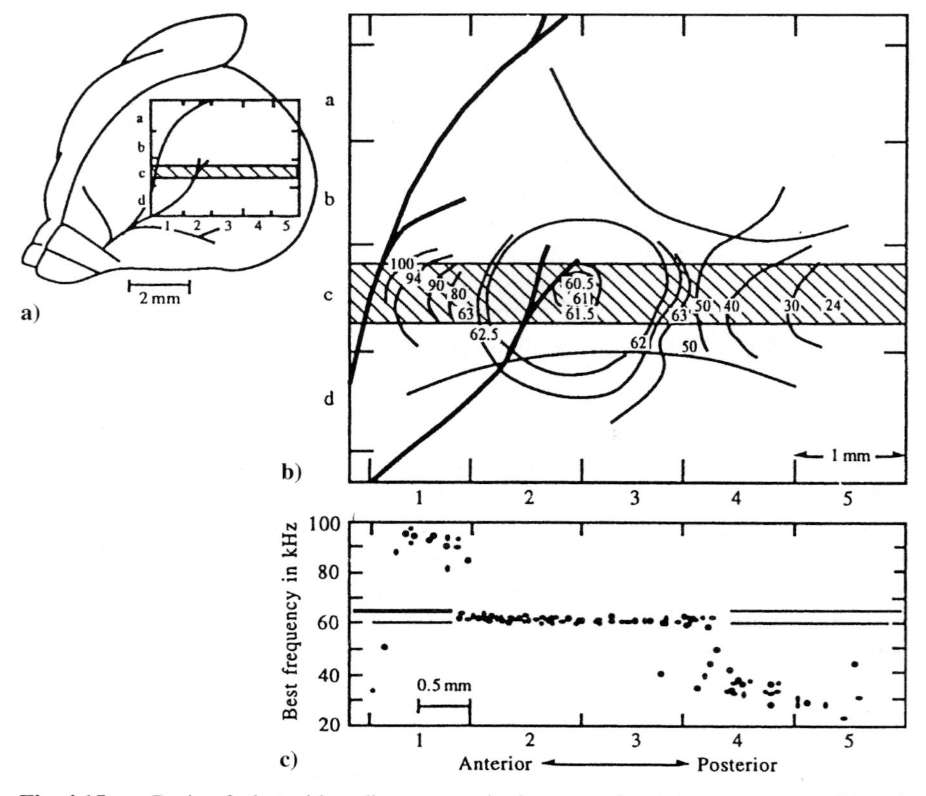

Fig. 4.15a–c. Brain of a bat with auditory cortex in the rectangle (**a**), its enlargement (**b**), and distribution of the best frequencies (**c**) [4.38]

rior to the anterior region of the auditive cortex. The frequency causing the highest excitation of a neuron is called the best frequency of that neuron. The region of the dotted lines is the primary auditory cortex. Figure 4.15c shows the distribution of the best frequencies in the hatched region of Fig. 4.15a. The majority of measured points are centered around the frequency of supersonic echo sounding. More than the half of the anterior-posterior region is used to analyze the Doppler effect in the supersonic echo sounding. It is remarkable that computer assisted simulations by self-organizing maps produce the actual representations of the auditory cortex in Fig. 4.15c [4.38].

The brain in primates consists of many regions with several neural net topologies. The retina, for instance, already develops during early ontogeny. It has a neural topology with five separate layers for photoreceptors, horizontal cells, bipolar cells, amacrine cells, and retinal ganglion cells. The photoreceptor layer in humans consists of about $120 \cdot 10^6$ receptor cells. The retinal output, which is represented by the spatio-temporal pattern of the impulse rates of all ganglion cells, travels along the optic nerve toward the thalamus. In humans, there are about $1.2 \cdot 10^6$ ganglion cells. Thus, the retina is a really complex system. Nevertheless, the complexity of more than $200 \cdot 10^6$ retinal neurons still is not understood completely. The cerebral cortex is the phylogenetically youngest brain region. The percentage of cerebral cortex to brain increased during evolution. Lower vertebrates like fishes did not evolve a cerebral cortex. Its magnitude increased from small parts in reptiles and birds to dogs, cats, and finally apes and humans. In primates, the cortex is divided into different regions with multilayered neural net topologies like visual, sensory, motor, and association cortex. The cerebellum consists of the cerebellar cortex with many multilayered subregions for specific sensorimotor functions.

The great variety of brain systems is described as densely packed sets of neurons with particular network topologies, communicating with each other via many nerves which consist of thousands of axons. In contrast to digital computers with separate central processing unit, memory, and registers, the brain and the central nervous system can be modeled as an ensemble with many special-purpose parallel processing networks. Each network is capable of independent processing and storage of information for sensory, motor, and associative functions.

Obviously, the biological brain does not apply principles known from program-controlled centralized digital computers. The process of network self-organization is fundamental to the structure of the brain. In the very long course of phylogenesis, complex structural forms, the purposes of which are sometimes not completely clear to us, have been produced. On the macroscopic scale, particular neural areas have been specialized for signals with different sensory functions, for information processing operations with different levels, for humans as well as for the animal and vegetative functions of the organism. Although they are distributed in different areas of the brain, they can be understood as self-organizing complex or collective effects.

Self-organization as a learning procedure demonstrates that organisms are not fully determined by genes containing a blueprint which describes the organism in detail. Each stage of brain organization involves some kind of self-organization.

Genes would not be able to store the complex structure of the brain. With a cerebral cortex of about 10^{14} synapses, ontogeny could not select the correct wiring diagram out of all alternatives if all were equally likely. Thus, ontogeny must use the self-organization of neural systems to handle their complexity. But the structure of the cortex is not understandable without knowing the principles of its ontogenesis.

In earlier chapters, we have studied the emergence of ordering patterns from complex systems in physics, chemistry, biology, metereology, and astronomy. Global order emerges in complex systems with a large number of locally interacting elements. There are interacting atoms or molecules in a liquid or crystal, subvolumes in an evolving star system, or neurons and synapses in a complex neural system like a brain. We remind the reader of Bénard convection ("rolling columns"), which arises by thermal fluctuations of a liquid.

How can global order be arranged by local interactions? The intermolecular forces, for instance, acting within a volume of liquid have a very short range, while the pattern of convective movement which is caused by the molecular interactions may be ordered on a large scale. This principle, which arises in physical, chemical, and biological evolution, has great importance to the brain, in which local interactions between neighboring cellular elements create states of global order leading to a coherent behavior of the organism. The ordering pattern is arranged by forces between elements of the complex system and by initial and boundary conditions. In the example of Bénard convection, the forces are hydrodynamic interactions, thermal conduction, expansion, and gravity. Boundary conditions are, for instance, the temperature which is given to the liquid. In the brain, the connection patterns are arranged by several rules for the interaction of cellular units. As neurons are connected by sometimes very long axons, a local interaction of two neurons does not imply their spatial proximity in the brain's anatomy, but only their immediate connection by axons.

Although the general structure is universal for all types of neurons and synapses, there can exist many qualitative and quantitative differences. The neural system of an invertebrate, for instance, is deterministic with a high degree of coded information in the specific location of individual neurons. For an associative system in the mammalian neocortex, the specific responses to specific input patterns are achieved by learning rules faciliating feedback of information from the output.

How adequate are complex system models to real neural networks? From a methodological point of view, we must be critically aware that models cannot be naively identified with each function and element of reality. Models are special-purpose abstractions which may explain and simulate some part of the central nervous system more or less, and other parts not at all. Sometimes model nets are criticized in that they only demonstrate more or less correct execution of an input-output function like a black box. But nothing could be revealed about how biological neural nets execute that function. Hidden units were only theoretical concepts like hidden variables in quantum physics which are assumed to be intrinsic elements of the system realizing the relation between the observed and measurable input and output values. Besides the architecture of a perhaps multi-layered network, the dynamics and learning procedures are an essential problem of simulations.

How realistic are the parameter-adjusting procedures of model nets to minimize error? Many learning rules of model nets take unacceptably long to converge to a set of weights that classify correctly. Although a successful weight-adjustment is sometimes found, its optimality is by no means decided. In 1960, Widrow and Hoff suggested a simple and elegant learning rule which was motivated by reasons of technical optimization, and not by biological insight into the brain's function [4.39]. The Widrow–Hoff-rule and its variants have been extensively used in technical networks in recent years (compare Chap. 6).

The rule assumes that there is an input pattern, and an output classification of the input pattern by an adaptive neuron, which can take values of either $a+1$ or $a-1$. Thus, a "teacher" is assumed knowing what the answer was supposed to be for that input. The adaptive neuron compute a weighted sum of activities of the inputs times the synaptic weight. The system is able to form an error signal between what the output is supposed to be and what the summer computed. According to the difference between them, the synaptic weights are adjusted, and the sum recomputed, so the error signal becomes zero.

Widrow and Hoff's strategy aims at reducing the square of the error signal to its smallest possible value. All possible values of the input weighting coefficients give rise to an error value. In Fig. 4.16, the situation is visualized by an error surface in the weight space [4.40].

The minimum of the error surface is not known exactly, because the entire surface cannot be seen. But the local topography can be measured. Thus, the directions of adjustment which decrease the error the most can be calculated. The so-called gradient descent method, which is well known in differential geometry and physics, always adjusts the weights so that changes in weights move the system down the error surface in the direction of the locally steepest descent.

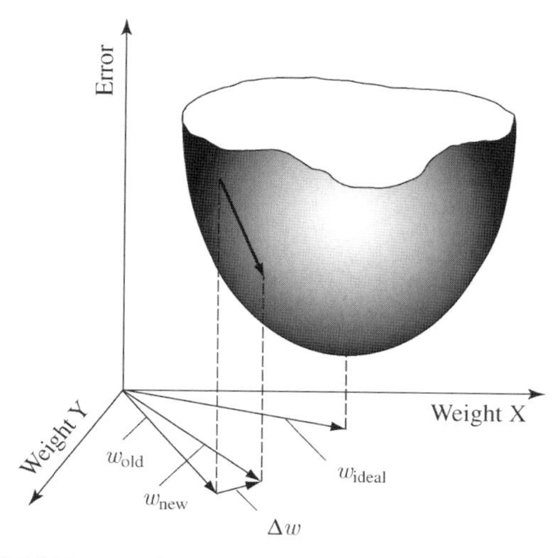

Fig. 4.16. Error surface and learning by gradient descent [4.40]

In Fig. 4.16, the gradient of the error surface is computed to find the direction of steepest descent. The weights are incrementally changed by a certain value Δw along this direction. The procedure is repeated until the weights reach w_{ideal} representing minimum error. For a nonlinear network the error surface may have many local minima. Problems with gradient descent in general involve getting trapped in local minima. Then the bottoms of valleys do not represent the lowest global error.

Widrow and Hoff proved that there is a simple quadratic error surface with only one global minimum. Mathematically, computing the gradient at a point means computing partial derivatives of the square of the error with respect to the weights. Widrow and Hoff proved that this derivative is proportional to the error signal. Thus, the measurement of the error signal provides the direction of movement, in order to correct the error. Technically, the existence of a "teacher" with perfect knowledge may be justified for special purposes. But the assumption of supervised learning procedures in nature seems to be rather unlikely.

In network models, so-called backpropagation is the best known supervised algorithm which is a generalization of the simple Widrow-Hoff rule. In short, backpropagation is a learning algorithm for adjusting weights in neural networks. The error for each unit, which is the desired minus the actual output, can be calculated at the output of the network and recursively propagated backward into the network. This method enables the system to decide how to change the weights inside the network to improve its overall performance. Figure 4.17 illustrates the backpropagation method through an entire net with several layers [4.41].

Although models with backpropagation can be as successful as biological networks, it is not assumed that the real brain is organized by backpropagation. Many values of parameters in a real network are sometimes known by measurement and experimentation in anatomy, physiology, and pharmacology. For instance, the number of cell types and cells themselves may be roughly estimated. The topology and architecture may be described, the question whether specific synapses are excitatory or inhibitory may be decided, and so on. But the specific weights are unknown.

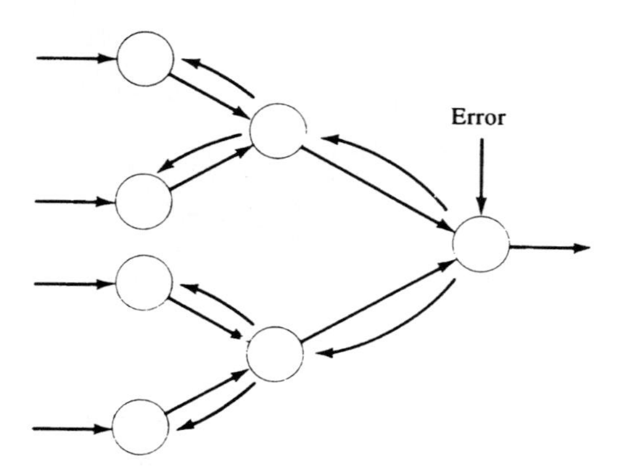

Fig. 4.17. Backpropagation

When there are thousands of them, the probability that the global minimum of a brain net and a model net are exactly the same, weight for weight, is rather low.

Thus, the error minimization strategies of model nets may in general be hypothetical, but they are necessary to deal with the complexity of perhaps thousands of unknown parameters. They allow one to predict some local or global properties of brain nets if the degree of similarity between the topology, architecture, and synaptical dynamics of model and brain nets is rather high. Backpropagation in neurobiology is justified as a search tool to find local minima, but not as a replacement for neurobiological analysis, which may reveal the real learning procedure of a neural network.

Error minimization strategies have a long tradition in the evolution of nature and have not emerged for the first time in the learning procedures of the brain. Natural selection of, for instance, an ecological population can often be modeled as a process of sliding down an error gradient to an error minimum representing an environmental survival niche. But the parameter-adjusting procedure may only find a local minimum, not necessarily the global minimum. As far as we know, evolution in general does not find the best possible solution, but only a satisfactory one which is good enough for survival. Only local minima of the entire evolution can be empirically evaluated with respect to their evolutionary survival value. This evaluation depends on the observed and measured constraints of the chosen model.

Thus, from a methodological point of view, the idea of an always globally and perfectly optimizing nature is a metaphysical fiction. It is the secularized idea of a godhead called "nature" or "evolution" which was born in the optimistic century of enlightenment in order to replace the Christian God with his plan of creation. It was already Kant who criticized the idea of a totally self-optimizing nature as a human fiction which cannot be empirically justified in any sense. There is no supercomputer with a separate central processing unit which can optimize the evolutionary strategy totally and in the long run. There are only locally more or less satisfactory solutions, even many failures, and imperfection in real evolutionary processes. Their complexity contradicts the simplified models of a perfect world à la Laplace.

4.3 Brain and the Emergence of Consciousness

How can cognitive features be explained by neural interactions in complex models of the brain? Leibniz already had the problem that consciousness, thoughts, and feelings cannot be found in the elements of the brain if it is interpreted as a mere machine. Kant underlined that an organizing force is necessary to animate a physical system. Until this century some physicists, biologists, and philosophers believed in a immaterial organizing life factor which was called "elan vital" (Bergson) or "entelechy" (Driesch). From the point of view of complex systems, Köhler's gestalt psychology was an interesting approach referring to the existence of physical systems in which complex psychic structures originate spontaneously from the system's own intrinsic dynamics. Popularly speaking, the macroscopic "gestalt" (form) of a perceived object is more than the sum of its atomic parts, and cannot be reduced to the microscopic scale.

Köhler had the idea that the emergence of visual phenomena can be explained in the framework of thermodynamic models. But in those days, he referred to Boltzmann's linear thermodynamics in equilibrium. He assumed: "The somatic processes underlying static visual fields are stationary equilibrium distributions developed from the inner dynamics of the optical system itself" [4.42]. Köhler even realized that an organism is not a closed system, and tried to explain the emergence of ordered states as a kind of intuitively understood synergy. In this respect, Köhler was already correct with his clear distinction between the microscopic level of elementary interactions and the macroscopic level of emerging ordered states in a synergetic system. But he still lacked an adequate framework of complex dynamical systems to provide the formalism for a thermodynamics far from thermal equilibrium.

The complex system approach offers the possibility for modeling the neural interactions of brain processes on the microscopic scale and the emergence of cognitive structures on the macroscopic scale. Thus, it seems to be possible to bridge the gap between the neurobiology of the brain and the cognitive sciences of the mind, which traditionally has been considered as an unsolvable problem.

Complex models consist of state spaces and nonlinear evolution equations describing a system's dynamics. With about 10^{11} nonsensory neurons, the human brain is represented by a state space of 10^{11} dimensions. Even a typical subsystem contains about 10^8 elements. In a state space with 10^8 dimensions and only 10 levels of neural activitation, there are at least 10^{10^8} distinct positions representing activitation vectors. If we assume 10^3 synaptic connections between each neuron and the other 10^8 neurons of a subsystem, then about 10^{11} synapses must be distinguished. Consequently, for only 10 distinct weights at each synapse, we get the huge number of $10^{10^{11}}$ weights in a subsystem alone. This complexity provides numerous possibilities for coding, representing, and processing information, which can be modeled mathematically by vector and tensor transformations [4.43].

In Kohonen's competitive learning network, the system self-organizes so that nearby vectors map onto nearby points of the net. It is assumed that similar impressions are represented by similar vectors with tiny distances to some prototype vectors. In the framework of complex systems, prototype vectors are interpreted as attractors. Thus, two distinct categories or classes are represented by two different attractors in the state space (Fig. 4.7b). The learning process of cognitive distinction is modeled by a training process of the network which involves adjusting the weights so that an input vector (for instance, a visual or acoustic pattern) is submitted to the prototype vector with the most similarity.

The concept of prototype in neural state spaces allows some interesting interpretations of cognitive processes. How can a network recognize a pattern when the input impression is only partially given? The task of vector completion is crucial for animals to survive in the wild. Imagine a coyote in the desert which detects the tail of a rat in the grass (Fig. 4.18a). The input to the retina of the coyote is limited to

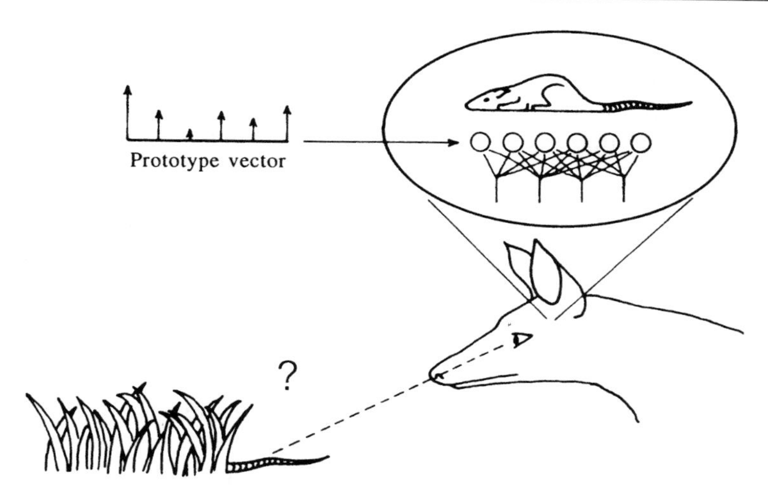

Fig. 4.18a. Recognition by prototype-activation in neural nets: a coyote recognizes the tail of a rat [4.44]

that detail of the rat. The assumption or "hypothesis" that there is a rat in the grass is enabled by the coyote's visual system completing the input vector with the learnt prototype vector of a rat. In this sense, we may say that the coyote has a "concept" of a rat represented by a corresponding prototype activation pattern in the brain [4.44].

Paul Churchland even suggests interpreting human high-level cognitive abilities by the prototype vector approach. So, explanatory understanding is reduced to the activation of a specific prototype vector in well-trained networks. Prototype vectors embody a huge amount of information which may differ for different people. The reason is that different people may not always have the same items satisfying the constraints of a prototype cluster. Indeed, people mostly have different degrees of explanatory understanding, although they classify an object or situation in nearly the same manner. A joiner, for example, has a higher degree of understanding of what may be a chair than most other people. Nevertheless, they all will agree in most cases. Thus, the prototype-activation model is rather realistic, because it considers the fuzziness of human concepts and understanding.

In epistemology and cognitive psychology, it is usual to distinguish between different kinds of explanation. There are classifying explanations ("Why is the whale a mammal?"), causal explanations ("Why does the stone fall down?"), functional explanations ("Why does a bird have wings?"), and others which correspond to prototype activations of clusters, causal relations, functional properties, and so on. Even mastering social situations is a matter of activating social-interaction prototypes which have been trained and taught during a lifetime.

In the complex system approach, mental states are correlated with neural activation patterns of the brain which are modeled by state vectors in complex state spaces. External mental states referring to perceptions of the outer world may be testable and correlated to neural activities of the brain. How can we test and explain internal states of consciousness which do not refer to events of the outer world, but to mental states themselves?

It is well known that we can even reflect about our self-reflections, and reflect about the reflections about our self-reflections, and so on in an iteration process which is in principle unlimited (Fig. 4.18b). Self-experience and self-reflection lead to the concept of self-consciousness, which traditionally was considered as the essential concept in the philosophy of mind and cognitive psychology. Self-consciousness was defined as the crucial feature of human personality. The definitions of self-consciousness which historically have been discussed are not only philosophically interesting. Obviously, these more or less speculative definitions have powerful consequences in medicine and law. Which criteria must be satisfied for a human being to be conscious and therefore responsible for his actions? Are there medical criteria for consciousness? How can consciousness be disturbed or even destroyed? What about the consciousness of animals? Can we feel like our neighbor or like an animal?

The fundamental questions arise (1) if there are particular brain processes causing the emergence of consciousness, and (2) if the emergence of consciousness from brain processes can be modeled by complex systems. The methodological difficulty

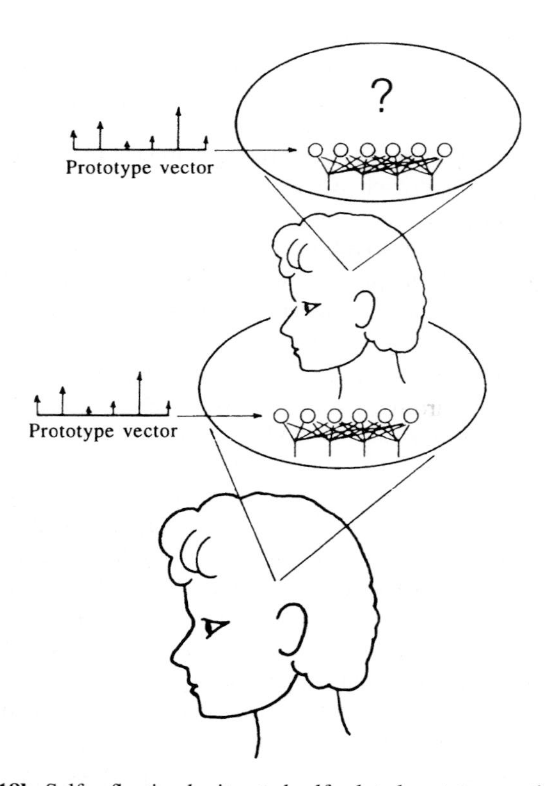

Fig. 4.18b. Self-reflection by iterated self-related prototype-activations

seems to be that subjective feelings like pains, smells, and so on are only accessible to introspection. These subjective states of feeling and consciousness are sometimes called phenomenal states. Some philosophers have criticized that a physical description of brain states fails to grasp the essence of what is a phenomenal state. Proponents of the opposite view have argued that notions of phenomenal states can be reduced to notions of neurophysiological states of the brain. These arguments are nothing more than modern variations of traditional positions which are known as physicalism and mentalism (or antiphysicalism). Both positions are ideological reductionisms and exaggerations which are neither justified by research nor very helpful in research [4.45].

According to the complex system approach, neurophysiological states and mental states are modeled by mathematical formalisms without reductionist ambitions. Some philosophers fail heavily with their prejudice against mathematics, because they seem to believe that formulas only can designate "physical" states. The reader may recall, for instance, a Hopfield system which contains an "energy" formula by analogy with the energy formula for a physical spin system. Nevertheless, in the framework of a Hopfield system, the so-called "energy" formula must not be identified with energy in solid state physics. The mathematical expression only determines the dynamics of a network, which may be simulated by neurobiological brains or silicon computers or angelic organisms from still unknown star systems.

The mathematical model is empirically corroborated if it fits the observed data. In other cases, it must be modified or dropped. We must be aware that a testable and corroborated theory of mental states and consciousness does not enable us to feel like our neighbor. A physician or a surgeon, for instance, who wishes to heal a patient's pain in the stomach does not need to feel the patient's stomach pain. He must have a good knowledge of the stomach based on anatomy, physiology, biochemistry, psychology, etc. In the terminology of the complex system approach, he must know the possible states of a stomach and their dynamics. In this sense, a model of mental states and consciousness should be developed and tested without any reductionist claims.

Obviously, there are many testable correlations between phenomenal states of consciousness and the neurobiological functioning of the brain. Everybody knows that a short period of oxygen deprivation causes unconsciousness. Electrical stimulations, psychotropic drugs, anesthesia, and lesions may influence the degree of consciousness too, which is not only experienced by self-experiment (autocerebroscopy) but also clinically testable by observations and measurements of functional deficits. The reason is that the brain is an open system whose states depends on the physical, chemical, and biological metabolism with its environment far from thermal equilibrium.

Conscious and unconscious states seem to depend on a rather complex neurophysiological system which contains feedback loops and interconnections at various levels. Figure 4.19 shows the network of the cerebral cortex with its subsystems of primary sensory cortex and association cortex. There are specific inputs ("afferents") from sense organs reaching the primary cortical projection areas through specific transmitting subsystems and pathways. Non-specific inputs reach the cor-

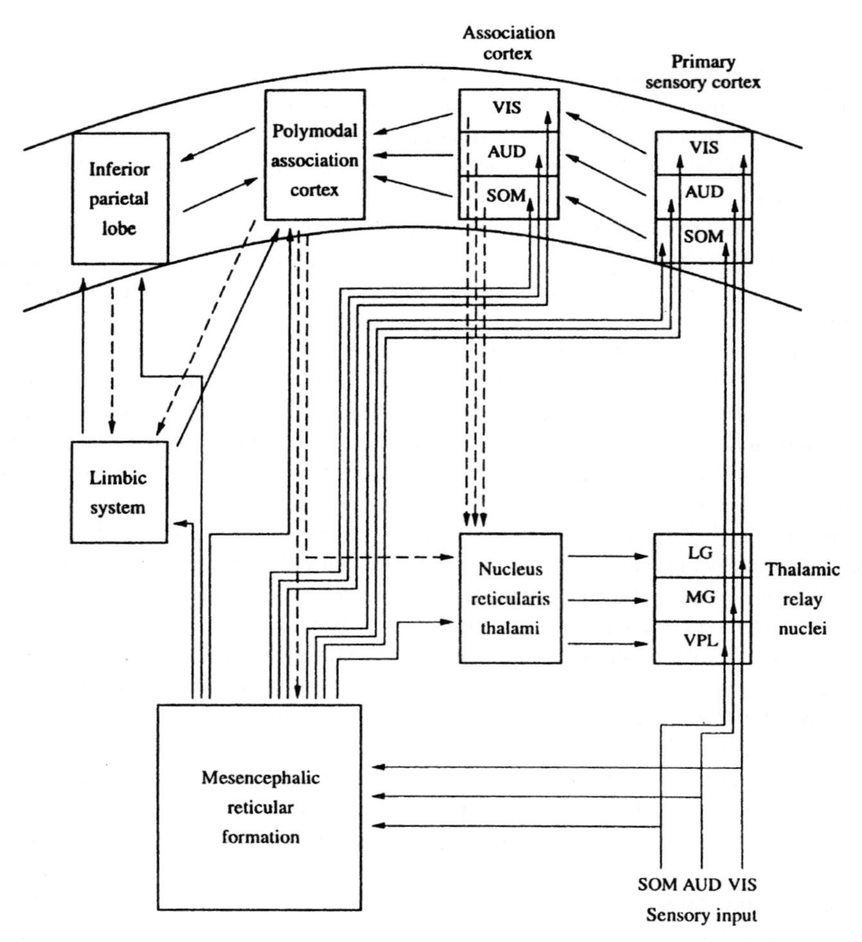

Fig. 4.19. Inputs to the cerebral cortex with somasensory pathways (SOM), auditory pathways (AUD), visual pathways (VIS), lateral geniculate (LG), medial geniculate (MG), nucleus ventralis posterolateralis (VPL) [4.46]

tex from a subsystem called the "mesencephalic reticular formation". The reticular formation designates a complex network of neurons and nerve fibers with widely distributed connections of synaptic contacts. It is known to play an essential role in arousal, wakefulness, and attention [4.46].

Lesions within the complex network lead to various disturbances of consciousness, which may be global or only local, with specific deficits of conscious experience during global wakefulness. Neurophysiology can experimentally demonstrate that degrees of consciousness depend on the two streams of specific and non-specific afferent signals processed in the cerebral cortex. But the question arises of how mental states of consciousness emerge from these networks. In the terminology of

Leibniz, we see the interacting elements like the cog wheels of a mill, but cannot bridge the gap between the neurophysiological machinery and the mental states of consciousness. Traditional neurophysiology has the conventional view that brain functions are made possible by electrical impulses spreading through a network of neurons connected by rigid synapses like the rigid connections of cog wheels in Leibniz' mechanistic model of a mill.

The complex system approach offers a view of self-organizing nets changing their synaptic connections, which are induced by synaptic activation and depend on the degree of activation. In the framework of neural complex systems, the microscopic level of interacting neurons is distinguished from the macroscopic level of global patterns produced as cell assemblies by self-organization. In earlier sections, it was already mentioned that the concept of self-organizing neural cell assemblies was introduced by Hebb. It was modified by Christoph von der Malsburg, Teuvo Kohonen, and others. If simultaneous activity is induced in some neurons of a net by a patterned input, then an assembly will be formed by synchronous activation according to a Hebb-like learning rule.

The modification suggested by von der Malsburg is that assembly formation is not a slow process, but produced by rapid synaptic changes [4.47]. These so-called "Malsburg synapses" are used to model networks with rapid weight adjustment dynamics. Today, there is empirical evidence of Hebb- and Malsburg-type synapses with high plasticity in the brain whose rule of interaction can be realized by molecular mechanisms. The formation of assemblies in a network depends on the degree of activation of its neurons.

But there is no "mother neuron" that can feel, think, or, at least, coordinate the appropriate neurons. The binding problem of pixels and features in perception is explained by cell assemblies of synchronously firing neurons dominated by learnt attractors of brain dynamics. The binding problem asked: How can the perception of entire objects be conceived without decaying into millions of unconnected pixels and signals of firing neurons? Barlow's theory [4.48] assumed single neurons for each property of a perceived object, other neurons for clusters of properties, and, finally, a neuron for the entire object ("grandmother neuron"). Thus, the brain needs an exploding number of specialized neurons which must be postulated in ad hoc hypotheses for every new perception of changing situations (Fig. 4.20a). Wolf Singer [4.49], and others confirmed Hebb's concept of synchronously firing neurons through observations and measurements (Fig. 4.20b). Thus, Barlow's theory is not necessary for the explanation of gestalt phenomena.

Concerning conscious and unconscious states, it is assumed that global activation of a cell population, as exerted by the reticular formation on the cortex (Fig. 4.19), would generally increase the probability of assemblies being formed. Thus, Hans Flohr has suggested that degrees of consciousness differ in the rate at which assemblies can be generated. The production rate of cell assemblies determines the amount, complexity, and duration of representations of sensory patterns from the outer world, for instance. Consciousness is a self-referential state of self-reflection (Fig. 4.18b). Thus, a conscious state is based on a cell assembly representing an internal state (and not only a state of the outer world). For example, I not

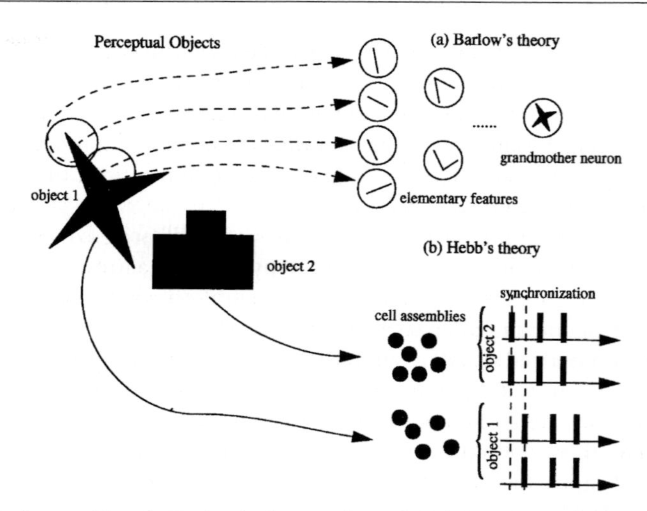

Fig. 4.20. Binding problem in Barlow's theory of grandmother neurons (**a**) and Hebb's theory of cell assemblies (**b**) [4.50]

only have the impression of a green tree, but I am conscious that I am looking at this tree. Furthermore, I can reflect on my state of being conscious of my looking at the green tree, and continue with an iterative production of meta-meta...representations reaching from phenomenal impressions and feelings to abstract and highly sophisticated states of self-reflection. Whenever a critical threshold rate of production is surpassed, phenomenal states must emerge. Deficits of consciousness occur below the critical threshold level.

This hypothesis is testable by particular EEG changes corresponding to an increasing formation rate of assemblies which represent a certain degree of attention. As the production rate of cell assemblies is based on particular synapses with changing weights, the degrees of consciousness may be tested by interventions on synaptic connections. Actually, patients anesthetized with chemical substances which influence synaptic plasticity experience vivid dreams, sensory illusions, visual and auditory hallucinations, and disorganized thoughts. In this sense, awareness can be considered as the result of a system's capacity to generate representations and metarepresentations.

Neural networks with a high rate of assembly formation can produce more complex representations than networks with a lower formation rate. Thus, at a sufficiently high formation rate, complex systems will develop self-referential and metacognitive activities. We may imagine a scale of more or less conscious systems corresponding to the degrees of consciousness in the evolution of living beings with more or less complex nervous systems from the worm to humans. It follows that in the framework of complex systems the emergence of consciousness is no epiphenomenon of evolution. It is a lawful occurrence of global states according to the dynamics of complex systems which produce macroscopic order patterns by microscopic interactions of their elements if certain critical conditions are satisfied.

If the complex systems approach is right, then the nervous systems of biological evolution are only particular realizations of self-referential systems, and other, perhaps technical systems with self-referential capacities based on materials different from the human brain's biochemistry cannot be excluded in principle (compare Chap. 6). We may even be able to translate representations from one complex system into an alien one. As the representations in both systems are not exactly the same, we would not exactly feel like our neighbor, an animal, or another alien system. But we would have a representation in the form of knowledge or a theory about their feelings or thoughts. In this sense, subjectivity is saved, and a wide field of hermeneutics in human communication remains even in the case of technical simulations.

Concerning the traditional mind-body problem, the complex system approach shows that cognitive activity is neither completely independent and different from brain activity nor simply identical, nor an epiphenomenon. Thoughts and feelings are assumed to be both product and producer of neural processes without being identical to them. In the framework of complex systems, the brain is modeled as a self-organizing system which operates far from thermal equilibrium and close to certain threshold values as instability points. During neural instability, different modes of collective excitations evolve to coherent macroscopic patterns which are neurophysiologically based on certain cell assemblies and psychologically expressed as certain feelings or thoughts [4.51].

We all know the experience that in a situation of emotional instability a certain feeling may dominate the other virtual ones and even guide our actions. In synergetics, the competition of stable and unstable modes is explained by the slaving principle. The reader may be reminded of decision situations in which one thought or concept begins to "enslave" the other possible ones. These nonequilibrium phase transitions are governed by very few order parameters in the sense of minimum information. Indeed, acting after a decision means an enormous reduction of complexity. Too much knowledge hinders action or to quote Goethe: "An acting person is always unscrupulous."

Cognitive phenomena are referred to macroscopic properties of the brain's dynamics and to order parameters which govern the underlying microscopic processes. Thus, the so-called mind-brain interaction is only an old-fashioned formulation of an inadequate and obsolete metaphysics that assumes some interacting substances like colliding balls in mechanics. The overlapping area of brain and cognitive sciences is modeled by the emergence of macroscopic properties from microscopic neural interactions during phase transitions in complex neural systems.

In synergetics, phase transitions are interpreted as a kind of symmetry breaking which can be visualized by an overdamped motion of a particle in a symmetric potential (Fig. 4.21a) [4.52].

At the maximum of the potential the position of the particle is symmetric, but unstable, and tiny initial fluctuations decide which of the two equal stable states of minima the particle will reach. In the complex system approach, the two valleys of Fig. 4.21a are interpreted as attractors. Obviously, the ambiguity of perceptions and the spontaneous decision of the visual system for one interpretation is a well-known psychological example of symmetry breaking. In Fig. 4.21b, there is an instability

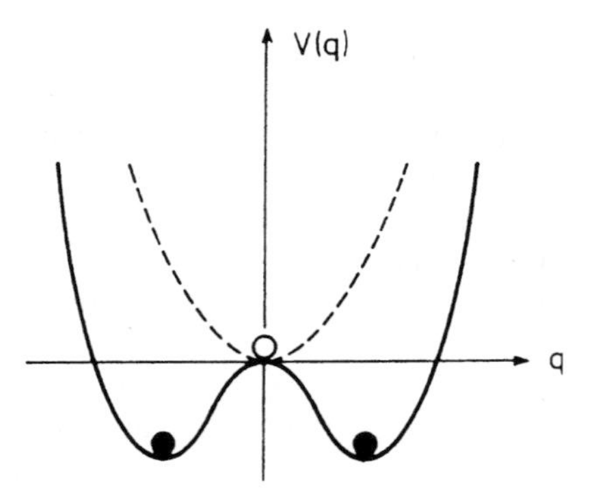

Fig. 4.21a. Symmetry breaking by an overdamped motion of a particle in a symmetric potential

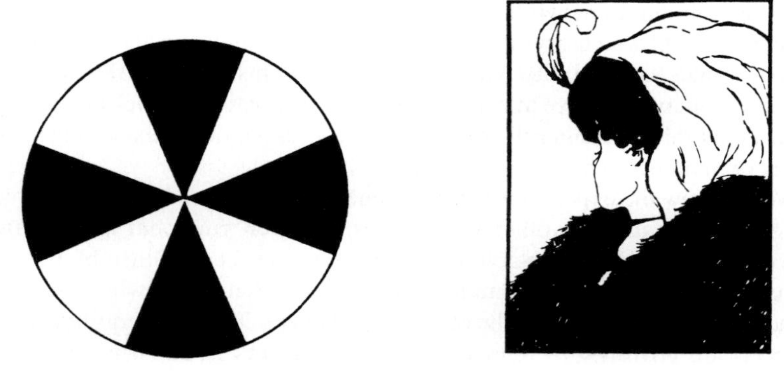

Fig. 4.21b,c. Ambiguity of meaning: (**b**) white or black cross? (**c**) old or young woman? [4.53]

of figure and ground. Do we see a white or black cross? Figure 4.21c shows an ambiguity of meaning. Is it a picture of a young lady or an old woman? [4.53]

Symmetry breaking in psychology is governed by the nonlinear causality of complex systems (the "butterfly effect"), which roughly means that a small cause can have a big effect. Tiny details of initial individual perspectives, but also cognitive prejudices, may "enslave" the other modes and lead to one dominant view. A neurophysiological model has to simulate the corresponding phase transitions of cell assemblies.

Phase transitions are well known in animal locomotion, for instance, in horse gait. With increasing speed horses fall into different movement patterns, from walking to trotting to galloping, in order to minimize the energy costs. This phenomenon of hysteresis is frequently observed in non-equilibrium phase transitions and interpreted as a sequence of stable states or attractors in the nervous system. Phase transitions appear also in thinking. The "aha-experience" and the sudden "insight" are surprising phenomena arising from a situation of fluctuations and instability. In history, there are many famous examples of scientists, engineers, artists, and composers who suddenly found a new problem solution, an invention, an idea for a painting, a melody, and so on in a situation of "creative" instability and confusion.

The complex system approach delivers no closed doctrine of psychology, but an interdisciplinary research program to explore old and new problems of cognitive science and to bring them nearer to an empirical and experimental analysis. Thus, an exploration of correlations between the rates of changing cell assemblies and intellectual abilities of learning, creativity, cognitive flexibility, and ability to visualize is suggested. Phenomena of cognitive instability are assumed to be macroscopic properties of the microscopic instability of nervous processes. Thoughts and expectations are interpreted as order parameters governing the activity of the whole system if it is operating close to instability points. A confirmation of this theory can be seen in psychological tests which produce hallucinations by suggestions that correspond to measurable physiological effects. By recording the regional cerebral blood flow it has been shown that even the thought or intention of acting increases the neuronal activity of the motor area.

Who will deny that thoughts can change the world and that they are not only mere interpretations of the world? In the field of psychosomatic phenomena the placebo effect, for instance, demonstrates that a mere belief or leading idea can alter not only the emotional state but also the physiological state. Obviously, psychosomatic states are close to instability points. The corresponding order parameters are not just theoretical concepts of psychologists, but real modes governing and dominating ("enslaving") the activity of the central nervous system.

The last examples show that the application of self-organizing complex systems in psychology cannot simply be evaluated by their forecasts and quantitative measurability. It is an intrinsic feature of a complex system that its nonlinear dynamics on the microscopic scale and its sensitive dependence on initial conditions do not allow one to forecast the system's final state. In the brain and cognitive research we are confronted with a huge degree of complexity excluding exact calculations or long term forecasts. Nevertheless, the complex system approach reveals essential qualitative features of the mind-brain system, like its high sensitivity with respect to tiny intrinsic fluctuations and changes in the outer world.

4.4 Intentionality and the Crocodile in the Brain

Besides consciousness, there is another fundamental feature of the human mind which was traditionally emphasized – intentionality.

Intentionality is the reference of mental states to objects or states of affairs in the outer world: I see *something*, I believe in *something*, I expect *something*, I am afraid of *something*, I want *something*, etc. Intentional mental states can be distinguished from non-intentional states without any reference object: I am nervous, I am afraid, I am tired, I am happy, I am depressed, etc.

The phenomenon of intentionality can also be visualized by simple examples. In Fig. 4.22, every observer sees a square, although there is physically not given

Fig. 4.22. Intentional object of perception (square) [4.54]

any part of a square's shape. The configuration of the lines suggests to the visual system that there is a particular closed object. An intentional reference between the observer and a configuration of stimuli is achieved [4.54].

Intentional objects or states may be fictional or real. Obviously, human culture is full of signs and symbols for intentional objects and states from traffic signals to religious symbols. Even buildings from memorials and churches to factories may represent intentional objects. The intentional meaning of languages has been constituted in the long development of human cultures. In traditional epistemology, some philosophers like Franz Brentano even proclaimed that intentionality is a particular ability of the human mind to refer to the world. Intentionality was understood as a feature of the mind which cannot be reduced to physical, chemical, or biological properties.

Some modern philosophers like John Searle maintain that intentionality is a distinctive feature of the human mind. But they agree that the biological evolution of the human brain somehow developed the intentional power of mental reference to the world [4.55].

Actually, intentionality is not reserved to brains. It is a feature of certain complex systems which can be modeled by the dynamics of attractors in the evolution of life. Nest construction by social insects is an example of a collective intentional dynamics. The specific feature of this complex system is the autocatalytic mechanism by which the goal-directed work of building nest ecosystems each consisting of a termite population with its environment is carried out. In the complex system approach, it is assumed that this social system already illustrates paradigmatic properties which can be observed in more highly developed systems like brains or central nervous systems [4.56].

The construction process of a nest involves the coordination of more than 5 million insects on the microscopic level, and results in an evolution of certain macroscopic building modes. African termites, for instance, build nests that stand more than 15 feet in height and weigh more than 10 tons. Each insect works independently of each other termite. But their actions are locally determined by distributions of some chemical substance being excreted by the termites themselves. The building material is marked by chemical substance. At first, the building material is distributed randomly, then in an increasingly regular way, until the architectural structure arises from the local interactions of the insects governed by the chemical distribution.

The pattern determines several centers as goals of collective activities, which can be interpreted mathematically as attractors of a diffusion field. In earlier chapters, an attractor was introduced as a solution shared by multiple trajectories originating from different initial conditions. The local trajectories either converge to or diverge from the attractor. In physical or chemical field models, the attractors define local regions in which the potential energy gradient degenerates, going to zero. The region surrounding the attractor is called the basin of attraction and is defined by the gradient flows converging to or diverging from the attractor. The flow pattern of the insects is globally organized by the layout of attractors in their work space, which is the phase portrait of the insects' dynamics. It is well known that attractors are not achieved for ever. If certain control parameters are changed, a pattern may become unstable and break down, being followed by a new pattern of attractors.

Figure 4.23a shows the chemical diffusion gradient surrounding two attractors which will be base of two pillars. As the two pillars act as competing attractors for

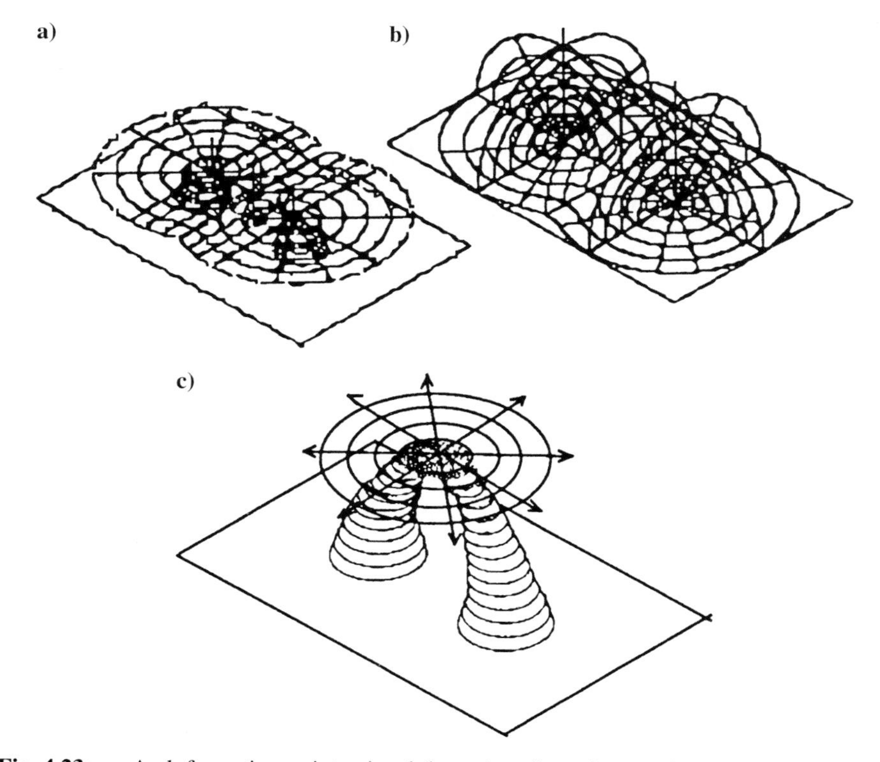

Fig. 4.23a–c. Arch formation as intentional dynamics of termites: (**a**) base of the two pillars as two attractors in a 2-dimensional diffusion gradient field, (**b**) 3-dimensional field governing the direction of the pillars' construction, (**c**) arc formation governed by an attractor of a diffusion gradient [4.57]

the termites, a saddle-point is determined between them. In a later building step, the initial two-dimensional field of Fig. 4.23a is followed by a three- dimensional one (Fig. 4.23b) governing the direction of the pillars' construction. In Fig. 4.23c, the arch formation is shown with one attractor of a chemical diffusion gradient [4.57].

Obviously, an intention at the ecological scale does not require that an individual component of a system must be aware of the global consequences of its actions. The intention is only globally manifested in the long range by the system's dynamics. Figure 4.23d shows the autocatalytic cycle of a nest-building intentional complex system. As it is not a supervised learning process, there is no "goal" or "plan" of some supervising authority like "God" or "Nature". That would be only a simplified anthropomorphic metaphor which does not correctly describe the nonlinear causality of the self-organizing complex system under consideration. Nevertheless, globally there is intentional collective behavior arising from complex nonlinear interactions.

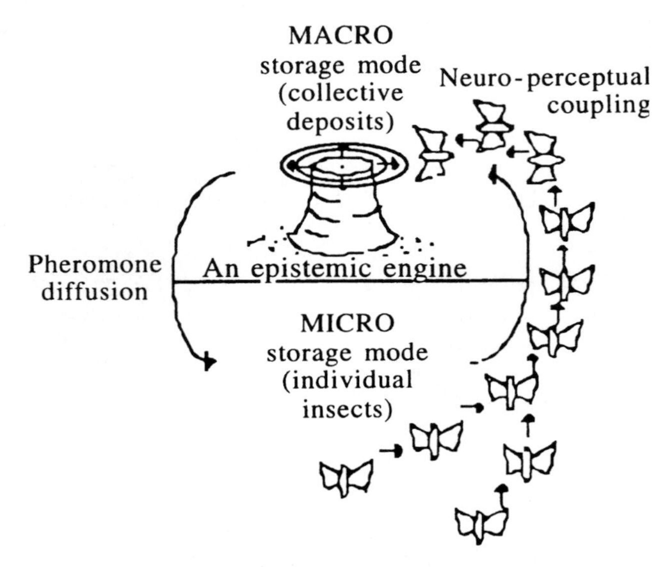

Fig. 4.23d. The autocatalytic cycle of a nest-building intentional complex system

As brains and central nervous systems are complex systems with a nonlinear dynamics governing their neurons and synapses, it is no wonder that they achieve intentional behavior patterns, too. Intentionality has not fallen from heaven as a miraculous feature to guide and distinguish human mind from nature. It is a global pattern emerging in particular complex systems under certain conditions. But there are different levels of intentionality depending on the increasing complexity of evolution.

Intentions must not necessarily be conscious. In Fig. 4.22, the intentional object of our visual system is a square without our exerting conscious will. The so-called perceptual illusions are also intentional patterns of our visual system emerging spontaneously without our conscious will. Figure 4.24 manifests a warping effect of manifolds which seems to be caused by the repellor gradients of different visual attractors. Two equidistant parallel lines seem to change their curvature by a pair of repellor gradients on the left and by a single repellor gradient on the right. The state space of the observer's visual system indicates different curvatures as a result of different visual gradient fields, although the lines remain equidistant and parallel in the physical figure.

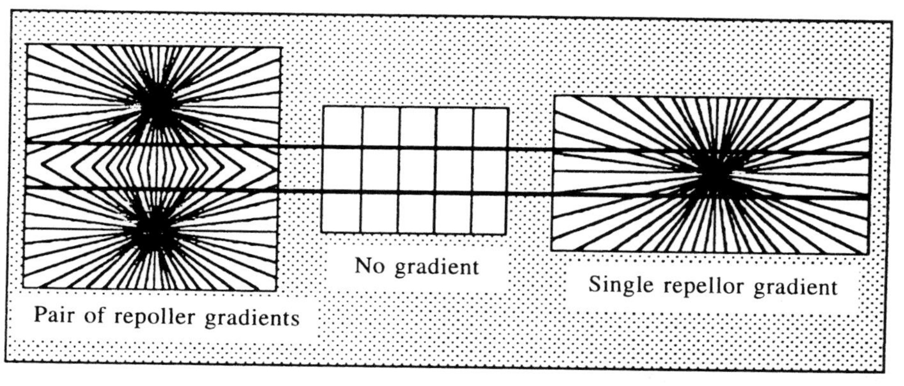

Fig. 4.24. Warping effects of two equidistant parallel lines by visual attractors

Even conscious intentional patterns of behavior are not exclusive to humans. A dog not only jumps, it jumps to catch prey, to greet its master, and so on. Intentionality in the sense of conscious goal-directedness is a property of more or less all animals. The question arises of how intentional behavior can be modeled by the complex system approach and how the model can be tested experimentally.

In this context, an intention is defined as an intended behavior pattern which may change the dynamical properties, such as stability, of intrinsic behavior pattern. Thus, psychologists can model the intrinsic dynamics of behavior patterns which may be changed by the dynamics of other, intended behavior patterns. Here we remind the reader of the intrinsic dynamics governing some patterns of behavior which can be modeled by nonequilibrium phase transitions and order parameters. Kelso, Haken, and others have analyzed the following simple examples: when persons are asked to move their fingers in parallel (Fig. 4.25a), they can easily perform this at low frequency. When the test persons are asked to increase the frequency of their finger movements, the fingers are suddenly moved in a symmetric and antiparallel fashion without conscious intention (Fig. 4.25b) [4.58].

In order to model this phase transition of behavior patterns, the frequency is interpreted as a control parameter, and the macroscopic variable describing the finger movement is the phase φ. The behavior can be modeled in an energy landscape relative to the changing phase. The landscape must be symmetric, as the left and right finger have equal functions. It must also be periodic in the phase angle (Fig. 4.26).

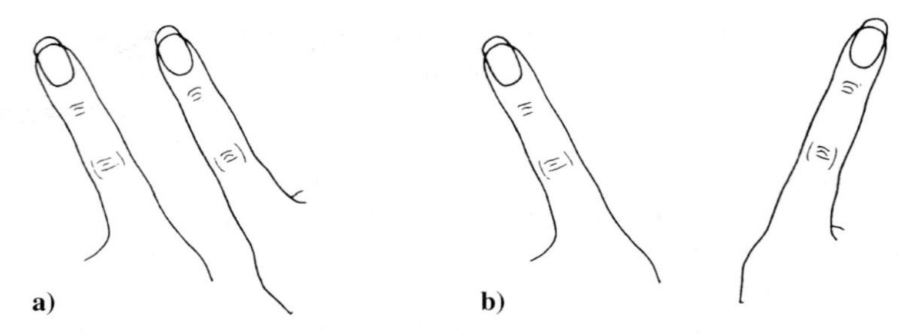

Fig. 4.25a,b. Two fingers moving in parallel (**a**) and antiparallel fashion (**b**) [4.58]

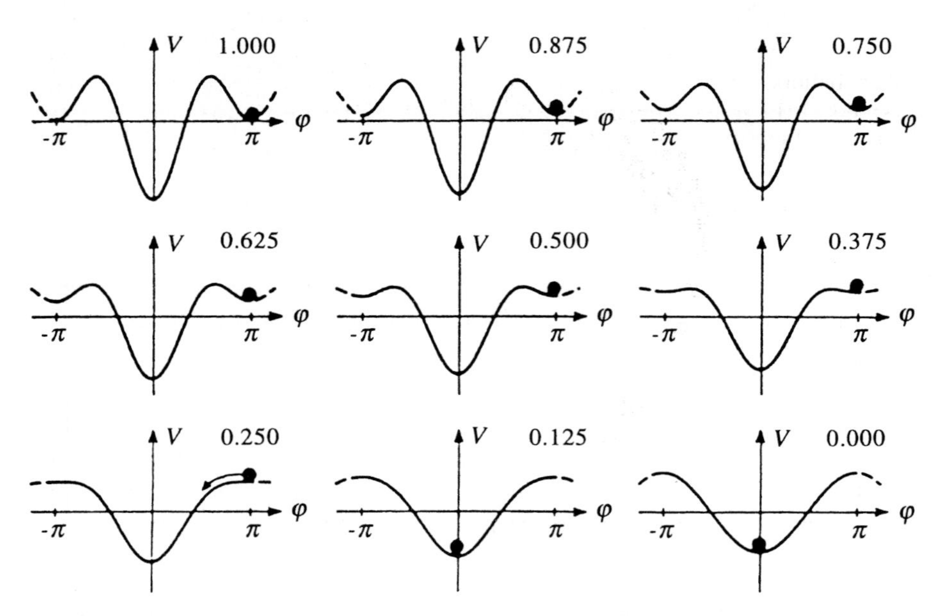

Fig. 4.26. Dynamics of moving fingers in an energy landscape with relative phase φ as order parameter

If the frequency increases, the landscape with its initially sharp valleys is deformed. In the beginning of slow movement the pattern is stable, corresponding to a stable phase at value π (Fig. 4.26a). Finally, the valley at π has disappeared, and a ball, initially in the valley at π, has run down to the deepest minimum, corresponding the symmetric movement of the fingers (Fig. 4.26c).

In some experiments, subjects were asked to switch intentionally between the two patterns of bimanual coordination. The duration of the transient corresponds to the switching time, which was measured. The stability of both patterns is measured by order parameter fluctuations. The relative phase dynamics is modeled by a nonlinear evolution equation.

Figure 4.27a visualizes the intrinsic dynamics according to the potential of this equation with two minima. The contribution of intentional information to the relative phase dynamics is shown by the potentials of Fig. 4.27b. The result of summing the intrinsic and the intentional dynamics to arrive at the full dynamics is shown by Fig. 4.27c. The ball in the landscape travels faster along the steeper slope at $\varphi = 0$ than $\varphi = 180$, corresponding to the empirically measured switching time. Obviously, an intention can change the intrinsic dynamics by destabilizing one pattern and stabilizing the other one. The intentional information is said to be a part of the pattern dynamics attracting the system toward the intended pattern. In this sense, intentional information defines an attractor in the same state space in which the intrinsic dynamics is modeled [4.59].

Intentionality and linguistic meaning are often proclaimed to be essential features of the human mind. Examples of intentional states are pains, tickles and itches, beliefs, fears, hopes, desires, perceptual experiences, experiences of acting, thoughts, feelings, etc., which are expressed by corresponding sentences like "I suffer from pain in the stomach", "I desire to get a car", "I believe in God", etc. Searle

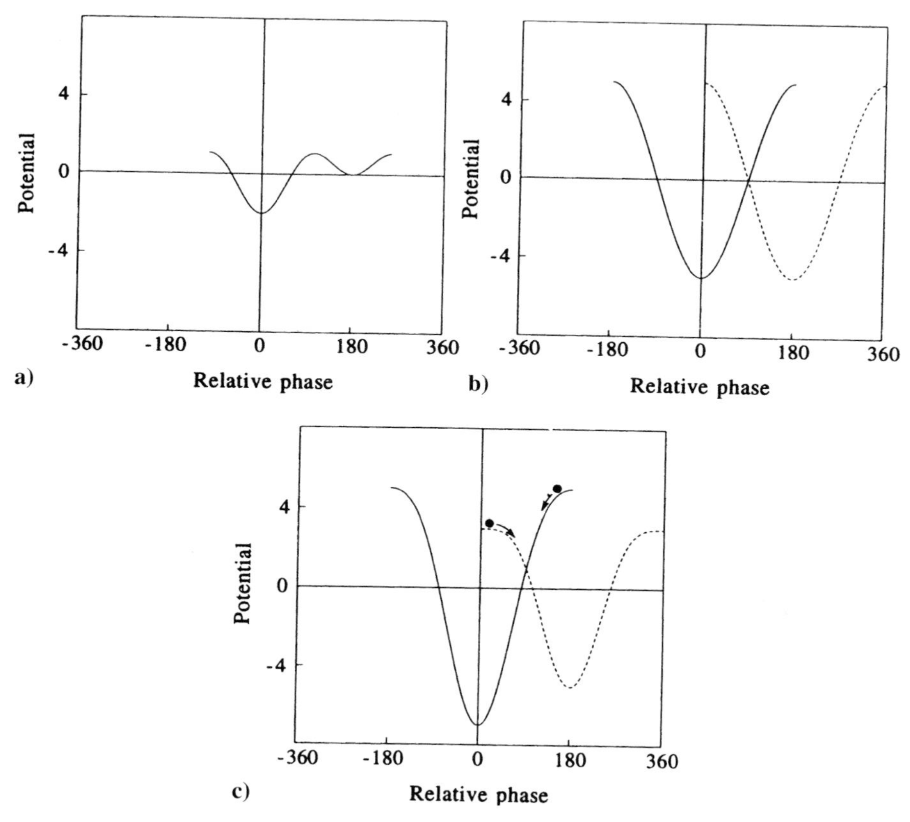

Fig. 4.27a–c. Relative phase dynamics with intentional information [4.59]

argues that mental states are as real as any other biological phenomena like lactation, photosynthesis, or digestion. He knows that mental states are macrostates of the biological brain which are caused by neurophysiological interactions between neurons on the microscopic scale. Thus, they cannot be identified with the neurophysiological states of single neurons.

The distinction between the micro- and macrostates of brains is illustrated by an analogy with, for instance, micro- and macrostates in liquids: the macrostate of liquidity cannot be reduced to single molecules or, in other words, single molecules cannot be liquid. In this sense, beliefs, desires, thirsts, and visual experiences are real causal features of the brain as much as the solidity of a table or the liquidity of water. Intentional states can themselves be caused by and realized in the structure of the brain. Searle declares that there is not, in addition, a metaphysical obstacle.

Nevertheless, he argues that no purely formal model will ever be sufficient by itself for intentionality because the formal properties are not by themselves constitutive of intentionality. His reason for holding this position is based on the thought experiment of the "Chinese room". A person who only understands English is locked in a room with a great store of Chinese symbols, and a set of complicated transformation rules, written in English, for performing operations on sequences of Chinese symbols. The person periodically receives sequences of Chinese symbols through a slot (Fig. 4.28). He applies the transformation rules in order to produce a further sequence of Chinese symbols which are put through the slot, again [4.60].

Fig. 4.28. Alice in the Chinese room

It is unknown to the person in the room that the store of sequences contains a large amount of information about certain topics, written in Chinese. The input sequences which are put through the slot are questions or comments on those topics.

The output sequences are reactions and comments to the received inputs. The transformation rules used are a formal program to simulate the conversational behavior of a native Chinese speaker. The person in the Chinese room applies the formal transformation rules correctly without understanding the sequences of Chinese symbols, which are meaningless to him.

Searle maintains that formal symbol manipulations by themselves do not have any intentionality, because they are quite meaningless to the user. Intentionality in this context is the feature of formal symbols like words, sentences, etc., that they refer to certain "meant" entities (semantic relation of the symbol) and to the user (pragmatic relation of the symbol). Searle asserts that this feature is intrinsic only for the mental states of the brain.

His arguments against "computer simulation" fail, if he restricts simulations to formal algorithms running on program-controlled Turing-type computers. But we have shown that a brain has the typical characteristics of a self-organizing and self-referential complex system which is quite different from a program-controlled computer (compare Chap. 6). Self-organization and self-referentiality of complex systems are not restricted to human or mammalian brains. They are only biochemical and neurophysiological realizations of particular complex structures which have been produced in biological evolution. Thus, in principle it cannot be excluded that these complex structures with their characteristic dynamics may be realized by quite different materials which may be produced via human technology. Consequently, as intentionality is made possible by the features of self-referentiality and self-organization, at least a partial simulation by complex models different from biological brains cannot be excluded in principle.

In traditional philosophies, intentionality is often founded on the so-called "self" of the human being, which is said to be able to refer to the world and to itself ("self-consciousness" as self-referentiality). But where is the self hidden in the brain? Traditional positions like kinds of platonism or spiritualism or materialism are even maintained by some modern researchers of the brain. For Sir John Eccles, for instance, the self seems to be a spiritual entity interacting with the brain, but completely different in nature [4.61]. But how should this hypothesis be defended or refuted? It is a mere postulate with high metaphysical costs, which one may believe or not believe.

Hypotheses must be criticizable, perhaps false, but fruitful for further research. Thus, the metaphysical price is too high. Ockham's razor from philosophy demands that we cut away superfluous hypotheses, remain economical with the postulation of metaphysical entities, and restrict hypotheses to the minimal number that seems indispensable for empirical research. The complex system approach is a mathematical research program of interdisciplinary models avoiding metaphysical dogmas. It may fail in the long run. But this strategy of modeling has been confirmed by an impressive number of successes in several sciences and technologies and, more important, it suggests some fruitful concepts for further empirical research. On the other hand, the traditional materialism which identifies mental states with neurophysiological processes in single neurons is simply false.

Nevertheless, in brain research, the question has arisen which part of the brain is the center of the "self". The cortex is the part of the brain which enables us to learn, to memorize, to think, and to create all the products of human culture and civilization. But if the cortex is mainly modeled as a complex associative memory store with certain learning procedures, then it is only a complex and highly sophisticated instrument which has been evolved in biological evolution for enhanced survival of the fittest.

Indeed, the cortex is the youngest part in the evolution of the human brain. There are some much older, but more primitive structures which also can be found in the brains of birds, reptiles, amphibians, and fishes. Some scientists assume that basic feelings like lust and pain and all the servo-mechanisms which were necessary to survive in a reptile's life are essentially realized in these early structures of the brain. This center would give the impulses for all kinds of activities, using the cortex only as huge and effective associative store. Thus, in this interpretation, the "self" is replaced by a little crocodile in the brain operating with some highly complex instruments like the cortex, in order to survive in a more and more complex environment [4.62]. Intentionality would be made possible by the cortex, but initiated by the basic instincts of the crocodile in the human brain.

The idea of crocodiles with highly effective neural instruments of survival seems to injure our vanity more than the popular Darwinistic motto of the last century that the ape is the ancestor of man. From a scientific point of view, of course, it should not be injured vanity which makes us criticize the concept of the "neural crocodile". The main objection is that our feelings have not rested at the level of a crocodile, but have developed during biological and cultural evolution, too.

Our feelings of lust and pain are rather complex, because they are influenced by the stimuli of a rather complex and sophisticated civilization which has been produced by human brains. Thus, there is a complex feedback which has shaped our feelings and desires from the crocodile until today. The history of literature, art, and psychology demonstrates that lust and pain have been highly sophisticated states of the human brain that are in permanent evolution. Thus, even the traditional concept of a human soul which is more or less sensitive still makes sense in the framework of complex system. But we must give up the traditional ideas of human mind and soul as strange substances controlling and interacting with the human body in a miraculous manner which cannot be conceived in principle.

4.5 Complexity and the Embodied Mind

The coordination of the complex cellular and organic interactions in an organism requires a special type of self-organizing control. This was made possible by the evolution of nervous systems that enabled organisms to adapt to changing living conditions and to learn bodily from experiences with their environments. We call this the emergence of the embodied mind [4.63]. The hierarchy of anatomical organization varies over different scales of magnitude, from molecular dimensions to those of the entire central nervous system (CNS). Research into these hierarchical levels concerns questions of (for example) how signals are integrated in dendrites, how neurons interact in a network, how networks interact in a system like that used in vision, how systems interact in the CNS, or how the CNS interacts with its environment. Each stratum can be characterized by some order parameters that determine its particular structure, which is caused by complex interactions of subelements with respect to the particular level of hierarchy.

At the micro level of the brain, there are (massively) many-body problems which need a reduction strategy to cope with the complexity. In the case of EEG pictures, a complex system of electrodes measures local states (electric potentials) of the brain. The whole state of a patient's brain at the micro level is represented by local time series. In the case of, say, petit mal epilepsy, these are characterized by typical cyclic peaks. The microscopic states determine the macroscopic electric field patterns during a cyclic period. Mathematically, the macroscopic patterns can be determined by spatial modes and order parameters – the amplitude of the field waves. In the corresponding phase space, they determine a chaotic attractor that characterizes petit mal epilepsy.

Neural self-organization at the cellular and subcellular level is determined by information processing in and between neurons. Chemical transmitters can effect neural information processing using direct and indirect mechanisms of great plasticity. The long-term potentiation (LTP) of synaptic interactions is an extremely interesting topic of recent brain research. LTP seems to play an essential role in the neural self-organization of cognitive features such as memory and learning. It is assumed that the information is stored in the synaptic connections of neural cell assemblies with typical macroscopic patterns.

However, while an individual neuron cannot see or reason or remember, brains can. Vision, reasoning, and memory are understood as being higher-level functions. Scientists who prefer a bottom-up strategy recommend that higher-level functions of the brain can be neither addressed nor understood until the particular properties of each neuron and synapse are explored and explained. An important insight gained from the complex system approach is that emergent effects of the whole system are synergetic system effects that cannot be reduced to single elements. They are due to nonlinear interactions. Therefore, the whole is more than the (linear) sum of its parts. Thus, from a methodological point of view, a purely bottom-up-strategy of exploring brain functions must fail. On the other hand, the advocates of a purely top-down strategy proclaiming that cognition is completely independent of the nervous system are caught in the old Cartesian dilemma: 'how does the ghost drive the machine?'

We can now distinguish several degrees of complexity in the CNS. These scales involve molecules, membranes, synapses, neurons, nuclei, circuits, networks, layers, maps, sensory systems, and the entire nervous system. Research into these hierarchical levels concerns questions of how signals are integrated in dendrites, how neurons interact in a network, how networks interact in a system like that of vision, how systems interact in the CNS, or how the CNS interacts with its environment. Each stratum can be characterized by some order parameters that determine its particular structures, which are caused by complex interactions of subelements at the particular level of hierarchy. Beginning at the bottom, we can distinguish the order associated with ion movement, channel configurations, action potentials, potential waves, locomotion, perception, behavior, feeling, and reasoning.

The different abilities of the brain require massively parallel information processing in a complex hierarchy of neural structures and areas. We have complex models of information processing in the visual and motor systems. The dynamics of the emotional system even interact in a nonlinear feedback manner with several structures of the human brain. These complex systems produce neural maps of cell assemblies. The self-organization of somatosensoric maps is well-known in the visual and motor cortices. They can be enlarged and changed by learning procedures, like that used when training an ape's hand.

PET (positron emission tomography) pictures show macroscopic patterns of neurochemical metabolic cell assemblies in different regions of the brain that are correlated with cognitive abilities and conscious states, such as looking, hearing, speaking, or thinking. Patterns formed by neural cell assemblies are even correlated with complex processes of psychic states [4.64]. Perturbations of metabolic cellular interactions (e.g., caused by cocaine intake) can lead to nonlinear effects that cause complex changes in behavior (e.g., addictions to drugs). These correlations between neural cell assemblies and order parameters (attractors) of cognitive and conscious states demonstrate the connection between neurobiology and cognitive psychology observed in recent research, depending on the precisions of the measuring instruments and procedures employed.

Many questions are still unanswered. We can only observe that someone is thinking and feeling, not what they are thinking and feeling. Also, there is no unique substance called consciousness, but complex macrostates of the brain that pay different degrees of attention to sensor, motor, or other types of function. Consciousness means not only that we look, listen, speak, hear, feel, think, etc., but also that we recognize when we are performing these cognitive processes. Our self is considered to be an order parameter of a state, which emerges from a recursive process of multiple self-reflections, self-monitoring, and supervising our conscious actions. Self-reflection is made possible by so-called mirror neurons (e.g., in the Broca area), which allow primates (especially humans) to imitate and simulate interesting behavior exhibited by their companions. Therefore, they can learn to see things from their own and their companion's perspectives, allowing them to understand their intentions and to empathize with them. The emergence of subjectivity is well understood neuropsychologically.

The brain observes, maps, and monitors both the external world and the internal states of the organism, especially its emotional states. To "feel" means to have an awareness of one's emotional states, which are mainly caused by the limbic system. In neuromedicine, the "Theory of Mind" (ToM) even analyzes the neural correlates of social feeling, which are situated in special areas of the neocortex [4.65]. Some people, such as those suffering from Alzheimer's disease, lose their feelings of empathy and social responsibility because the associated neural areas have been destroyed. Therefore, our moral reasoning and decision-making has a clear basis in brain dynamics.

From a neuropsychological point of view, the old philosophical problem of "qualia" is also solvable. Qualia are properties that are consciously experienced by a person. In a thought experiment, a neurobiologist is assumed to be caught in a black and white room. Theoretically, she knows everything about the processing of colors by neurons. However, she has never had a chance to experience colors. Therefore, exact knowledge says nothing about the quality of conscious experience. Qualia in this sense emerge through the interactions of self-conscious organisms bodily with their environment, which can be explained via nonlinear dynamics of complex systems. Therefore, we can explain the dynamics of subjective feelings and experiences, but, of course, the actual feeling is an individual experience. In medicine, the dynamics of a certain pain can often be completely explained by a physician, although the actual feeling of pain is an individual experience for the patient [4.66].

In order to model the brain and its complex abilities, it is adequate to distinguish the following categories. In neuron-level models, studies concentrate on the dynamic and adaptive properties of each nerve cell or neuron, in order to describe the neuron as a unit. In network-level models, identical neurons are interconnected, resulting in basic system functions. In nervous-system-level models, several networks are combined to demonstrate some of the more complex functions of sensory perception, motor functions, stability control, etc. In mental-operation-level models, the basic processes of cognition, thinking, problem-solving, etc., are described.

In the complex systems approach, the microscopic level of interacting neurons should be modeled by coupled differential equations that model the transmission of nerve impulses by each neuron. The Hodgekin–Huxley equation is an example of a nonlinear diffusion reaction equation with an exact solution of a traveling wave, which provides a precise prediction of the speed and shape of the nerve impulse of electric voltage. In general, nerve impulses emerge as new dynamical entities like ring waves in BZ reactions or fluid patterns in non-equilibrium dynamics. In short, they are the "atoms" of complex neural dynamics. At the macroscopic level, they generate a cell assembly whose macrodynamics are dominated by order parameters. For example, a synchronously firing cell assembly represents a visual perception of a plant which is not only the sum of its perceived pixels, but is characterized by some typical macroscopic features like form, background or foreground. At the next level, cell assemblies of several perceptions interact in a complex scenario. In this case, each cell assembly is a firing unit, generating a cell assembly of cell assemblies whose macrodynamics are characterized by some order parameters. The order parameters may represent similar properties of the perceived objects.

In this way, we obtain a hierarchy of emerging levels of cognition, starting with the microdynamics of firing neurons. The dynamics of each level are assumed to be characterized by differential equations with order parameters. For example, at the first level of macrodynamics, order parameters characterize a visual perception. At the following level, the observer becomes conscious of the perception. Then the cell assembly of perception is connected with the neural area that is responsible for states of consciousness. In a next step, a conscious perception may be the goal for planning activities. In this case, cell assemblies of cell assemblies are connected with neural areas in the planning cortex, and so on. They are represented by coupled nonlinear equations with firing rates of corresponding cell assemblies. Even high-level concepts like self-consciousness can be explained by self-reflections of self-reflections, connected with a personal memory which is represented in the corresponding cell assemblies of the brain. Brain states emerge, persist for a small fraction of time, then disappear and are replaced by other states. It is the flexibility and creativeness of this process that makes a brain so successful at enabling animals to adapt to rapidly changing and unpredictable environments.

5 Complex Systems and the Evolution of Computability

The evolution of complexity in nature and society can be understood as the evolution of computational systems. In the beginning of modern times, Leibniz already had the idea that the hierarchy of natural systems from stones and plants up to animals and humans corresponded to natural automata with increasing degrees of complexity (Sect. 5.1). The present theory of computability enables us to distinguish complexity classes of problems, meaning the order of corresponding functions describing the computational time of their algorithms or computational programs. But we can also consider the size of a computer program when defining the algorithmic complexity of symbolic patterns (Sect. 5.2).

Information dynamics in complex systems are analyzed by Shannon's concept of information entropy and Kolmogorov-Sinai entropy. Thus, the information flow in complex systems with stable, oscillating, chaotic, or random dynamics can be distinguished by well-defined methods. The degree of complexity of $1/f^b$ noise can be linked to attractors in nonlinear dynamics (Sect. 5.3). In general, any stochastic process can be classified according to the degree of complexity of the probabilistic attractor. This offers deep insights into the power laws of complex systems, indicating the self-organization and emergence of order in nature and society (Sect. 5.4). Further on, we ask if more efficient information processing can be expected from quantum computers and quantum complexity theory. Is matter nothing more than "condensed" quantum information with different degrees of complexity (Sect. 5.5)? Leibniz's idea of natural automata has been made mathematically precise by John von Neumann's concept of cellular automata. Pattern formation in complex systems can be analyzed in the framework of cellular automata. Even chaos and randomness can be generated by simple rules of cellular automata, as demonstrated by Stephen Wolfram's computer experiments (Sect. 5.6).

5.1 Leibniz and Mathesis Universalis

One of the most speculative applications of complex systems is the evolution of artificial intelligence (AI) [5.1]. In the tradition of classical AI, the brain has been understood as computer hardware of the most advanced machinery, while the mind is the corresponding software program with deterministic algorithms. Even knowledge-based expert systems are conceived by algorithmical representations of highly developed AI programming languages. But theoretical results of mathematical logic

(Church, Turing, Gödel, etc.) and practical problems of programming limit the mechanization of thought in the framework of classical AI.

A theory of the "cerebral computer" as a product of natural evolution has been suggested to model the nature of the brain and its mental states by the non-linear dynamics ("self-organization") of complex neural networks. The question arises of wether the insight into their dynamics delivers the "blue-prints" of a new revolutionary technology which will pursue the natural evolution of brain and mind. Actually, the development of human knowledge and knowledge technology seems to be a kind of technical evolution which has led to technical innovations like mutations in biological evolution.

The first level was realized by simple tools like the hammer, the lever, and so on. On the next level, machines using force and energy were invented. Today program-controlled computers and information-processing automata have become tools of everyday life. Computer scientists distinguish several generations of hardware and software in the historical development of their machines. In artifical-intelligence research one speaks of the "second computer age", meaning the transition from number-processing machines to knowledge-processing systems such as expert systems, which are said to simulate human experts, at least partially [5.2].

The early historical roots of computer science stem back to the age of classical mechanics. The mechanization of thoughts begins with the invention of mechanical devices for performing elementary arithmetic operations automatically. A mechanical calculation machine executes serial instructions step by step. Thus, its dynamics is determined by mechanical mono-causality, differing essentially from the parallelism and self-organization of complex systems. In general, the traditional design of a mechanical calculation machine contains the following devices.

First, there is an input mechanism by which a number is entered into the machine. A selector mechanism selects and provides the mechanical motion to cause the addition or subtraction of values on the register mechanism. The register mechanism is necessary to indicate the value of a number stored within the machine, technically realized by a series of wheels or disks. If a carry is generated because one of the digits in the result register advances from 9 to 0, then that carry must be propagated by a carry mechanism to the next digit or even across the entire result register. A control mechanism ensures that all gears are properly positioned at the end of each addition cycle to avoid false results or jamming the machine. An erasing mechanism has to reset the register mechanism to store a value of zero.

Wilhelm Schickard (1592–1635), professor of Hebrew, oriental languages, mathematics, astronomy, and geography, is presumed to be the first inventor of a mechanical calculating machine for the first four rules of arithmetic. The adding and subtracting part of his machine is realized by a gear drive with an automatic carry mechanism. The multiplication and division mechanism is based on Napier's multiplication tables. Blaise Pascal (1623–1662), the brilliant French mathematician and philosopher, invented an adding and subtracting machine with a sophisticated carry mechanism which in principle is still realized in our hodometers of today [5.3].

But it was Leibniz' mechanical calculating machine for the first four rules of arithmetic which contained each of the mechanical devices from the input, selector, and register mechanism to the carry, control, and erasing mechanism. The Leibniz machine became the prototype of a hand calculating machine. If we abstract from the technical details and particular mechanical constructions of Leibniz' machine, then we get a model of an ideal calculating machine which in principle is able to calculate all computable functions of natural numbers.

Figure 5.1 is a scheme of this ideal machine with a crank C and three number stores SM, TM, RM [5.4]. Natural numbers can be entered in the set-up (input) mechanism SM by the set-up handles SH. If crank C is turned to the right, then the contents of SM are added to the contents of the result mechanism RM, and the contents of the turning mechanism TM are raised by 1. A turn to the left with crank C subtracts the contents of SM from the contents of RM and diminishes the contents of TM by 1.

Addition means the following. At the beginning of the calculation, the erasing procedure is implemented by setting TM and RM to zero. Then the first number is set up in SM by SH. A turn to the right of crank C transports this number into RM. In other words, the number is added to the zero 0 in RM. Now the second number is set up in the SM and added to the contents of RM by a turn to the right. The sum of both numbers can be read in the RM. After turning the crank twice to the right, the TM shows 2. Multiplication only means a repeated addition of the same number. The product b · a results from adding the number a to itself b times.

Leibniz even designed a mechanical calculating machine for the binary number system with only two digits 0 and 1, which he discovered some years earlier. He described a mechanism for translating a decimal number into the corresponding binary number and vice versa. As modern electronic computers only have two states 1 (electronic impulse) and 0 (no electronic impulse), Leibniz truly became one of the pioneers of computer science [5.5].

Leibniz' historical machines suffered from many technical problems, because the materials and technical skills then available were not up to the demands. Nevertheless, his design is part of a general research program for a *mathesis universalis* intended to simulate human thinking by calculation procedures ("algorithms") and to implement them on mechanical calculating machines. Leibniz proclaimed two basic disciplines of his *mathesis universalis.*

An *ars iudicandi* should allow every scientific problem to be decided by an appropriate arithmetic algorithm after its codification into numeric symbols. An *ars inveniendi* should allow scientists to seek and enumerate possible solutions of sci-

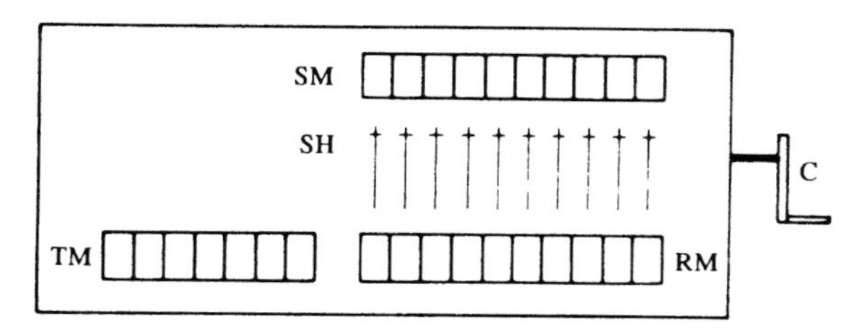

Fig. 5.1. Hand calculating machine

entific problems. Leibniz' *mathesis universalis* seems already to foreshadow the famous Hilbert program in our century with its demands for formalization and axiomatization of mathematical knowledge. Actually, Leibniz developed some procedures to formalize and codify languages. He was deeply convinced that there are universal algorithms to decide all problems in the world by mechanical devices [5.6].

Consequently, he proposed that natural systems like cells, plants, animals, and even humans are more or less complex automata. In his *Discourse on Metaphysics* (1686), Leibniz underlines that the mechanistic description and causal explanation of living systems is not in contradiction to a teleological consideration that has great heuristic value in science (§22). In his *Monadology* (§18) he introduced an individual substance (*monade*) as an elementary automaton (*automates incorporels*) which is characterized by a (continuous) series of states ("perceptions"). The elementary automata constitute aggregations of more or less complexity which are characterized by different correlations and which can be interpreted as composite automata. In his *Theodicée* (§200), Leibniz discusses the hierarchical structure and subordination in living systems:

> ... the connection and order of things brings it about that the body of every animal and of every plant is comprised of other animals and of other plants, or of other living organic beings: consequently there is subordination, and one body, one substance, serves the other.

The unity of a living system is guaranteed by its form of organization, which Leibniz, adapting an idea of Aristotle, called "entelechy". But Leibniz only used an old metaphysical term in order to introduce his own new concept. For Leibniz a system can only be more or less unified in the sense of higher or lower degrees of subordination and hierarchy. An aggregation with the same correlation between all its substances has no hierarchical order and is less structured than a primitive cellular organism, while in plants, animals, and humans we can observe a growing degree of subordination.

For Leibniz the teleological terminology has a heuristical value, although in principle nature can be explained by mechanistic causes. But it is a fundamental error and misunderstanding when disciples of vitalism refer to Leibniz. The main difference is that for Leibniz no new principle or *force vitale* is necessary to explain living systems. At a certain degree of complexity, it is only heuristically suitable to describe natural systems in the terminology of teleology. But, unlike natural systems, artificial mechanical automata are constructed by humans in finite steps. Only an infinite analysis could demonstrate the complexity of a natural automaton, which is correlated with each individual automaton ("substance") in the world. Obviously, Leibniz designed a theory of complex systems, but still in the framework of classical mechanics and the belief in decidable universal algorithms.

In the 19th century it was the English mathematician and economist Charles Babbage who not only constructed the first program-controlled calculation machine (the "analytical engine") but also studied its economic and social consequences [5.7]. A forerunner of his famous book *On the economy of machinery and manufactures* (1841) was Adam Smith's idea of economic laws, which paralleled Newton's mechanical laws (compare Sect. 6.2). In his book *The Wealth of Nations*,

Smith described the industrial production of pins as an algorithmic procedure and anticipated Henry Ford's idea of program-controlled mass production in industry.

5.2 Computability and Algorithmic Complexity

The modern formal logic of Frege and Russell and the mathematical proof theory of Hilbert and Gödel have been mainly influenced by Leibniz' program of *mathesis universalis*. The hand calculating machine (Fig. 5.1) which was abstracted from the Leibniz machine in Sect. 5.1 can easily be generalized to Marvin Minsky's so-called register machine [5.8]. It allows the general concept of computability to be defined in modern computer science.

A hand calculating machine had only two registers TM and RM, and only rather small natural numbers can be input. An ideal register machine has a finite number of registers which can store any finite number of a desired quantity. The registers are denoted by natural numbers $i = 1, 2, 3, \ldots$. The contents of register i are denoted by $\langle i \rangle$. As an example, the device $\langle 4 \rangle$:=1 means that the content of the register with number 4 is 1. The register is empty if it has the content 0.

In the hand calculating machine an addition or subtraction was realized only for the two registers ⟨SM⟩ and ⟨RM⟩, with ⟨SM⟩+⟨RM⟩ or ⟨RM⟩ − ⟨SM⟩ going into the register RM. In a register machine the result of subtraction $\langle i \rangle - \langle j \rangle$ should be 0 if $\langle j \rangle$ is greater than $\langle i \rangle$. This modified subtraction is denoted by $\langle i \rangle \dot{-} \langle j \rangle$. In general, the program of an ideal register machine is defined using the following elementary procedures as building blocks:

1) Add 1 to $\langle i \rangle$ and put the result into register i, in short: $\langle i \rangle := \langle i \rangle + 1$
2) Subtract 1 from $\langle i \rangle$ and put the result into register i, in short: $\langle i \rangle := \langle i \rangle \dot{-} 1$

These two elementary procedures can be composed using the following concepts:

3) If P and Q are well-defined programs, then the chain $P \to Q$ is a well-defined program. $P \to Q$ means that a machine has to execute program Q after program P.
4) The iteration of a program, which is necessary for multiplication, for instance, as iterated addition is controlled by the question of whether a certain register is empty.

A diagram illustrates this feedback:

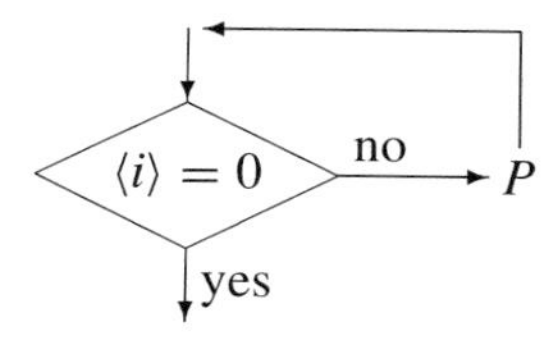

If P is a well-defined program, then execute P until the content of the register with number i is zero.

Each elementary operation (1) and (2) of a program is counted as a step of computation. A simple example is the following addition program:

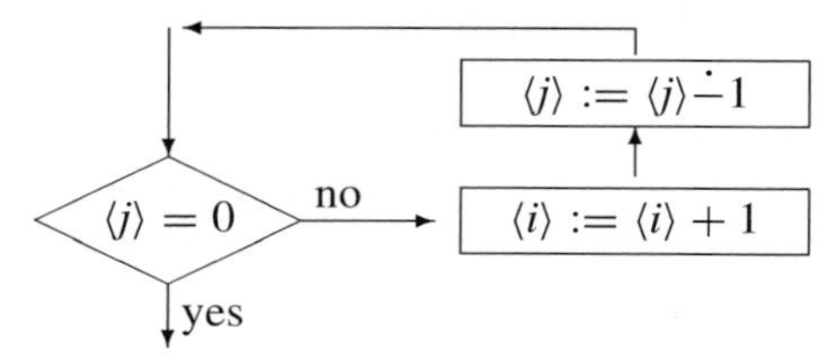

Each state of the machine is illustrated by the following matrix, which incrementally adds the content y of register $\langle j\rangle$ to the content x of register $\langle i\rangle$ and simultaneously decrements the content of $\langle j\rangle$ to zero. The result $x + y$ of the addition is shown in register $\langle j\rangle$:

$$\begin{array}{cc} \langle i\rangle & \langle j\rangle \\ x & y \\ x+1 & y \dot{-} 1 \\ \vdots & \vdots \\ x+y & y \dot{-} y \end{array}$$

A register machine with program F is defined to compute a function f with n arguments if for arbitrary arguments $x_1, \ldots, x_n$ in the registers $1, \ldots, n$ (and zero in all other ones) the program F is executed and stops after a finite number of steps with the arguments of the function in the registers $1, \ldots, n$ and the function value $f(x_1, \ldots, x_n)$ in register $n + 1$:

The program

$$\begin{array}{c} \langle 1\rangle := x_1;\ \ldots;\ \langle n\rangle := x_n \\ \downarrow \\ F \\ \downarrow \\ \langle n+1\rangle := f(x_1, \ldots, x_n) \end{array}$$

works according to a corresponding matrix. A function f is called computable by a register machine RM (RM-computable) if there is a program F computing f.

The number of steps which a certain program F needs to compute a function f is determined by the program and depends on the arguments of the function. The complexity of program F is measured by a function $s_F(x_1, \ldots, x_n)$ counting the steps of computation according to program F. For example, the matrix of the addition program for $x + y$ shows that y elementary steps of adding 1 and y elementary steps of subtracting 1 are necessary. Thus, $s_F(x, y) = 2y$. As an RM-computable function f may be computed by several programs, a function g is called the step counting function of f if there is a program F to compute f with $g(x_1, \ldots, x_n) = s_F(x_1, \ldots, x_n)$ for all arguments $x_1, \ldots, x_n$. The complexity of a function is defined

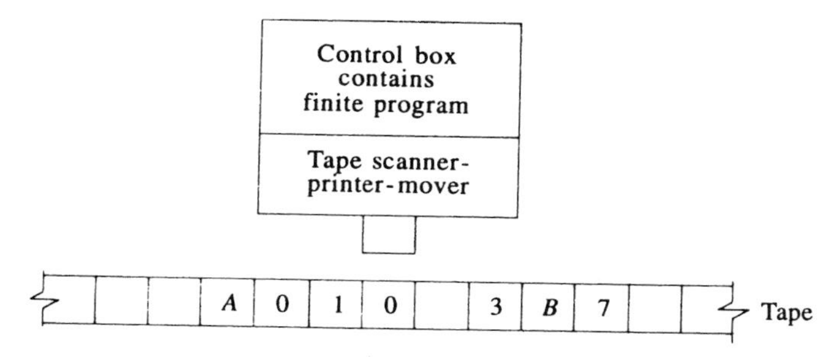

Fig. 5.2a. Turing machine with one tape

as the complexity of the best program computing the function with the least number of steps.

Obviously, Minsky's register machine is an intuitive generalization of a hand calculating machine à la Leibniz. But, historically, some other, but equivalent formulations of machines were at first introduced independently by Alan Turing and Emil Post in 1936. A Turing machine (Fig. 5.2a) can carry out any effective procedure provided it is correctly programmed [5.9]. It consists of

a) a control box in which a finite program is placed,
b) a potentially infinite tape, divided lengthwise into squares,
c) a device for scanning, or printing on one square of the tape at a time, and for moving along the tape or stopping, all under the command of the control box.

If the symbols used by a Turing machine are restricted to a stroke | and a blank $*$, then an RM-computable function can be proved to be computable by a Turing machine and vice versa. We must remember that every natural number x can be represented by a sequence of x strokes (for instance 3 by |||), each stroke on a square of the Turing tape. The blank $*$ is used to denote that the square is empty (or the corresponding number is zero). In particular, a blank is necessary to separate sequences of strokes representing numbers. Thus, a Turing machine computing a function f with arguments $x_1, \ldots, x_n$ starts with tape $\cdots * x_1 * x_2 * \cdots * x_n * \cdots$ and stops with $\cdots * x_1 * x_2 * \cdots x_n * f(x_1, \ldots x_n) * \cdots$ on the tape.

From a logical point of view, a general purpose computer – as constructed by associates of John von Neumann in America and independently by Konrad Zuse in Germany – is a technical realization of a universal Turing machine which can simulate any kind of Turing program. Analogously, we can define a universal register machine which can execute any kind of register program. Actually, the general design of a von-Neumann computer consists of a central processor (program controller), a memory, an arithmetic unit, and input-output devices. It operates step by step in a largely serial fashion. A present-day computer à la von Neumann is really a generalized Turing machine. The efficiency of a Turing machine can be increased by the introduction of several tapes, which are not necessarily one-dimensional, each acted on by one or more heads, but reporting back to a single control box which coordinates all the activities of the machine (Fig. 5.2b) [5.10]. Thus, every

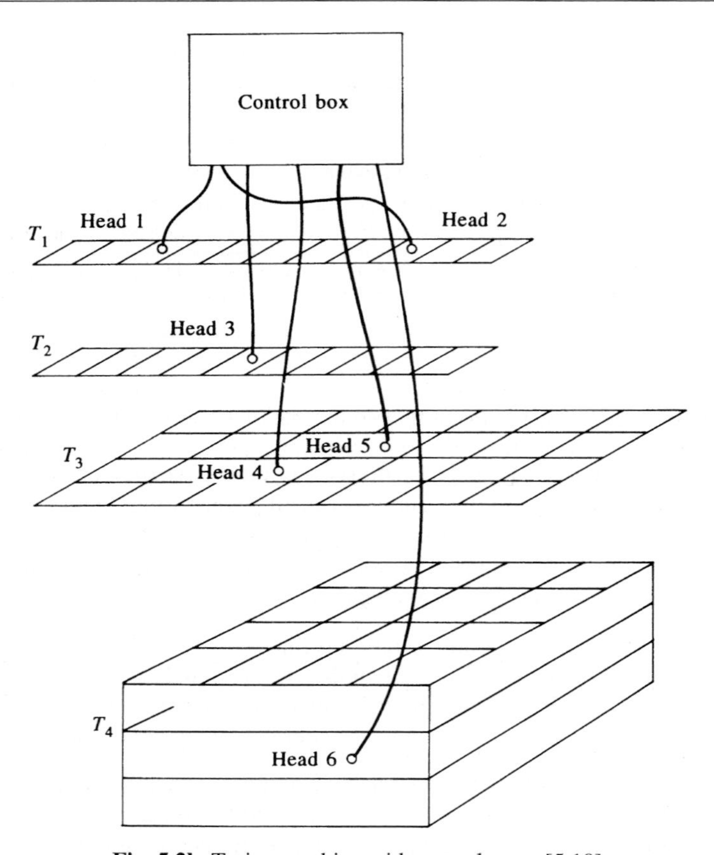

Fig. 5.2b. Turing machine with several tapes [5.10]

computation of such a more effective machine can be done by an ordinary Turing machine. Concerning the complex system approach, even a Turing machine with several multidimensional tapes remains a sequential program-controlled computer, differing essentially from self-organizing systems like neural networks.

Besides Turing- and register machines, there are many other mathematically equivalent procedures for defining computable functions. Recursive functions are defined by procedures of functional substitution and iteration, beginning with some elementary functions (for instance, the successor function $n(x) = x + 1$) which are obviously computable. All these definitions of computability by Turing machines, register machines, recursive functions, etc., can be proved to be mathematically equivalent. Obviously, each of these precise concepts defines a procedure which is intuitively effective.

Thus, Alonzo Church postulated his famous thesis that the informal intuitive notion of an effective procedure is identical with one of these equivalent precise concepts, such as that of a Turing machine. Church's thesis cannot be proved, of course, because mathematically precise concepts are compared with an informal intuitive notion. Nevertheless, the mathematical equivalence of several precise concepts of computability which are intuitively effective confirms Church's thesis. Consequently, we can speak about computability, effectiveness, and computable functions without referring to particular effective procedures ("algorithms") like Turing machines, register machines, recursive functions, etc. According to Church's thesis, we may in particular say that every computational procedure (algorithm) can be calculated by a Turing machine. So every recursive function, as a kind of machine program, can be calculated by a general purpose computer [5.11].

Now we are able to define effective procedures of decision and enumerability, which were already demanded by Leibniz' program of a *mathesis universalis*. The characteristic function f_M of a subset M of natural numbers is defined as $f_M(x) = 1$ if x is an element of M, and as $f_M(x) = 0$ otherwise. Thus, a set M is defined as effectively decidable if its characteristic function saying whether or not a number belongs to M is effectively computable (or recursive).

A set M is defined as effectively (recursively) enumerable if there exists an effective (recursive) procedure f for generating its elements, one after another (formally $f(1) = x_1, f(2) = x_2, \ldots$ for all elements $x_1, x_2, \ldots$ from M). It can easily be proved that every recursive (decidable) set is recursively enumerable. But there are recursively enumerable sets which are not decidable. These are the first hints that there are limits to Leibniz' originally optimistic program, based on a belief in universal decision procedures.

Concerning natural and artificial intelligence, the paradigm of effective computability implies that mind is represented by program-controlled machines, and mental structures refer to symbolic data structures, while mental processes implement algorithms. Historically the hard core of AI was established during the Dartmouth Conference in 1956 when leading researchers such as John McCarthy, Alan Newell, Herbert Simon, and others from different disciplines, formed the new scientific community of AI. They all were inspired by Turing's question "Can machines think?" in his famous article "Computing machinery and intelligence" (1950).

In the tradition of Leibniz' *mathesis universalis* one could believe that human thinking could be formalized with a kind of universal calculus. In a modern version one could assume that human thinking could be represented by some powerful formal programming language. In any case, formulas are sequences of symbols which can be codified by natural numbers. Then assertions about objects would correspond to functions over numbers, conclusions would follow from some kind of effective numerical procedure, and so on. Actually, the machine language of a modern computer consists of sequences of numbers, codifying every state and procedure of the machine. Thus, the operations of a computer can be described by an effective or recursive numerical procedure.

If human thinking can be represented by a recursive function, then by Church's thesis it can be represented by a Turing program which can be computed by a universal Turing machine. Thus, human thinking could be simulated by a general purpose computer and, in this sense, Turing's question must be answered with "yes". The premise that human thinking can be codified and represented by recursive procedures is, of course, doubtful. Even processes of mathematical thinking can be more complex than recursive functions. Recursiveness or Turing computability is only a theoretical limit of computability according to Church's thesis.

In the following we want to consider problems with a degree of complexity both below and beyond this limit. Below this limit there are many practical problems concerning certain limitations on how much the speed of an algorithm can be increased. Especially among mathematical problems there are some classes of problems that are intrinsically more difficult to solve algorithmically than others. Thus, there are degrees of computability for Turing machines which are made precise in complexity theory in computer science [5.12].

Complexity classes of problems (or corresponding functions) can be characterized by complexity degrees, which give the order of functions describing the computational time (or number of elementary computational steps) of algorithms (or computational programs) depending on the length of their inputs. The length of inputs may be measured by the number of decimal digits. According to the machine language of a computer it is convenient to codify decimal numbers into their binary codes with only binary numbers 0 and 1 and to define their length by the number of binary digits. For instance, 3 has the binary code 11 with the length 2. A function f has linear computational time if the computational time of f is not greater than $c \cdot n$ for all inputs with length n and a constant c.

The addition of two (binary) numbers has obviously only linear computational time. For instance, the task 3+7=10 corresponds to the binary calculation

$$\begin{array}{r} 0\,1\,1 \\ 1\,1\,1 \\ \hline 1\,0\,1\,0 \end{array}$$

which needs 5 elementary computational steps of adding two binary digits (including carrying). We remind the reader that the elementary steps of adding binary digits are 0+0=0, 0+1=1, 1+0=1, 1+1=10, and carry. It is convenient to assume that the two numbers which should be added have equal length. Otherwise we simply start the shorter one with a series of zeros, for instance 111 and 011 instead of 11. In general, if the length of the particular pair of numbers which should be added is n, the length of a number is $\frac{n}{2}$, and thus, we need no more than $\frac{n}{2} + \frac{n}{2} = n$ elementary steps of computation including carrying.

A function f has quadratic computational time if the computational time of f is not greater than $c \cdot n^2$ for all inputs with length n and a constant c.

A simple example of quadratic computational time is the multiplication of two (binary) numbers. For instance, the task $7 \cdot 3 = 21$ corresponds to the binary calculation:

$$\begin{array}{r} 1\,1\,1 \cdot 0\,1\,1 \\ \hline 0\,0\,0 \\ 1\,1\,1 \\ 1\,1\,1 \\ \hline 1\,0\,1\,0\,1 \end{array}$$

According to former conventions, we have $n = 6$. The number of elementary binary multiplications is $\frac{n}{2} \cdot \frac{n}{2} = \frac{n^2}{4}$. Including carrying, the number of elementary binary additions is $\frac{n}{2} \cdot \frac{n}{2} - \frac{n}{2} = \frac{n^2}{4} - \frac{n}{2}$. In all, we get $\frac{n^2}{4} + \frac{n^2}{4} - \frac{n}{2} = \frac{n^2}{2} - \frac{n}{2}$, which is smaller than $\frac{n^2}{2}$.

A function f has polynomial computational time if the computational time of f is not greater than $c \cdot n^k$, which is assumed to be the leading term of a polynomial $p(n)$. A function f has exponential computational time if the computational time of f is not greater than $c \cdot 2^{p(n)}$. Many practical and theoretical problems belong to the complexity class P of all functions which can be computed by a deterministic Turing machine in polynomial time.

In the history of mathematics, there have been some nice problems of graph theory to illustrate the basic concepts of complexity theory [5.13]. In 1736, the famous mathematician Leonhard Euler (1707–1783) solved one of the first problems of graph theory. In the city of Königsberg, the capital of eastern Prussia, the so-called old and new river Pregel are joined in the river Pregel. In the 18th century, there were seven bridges connecting the southern s, northern n, and eastern e regions with the island i (Fig. 5.3a). Is there a route which crosses each bridge only once and returns to the starting point?

Euler reduced the problem to graph theory. The regions n, s, i, e are replaced by vertices of a graph, and the bridges between two regions by edges between the corresponding vertices (Fig. 5.3b).

In the language of graph theory, Euler's problem is whether for every vertex there is a route (an "Euler circuit") passing each edge exactly once, returning finally to the starting point. For arbitrary graphs Euler proved that an Euler circuit exists if and only if each vertex has an even number of edges (the "Euler condition"). As the graph of Fig. 5.3b does not satisfy this condition, there cannot be a solution of Euler's problem in this case. In general, there is an algorithm testing an arbitrary graph by Euler's condition if it is an Euler circuit. The input of the algorithm consists of the set V of all vertices $1, \ldots, n$ and the set E of all edges, which is a subset of the set with all pairs of vertices. The computational time of this algorithm depends linearly on the size of the graph, which is defined by the sum of the numbers of vertices and edges.

In 1859, the mathematician William Hamilton (1805–1865) introduced a rather similar problem that is much more complicated than Euler's problem. Hamilton

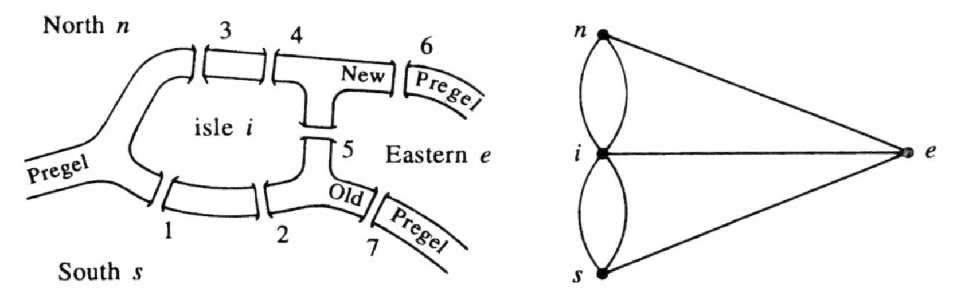

Fig. 5.3a,b. Euler's Königsberg river problem (**a**). Graph of Euler's river problem (**b**)

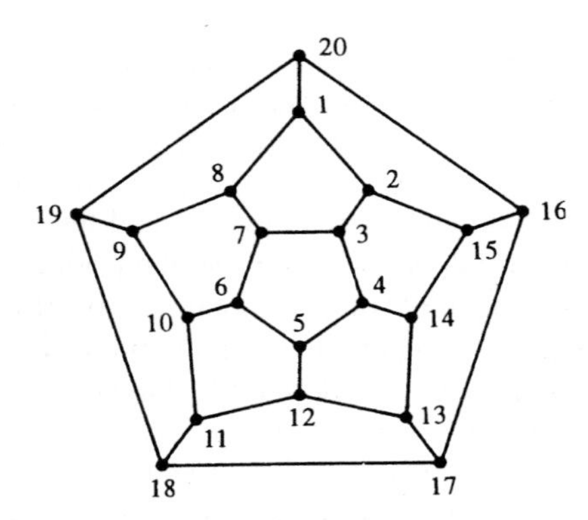

Fig. 5.3c. Hamilton's problem

considered an arbitrary graph, which means nothing else than a finite collection of vertices, a certain number of pairs of which are connected together by edges. Hamilton's problem is whether there is a closed circuit (a "Hamilton circuit") passing each vertex (not each edge as in Euler's problem) exactly once. Figure 5.3c shows a graph with a Hamilton circuit passing the vertices in the order of numbering.

However, unlike the case of Euler's problem, we do not know any condition which exactly characterizes whether a graph contains a Hamilton circuit or not. We only can define an algorithm testing whether an arbitrary graph contains a Hamilton circuit or not. The algorithm tests all permutations of vertices to see if they form a Hamiltonian circuit. As there are $n!$ different permutations of n vertices, the algorithm does not need more than c $\cdot$ $n!$ steps with a constant c to find a solution. It can easily be proved that an order of $n!$ corresponds to an order of n^n. Consequently, an algorithm for the Hamilton problem needs exponential computational time, while the Euler problem can be solved algorithmically in linear computational time. Thus, Hamilton's problem cannot practically be solved by a computer even for small numbers n.

The main reason for a high computational time may be a large number of single subcases which must be tested by a deterministic computer step by step. It is more convenient to use a non-deterministic computer which is allowed to choose a computational procedure at random among a finite number of possible ones instead of performing them step by step in a serial way. Let us consider Hamilton's problem again. An input graph may have n vertices $\nu_1, \ldots, \nu_n$. A non-deterministic algorithm chooses a certain order $\nu_{i_1}, \ldots, \nu_{i_n}$ of vertices in a non-deterministic, random way. Then the algorithm tests whether this order forms a Hamilton circuit. The question is whether for all numbers j ($j = 1, \ldots, n-1$) the successive vertices ν_{i_j} and $\nu_{i_{j+1}}$ and the beginning and starting vertices ν_{i_n} and ν_{i_1} are connected by an edge. The computational time of this non-deterministic algorithm depends linearly on the size of the graph.

In general, NP means the complexity class of functions which can be computed by a non-deterministic Turing machine in polynomial time. Hamilton's problem is an example of an NP-problem. Another NP-problem is the "travelling salesman problem", which is rather like Hamilton's problem except that the various edges have numbers attached to them. One seeks that Hamilton circuit for which the sum of the numbers, or more intuitively the distance travelled by the salesman, is a minimum.

By definition every P-problem is an NP-problem. But it is a crucial question of complexity theory whether P = NP or, in other words, whether problems which are solved by non-deterministic computers in polynomial time can also be solved by a deterministic computer in polynomial time [5.14].

Hamilton's problem and the travelling salesman problem are examples of so-called NP-complete problems. This means that any other NP-problem can be converted into it in polynomial time. Consequently, if an NP-complete problem is actually proved to be a P-problem (if for instance a deterministic algorithm can be constructed to solve Hamilton's problem in polynomial time), then it would follow that all NP-problems are actually in P. Otherwise if P $\neq$ NP, then no NP-complete problem can be solved with a deterministic algorithm in polynomial time.

Obviously, complexity theory delivers degrees for the algorithmic power of Turing machines or Turing-type computers. The theory has practical consequences for scientific and industrial applications. But does it imply limitations for the human mind? The fundamental questions of complexity theory (for example N = NP or N $\neq$ NP) refer to the measurement of the speed, computational time, storage capacity, and so on, of algorithms. It is another question how one sets out to find more or less complex algorithms. This is the creative work of a computer scientist which is not considered in the complexity theory of algorithms.

On the other hand, Gödel's famous theorems are sometimes said to limit the mathematical power of computers and the human mind. His incompleteness theorem says that in every consistently axiomatized enlargement of formal number theory there is a (closed) formula which is not decidable. Actually, his theorem states that any adequate consistent arithmetical logic is incomplete in the sense that there exist true statements about the integers that cannot be proved within such a logic. Even if we enlarge our axiomatization by the undecidable formula, then there is another formula which is not decidable in the enlarged formalism. Gödel's result showed that the formalistic search for a complete consistent arithmetical logic in the tradition of Leibniz and Hilbert must fail [5.15].

Furthermore, Gödel proved that it is impossible to show that arithmetical logic, which may be incomplete, is consistent by methods that could be represented in the logic itself. Some years after Gödel's famous result, Gerhard Gentzen (1909–1945) proved the consistency of elementary number theory using so-called ε_0-induction, which is an infinitary extension of the usual arithmetical induction over natural numbers. But the consistency of Gentzen's extended proof method is as open to doubt as that of the system to be justified. In other words, the complexity of the justifying method is no less than that of the system to be justified. So there are only relative consistency proofs using methods which have to be justified by methods which have to be justified, and so on. For human thinking there is no absolute foundation of self-consistency which can be delivered by formal algorithms.

From Gödel we know that a consistent axiomatic system for arithmetic cannot be complete [5.16]. But there could still be a decision procedure that would enable us to decide if a given assertion is true or not. It was Turing who in 1936 proved that there cannot be such a universal decision procedure, a claim made in the tradition of Leibniz and Hilbert [5.17]. Turing's argument is in some sense deeper than Gödel's, because he reduced Hilbert's Entscheidungsproblem to the so-called halting problem, a basic problem of computabilty and algorithmic complexity: A universal decision procedure would be able to determine whether an arbitrary computer program stops after finite steps. Turing proved that the halting problem is in principle unsolvable. Then, Gödel's incompleteness is only a corollary of Turing's proof.

Turing started his proof with the question, are real numbers computable? A real number like $\pi = 3.1415926\ldots$ has an infinite number of digits that seem to be randomly distributed behind the decimal point. Nevertheless, there are simple finite programs for calculating the digits step by step with increasing precision of π. In this sense, π is called a computable real number. In a first step, Turing constructed an uncomputable real number. Remember that a computer program of a Turing machine, for example, consists of a finite list of symbols. Thus, it can be coded by a natural number called the program number. Imagine a list of all possible computer programs that are ordered according to their increasing program numbers $p_1, p_2, p_3, \ldots$. If a program computes a real number with an infinite number of digits behind the decimal point (e.g., π), then they should be written down behind the corresponding program number. Otherwise, there is a blank line in the list:

$$\begin{array}{ll}
p_1 & -.\,\underline{d_{11}}d_{12}d_{13}d_{14}d_{15}d_{16}d_{17}\ldots \\
p_2 & -.\,d_{21}\underline{d_{22}}d_{23}d_{24}d_{25}d_{26}d_{27}\ldots \\
p_3 & -.\,d_{31}d_{32}\underline{d_{33}}d_{34}d_{35}d_{36}d_{37}\ldots \\
p_4 & \\
p_5 & -.\,d_{51}d_{52}d_{53}d_{54}\underline{d_{55}}d_{56}d_{57}\ldots \\
\vdots &
\end{array}$$

Following Cantor's diagonal procedure, Turing changed the underlined digits on the diagonal of the list and put these changed digits together into a new number with a decimal point in front:

$$-.\neq d_{11} \neq d_{22} \neq d_{33} \neq d_{44} \neq d_{55} \ldots$$

This new number cannot be in the list because it differs from the first digit of the first number behind p_1, the second digit of the second number behind p_2, etc. Therefore, it is an uncomputable real number. With this number Turing got the unsolvability of the halting problem. If we could solve the halting problem, then we could decide if the n-th computer program ever puts out an n-th digit behind the decimal point. In this case, we could actually carry out Cantor's diagonal procedure and compute a real number, which, by its definition, has to differ from any computable real.

The unsolvability of the halting problem refutes Hilbert's Entscheidungsproblem. If there is a complete formal axiomatic system from which all mathematical truth follows, then it would give us a procedure to decide if a computer program will ever halt. We just run through all the possible proofs until we either find a proof that the program halts, or we find a proof that it never halts. So if Hilbert's finite set of axioms from which all mathematical truth should follow were possible, then by running through all possible proofs while checking which ones are correct, we would be able to decide if computer program halts. That is impossible using Turing's proof.

A formal axiomatic system has the great advantage of compressing a lot of theorems into a set of a few axioms. Thus, it delivers a shorter description of mathematical truth. Even a physical theory can be understood as a shorter description of many empirical data. In general, a formal theory can be considered a computer program that calculates true theorems or data. The smaller the program is relative to the output, the better the theory. Obviously, besides running time, the size of a computer program is an important measure of computational complexity. As a program is a finite list of symbols, its length can be measured by its number of symbols in binary coding. For example, consider the following sequences of binary digits:

$$s_1 = 11111111111111111$$
$$s_2 = 0101010101010101$$
$$s_3 = 0110100011011110100$$

For s_1 and s_2, there are shorter descriptions or printing programs than the actual output: "14 times 1" for s_1 and "8 times 01" for s_2. But for s_3, there seems to be no shorter description than the actual output itself. Gregory J. Chaitin and Andrej N. Kolmogorov came up with the idea that the algorithmic complexity of a symbolic s sequence should be defined by the length of the shortest computer program for generating s (measured in bits) [5.18]. Algorithmic complexity is sometimes called the algorithmic information content of a symbolic sequence, which is the subject of the algorithmic information theory. As random sequences have no regularities, they cannot be described by shorter programs. They are incompressible with an algorithmic complexity equivalent to their length. But, again, we are confronted with incompleteness and undecidability. The reason is that we can never decide if an individual string of digits satisfies this definition of randomness and incompressibility. We can never calculate the program-size complexity, because, in general, it is not decidable if a certain program is the shortest one. If we have a program generating a sequence, its size is only an upper bound on the program-size complexity of the sequence. But we can never prove lower bounds, which means a first incompleteness result in algorithmic information theory.

In the theory of computational complexity, with respect to the running time of programs, lower bounds are much harder than upper bounds. If we find a fast program, we only get an upper bound on the calculating time. At least in some cases, it can be proved that a certain program is the fastest possible one. But in algorithmic information theory, we can never prove any lower bounds. Nevertheless, there are some relativizing results. The program-size complexity of formal theories and programs can be related to programming languages in which they are written. Chaitin preferred the AI-programming language LISP [5.19]. In LISP, a formal axiomatic system with program-size complexity N cannot be used to prove that for any LISP-expression more than $N + 356$ characters long there is no smaller program with the same output. So this formal axiomatic system can only prove that for many finite expressions no smaller program has the same output. In principle, the randomness of a formal sequence cannot be decided. But for practical applications we can at least refer to standard procedures for detecting regularities in a sequence. If we are not successful, a sequence is called random with respect to these algorithms.

5.3 Information, Probability, and $1/f$-Complexity

Computational systems can be described as information processing machines. Algorithmic information theory refers to the size of a computer program in order to determine the algorithmic information content of a message. According to Shannon's information theory [5.20], a message from a sender (e.g., phone, PC) is sent to a recipient by coding the signs of the message into binary digits ("bits"), representing binary technical signals (e.g., electrical pulses), and decoding them when the message arrives. Communication means the exchange of information. The information content of a symbol is the number of binary decisions leading to it. For N symbols, there are $N = 2^I$ selecting procedures with I binary decisions, i.e., $I = \mathrm{ld}\, N$ bit. If the symbols s_i $(1 \leq i \leq N)$ occur with different probabilities p_i, then their information content is $I(s_i) = \mathrm{ld}\, p_i^{-1} = -\mathrm{ld}\, p_i$ bit. A more probable symbol has less information content than an improbable one. In this sense, the information content of a symbol can be considered a measure of news for the receiver.

The mean information content of a sender with symbols s_i is the expectation value of the information contents $I(s_i)$ of its symbols s_i, i.e., $H = \sum_i p_i I(s_i) = -\sum_i p_i \mathrm{ld}\, p_i$ with $\sum_i p_1 = 1$. The mean information content H can be considered a measure of uncertainty for the probabilistic distribution of the symbols of a source (Fig. 5.4). The reason being that in the case of the uniform distribution of probabili-

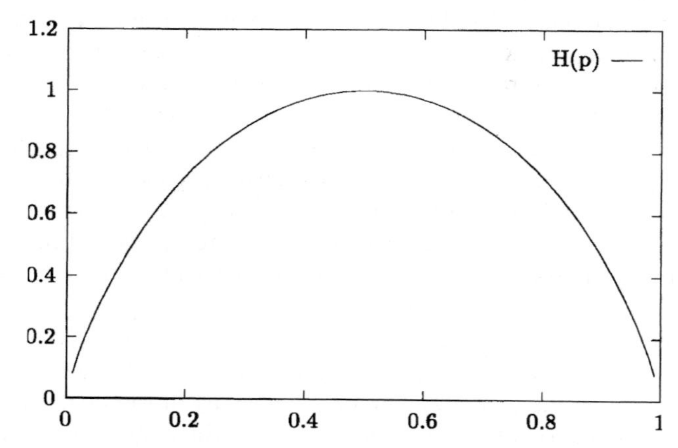

Fig. 5.4. Mean information content (information entropy) of a system with two symbols (states) with $p_1 = p$ and $p_2 = 1 - p$

ties, the mean information content $H_{\max}$ of a source is maximal, i.e., the uncertainty of a symbol is maximal. For $H = 0$ is $p_i = 1$, i.e., symbol s_i is determined by the source.

Shannon's concept of information is not only applicable to technology. In evolution, chemical and biological information is molecularly coded and can be recognized (decoded) by appropriate molecules, cells, or organisms (molecular pattern recognition). The genetic information of an organism is coded by the four chemical compounds adenine (A), cytosine (C), guanine (G) and uracil (U). With binary coding A = 00, U = 11, G = 01, and C = 10, we get a genetic code in bits. Sensorial stimuli of the human organism are analogous signals (e.g., mechanical pressure of skin or muscles, acoustic waves in the ear, electromagnetic waves of the retina, chemical stimuli in the nose) which are received by sensorial cells, coded into digital action potentials, and sent as binary codes (firing and non- firing of neurons) in the central nervous system (CNS) to the brain. Specific nervous signals (neural information) are decoded as sensorial perceptions, emotions, imaginations, or thoughts by specific areas of the brain. A mechanical stimulus (e.g., stretch of a muscle) is received by a sensorial cell as an analogous signal and transformed into digital action potentials. The intensity of the stimulus is coded by the number of equal action potentials. According to information theory, information can be reduced to bits, the smallest units of binary states 0 and 1. According to quantum theory, elementary particles (e.g., photons) have binary spin-states $\uparrow$ (up) and $\downarrow$ (down) that can be superposed in coherent states, called quantum bits [5.21]. Thus, each state of matter can be considered a kind of "condensed" quantum information.

Information storage and information flow in matter, life, and the brain depend on the dynamics of complex systems. According to L. Boltzmann, entropy S is a measure of the probable distribution of microstates of elements (e.g., molecules of a gas) in a complex dynamical system, generating a macrostate (e.g., temperature of a gas), i.e., $S = k_B \ln W$ with k_B Boltzmann-constant and W number of probable distributions of microstates, generating a macrostate. According to the 2nd law of thermodynamics, entropy is a measure of increasing disorder in isolated systems. The reversible process is extremely improbable. In information theory, entropy can be introduced as a measure of uncertainty of random variables. The information entropy $H(X)$ of random variable X is the expectation value of the probabilistic distribution of its values x, i.e., $H(X) = -\sum_x p(x) \log p(x)$. Thus, in thermodynamic systems, $H(X)$ is the expectation value of the probabilistic distribution of their microstates. For $H(X) = 0$, the process X is deterministic. For $H(X)$ maximal, there is uniform distribution with maximal uncertainty of x. Information entropy is considered a measure of uncertainty.

According to Shannon, further concepts of information can be introduced in order to measure the information flow in a dynamical system. The joint entropy $H(X, Y)$ of random variables X and Y is the expectation value of the distribution of joint probabilities $p(x, y)$ of values x of X and y of Y. The conditional entropy $H(Y \mid X)$ of X and Y is the average outcome of the degree of uncertainty of Y over all concrete outcomes of X. The relative entropy or cross-entropy is a measure of the difference ("distance") between two distributions $p(x)$ and $q(x)$. Mutual information $I(X; Y)$ measures the statistical independence of random variables

X and Y with associated probability distributions $p(x)$ and $p(y)$: If X and Y are independent, then $I(X;Y) = 0$. Mutual information is a symmetric measure, because $I(X;Y) = I(Y;X)$, $I(X;X) = H(X)$. Mutual information can be considered measure of correlations between X and Y. When X is the input and Y is the output of a stochastic channel, then $I(X;Y)$ is the amount of information transmitted in the stochastic channel. There is a remarkable application of mutual information in brain research: In a self-organizing learning process, the brain responds to different stimuli with different clusters of synchronously firing neurons. According to Hebb's theory (compare Sect. 4.2), these cell assemblies code the binding of single features in a perceptual object. The reliability of discrimination between different stimuli and different clusters is measured by the mutual information between the corresponding random variables.

An information system produces a time series of N different symbols s_i ($1 \leq i \leq N$). Let β be a partition of the symbolic dynamics and p_i^β the probability of observing symbols s_i of the partition β. The entropy of the symbolic sequence with partition β is defined by $H^\beta = -\sum_i p_i^\beta \log p_i^\beta$. The flow of information I_p^β measures the predictability of a dynamical step p steps into the future, given the whole past of $n \to \infty$ steps, with $I_p^\beta = \lim_{n\to\infty} I^\beta(n;p)$, where $I^\beta(n;p)$ is the mutual information between a word of n subsequent symbols and the symbol that is p steps ahead [5.22]. Therefore, this concept of information flow is an extension of the Kolmogorov-Sinai-entropy (Table 2.1, measuring the predictability only one step ahead [5.22]. It follows $0 \leq I^\beta(n;p) \leq H^\beta$, where the minimal value (0) corresponds to statistical independence and the maximal value (H^β) to perfect predictability. For a chaotic time series, we have $I^\beta(n;p) > I^\beta(n;p+1)$, which expresses the loss of information in the prediction horizon (Fig. 5.5).

A dynamical system can be considered an information processing machine, computing a present or future state as output from an initial past state of input. Thus, the computational efforts to determine the states of a system characterize the computational complexity of a dynamical system. The transition from regular to chaotic

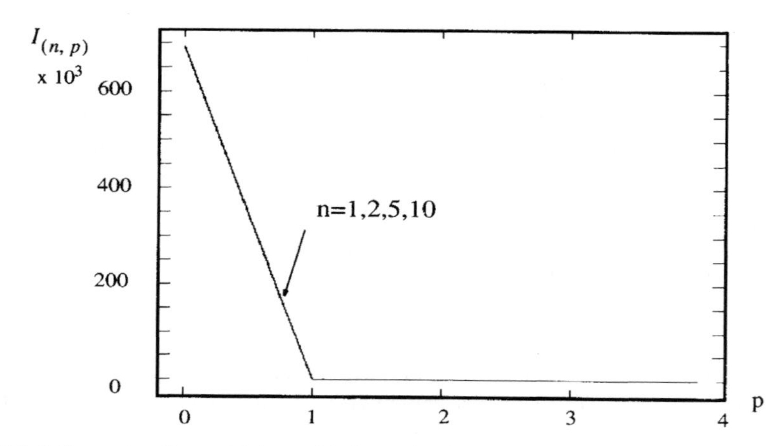

Fig. 5.5. Information flow with loss of information in one step ($p = 1$) for a chaotic logistic map $x_{n+1} = 4x_n(1 - x_n)$ with $n = 1, 2, 5, 10$ and bipartition [5.23]

systems corresponds to increasing computational problems, according to the computational degrees in the theory of computational complexity. In statistical mechanics, the information flow of a dynamical system describes the intrinsic evolution of statistical correlations between its past and future states. The Kolmogorov-Sinai (KS) entropy is an extremely useful concept in studying the loss of predictable information in dynamical systems, according to the complexity degrees of their attractors (Table 2.1). Actually, the KS-entropy yields a measure of the prediction uncertainty of a future state provided the whole past is known (with finite precision).

In the case of fixed points and limit cycles, oscillating or quasi-oscillating behavior, there is no uncertainty or loss of information, and the prediction of a future state can be computed from the past. In chaotic systems with sensitive dependence on the initial states, there is a finite loss of information for predictions of the future, according to the decay of correlations between the past states and the future state of prediction. The finite degree of uncertainty of a predicted state increases linearly to its number of steps in the future, given the entire past. But in the case of noise, the KS-entropy becomes infinite, which means a complete loss of predicting information corresponding to the decay of all correlations (i.e., statistical independence) between the past and the noisy state of the future. The degree of uncertainty becomes infinite.

The degree of complexity of noise can also be classified via Fourier analysis of time series in signal theory. Early in the nineteenth century, the French mathematician Jean-Baptiste-Joseph Fourier (1768–1830) proved that any continuous signal (time series) of finite duration can be represented as a superposition of overlapping periodic oscillations of different frequencies and amplitudes. The frequency f is the

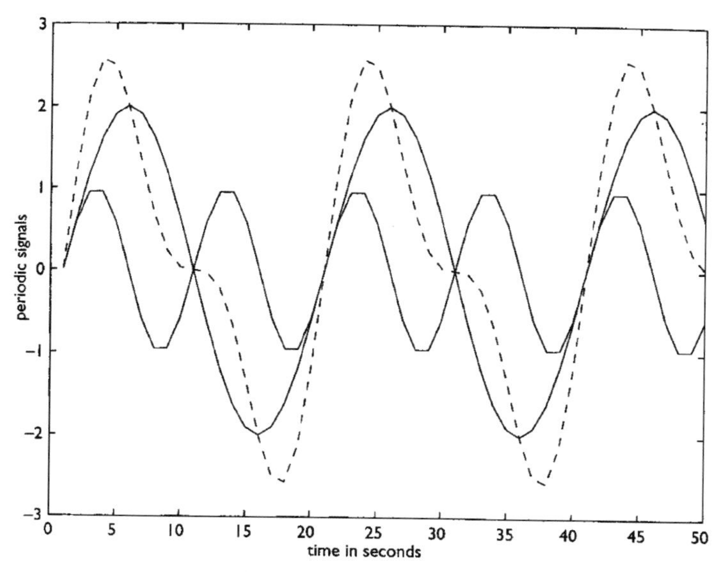

Fig. 5.6. Fourier analysis with two periodic signals and their superposition [5.24]

reciprocal of the length of the period, which means the duration $1/f$ of a complete cycle. This measures how many periodic cycles there are per unit time. Figure 5.6 shows a portion of two periodic signals (solid lines) with different oscillation amplitudes. The smaller fluctuation has a period of 10 seconds, while the larger fluctuation has a period of 20 seconds. The frequencies are therefore 0.10 and 0.05 cycles per second, respectively. Their sum (dashed line) is superimposed on the two oscillations.

Each signal has a spectrum (Table 2.1), which is a measure of how much variability the signal exhibits in each of its periodic components. The spectrum is usually expressed as the square of the magnitude of the oscillations at each frequency. This indicates the extent to which the magnitude of each periodic oscillation contributes to the total signal. If the signal is periodic, with a period of $1/f$, then its spectrum is zero except at the isolated value f. In the case of a signal that is a finite sum of periodic oscillations, the spectrum will exhibit a finite number of values at the frequencies of the oscillations that comprise the signal. For example, the spectrum of the dashed curve in Fig. 5.6 consists of two isolated values at the frequencies 0.05 and 0.10.

The opposite of periodicity is a signal whose values are statistically independent and uncorrelated. In signal theory, a distribution of independent and uncorrelated values is called white noise. It contains contributions from oscillations whose amplitudes are uniform over a wide range of frequencies. In this case the spectrum has a constant value, flat throughout the frequency range. The contributions of periodic components cannot be distinguished. Examples of periodicity and white noise are given by the sequences of binary digits in Sect. 5.2: the sequence 010101 ... is obviously an example of a periodic signal, while a random string gives white noise.

However, in nonlinear dynamics of complex systems we are mainly interested in complex series of data that conform to neither of these extremes. They consist of many superimposed oscillations at different frequencies and amplitudes, with a spectrum that is approximately proportional to $1/f^b$ for some b greater than zero. In that case, the spectrum varies inversely with the frequency. Such signals are called $1/f$ noise. Figure 5.7 illustrates examples of signals with spectra from pink noise ($b = 1$), red noise ($b = 2$), and black noise ($b = 3$). White noise is designated by $b = 0$. The degree of irregularity in the signals decreases as b increases.

When b exceeds 2 the correlations become persistent, because upwards and downwards trends tend to maintain themselves. A large excursion in one time interval is likely to be followed by another large excursion in the next time interval of the same length. The time series seem to have a long-term memory, which is sometimes called the "Joseph effect." In Sect. 7.4, we will remind the reader of Joseph's biblical story of seven years of plenty followed by seven years of famine. When b is less than 2 the correlations are antipersistent, in the sense that an upswing is now likely to be quickly followed by a downturn, and vice versa. When b increases from the antipersistent to the persistent case, the curves in Fig. 5.7 become less jagged. In Sect. 7.4, the change from uniform and antipersistent to persistent behavior is mathematically characterized by the Hurst parameter.

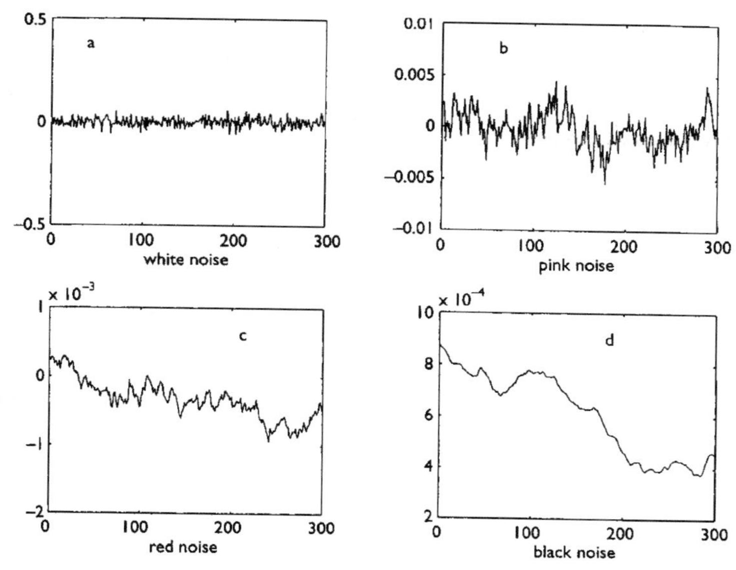

Fig. 5.7. Degrees of complexity of $1/f^b$ noise, with white noise ($b = 0$), pink noise ($b = 1$), red noise ($b = 2$), and black noise ($b = 3$) [5.25]

The spectrum gets progressively smaller as the frequency increases. Therefore, large-amplitude fluctuations are associated with long-wavelength (low-frequency) oscillations, and smaller fluctuations correspond to short-wavelength (high-frequency) cycles. In nonlinear dynamics, pink noise with b roughly equal to 1 is particular interesting, because it characterizes processes that lie between the regular order of black noise and the complete disorder of white noise. For pink noise, the fraction of total variability in the data between two frequencies $f_1 < f_2$ equals the percentage variability within the interval $cf_1 < cf_2$ for any positive constant c. Therefore, there must be fewer large-magnitude fluctuations at lower frequencies than there are small-magnitude oscillations at high frequencies. As the time series increases in length, more and more low-frequency but high-magnitude events are uncovered because cycles of longer periods are included. The longest cycles have periods comparable to the duration of the sampled data. Like all fractal patterns, small changes in signals are superimposed on larger ones with self-similarity at all scales (compare the fluctuation of the information packet from the World Wide Web in Fig. 8.16).

In electronics, $1/f$ spectra are known as flicker noise, since they differ from the uniform sound of white noise due to the individual signals [5.26]. The high-frequency occurrences are hardly noticed compared to the large-magnitude events. One remarkable application of $1/f$ spectra involves its use in different kinds of music. The fluctuations in loudness as well as the intervals between successive notes in the music of Bach have a $1/f$ spectrum. In contrast to Bach's pink noise music, white noise music consists of successive uncorrelated values. The brain fails to find any pattern in a structureless and irritating sound. On the other hand, black noise music seems too predictable and boring, because the persistent signals depend strongly on previous values. Obviously, creating impressive music involves finding a balance between order and disorder, regularity and surprise.

$1/f$ spectra are typical of processes that organize themselves to a critical state at which many small interactions can trigger the emergence of a new, unpredicted phenomenon. Earthquakes, atmospheric turbulence, stock market fluctuations, and physiological processes of organisms are typical examples. Self-organization, emergence, chaos, fractality, and self-similarity are features of complex systems with nonlinear dynamics [5.27]. The fact that $1/f$ spectra are measures of stochastic noise again emphasizes the deep relationship between information theory and systems theory: any complex system can be considered to be an information processing system. In the following section, distributions of correlated and unrelated signals are analyzed according to the theory of probability. White noise is characterized by the normal distribution of the Gaussian bell curve. Pink noise with a $1/f$ spectrum is decidedly non-Gaussian. Its patterns are footprints of complex self-organizing systems.

5.4 Stochastic Processes, Probabilistic Attractors, and Probabilistic Complexity

In complex systems, the behavior of a single element is often completely unknown and therefore considered to be a random process. In this case, it is not necessary to distinguish between chance that occurs because of some hidden order that may exist and chance that is the result of blind lawlessness. A stochastic process is assumed to be a succession of unpredictable events. Nevertheless, the whole process can be characterized by laws and regularities, or in the words of A.N. Kolmogorov, the founder of the modern theory of probability: "The epistemological value of probability theory is based on the fact that chance phenomena, considered collectively and on a grand scale, create non-random regularity" [5.28]. When tossing a coin, for example, heads and tails are each assigned a probability of $1/2$ whenever the coin appears to be balanced. This is because one expects that an outcome of heads or tails is equally likely in each flip. Therefore, the average number of heads or tails in a large number of tosses should be close to $1/2$, according to the law of large numbers. This is what Kolmogorov meant.

The outcomes of a stochastic process can also have different probabilities of occurring. Binary outcomes are designated by probabilities of p and $1 - p$. In the simplest case of $p = 1/2$, there is no propensity for one outcome to occur more than another, and the outcomes are said to be uniformly distributed. For instance, the six faces of a balanced die are all equally likely to land face-up after a toss, and so the probability of each face is $1/6$. In this case, a random process is thought of as a succession of independent and uniformly distributed outcomes. In order to turn this intuition into a more precise statement, we consider coin tossing with two possible outcomes, labeled zero or one. The number of ones in n trials is denoted by

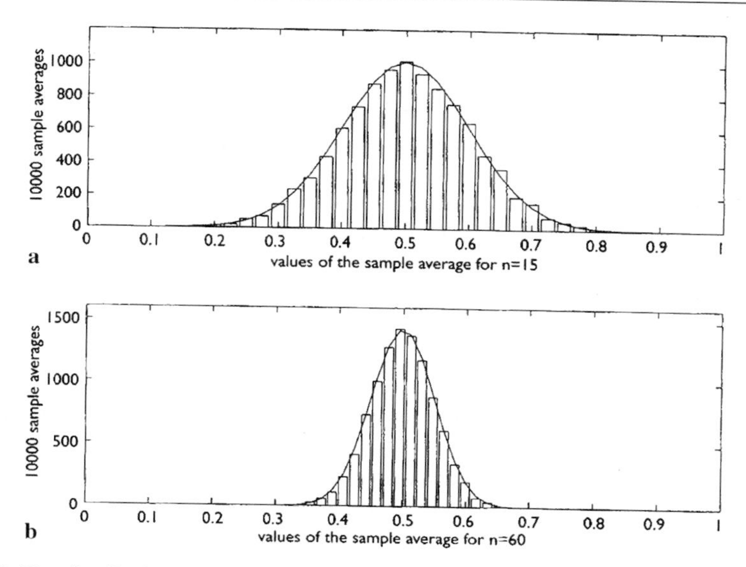

Fig. 5.8. The distributions of values of r_n/n when (**a**) $n = 15$ and (**b**) $n = 60$ after 10 000 samples

r_n, and the sample average r_n/n represents the fraction of the total number of trials n that result in ones. Then, according to the law of large numbers, the probability that r_n/n is within some fixed interval around $1/2$ will tend to one as n increases without bound.

In Fig. 5.8a, the distribution of values of r_n/n for $n = 15$ obtained after 10 000 samples is plotted for a probability $p = 1/2$. Obviously, the values cluster about $1/2$, with a dispersion that appears roughly bell-shaped. The height of each rectangle in the figure indicates the number of all sample averages that lie in the indicated interval along the horizontal axis. Figure 5.8b shows the distribution of values of r_n/n for $n = 60$ for 10 000 samples with a probability $p = 1/2$. The distribution of values also appears to follow a bell-shaped curve, but the curve is narrower than for $n = 15$ and it has a higher peak. The bell-shaped Gaussian curve illustrates Kolmogorov's statement that regularity emerges when large ensembles of random events are considered.

The same general bell shape appears for several games with different average outcomes, like playing with coins, throwing dice, or dealing cards. Some bells are squatter and some narrower, but each can be described as a Gaussian curve. In fact, the values of just two curve parameters are required to differentiate each curve: the mean or average error and the variance or standard deviation, which expresses how widely the bell spreads.

Another example of a stochastic process is a random walk. A single walker may perform a number of independent identically distributed steps. If n is the number of steps performed and Δt the time interval required to perform one step, the position

$x(t)$ of the walker at time $t = n\Delta t$ can be considered to be the sum $S_n = x_1 + \ldots + x_n$ of n independent identically distributed random variables x_i with $1 \leq i \leq n$. For a random walk, the variance of the stochastic process $x(t)$ grows linearly with the number of steps. Starting from a discrete random walk, a continuous limit can be obtained by making the limit $n \to \infty$ and $\Delta t \to 0$ such that $t = n\Delta t$ is finite. The linear dependence of the variance on t is characteristic of a diffusive process, known as a Wiener process. A random walk is only a Gaussian distribution for $n \to \infty$; the Gaussian shape is assumed asymptotically. The probability distribution (density) function $P(S_n)$ depends on n and its shape changes with time. $P(x_i)$ is arbitrary. Figure 5.9 shows four different probability distribution functions, where (i) is a delta distribution, (ii) a uniform distribution, (iii) a Gaussian distribution, and (iv) is a Cauchy distribution. When one of these distributions characterizes the random variables x_i, the probability distribution function $P(S_n)$ changes as n increases (Fig. 5.10).

From Fig. 5.10, the delta and uniform distributions behave in a different way to the Gaussian and Cauchy distributions as n is increased. The function $P(S_n)$ changes both in scale and in functional form as n increases for the delta and the uniform distributions, while the Gaussian and the Cauchy distributions do not change in shape, only in scale; they become broader when n increases. When the functional form of $P(S_n)$ is the same as the functional form $P(x_i)$, the stochastic process is said to be stable. Therefore, while Gaussian and Cauchy processes are stable, stochastic processes generally are not stable.

As long as the random variables x_i exhibit both independence and finite variance, the central limit theorem [5.31] holds: the distribution $P(S_n)$ gradually converges to the Gaussian shape as n increases. For example, in Fig. 5.11, the stochastic process S_n is simulated under the assumption that x_i is characterized by a uniform $P(x_i)$. Obviously, the distribution $P(S_n)$ broadens when n increases. The convergence to the Gaussian asymptotic distribution can be emphasized by plotting the probability density function using scaled units with $\tilde{x} = x/n^{1/2}$ and $\tilde{P}(\tilde{x}) = P(\tilde{x})\ n^{1/2}$. In this case, the distribution rapidly converges to the functional form of a Gaussian of unit variance, which is a smooth curve for large n.

If the conditions (independence and finite variance of the random variables) are not satisfied, other limit theorems must be considered. Studies of limit theorems use the concept of the basin of attraction of a probability distribution. This concept relates to the changes that occur in the functional form of $P(S_n)$ as n changes. In the case of independent identically distributed random variables x_i, $P(S_1)$ coincides with $P(x_i)$ and is characterized by the choices made when selecting the random variables x_i. As n increases, $P(S_n)$ changes its functional form and assumes the Gaussian functional form for an asymptotically large value of n if the conditions of the central limit theorem are satisfied. All of the probability density functions define a functional space. The Gaussian probability function is a fixed-point attractor for stochastic processes in that functional space. The set of probability density functions that fulfill the requirements of the central limit theorem (independence and finite variance of random variables) constitutes the basin of attraction of the Gaussian distribution.

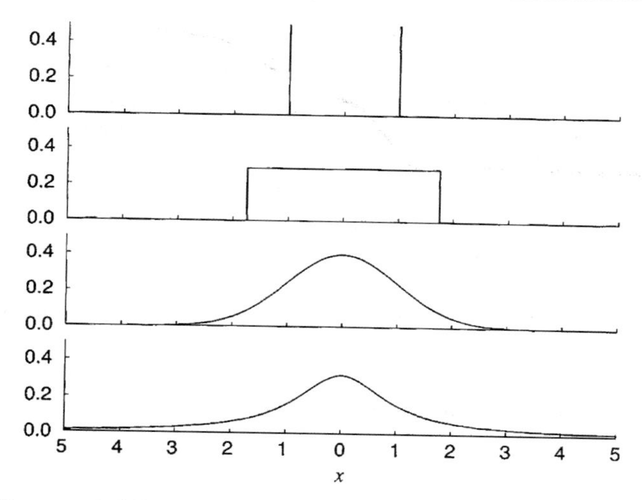

Fig. 5.9. Different probability density functions: (i) a delta distribution, (ii) a uniform distribution, (iii) a Gaussian distribution, and (iv) a Cauchy distribution [5.29]

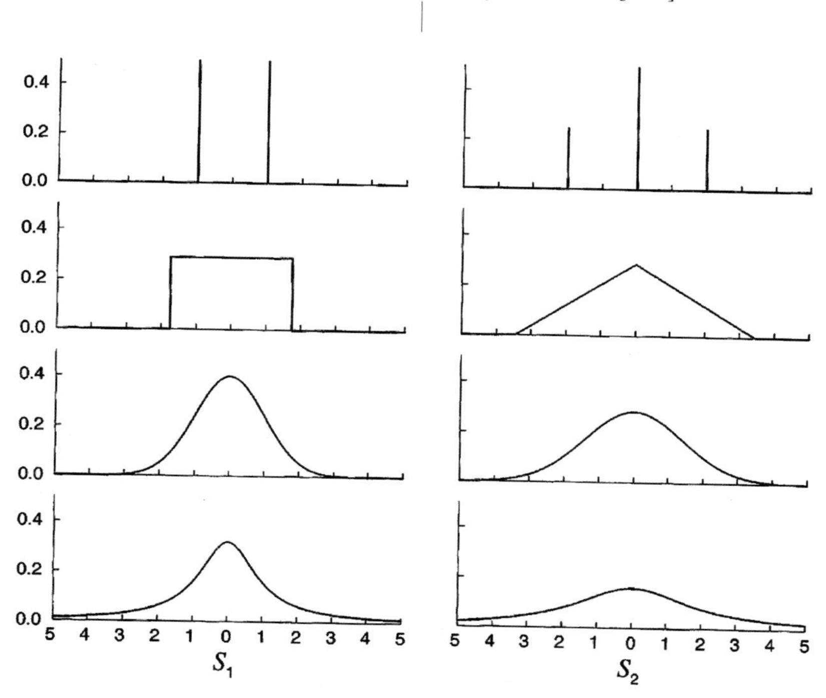

Fig. 5.10. $P(S_n)$ for independent identically distributed random variables ($n = 1, 2$) for the probability density functions shown in Fig. 5.9 [5.30]

In that functional space, we can imagine how two stochastic processes S_n converge to the Gaussian attractor (Fig. 5.12). Both stochastic processes are obtained by summing n independent identically distributed random variables x_i and y_i. If the two processes x_i and y_i differ in their probability density functions, they start from different regions of the functional space. With increasing n, both probability density functions $P(S_n)$ become progressively closer to the Gaussian attractor $P_G(S_\infty)$. The number of steps required to observe the convergence of $P(S_n)$ to $P_G(S_\infty)$ reflects the speed of convergence of the two approximations. The Gaussian attractor is the most important attractor in this functional space, but other attractors also exist.

Gaussian and Cauchy distributions are examples of stable distributions. A stable distribution of the sum of n independent identically distributed random variables is encountered when the distribution does not change its functional form for different values of n. The French mathematician Paul Lévy (1886–1971) determined the

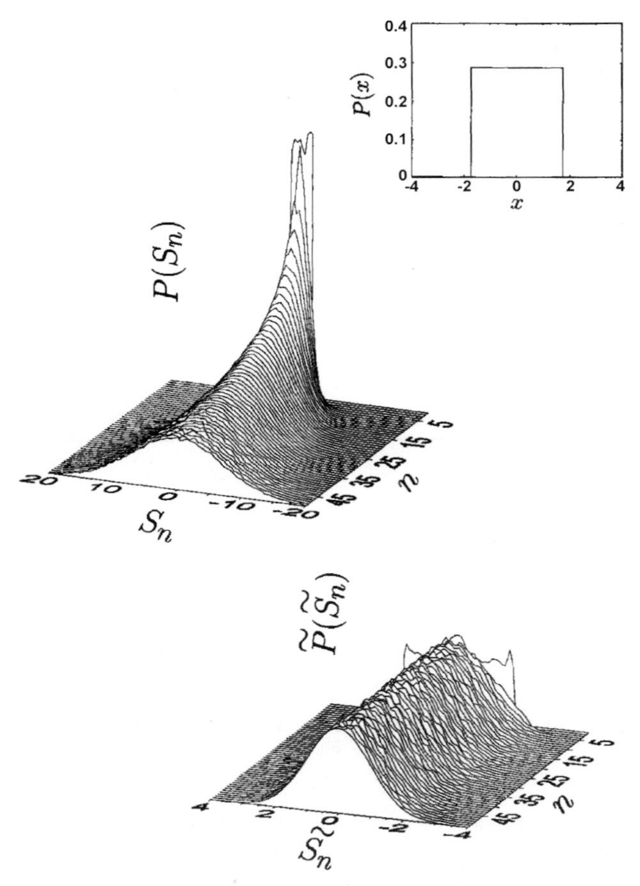

Fig. 5.11. Simulation of $P(S_n)$ for n ranging from $n = 1$ to $n = 50$ for the case when $P(x)$ is uniformly distributed (*top*), and the same distribution in scaled units (*bottom*) [5.32]

entire class of stable distributions [5.33]. In general, they are characterized by a parameter $\alpha (0 < \alpha \leq 2)$, with $\alpha = 2$ for the Gaussian distribution and $\alpha = 1$ for the Cauchy distribution. In contrast to the Gaussian distribution, non-Gaussian ("Lévy") stable stochastic processes with $\alpha < 2$ have infinite variance. Their asymptotic behavior is characterized by distributions of the form $P_L(x) \sim x^{-(1+\alpha)}$ that show power-law behavior for large values of x. Unlike the smooth Gaussian bell curve, their ("fat") tails indicate fluctuations with a leptokurtic shape. Thus, they do not have a characteristic scale, but they can be rescaled with self-similarity. Just like the Gaussian distribution, non-Gaussian stable distributions can be attractors in the functional space of probability density functions. There is a limit theorem which states that the probability density function $P(S_n)$ of a sum S_n of n independent identically distributed random variables x_i converges, in probability, to a stable Lévy distribution $P_L(x)$ provided that certain conditions on the probability density function of the random variable x_i are upheld. $P(S_n)$ belongs to the attraction basin of $P_L(x)$.

The functional space of probability density functions is characterized by the continuous parameter α with $0 < \alpha \leq 2$. Therefore, there are an infinite number of attractors that comprise the set of all stable distributions. Figure 5.12 illustrates several such attractors with the convergence of some stochastic processes. Attractors classify the functional space of probability density functions into regions with different complexities. The complexity of the stochastic process is different for the Gaussian attractor and stable non-Gaussian attractors. In the Gaussian basin of attraction, finite-variance random variables are present. However, in the basins of attraction of stable non-Gaussian distributions, random variables with infinite variance can be found. Therefore, distributions with power-law tails are present in the stable non-Gaussian basins of attraction.

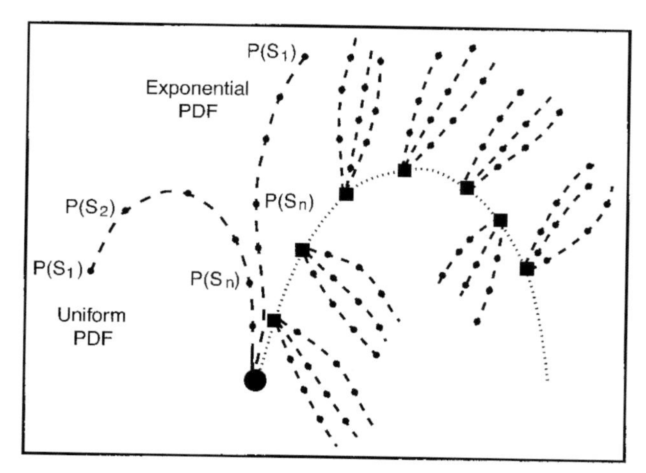

Fig. 5.12. Convergence (in probability) to some of the stable attractors of the sum of independent identically distributed random variables. The black circle is the Gaussian attractor $P_G(S_\infty)$ with $\alpha = 2$, and the black squares are the Lévy-stable non-Gaussian attractors characterized by different values of $\alpha < 2$ [5.34]

Power-law distributions and infinite variance indicate highly complex stochastic behavior [5.35]. Stochastic processes with infinite variance, although well-defined mathematically, are extremely difficult to use and, moreover, raise fundamental questions when applied to real systems. In the closed physical systems of equilibrium statistical mechanics, variance is often related to the system temperature. In this case, infinite variance implies an infinite or undefined temperature. Nevertheless, power-law distributions are used to describe open systems. They are assuming increasing importance in descriptions of, for example, complex economic and physiological systems. Actually, a power-law distribution was first introduced in economics, as Pareto's law of incomes. Turbulence in complex financial markets is also characterized by power-law distributions with fat tails. In financial systems, infinite variance would complicate the important task of risk estimation (see Sect. 7.4).

5.5 Quantum Information, Quantum Computers, and Quantum Complexity

In general, dynamical systems can be represented by computational models with different degrees of complexity. Computational models permit information about present or future states to be computed from initial conditions using the corresponding dynamical equations. However, in the case of deterministic systems, the computability is limited by the degree of algorithmic complexity (Sect. 5.2). The computability of stochastic systems is limited by probabilistic measures (Sects. 5.3–5.4). In any case, computational models of complex dynamical systems are not always computable. With these limitations in mind, dynamical systems can still be thought of as computers that sometimes cannot deliver results in a reasonable time. How far can we go with this assumption? Is the world a complex computer in the sense of Leibniz, with the corrections and limitations of modern algorithmic and probabilistic theories?

Obviously, a Turing machine can be interpreted in the framework of classical physics (Fig. 5.13). Such a computing machine is a physical system, the dynamical evolution of which takes it from one of a set of input states to one of a set of output states. The states are labeled such that they form a series. The machine is initialized to a state with a given input value and then, following some deterministic evolution, the output state is measured. For a classical deterministic system, the measured output label is a definite function f of the input label. In principle, the value of that label can be measured by an outside observer, and the machine is said to compute the function f. But classical stochastic computing machines do not compute functions in the above sense. The output state of a stochastic machine is random; the output corresponds to a probability distribution depending on the input state.

From a modern physical point of view, quantum systems are the fundamental dynamical systems of nature. In that case, the output state of a quantum machine, although fully determined by the input state, is not an observable and so in general the observer cannot discover its label. Why is this? We must now recall some basic concepts of quantum mechanics, which were introduced in Sect. 2.3. In quantum me-

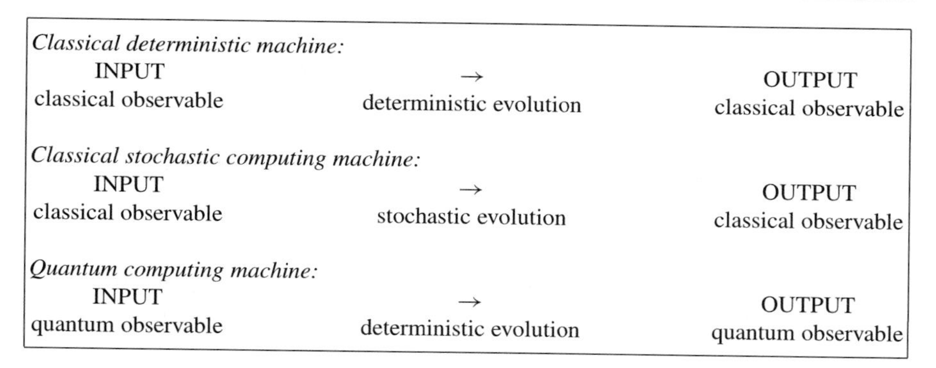

Fig. 5.13. Classical and nonclassical computing machines

chanics, vectors like momentum or position must be replaced by operators satisfying a non-commutative relation depending on Planck's quantum (Fig. 2.18). Classical systems described by a Hamiltonian function are replaced by quantum systems, for instance, electrons or photons described by a Hamiltonian operator. States of a quantum system are described by vectors of a Hilbert space spanned by the eigenvectors of its Hamiltonian operator. The causal dynamics of quantum states is determined by a partial differential equation called the Schrödinger equation. While classical observables commute and always have definite values, non-classical oberservables of quantum systems do not commute and in general have no common eigenvector and consequently no definite eigenvalues. For observables in a quantum state only statistical expectation values can be calculated.

A major difference from classical mechanics is given by the superposition principle demonstrating the linearity of quantum mechanics. In an entangled pure quantum state of superposition an observable can only have indefinite eigenvalues. In short, the superposition or linearity principle of quantum mechanics delivers correlated ("entangled") states of combined systems which are highly confirmed by the EPR experiments (Alain Aspect 1981). Philosophically, the (quantum) whole is more than the sum of its parts.

The superposition principle has severe consequences for the measurement of quantum systems. In the quantum formalism a quantum system and a measuring apparatus are represented by two Hilbert spaces which are combined in a tensor product $H = H_1 \otimes H_2$. In the initial state $\phi(0)$ of measurement at time 0 the systems H_1 and H_2 are prepared in two separated states ψ and φ respectively, with $\phi(0) = \psi \otimes \varphi$. The causal development of both systems is determined by the Schrödinger equation, i.e., $\phi(t) = U(t)\phi(0)$ with the unitary operator $U(t)$. Because of the linearity of $U(t)$, the state $\phi(t)$ is entangled with indefinite eigenvalues while the measuring apparatus at time t shows definite measurement values. Thus, the linear quantum dynamics cannot explain the measurement process.

In a more popular way the measurement process is illustrated by Schrödinger's thought experiment of a cat in a linear superposition of the two states "dead" and "alive" (Fig. 5.14a). Imagine a cat which is locked in a closed box with a sample

Fig. 5.14a. Schrödinger's cat

of radium. The radium is chosen in such a way that during one hour a single decay takes place with a probability 1:2. If a decay happens, then an electrical circuit is closed, causing a mechanism with a hammer to destroy a bottle of prussic acid and thus killing the cat. The box remains closed for one hour [5.36].

According to quantum mechanics the two possible states of the cat, dead and alive, remain undetermined until the observer decides them by opening the box. For the cat's state in the closed box, quantum mechanics as interpreted by Erwin Schrödinger forecasts a correlated ("entangled") state of superposition, i.e., the cat is both dead and alive with equal parts. According to the measurement process, the states "dead" and "alive" are interpreted as measurement indicators representing the states "decayed" or "not decayed" of the radium.

In the Copenhagen interpretation of Bohr, Heisenberg, and others, the measurement process is explained by the so-called "collapse of the wave-packet", i.e., splitting up of the superposition state into two separated states of measurement apparatus and measured quantum system with definite eigenvalues. Obviously, we must distinguish the linear dynamics of quantum systems from the nonlinear act of measurement. The reason for nonlinearity in the world is sometimes explained as the emergence of human consciousness.

Eugene Wigner (1961) suggested that the linearity of Schrödinger's equation might fail for conscious observers, and be replaced by some nonlinear procedure, according to which either one or the other alternative would be resolved out (Fig. 5.14b). But Wigner's interpretation forces us to believe that the complex quantum linear superpositions would be resolved into separated parts only in those corners of the universe where human-like consciousness emerges. In the macroscopic world of billiard balls, planets, or galaxies, EPR correlations are not measured, and appear only in the microscopic world of elementary particles like photons. It seems to be rather strange that the separated states of systems in the macroscopic world which can be described in classical physics with definite measurement values are caused by human-like consciousness.

The Everett "many-worlds" interpretation of quantum mechanics seems to avoid the problems of nonlinear reductions by splitting up human consciousness into branching paths inhabiting different, mutually incompatible worlds (Fig. 5.14c).

In the measurement process the dynamics of measurement instrument and quantum system is described by the equation $\phi(t) = \sum_i c_i(t)\psi_i \otimes \varphi_i$ with states (φ_i) referring to the measurement values of the measuring instrument. Everett argues that the state vector $\phi(t)$ never splits up into partial states, but all branches $\psi_i \otimes \varphi_i$ are actualized. The state $\phi(t)$ describes a manifold of simultaneously existing real worlds with $\psi_i \otimes \varphi_i$ corresponding to the state of the i-th parallel world. Thus, the measured partial system is never in a pure state. In Everett's sense, ψ_n may be interpreted as a relative state depending on the state of an observer or measuring instrument with $\psi_n = c_n^{-1}(\varphi_n, \phi)_{H_2}$. If φ_n are accepted as memory states, then an observer with a definite memory can only be aware of his own branch of the world $\psi_n \otimes \varphi_n$. But he can never observe the other partial worlds.

The advantage of Everett's interpretation is that a nonlinear reduction of superposition does not need to be explained. But the disadvantage is his ontological belief in myriads of worlds which are unobservable in principle. Thus, Everett's interpretation (if mathematically consistent) needs Ockham's razor.

In the history of science, anthropic or teleological arguments often showed that there were gaps or failures of explanation in science. Thus, some scientists, such as Roger Penrose, suppose that the linear dynamics of quantum mechanics is inconvenient (Einstein said it was "incomplete") for explaining cosmic evolution with the emergence of consciousness. He argues that a unified theory of linear quantum mechanics and nonlinear general relativity could at least explain the separated states of macroscopic systems in the world without reference to anthropic or teleological principles. In Penrose's proposed unified theory a linear superposition of a physical system splits into separated states when the system is large enough for the effects of relativistic gravitation. Penrose calculates a level of one graviton as the smallest unit of curvature for such an effect [5.37]. The idea is that the level should lie comfortably between the quantum level of atoms, molecules, etc., with linear laws of quantum mechanics and the classical level of everyday experiences. The advantage of Penrose's argument is that the linearity of the quantum world and the nonlinearity of the macroscopic world would be explained by a unified physical theory without

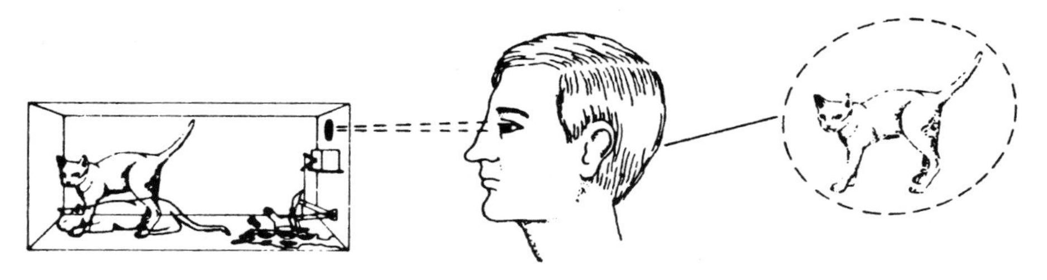

Fig. 5.14b. Wigner's interpretation of Schrödinger's cat

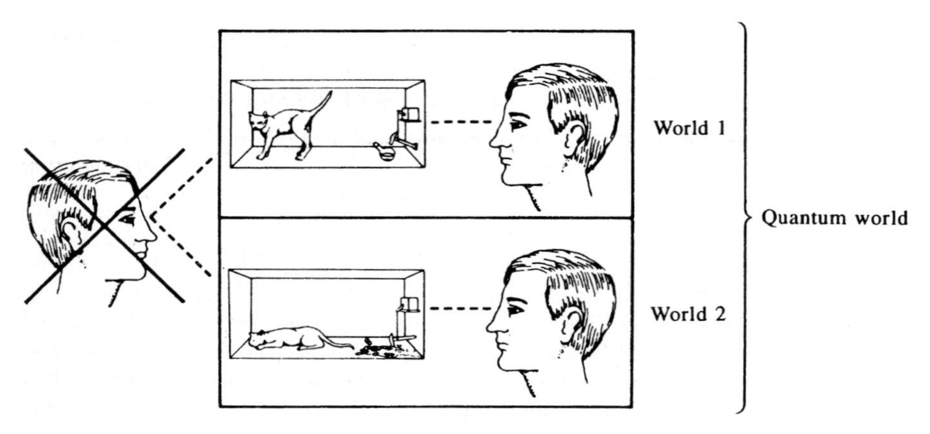

Fig. 5.14c. Everett's interpretation of Schrödinger's cat

reference to any human intervention. But, of course, we are still missing a testable unified theory (compare Sect. 2.4).

Concerning the human brain, we want to argue that the quantum level of elementary particles, atoms, and molecules was necessary for its evolution, but not the other way that mental states of the brain are necessary for the reduction of correlated states in physics. Actually, there is a significant number of neurons sensitive to single quanta with their superpositions and reductions of entangled states. But, of course, these quantum states cannot be identified with mental states of the brain. We have no consciousness either of superpositions or of their separation into single states initiated by nonlinear random events. Nevertheless, quantum effects are involved in the emergence and interaction of mental states of the brain in a way which is still far from being satisfactorily understood.

Nevertheless, the question arises of whether quantum mechanics delivers a framework for the evolution of the human brain, or at least for a new computer technology to replace classical computing systems. The basic idea of quantum mechanics is the superposition of quantum states as a result of linear quantum dynamics and the reduction of superpositions by some kind of measurement [5.38]. Thus, a quantum computer would need a quantum version of a logic gate, where the output would be the results of some unitary operator applied to the input and a final act of measurement. The superposition of quantum systems (for instance photons) reminds us of the parallelism of computations. A quantum computer would become useful if we were interested in some suitable combination of many computational results and not in their partial details. In this case a quantum computer could deliver the superposition of perhaps myriads of parallel computations in a rather short time, overcoming the efficiency of classical computing systems. But quantum computers would still work in an algorithmic way, because their linear dynamics would be deterministic. The non-deterministic aspect comes in via the nonlinear act of measurement. Thus, it cannot be expected that quantum computers will perform non-algorithmic operations beyond the power of a Turing machine. So quantum computers (if they are ever built) may be more interesting for complexity theory and for overcoming practical constraints of computation.

Quantum computers open new avenues of information processing, computation, and communication. An essential feature of the quantum world is the superposition of quantum states and the possibility of entangled states. If the binary information units 0 and 1 are considered to be the alternative states of a computing machine, then quantum physics allows a third state of quantum superposition. Bits that permit such a state are called quantum bits or "qubits." The laws of quantum mechanics have enormous practical consequences for computing [5.39]. If, for example, two partial subproblems of a problem are to be solved, then a classical computer must solve them step by step in a sequential way. For a quantum computer, both subproblems can be superposed and processed simultaneously. Since it is analogous to using parallel computers containing several processors, the information processing of superposed quantum information is called quantum parallelism. Consider a computer that must find an integer with a certain property. A classical computer counts the integers $1, 2, 3, \ldots$ and tests, step by step, whether a number satisfies the demanded property. If the number n sought is very large, then the property must be tested n times, which involves considerable computational time. A quantum computer could test the property for a large number of test numbers simultaneously in one step. Decimal numbers are represented by sequences of binary digits. In quantum computers, a bit corresponds to an alternative quantum state. One example is the spin of an elementary particle, which can adopt states of 0 and 1. In this case, a bit sequence represents a sequence of spin states. For seven particles, for example, there are 2^7 potential combinations, such as 0000000 (for decimal number 0), 0000001 (for decimal number 1), 0000010 (for decimal number 2), etc., including every number between 0 and 127.

In a classical computer, the binary numbers $0000000, 0000001, 0000010, \ldots$ must be tested sequentially. In a quantum computer, the spin states are changed using an appropriate energetic impulse. When the impulse is too weak, the particle only has a particular probability of changing its spin state. In analogy to Schrödinger's cat, which is simultaneously dead and alive in a closed box, an elementary particle is in a superposed state of alternative spin states as long as it is not measured or observed. If every particle receives a weak impulse, then all seven particles enter superposed states as long as they are not observed or measured. In this superposition, all 128 different states and all corresponding numbers can be simultaneously represented and tested in one computational step. However, it is a technical challenge to maintain a superposed state during a calculation. The reason for this is that tiny perturbations and interactions with the environment can lead to the collapse of a superposition or coherent state. This phenomenon is called the *decoherence* problem in quantum computing.

The registers of a classical computer store classical yes(1)/no(0) bits. In a quantum computer, the registers contain quantum systems whose states can be entangled. The gates of a classical computer are the elementary logical operations, such as the NOT gate, which transforms a 0 into a 1 and a 1 into a 0, or the AND gate, which transforms a pair of 1 bits (11) into another 1, but any other pair combination (01,

10, and 00) into a 0 – corresponding to the logic that the whole proposition is only true if both of its partial propositions are true. An OR proposition is only true if at least one partial proposition is true. Therefore, an OR gate transforms all pairs (01, 10 and 11) into a 1 except for a 00 pair, which it converts into a 0. All of the other logical connections can be reduced to these types of gates. A classical program is a sequence of gate operations that can be illustrated in a logical network.

The input of a classical computer is a sequence of bits, either 0 or 1. The input of a quantum computer is an initialized group of registers. Sometimes it is a superposition of bit states in the registers. The output is uniquely determined in a classical computer. A quantum computer can generate a superposition of bit states. In this case, one of the two superposed values 0 and 1 is determined by measurement. During the measurement, the superposition jumps to state 0 or 1 according to the laws of quantum mechanics. Therefore, this different type of processing can give different results. Only probabilities of results can be forecast. Quantum registers differ from classical registers due to the additional possibility of superpositions. In quantum mechanics, superpositions are transformed by unitary operators, but they cannot be realized by classical gates. Unitary operators are reversible, due to the time symmetry of quantum laws. Classical gates are largely irreversible. For example, the classical OR gate delivers the value 1 in three different cases, and the AND gate delivers the value 0 in three different cases; in such cases it is not possible to derive the input directly from the output. However, reversibility can be ensured if the gate "remembers" the input and generates it along with the computed result. The computation of function f with input x and output $f(x)$ is replaced by the transformation of the input x into the output $(x, f(x))$. In Fig. 5.15, the input and output are designated as quantum mechanical state vectors. In the following diagrams, quantum states are distinguished from classical states by using straight lines for classical and wavy lines for quantum states.

Classical information can be transferred between senders and receivers realized using different physical, chemical, or biological systems. But problems arise if they are miniaturized to the quantum scale. In the quantum world, a sender corresponds to an initial quantum system, while a receiver corresponds to a measurement M (see Fig. 5.16). Quantum systems (e.g., elementary particles) that evolve from an initial state to a measurement transfer quantum information. The individual result of a quantum transfer is random, because in quantum mechanics only statistical statements about future events are possible. If the same experiment is repeated different results may be obtained, but – as in the case of coin tossing – with a reproducible

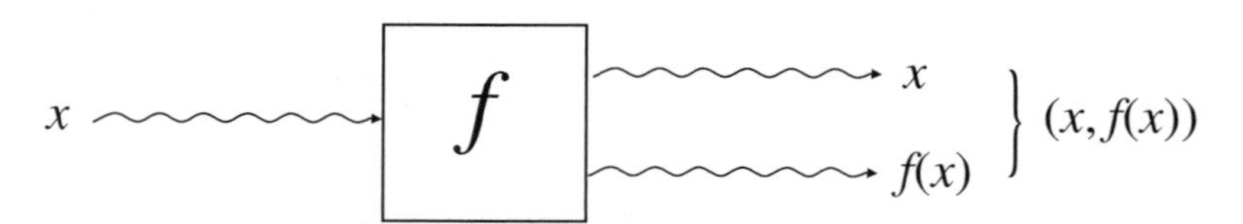

Fig. 5.15. Reversible computation of quantum gates

frequency. Statistical frequencies can be used for probabilistic forecasts. Figure 5.16 illustrates the statistical transfer of quantum information.

Therefore, quantum information depends on the statistical laws of quantum mechanics. It is a new type of information that cannot be translated into classical information without loss. Why is this? Figure 5.17 illustrates the translation of quantum information into classical information and vice versa. A measurement M of a quantum system (e.g., an elementary particle) containing certain quantum information (the first wavy line) is obtained. This measurement yields classical information (straight line) which is passed to a receiver P and used to prepare a new quantum system (wavy line). The fact that it is impossible to fully translate quantum information into classical information results from another statement of impossibility, the no-cloning theorem, according to which no quantum information can be copied. An impossible quantum-copying machine is illustrated in Fig. 5.18. This is a device that takes one quantum system as input and produces two systems of the same type as output. The two copies should be indistinguishable from the input in a statistical sense. If such a machine is not possible, important consequences follow: data cannot be secured by copying it, as done in classical computers. Further, it is not possible to read data in a quantum database without changing them. However, the fact that quantum copying machines are impossible would also prevent secret attacks on quantum data, because such attacks would change the quantum information.

The impossibility of quantum copying machines derives from Heisenberg's uncertainty relation, which states that pairs of quantities, such as the location and energy of a quantum particle, cannot be determined simultaneously with certainty.

Let us assume, for the sake of argument, that it was possible to construct a quantum copying machine. Then, two copies of a quantum particle could be produced in order to measure the energy of one copy and the location of the other, both with certainty. Due to the impossibility of a quantum copying machine, we can immediately see that classical teleportation is impossible. If classical teleportation was

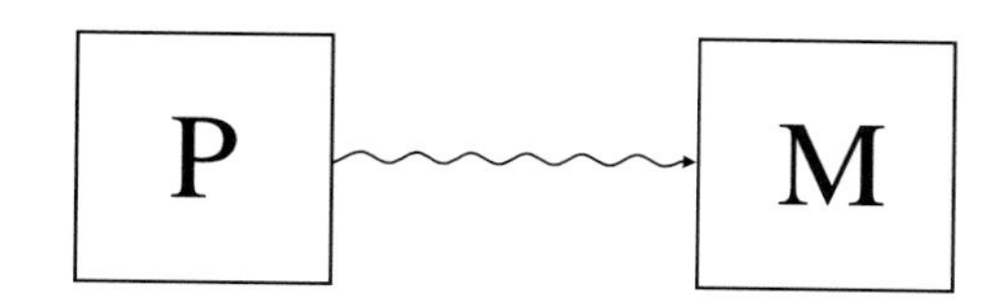

Fig. 5.16. The statistical transfer of quantum information

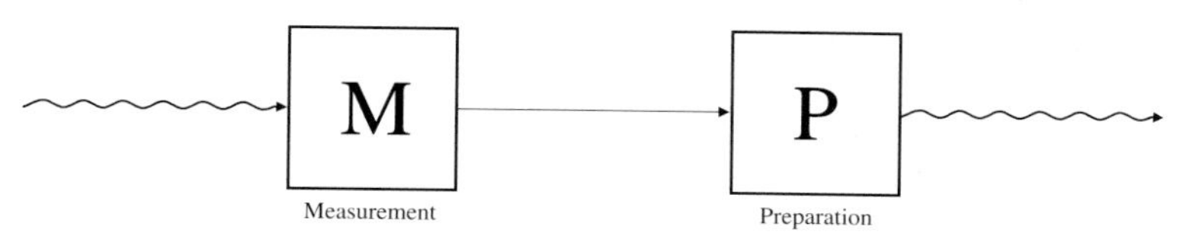

Fig. 5.17. Classical translation of quantum information into classical information and vice versa [5.40]

possible we could easily construct a quantum copying machine: if we used classical teleportation in Fig. 5.17, the transferred classical information could be copied with a classical computer. These copies could be transferred to the receivers P and P′ (Fig. 5.19) in order to prepare new quantum particles. The system of Fig. 5.19 would perform the task of a quantum copying machine in Fig. 5.18.

Quantum information is not only associated with the impossibility of classical teleportation, but also with the possibility of quantum teleportation. In science fiction movies, teleportation means the almost instantaneous transfer of objects across large distances. In order to do this, it is assumed that information (software) about the atomic structure of an object can be separated from its material substance (hardware); this information is then "beamed" to the desired location, where it is used to "rebuild" the object. However, this assumption ignores the fact that the quantum world adheres to the uncertainty principle, which states that at any particular instant, it is not possible to measure all of the properties of elementary particles precisely. Quantum information can only be transported without changing it if it is not measured or observed during the information transfer.

An amazing approach to instantaneous quantum teleportation can be realized via entangled quantum states. According to EPR (Einstein–Podolsky–Rosen) experiments, pairs of elementary particles (e.g., photons) that are emitted from a central source in opposite directions remain correlated in the superposition of an entangled quantum state. If one of two entangled quantum properties is measured at the

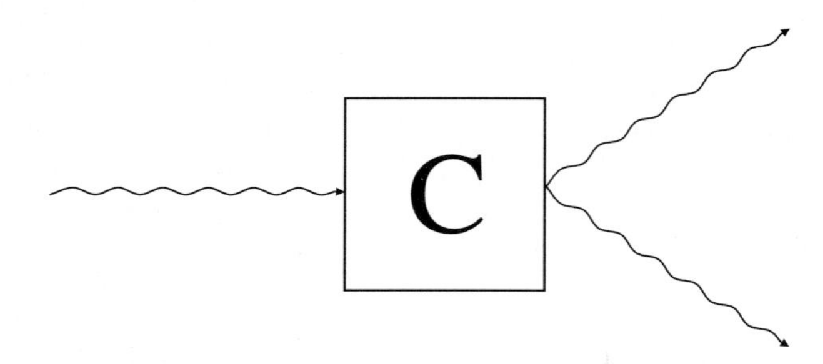

Fig. 5.18. Scheme of a "quantum copying machine"

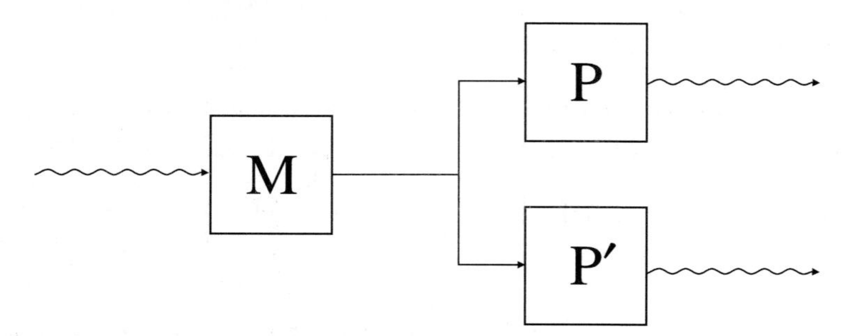

Fig. 5.19. Quantum copying machine utilizing classical teleportation [5.41]

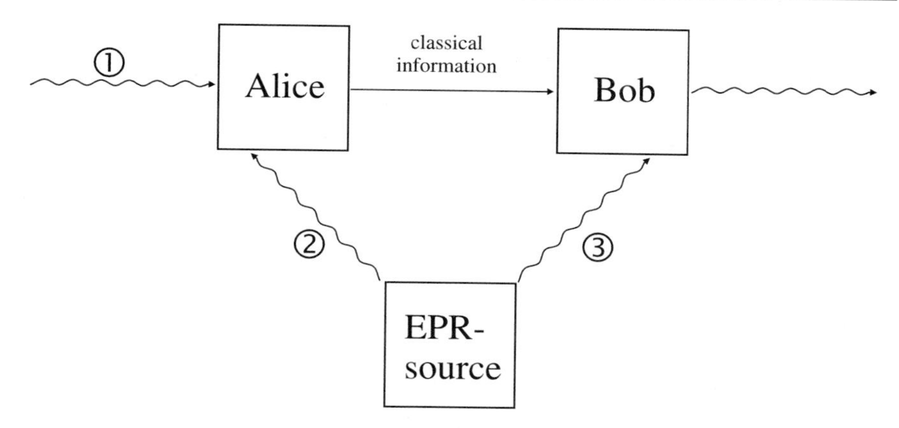

Fig. 5.20. Quantum teleportation

location of one particle, then the value of the other quantum property is instantly determined at the location of the other particle, which has traveled the same distance in the opposite direction. In Fig. 5.20, the sender of quantum information is called Alice, while the receiver is called Bob. The teleported quantum particle is designated 1 (the wavy line). To achieve teleportation, Alice and Bob use an EPR source that generates entangled pairs of quantum particles. An entangled pair of particles consists of particle 2 (for Alice) and particle 3 (for Bob). Alice and Bob do not know the individual states of the particles, but they do know that the pairs are correlated: we assume they are in alternative and opposite states, like the "dead" and "alive" states of Schrödinger's cat in Fig. 5.14a. Alice does not know the teleported state of particle 1 or the state of particle 2. For the teleportation, particle 1 is entangled with particle 2.

In an EPR experiment, a quantum particle (e.g., a photon) can have alternative quantum states, such as vertical or horizontal polarization. In quantum mechanics, a quantum state ψ is represented by a vector $|\psi\rangle$. While classical bit values are 0 and 1, the alternative quantum bit (qubit) values are $|0\rangle$ and $|1\rangle$, which can be entangled. In the case of entanglement, a quantum system simultaneously transports the qubit value $|0\rangle$ with a certain probability and the qubit value $|0\rangle$ with the remaining probability. The qubit only takes a value of $|0\rangle$ or $|1\rangle$ (at random) when the quantum system is measured [5.42].

Alice prepares an entangled pair of particles, particle 1 (which she wishes to teleport) and particle 2. Entangled pairs can be obtained for pairs of qubits when 0 and 1 occur with the same probability for each qubit. Based on the combinatorial possibilities of the classical bit pairs 00, 01, 10, and 11, there are four possible entangled states (Bell states) for pairs of qubits in this scenario [5.43]. Alice determines that her pair of particles 1 and 2 is in one of these four possible entangled states (at random) without measuring their states. For example, she determines the particular entangled Bell state where both qubits are in opposite orientations. However, we assumed that the entangled pair 2 and 3 (Bob's particle) of the EPR source are also in opposite orientations. It follows therefore that Bob's particle (3) must be in the same state as Alice's teleported particle (1). Therefore, classical measurement of the teleported particle gives the value of the qubit for Bob's particle. In this sense, it is possible to instantaneously teleport quantum information between the sender Alice and the receiver Bob.

Instantaneous quantum teleportation is also possible if another entangled Bell state is randomly determined by Alice. The sequentially chained correlations of particles 1, 2, and 3 in Fig. 5.20 allow Bob to adjust the qubit value of his particle appropriately, by reversing the direction of the photon for instance. Bob and Alice know the EPR-entanglement of particles 2 and 3. Thus, Bob's manipulation only depends upon the entangled state of particles 1 and 2. The correlation between particle 1 and 2 is sent from Alice to Bob as classical information, by phone for example (Fig. 5.20). However, unlike instantaneous quantum teleportation, this classical information cannot be transferred at velocities faster than that of light.

The disadvantage of quantum teleportation is the fact that the transported quantum information is unknown until it is determined randomly by measurement. Therefore, quantum teleportation cannot be used for direct information transfer. Thus, there is no conflict with Einstein's theory of relativity and his postulate on the maximum velocity of signals. However, so long as quantum information is not measured or observed, it can be transferred instantaneously via entanglement. Quantum entanglement makes quantum parallelism and thus increased computational velocity possible. Reducing computational time through massive quantum parallelism also means reducing computational complexity.

Quantum computing does not only lead to the exponential growth of computational capacity and a reduction in computational complexity. Any form of matter stores quantum information. Therefore, any elementary particle is a processor of quantum information. The computational rules of these processors are symmetric due to the principles of quantum symmetry. Any computational step is also reversible due to the quantum symmetry of time (microreversibility) [5.44]. Phase transitions of matter are quantum information processing. The universe is an expanding quantum computer that produces quantum information. Furthermore, it is an immense database that conserves all quantum information via symmetry. We must not forget that the concept of a computing machine is not restricted to human technology with symbolic data dynamics. Symbols are only used to represent states of dynamical systems for human purposes. Information processing does not depend on human purposes and interests. Human knowledge only relates to a tiny part of the information in the world. In principle, quantum information does not depend on the presence of an observer or measurement process. Observing and measuring quantum systems is only a special example of an interaction of a quantum system with another system.

Quantum computers are quantum systems, and quantum information corresponds to quantum states. If quantum systems are considered to be quantum computers, then the whole universe is a quantum computer, a concept analogous to that of Leibniz for classical mechanics. But in contrast to Leibniz, the quantum randomness of the universe is an integral part of this system. Looking deeper, conflicts with

Einstein's deterministic view of the world arise. According to Einstein's theory of general relativity, black holes with extremely strong curvatures of space–time are possible, which attract and swallow all forms of matter and light. The center of a black hole is a point singularity without return. Does it also destroy information about the architecture and structure of the material systems swallowed? Does the black hole of an imploding star erase the information about this part of the universe? Are black holes irreversible "memory holes" in space–time that increase in number as the universe ages, just as Alzheimer's disease leads to decay and loss of information in the human brain?

According to the laws of quantum physics, quantum information must remain constant. The wavefunction of a quantum system contains all of the information on its state. The time-dependent development of a quantum state is determined by a unitary transformation, which allows the initial state of the system to be reconstructed from its end state without loss. Thus, no quantum information is lost in quantum physics. But according to Einstein's theory of relativity, information is definitely lost in the point singularity of a black hole. This conflict between quantum and relativistic physics is called the *paradox of quantum information*. Hawking suggested that the quantum vacuum around a black hole should be considered. According to Heisenberg's uncertainty relation, this vacuum is not completely "empty" but is actually filled with quantum fluctuations that cause the spontaneous creation of pairs of particles and antiparticles that are annihilated after very small periods of time. However, some of the particles fall into the black hole while their partners escape from the black hole as thermal radiation (known as Hawking radiation). In Hawking's model, however, the escaping particles are completely independent of their swallowed partners. Thus, no information can escape from the center of the black hole in order to save the quantum information from being erased.

A potential solution to this problem is obtained by assuming entangled quantum states. In the Horowitz–Maldacena–model [5.45], the pairs of particles are entangled. Therefore, an escaping particle does not only transports mass but also information. It is entangled with its partner, which falls into the black hole. Thus, information about the swallowed matter can be "beamed" from inside to outside of the black hole by quantum teleportation. The quantum world does not forget anything. Only the Einsteinian deterministic model of the universe suffers from cosmic Alzheimer's disease.

5.6 Cellular Automata, Chaos, and Randomness

The dynamics of life and brain have been a challenge of traditional computer science and artificial intelligence. It is obvious that algorithmic mechanization with Turing-like machines confront severe obstacles which cannot be overcome by growing capacities of classical or quantum computers. For example, pattern recognition and other complex tasks of human perception cannot be mastered by program-controlled computers. The structure of the human brain seems to be completely different.

In the history of science, the brain was illustrated by technical models of the most advanced machinery [5.46]. Thus, during the age of mechanization, the brain functions were thought of as hydraulic pressures which are conducted along the nerves to operate on the muscles. With the beginning of electrical technology, the brain was compared with telegraphs or telephone switchboards. Since the development of computers, the brain has been identified with the most advanced hardware generations. In the last chapter, we saw that even quantum computers (if they are ever constructed) could not increase their power beyond the complexity of Turing-like algorithms.

Unlike program-controlled serial computers, the human brain and mind are characterized by contradictions, incompleteness, robustness, and resistance to noise, but also by chaotic states, dependence on sensitive initial conditions, and last but not least by learning processes. These features are well known in the complex system approach. Concerning the architecture of Turing-like and complex systems, an essential limitation derives from the sequential and centralized control of classical systems, but complex dynamical systems are intrinsically parallel and self-organized.

Nevertheless, historically, the first designs for neural network computers were still influenced by Turing's concept of a machine. In their famous paper "A logical calculus of the ideas immanent in nervous activity" (1943), McCulloch and Pitts offered a complex model of neurons as threshold logic units with excitatory and inhibitory synapses, which applied concepts of the mathematical logic of Russell, Hilbert, Carnap, and others, and the Turing machine. A McCulloch–Pitts neuron fires an impulse y along its axon at time $n+1$ if the weighted sum of its inputs $x_1, \ldots, x_m$ and weights $w_1, \ldots, w_m$ at time n exceeds the threshold Θ of the neuron (Fig. 5.21a) [5.47].

Particular applications of McCulloch–Pitts neurons are the following models of logical connections: the OR-gate (Fig. 5.21b) models the logical disjunction x_1 OR x_2 of sentences x_1 and x_2 (formally, $x_1 \vee x_2$) which is only false if x_1 and x_2 are false sentences, and true otherwise. The truth-values are binarily represented by 0 for "false" and 1 for "true". For threshold $\Theta = 1$ and weights $w_1 = 1$ and $w_2 = 1$, the OR-gate fires with $w_1x_1 + w_2x_2 \geq \Theta$ so long as x_1 or x_2 or both x_1 and x_2 are 1.

The AND-gate (Fig. 5.21c) models the logical conjunction x_1 AND x_2 (formally, $x_1 \wedge x_2$) which is only true if x_1 and x_2 are true sentences, and false otherwise. For the threshold $\Theta = 2$ and weights $w_1 = 1$ and $w_2 = 1$, the AND-gate fires with $w_1x_1 + w_2x_2 \geq \Theta$ only if both x_1 and x_2 are 1.

The NOT-gate (Fig. 5.21d) models the logical negation NOT x_1 (formally, $\neg x_1$) which is only true if x_1 is false, and false otherwise. For threshold $\Theta = 0$ and weight $w_1 = -1$ the NOT-gate fires $w_1x_1 \geq \Theta$ only if x_1 is 0. Thus, if x_1 is 1, then the NOT-gate does not fire, which means that the ouput $y = \neg x_1 = 0$.

A neural net à la McCulloch–Pitts is a system of McCulloch–Pitts neurons, interconnected by splitting the output of each neuron into lines and connecting some of these to the inputs of other neurons (Fig. 5.22). Although this concept of a system may be very simplified, any "classical" von-Neumann computer can be simulated by a network of such neurons. In 1945, John von Neumann wrote a draft report which has become famous as the first place where the idea of a stored program, which resided in the computer's memory along with the data it was to operate on, was clearly stated. This historic document shows that von Neumann was completely aware of the possibility of computation by a McCulloch–Pitts network.

Mathematically, a von-Neumann computer can be conceived as a finite automaton, consisting of a finite set X of inputs, a finite set Y of outputs and a finite set Q of states. The dynamics of a finite automaton is defined by a next-state function δ, transforming the state q and input x at time t into a state $\delta(q, x)$ at the following time $t + 1$, and an output function β, connecting a state q with an output $\beta(q)$.

The components of a von-Neumann computer, such as an input-output unit, a store, a logical control unit, and an arithmetic unit, can easily be shown to be finite automata. Even a modern digital computer which is made up of a network of thousands of elements and integrated onto chips can be understood as a McCulloch–Pitts-like neural network. In general, every register machine, Turing machine, or recursive function can be simulated by an appropriate network of finite automata. But these applications of McCulloch–Pitts networks still work in the framework of program-controlled serial computers.

It was again John von Neumann who first tried to extend Turing's concept of a universal computer to the idea of a self-reproducing automaton [5.48]. He noted that a machine building other machines decreased its complexity, because it seems to be impossible to use more material than is given by the building machine. Contrary to this traditional mechanistic view, living organisms in biological evolution seem to be at least as complex as their parents, with increasing complexity in the long run of evolution (Herbert Spencer).

Von Neumann's concept of cellular automata gave the first hints of mathematical models of living organisms conceived as self-reproducing networks of cells. The state space is a homogeneous lattice which is divided into equal cells like a chess board. An elementary cellular automaton is a cell which can have different states, for instance "occupied" (by a mark), "free", or "colored". An aggregation of elementary automata is called a composite automaton or configuration. Each automaton is characterized by its environment, i.e., the neighboring cells. The dynamics of the

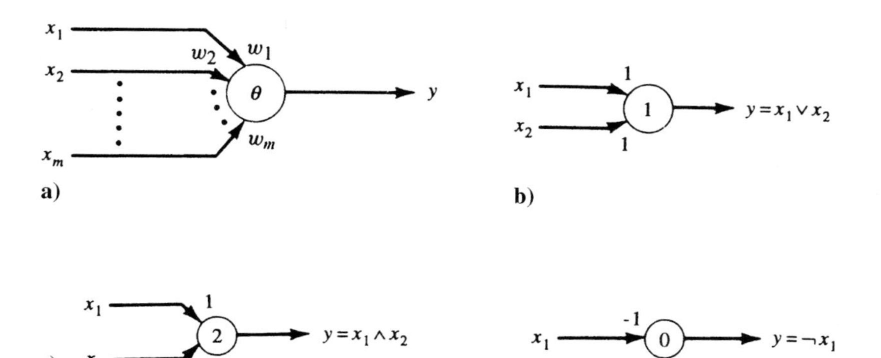

Fig. 5.21a–d. McCulloch-Pitts neuron (**a**), OR-gate (**b**), AND-gate (**c**), NOT-gate (**d**) [5.34]

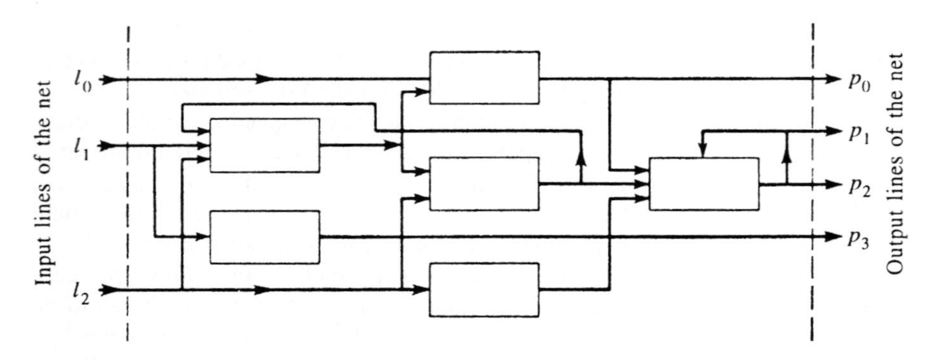

Fig. 5.22. McCulloch-Pitts network

automata is determined by synchronous transformation rules. Von Neumann proved that the typical feature of living systems, their tendency to reproduce themselves, can be simulated by an automaton with 200 000 cells (in the plane), where each cell has 29 possible states and the four orthogonal neighboring cells as environment [5.49].

This idea was developed by John Conway, whose cellular automata can simulate growth, change, and death of populations of living systems. A simple example is defined by the following synchronous rules for cells with two possible states "occupied" (by a mark) or "free":

1) Rule of survival: An occupied cell with 2 or 3 occupied neighboring cells remains unchanged.
2) Rule of death: A cell loses its mark if it either has more than 3 neighboring cells ("overpopulation") or has less than 2 neighboring cells ("isolation").
3) Rule of birth: If an empty cell has exactly 3 occupied neighboring cells, then it gets a mark.

Figure 5.23a illustrates the "death" of a configuration in the third generation and Fig. 5.23b shows "survival" after the second generation. Conway's theory has some more surprising results, which were discovered via computer experiments.

Cellular automata are not only nice computer games. They have turned out to be discrete and quantized models of complex systems with nonlinear differential equations describing their evolution dynamics. Imagine a chessboard-like plane with cells, again. A state of a one-dimensional cellular automaton consists of a finite string of cells, each of which can take one of two states ("black" (0) or "white" (1)) and is connected only to its two nearest neighbors, with which it exchanges information about its state. The following (later) states of a one-dimensional automaton are the following strings on the space-time plane, each of which consists of cells taking one of two states, depending on their preceding (earlier) states and the states of their two nearest neighbors. Figure 5.24b–e illustrates the time evolution of four automata in 60 time steps. Thus, the dynamics of an one-dimensional cellular au-

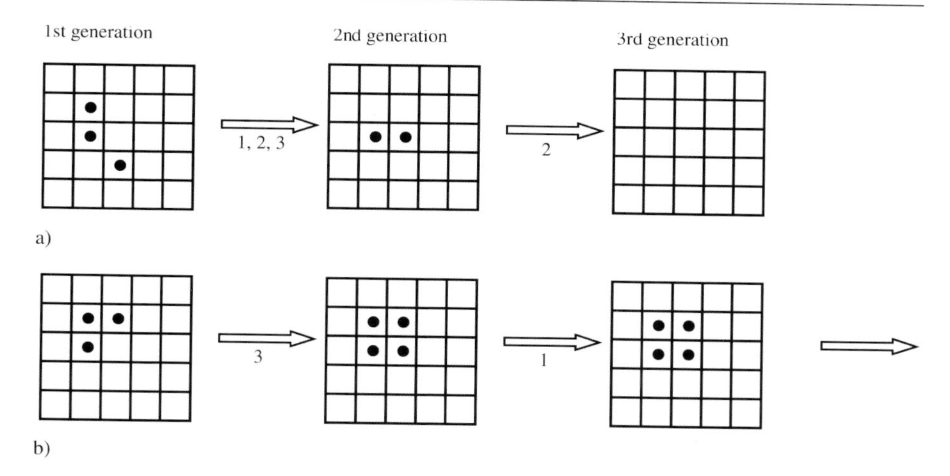

Fig. 5.23a,b. Conway's cellular automata modeling "death" (**a**) and "survival" (**b**)

tomaton is determined by a Boolean function of three variables, each of which can take either the value 0 or 1 [5.50].

For three variables and two values, there are $2^3 = 8$ possibilities for three nearest neighbor sites. In Fig. 5.24a, they are ordered according to the corresponding three-digit binary number. For each of the three nearest neighbor sites, there must be a rule determining the following state of the middle cell. For eight sequences and two possible states, there are $2^8 = 256$ possible combinations. One of these rule combinations, determining the dynamics of a one-dimensional cellular automaton, is shown in Fig. 5.24a.

Each rule is characterized by the eight-digit binary number of the states which each cell of the following string can take. These binary numbers can be ordered by their corresponding decimal numbers.

The time evolution of these simple rules, characterizing the dynamics of a 1-dimensional cellular automaton, produces very different cellular patterns, starting from simple or random initial conditions. According to Stephen Wolfram, computer experiments give rise to the following classes of attractors the cellular patterns of evolution are aiming at. After a few steps, systems of class 1 reach a homogeneous state of equilibrium independently of the initial conditions. This final state of equilibrium is visualized by a totally white plane and corresponds to a fixed point as attractor (Fig. 5.24b).

Systems of class 2, after a few time steps, show a constant or periodic pattern of evolution which is relatively independent of the initial conditions. Specific positions of the pattern may depend on the initial conditions, but not the global pattern structure itself (Fig. 5.24c).

In a 3rd class, cellular automata produce patterns that seem to spread randomly and irregularly over a grid (Fig. 5.24d), while in a 4th class, evolutionary patterns with ocassional quasi-organic and locally complex structures can be ob-

111	110	101	100	011	010	001	000
0	0	1	0	1	1	0	1

Fig. 5.24a. Dynamics of 1-dimensional cellular automata

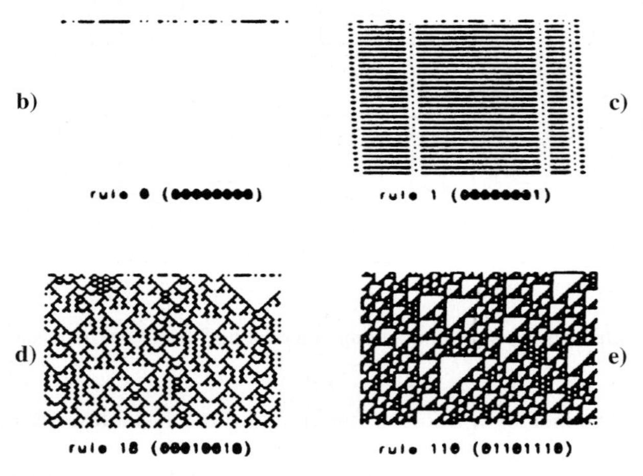

Fig. 5.24b–e. Attractors of 1-dimensional cellular automata

served (Fig. 5.24e). Contrary to 1st and 2nd class automata, patterns in the 3rd and 4th class sensibly depend on their initial conditions. Obviously, these four classes of cellular automata model attractor behavior of nonlinear complex systems, a fact well-known from self-organizing processes. They remind us of the familiar classifications of materials into solids, liquids, and gases, or living organisms, such as plants and animals. In general, the cellular automata approach confirms the intuitive idea that complex systems lie somewhere between regular order (like ice crystals and Buckminsterfullerens) and complete irregularity or noise (like molecules in a heated gas). Organisms and brains are highly complex, but neither is completely ordered nor completely random and disordered [5.51].

Obviously, these four classes of cellular automata model the attractor behavior of nonlinear complex systems which is well known from self-organizing processes. In the preceding chapters, we have seen many examples in the evolution of matter, life, and mind–brain. In Chaps. 7 and 8, we will consider many analogies with the evolution of human societies. In general, self-organization has been understood as a phase transition in a complex system. Macroscopic patterns arise from complex nonlinear interactions of microscopic elements. There are different final patterns of phase transitions corresponding to mathematically different attractors.

In Fig. 2.24a–e, a survey was given of the different attractors of a stream, the velocity of which is accelerated step by step. These patterns of the stream have many analogies with corresponding evolution patterns of cellular automata. At a first level, the stream reaches a homogeneous state of equilibrium ("fixed point"). At a higher velocity, the bifurcation of two or more vortices can be observed corresponding to periodic and quasi-periodic attractors. Finally, the order decays into deterministic chaos as a fractal attractor of complex systems. Classes 3 and 4 of the cellular automata are extremely interesting for modeling processes. Class 3 delivers evolution patterns which seem to be irregular, random, and noisy. Class 4 shows evolution patterns of dissipative systems with sometimes quasi-organic forms which can be observed in the evolution of organisms and populations.

From the information point of view, there are basic differences in the information processing of each class of automata. As pattern formation of 1st class automata is completely independent of initial patterns, the information about intial conditions is rapidly forgotten. The final uniform pattern of equilibrium shows no trace of them. In the oscillating patterns of 2nd class automata, some information about initial conditions is retained, but only in localized substructures. The global oscillating behavior is independent of initial conditions. 3rd class automata are highly sensitive to tiny changes in initial conditions. Thus, they show long-range communication of information. Any local change in pattern formation is communicated globally to the most distant parts, according to the butterfly effect. In 4th class automata, long-range communication of information is possible, but is sometimes restricted to the localized structures of their patterns. Another aspect of information processing is the loss of predictable information, or the degree of prediction uncertainty in each class of cellular automata measured by the Kolmogorov–Sinai-entropy. In the case of 1st and 2nd class automata, the final states with uniform or oscillating patterns are well-known and predictable from past states without loss of information. In the case of 3rd and 4th class automata, the prediction of random patterns has an infinite degree of uncertainty, while in the case of chaotic structures, predictions of a future state can be computed from past states with a finite degree of uncertainty.

Predictions of future development are easy for cellular automata of the first two classes. In the 1st class, cellular automata always evolve after finite steps to a uniform pattern of rest, which is repeated for all further steps in the sense of a fixed point attractor. As they preserve no information about the arrangement of cells on earlier steps, the evolution is irreversible: We have no chance to go backwards and reconstruct the initial conditions from which the automata actually started. In the 2nd class, the development of repeated patterns is obviously reversible for all future developments. It preserves sufficient information to allow one to go backwards or forwards from any particular step. In random patterns of the 3rd class, all correlations have decayed, and, therefore, the evolution is irreversible. For localized complex structures of the 4th class, we perhaps have a chance to recognize strange or chaotic attractors, which are highly complex and correlated patterns, contrary to the complete loss of structure in the case of randomness.

Of the 256 simplest 1-dimensional cellular automata with nearest neighbors and binary cellular states (or two colors), only six have reversible behavior. They only generate simple repetitive changes in the initial conditions. In these cases, it is always possible to reproduce the configurations of all previous steps, starting from any given configuration. In other words, it is possible to interchange the past and future. If we increase the number of cellular states to three, instead of two, colors, we

1	1	1	1	1	1	1	1
111	110	101	100	011	010	001	000
1	0	0	0	0	1	0	1

0	0	0	0	0	0	0	0
111	110	101	100	011	010	001	000
0	1	1	1	1	0	1	0

Fig. 5.25. Rule of the reversible cellular automaton 122R

get $3^3 = 27$ possibilities for three nearest neighbor sites and the gigantic number of $3^{27} = 7.625.597.484.987$ 1-dimensional cellular automata. Among them, there are 1800 reversible automata, so starting from any configuration of cells, it is possible to generate the configurations of all previous steps. But some of these 1800 reversible 1-dimensional automata no longer only deduce simple repetitive transformations of initial conditions, but show complex, scrambled patterns. Thus, reversible micro rules can generate complex macro behavior.

For example, we can construct reversible rules that remain the same even when turned upside-down. Therefore, the rules of a 1-dimensional cellular automaton are affected by the dependence on colors two steps back. In Fig. 5.25, we take the elementary rule 122 of the 256 simplest 1-dimensional automata with nearest neighbors and binary cellular states (or two colors). We add the restriction that the new state (colour) of a cell should be inverted if the the cell is black (1) two steps back. With knowledge of not one but two successive steps, it is always possible to determine the cellular configurations of future or past steps.

The reversibility and irreversibility of temporal development are important topics in natural science. All fundamental laws of classical, relativistic, and quantum physics are reversible: They are invariant with respect to the two possible directions of time, t or $-t$. Our everyday experience seems to support an irreversible development with one direction of time. According to the 2nd law of thermodynamics, increasing disorder and randomness ("entropy") is generated from simple and ordered initial conditions of closed dynamical systems (compare Sect. 3.2). Irreversibility is highly probable inspite of the microreversibility of molecular laws. Some cellular automata with reversible rules generate patterns of increasing randomness, starting from simple and ordered initial conditions. In Fig. 5.26, the reversible cellular automaton of rule 122R can start from an initial condition in which all black cells or particles lie in a completely ordered pattern at the center of a box. Running downwards, the distribution seems to become more and more random and irreversible, in accordance with the 2nd law.

In principle, reversibility is possible, analogous to Poincaré's famous theorem of reversibility in statistical mechanics, but extremely improbable. By starting with a simple state and tracing the actual evolution, one can find initial conditions that will lead toward to decreasing randomness (Fig. 5.26). But for cellular automata, the computational amount to go backwards and find these conditions cannot be reduced to the actual evolution from simple to random patterns: Computational irreducibility corresponds to temporal irreducibility and improbability. Thus, in computer experi-

Fig. 5.26. Reversible cellular automaton with random pattern formation illustrating the 2nd law of thermodynamics [5.52]

ments with cellular automata, we get a computational equivalence of the 2nd law of thermodynamics.

Different increasingly complex and random patterns can be generated by the same simple rules of cellular automata with different initial conditions. In many cases, there is no finite program to forecast the development of complex and random patterns. The algorithmic complexity (compare Sect. 5.2) is incompressible due to its computational irreducibility. In this case, the question of how the system will behave in the future is undecidable, because there can be no finite computation that will decide it. Obviously, computational irreducibility is connected to Turing's fundamental problem of undecidability. Whether a pattern of a cellular automaton ever dies out can be considered analogous to the halting problem of Turing machines.

Computational irreducibility means that there is no finite method of predicting how a system will behave except by going through nearly all the steps of actual development. In the history of science, one assumes that the precise knowledge of laws allows for precise forecasting of the future. Even in the case of chaos theory, there are methods of time series analysis (compare Sect. 2.6) that determine, at least, future trends and attractors of behavior. But in the case of randomness, there is no shortcut to the actual evolution. Stephen Wolfram supposes that the sciences of complexity are basically characterized by computational irreducibility [5.53]. Even if we know all the laws of behavior on the micro level, we cannot predict the development of a random system on the macro level. The brain, as a complex system, is determined by simple synaptic rules (e.g., Hebb's rule) on the micro level of neurons that are more or less well- known. Nevertheless, there is no chance of computing pattern formation of neural cell assemblies in all its details. In a philosophical sense, computational irreducibility seems to support personal individuality: Our personal life is influenced by many unexpected and random events. The pattern of our way of life is highly nonlinear, complex, and random. Thus, there is no shortcut to predicting life: If we want to experience our life, we have to live it.

From a methodological point of view, a 1-dimensional cellular automaton delivers a discrete and quantized model of the phase portrait which describes the dynamical behavior of a complex system with a nonlinear differential equation of evolution, depending on one space variable. There are many reasons for restricting oneself to discrete models. The complexity of nonlinear systems is often too great to calculate numerical approximations within a reasonable computation time. In that case, computer experiments with discrete models give, at least, a rough idea and feeling of what is going on, similar to laboratory experiments.

Two-dimensional cellular automata which have been used in Conway's game of Life can be interpreted as discrete models of complex systems with nonlinear evolution, depending on two space variables. Obviously, cellular automata are a very flexible and effective modeling instrument when the complexity of nonlinear systems is increasing and the possibility of determining their behavior by solving differential equations or even by calculating numerical approximations becomes more and more hopeless.

6 Complex Systems and the Evolution of Artificial Life and Intelligence

All kinds of complex dynamical systems can be modeled by computational systems. Therefore, the natural evolution of life and intelligence could become an important paradigm for computational models. They are no longer restricted to symbolic knowledge representation and artificial intelligence (AI) (Sect. 6.1). Their concepts are inspired by the successful technical applications of nonlinear dynamics to solid-state physics, spin-glass physics, chemical parallel computers, optical parallel computers, laser systems, and the human brain (Sect. 6.2). The cellular neural network (CNN) model has recently become an influential paradigm in complexity research and is currently being realized in information and chip technology. CNNs have resulted in a breakthrough in analog neural computing for visual computing and pattern formation. A CNN is a highly complex computational system, because it consists of a massively parallel focal-plane array with the computational power of a supercomputer (Sect. 6.3). Like the universal Turing machine model for digital computers, there is a universal CNN machine for modeling analog neural computers. CNNs are used not only for pattern recognition, but to simulate various types of pattern formation. The degree of dynamic complexity is found through empirical observations in computer experiments, and it is also rigorously defined via mathematical methods (Sect. 6.4). Exciting applications of artificial neural networks already exist in the fields of organic computing, neurobionics, medicine, and robotics (Sect. 6.5). Natural life and intelligence depends decisively on the evolution of organisms and brains. Therefore, embodied life and mind lead to embodied artificial intelligence and embodied artificial life of embodied robotics (Sect. 6.6).

6.1 Turing and Symbolic Artificial Intelligence

Symbolic knowledge representation, which is currently used in database applications, artificial intelligence, software engineering, and many other disciplines of computer science, has its roots in logic and philosophy. Aristotle (384–322 B.C.) developed logic as a precise approach to reasoning in the search for knowledge. Syllogisms were introduced as formal patterns that represent special forms of logical deduction. According to Aristotle, the subject of ontology is the study of categories of things that exist or may exist in some domain.

More recently, Descartes considered the human brain to be a store of knowledge. Recognition was made possible by an isomorphic correspondence between internal geometrical representations (*ideae*) and external situations and events. Leibniz was deeply influenced by these traditions. His *mathesis universalis* required a universal formal language (*lingua universalis*) that could be used to represent human thinking in calculation procedures which would then be implemented in mechanical calculating machines. An *ars iudicandi* would permit every problem to be solved by an algorithm after representating it in numerical symbols, and an *ars inveniendi* would enable users to seek out and enumerate desired data and solutions to problems. In the age of mechanics, knowledge representation was reduced to mechanical calculation procedures.

In the twentieth century, computational cognitivism arose along with Turing's theory of computability. In his functionalism, the hardware of a computer is related to the wetware of the human brain. The mind is considered to be the software of a computer. Turing argued that if a human mind is computable, it can be represented by a Turing program (Church's thesis), which can be computed by a universal Turing machine, i.e., technically by a general-purpose computer. Even people that do not believe in Turing's strong AI thesis often claim that classical computational cognitivism holds as follows. Computational processes operate on symbolic representations that refer to situations in the outside world (Fig. 6.1). These formal representations should obey Tarski's correspondence theory of truth, as described below. Imagine a real-world situation X_1 (e.g., some boxes on a table) which is encoded by a symbolic representation R_1=encode(X_1) (i.e., a description of the boxes on the table). If the symbolic representation R_1 is decoded, then we get the real world situation X_1 as its meaning, i.e., we decode(R_1)=X_1. A real-world operation T (e.g.,

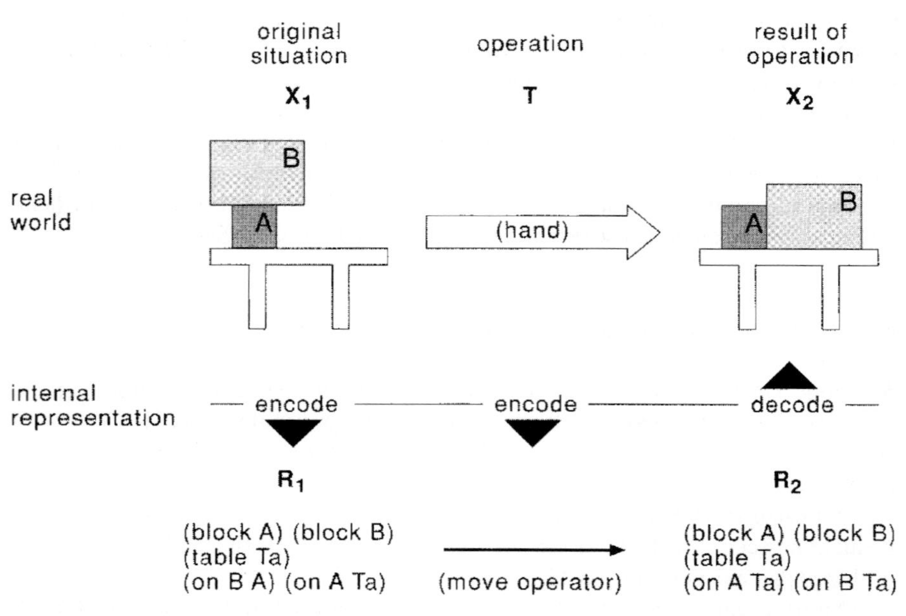

Fig. 6.1. Symbolic representation of real-world situations [6.1]

a manipulation of the boxes on the table by hand) should produce the same real-world result X_2, whether it is performed in the real world or simply on the symbolic representation: decode(encode(T)(encode(X_1)))=T(X_1)=X_2. Thus, there is an isomorphism between the outside situation and its formal representation in Cartesian tradition. As the symbolic operations are completely determined by algorithms, the real-world processes are assumed to be completely controlled. Therefore, classical robotics operate with completely determined control mechanisms.

Symbolic knowledge representation and problem solving have become key concepts in the development of Artificial Intelligence (AI). The first period of AI (1957–1962) was dominated by questions of heuristic programming, which means the automated search for human problem solutions in trees of possible derivations, controlled and evaluated by heuristics [6.2]. An example was the "Logical Theorist" (1957) of Newell, Shaw, and Simon, which delivered proofs for the first 38 theorems of Russell and Whitehead's *Principia Mathematica*. Its heuristics were extracted from the rules of thumb used by several persons in psychological tests.

In 1962 these simulative procedures were generalized and enlarged for the so-called "General Problem Solver" (GPS), which was assumed to be the heuristic framework of human problem solving. But GPS could only solve some insignificant problems in a formalized micro-world. Another example of heuristic programming was the search for winning strategies in games (chess, checkers). The first programs of pattern recognition (for instance lexical and syntactical lists of words and symbols) were based on statistical methods. But in the long run the euphoric belief in general cognitive simulation procedures was not justified by any program of this early period. At least its metaphysics inspired the invention of McCarthy's programming language LISP, which was introduced as a functional programming language for the comfortable processing of symbolic lists and which has become the most powerful programming language for knowledge based systems today.

After the failure of general methods, the AI researchers propagated ad hoc procedures of "semantic information processing". The second period of AI (1963–1967) was characterized by the development of specialized programs like STUDENT for solving simple algebraic problems, ANALOGY for pattern recognition of analogical objects, and so on. Marvin Minsky, who was the leading figure at the MIT during this period, gave up the claim of psychological simulation: "The current approach", he said, "is characterized by ad hoc solutions of cleverly chosen problems which give the illusion of complex intellectual activity." For the first time it was underlined that successful practical programming depends on special knowledge, which became the central idea of knowledge based systems later on.

The search for general principles of problem solving still survived in theoretical computer science: J.A. Robinson introduced the so-called resolution principle based on the calculus of predicate logic and Herbrand's completeness theorem permitted the finding of proofs by logical refutation procedures.

The drive to practical and specialized programming in AI was accelerated during the third period (1967–1972). It was characterized by the construction of specialized systems, methods for representation of knowledge, and interest in natural languages. J. Moses, who invented the successful MACSYMA program for mathematical applications, described the change of paradigms in AI: "In fact, 1967 is the turning point in my mind when there was enough feeling that the old ideas of general principles had to go ... I came up with an argument for what I call the primacy of expertise."

Another famous example of this period is the DENDRAL program, which uses the specialized knowledge of a chemist in mass spectroscopy in order to find structural formulas for molecules. A paradigmatic example of this period became the SHRDLU program for a robot who could manipulate a mini-world of different blocks. The system could understand and answer questions in English concerning the block-world, carry out orders for manipulating the block-world, divide orders into a series of operations, understand what is done and why, and describe its actions in English.

In the fourth period (1972–1977), description, organization, and processing of knowledge became the central paradigm combining engineering and the philosophy of AI. Mitchell Feigenbaum introduced the term "knowledge engineering" for the development of so-called expert systems. An example was the MYCIN program for medical diagnosis, which simulates a physician with special medical knowledge of bacterial infections.

A new method of knowledge representation was the conception of frames by Marvin Minsky. A new programming language for symbolic knowledge processing was PROLOG ("Programming in Logic"), which can be compared with LISP.

The fifth period of AI (1977–1986) is, so to say, "normal" in the sense of Thomas Kuhn, meaning the paradigm of expert systems is worked out and commercialized. Tools are developed in order to build new expert systems like automobiles in mass production. AI is emerging from the laboratory and philosopher's study and is becoming the key technology of a world wide knowledge industry.

In the following, emphasis is put on expert systems, because they seem to be of most interest for philosophical questions [6.3]. An expert system is a computer program that has built into it the knowledge and capability that will allow it to operate at an expert's level (for instance DENDRAL in chemistry, MYCIN in medicine). The reasoning of a human expert is illustrated in Fig. 6.2.

Some expert systems can even explain why they rejected certain paths of reasoning and chose others. Designers work hard to achieve this because they understand that the ultimate use of an expert system will depend upon its credibility to its users, and its credibility will rise if its behavior is transparent and explainable.

But unlike that of human beings, the knowledge of an expert system is restricted to a specialized information base without generalized and structuralized knowledge of the world. Therefore, expert systems have an intermediate function between the conventional programs of numerical computers and human beings (Fig. 6.3).

The architecture of an expert system consists of the following components: knowledge base, problem solving component (interference system), explanation component, knowledge acquisition and dialogue component. Their coordination is demonstrated in Fig. 6.4 [6.4].

Knowledge is the key factor in the performance of an expert system. The knowledge is of two types. The first type is the facts of the domain that are written in

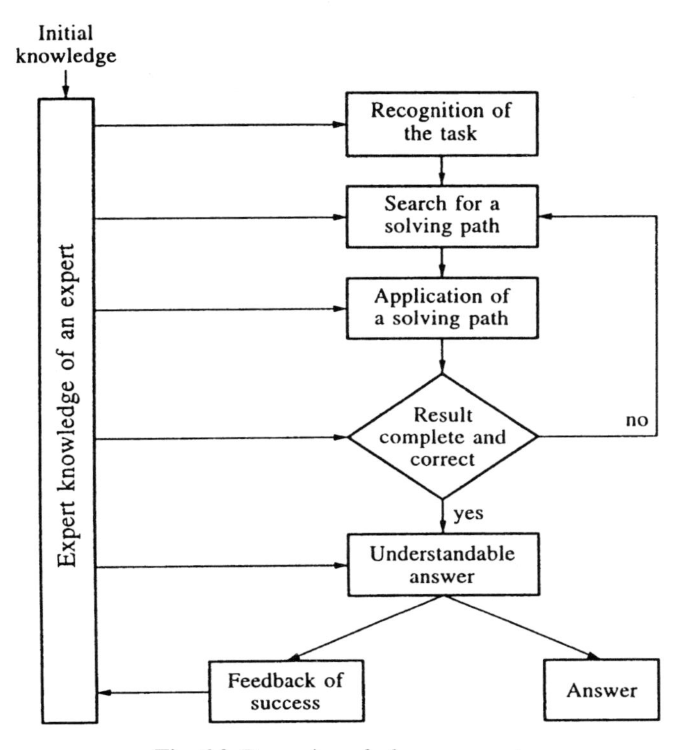

Fig. 6.2. Reasoning of a human expert

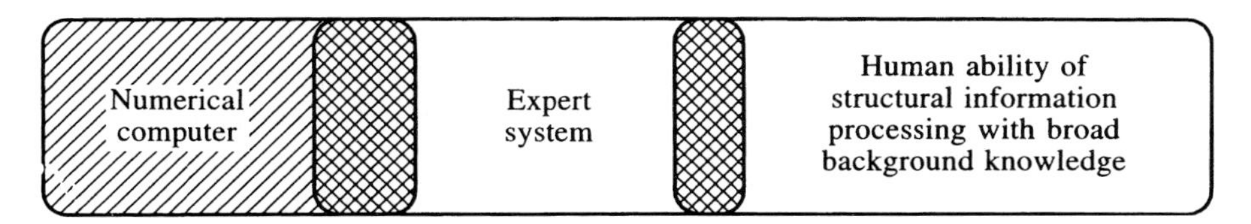

Fig. 6.3. Expert systems between numerical computers and humans

textbooks and journals in the field. Equally important to the practice of a field is the second type of knowledge, called heuristic knowledge, which is the knowledge of good practice and judgement in a field. It is experimental knowledge, the art of good guessing that a human expert acquires over years of work.

By the way, knowledge bases are not the same as data bases. The data base of a physician, for instance, is the patient's record, including patient history, measurements of vital signs, drugs given, and response to drugs. These data must be interpreted via the physician's medical knowledge for purposes of continuing diagnosis and therapy planning. The knowledge base is what the physician learned in his medical education and in the years of internship, residency, specialization, and practice. It consists of facts, prejudices, beliefs, and heuristic knowledge.

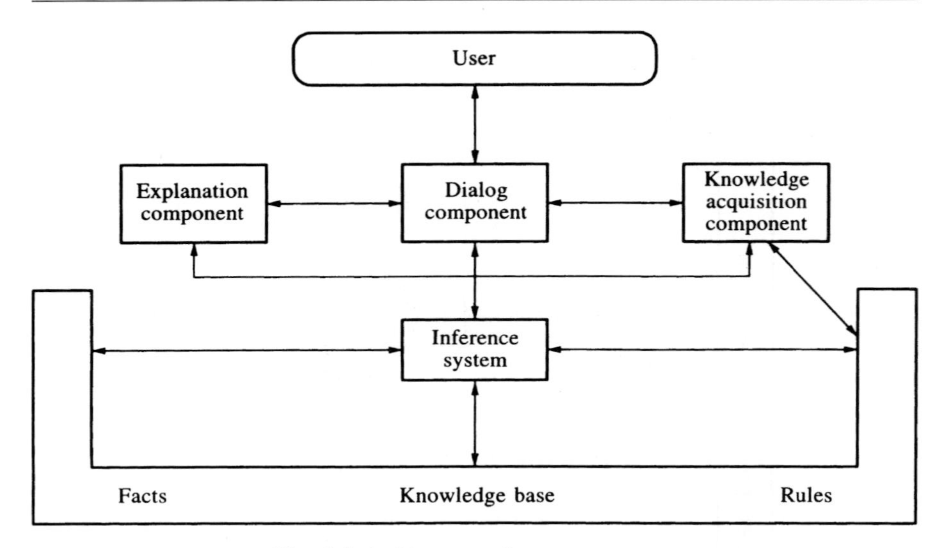

Fig. 6.4. Architecture of an expert system

The heuristic knowledge is hardest to get at because experts rarely have the self-awareness to recognize what it is. Therefore knowledge engineers with interdisciplinary training have to acquire the expert's rules, to represent them in programming language, and to put them into a working program. This component of an expert system is called knowledge acquisition. Its central function in the knowledge processing of an expert system is illustrated in Fig. 6.5.

The most important methods of knowledge representation are production systems, logic, frames, and semantical networks. In addition to knowledge, an expert system needs an inference procedure, a method of reasoning used to understand and act upon the combination of knowledge and problem data. Such procedures are independent of the special knowledge base and are founded upon different philosophical methodologies, which will be analyzed for several examples of expert systems later on.

The explanation component of expert systems has the task of explaining the steps of the procedure to the user. The question "how" aims at the explanation of facts or assertions which are derived by the system. The question "when" demands the reasons for the questions or orders of a system.

The dialogue component handles the communication of expert system and user. A natural-language processor could, of course, increase acceptance even for untrained users.

From a technological point of view the limits of expert systems are obvious. First is the problem of knowledge representation. How shall the knowledge of a domain of work be represented as data structures in the memory of the computer and accessed for problem solving? Second is the problem of knowledge utilization. How should the inference engine be designed? Third is the question of knowledge acqui-

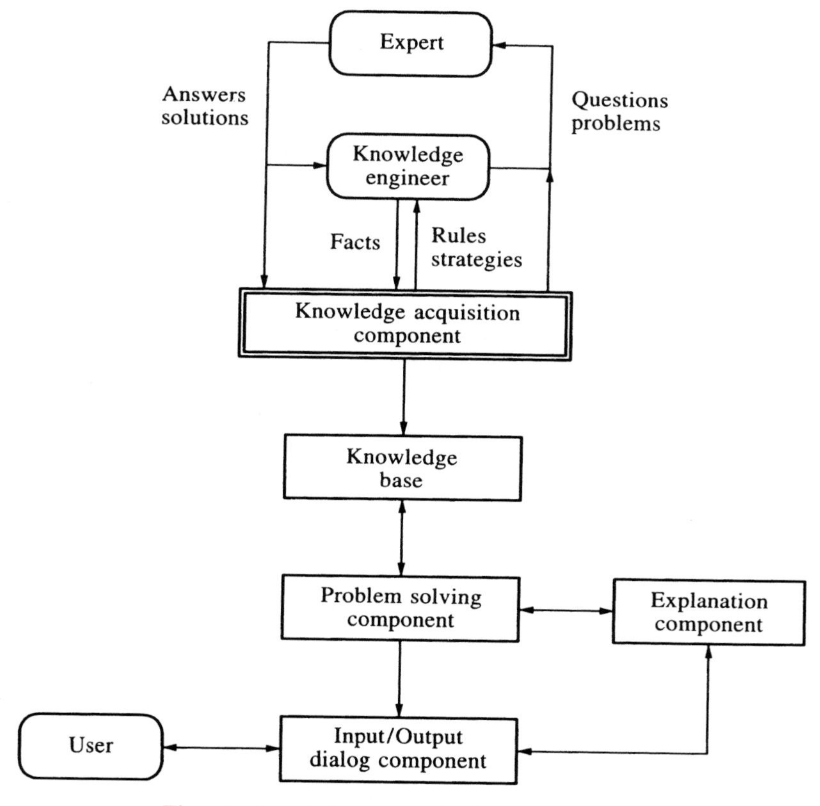

Fig. 6.5. Knowledge processing in an expert system

sition. How is it possible to acquire the knowledge so important for problem solving automatically so that the computer eases the transfer of expertise from humans to the symbolic data structures?

The last and most important problem of expert systems is philosophical. How should the specialized knowledge base of an expert system be combined with the generalized and structuralized background knowledge of the world which influences the decisions and actions of a human expert?

Thus, when deciding on an operation, a physician will also take into consideration non-objective impressions he has concerning the patient's living conditions (family, job, etc.) and his attitude towards life. Especially in fundamental questions of life and death, for example in connection with the present dispute over the dignity of dying, the whole attitude and horizon of a physician flows into his decisions in a manner that cannot be codified, although legislation seeks to lay down general standards of behavior. In expert systems of law, for example, the same aspect could be shown. Despite all consistent systems of norms a judge will in the end find a formal scope for possible decisions where he will orient towards his personal outlook on life and the world. This invasion into subjectivity should not be complained of as a lack of objectivity, but should be taken as a chance for a more humane medicine and jurisdiction. With that, however, it is not excluded that in the future computer science should aim at an expansion of knowledge based expert systems, which are nowadays still very specialized. Yet essential limits are evident, resulting from the nature of expert systems.

Expert systems are technological realizations of problem solving procedures. Therefore the factually existing expert systems can be classified by the special problems which they are intended to solve. Figure 6.6 illustrates the most important problem classes of expert systems.

A well analyzed problem class concerns "diagnosis", for instance in medicine. The input of such an expert system consists of measurement data, symptoms, etc, and it delivers patterns recognized in the data regularities as output. Another problem class concerns "design". The problem is how to find a product which conforms to some constraint. The solution of a planning problem demands a sequence of actions which transform an initial state into a goal state. A simulation problem starts with the initial state of a model, the following states of which must be calculated and evaluated [6.5].

The problem-solving strategies are inferred by production rules which must be chosen by a so-called rule interpreter. If several rules are applicable, a conflict-solving strategy decides which rule is appropriate. The possible rules can be ordered, for instance, by degrees of priority or generality. Then it may be suitable to choose the rule with the highest degree of priority or specificity.

The combination of rules in an inference can be realized by so-called forward and backward chaining. Forward chaining starts with given data and facts A and applies the deduction machinery until the given goal D is deduced (Fig. 6.7).

In contrast to the data-driven forward chaining method, the backward chaining method is goal-directed, which means that it starts with the given goal and tries

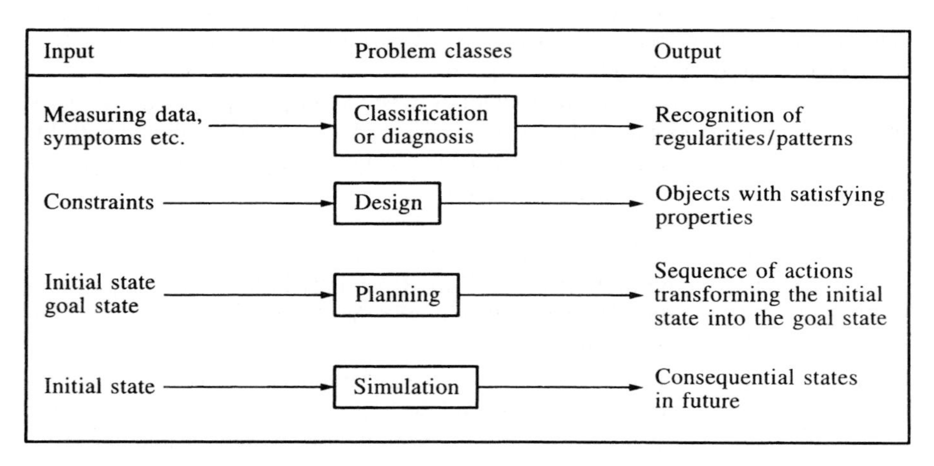

Fig. 6.6. Problem classes of expert systems

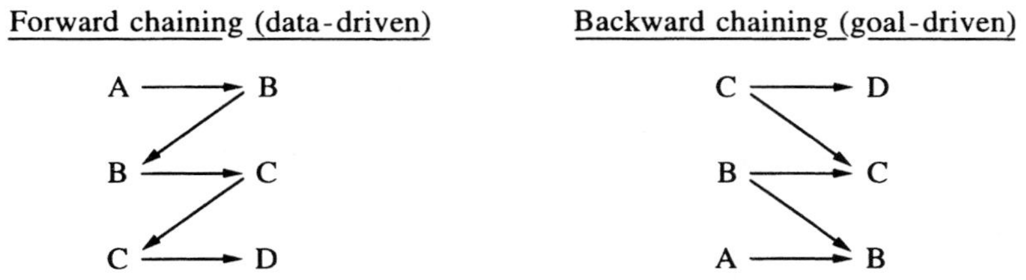

Fig. 6.7. Problem solving strategies

to find the premises of rules which make the goal deducible. Therefore a premise A must be found which is "true" or realizable (Fig. 6.7).

From a methodological point of view, the forward and backward chaining procedures of expert systems are nothing more than the well known methods of the antique logician and philosopher Pappos for finding necessary or sufficient reasons for affirmations. It is not surprising that nearly all inference strategies of expert systems are founded upon well known philosophical methodologies.

Nowadays, most of the philosophical theories used in AI are not taken from the philosophical literature directly, but that does not make them any less interesting philosophically. Nevertheless, some authors of famous expert systems were influenced directly by philosophers [6.6].

To see that AI is philosophical logic and methodology, one need only consider some expert systems in detail. Their problem class determines which strategy is suitable for problem solving. In general, a strategy aims at a reduction of the problem's complexity.

The task that the DENDRAL program addresses is the determination of molecular structure from data consisting of the molecular formula of a compound, and the mass spectrum of the compound [6.7]. The output is an ordered list of more or less probable structural formulas. Its strategy of problem solving is called "Generate-and-Test" and is an algorithm for generating the topological structures of organic molecules consistent with a given molecular formula and rules about which molecular bonds in a molecule are most likely to break. In short, we may say that the program reduces the complexity of the generated tree of solutions by early pruning of bad branches. Methodologically it involves a confirmation criterion.

In general, the following points are of importance, regardless of the chemical application:

a) There is a set of formal objects in which the solution is contained.
b) There is a generator, i.e., a complete enumeration process of this set.
c) There is a test, i.e., a predicate identifying a generated element as part or not part of a set of solutions.

This general method is defined by the following algorithm, i.e., by the following recursive function according to Church's thesis:

Function GENERATE_AND_TEST (SET):
If the set SET to be examined is empty,
then failure,
otherwise let ELEM be the "following" element of SET;
If ELEM is goal element,
then give it as solution, otherwise repeat this function with the set SET without the element ELEM.

For a translation into the AI programming language LISP [6.8], some recursive auxiliary functions have to be introduced, such as GENERATE (generates an element of a given set), GOALP (is a predicate function delivering T(true), if the argument is part of the set of solution, otherwise NIL), SOLUTION (prepares the solution element for "output"), and REMOVE (delivers the set minus the given element). Considering the common abbreviations in LISP, for example DE (Definition), COND (Condition), EQ (Equation), T (true), and the LISP conventions (e.g., brackets rules) when working out a list of symbols, the following algorithm in LISP is received:

```
(DE GENERATE_AND_TEST(SET)
 (COND((EQ SET NIL)'FAIL)
      (T(LET(ELEM(GENERATE SET))
        (COND((GOALP ELEM)(SOLUTION ELEM))
             (T(GENERATE_AND_TEST
               REMOVE ELEM SET)))))))
```

All chemical structures are systematically produced out of a given chemical sum formula, for example C_5H_{12}, in a first step:

```
 (C2H7)                       H                        H
   |    H                     |                        |
   |    |                 H — C — H                H — C — H
   C == C                     |                        |
   |    |            (C2H6) == C              (C2H5) — C
   |    H                     |                        |
H — C — H                 H — C — H                H — C — H
    |                         |                        |
    H                         H                        H

                                          H
                                          |
        H                        H ——— C ——— H
        |                        H     |     H
(C2H5) — C — (C2H5)              |     |     |
        |                    H — C ——— C ——— C — H
        H                        |     |     |
                                 H  H — C — H  H
                                       |
etc.                                   H
```

Some chemical structures are eliminated because they are unstable or contradictory. In the next step the corresponding mass spectrograms are computed and compared with mass spectrograms empirically obtained. This comparison is the test step of the process. GENERATE_AND_TEST thus technically realizes a methodology for generating presumptions with elimination of the impossible and test of the probable variants.

The META-DENDRAL program is designed to improve the rules of the DENDRAL program regarding which molecular bonds will break in the mass spectrometer. So META-DENTRAL uses the DENDRAL program plus the prediction criterion of confirmation which was critically analyzed by Hempel.

The MYCIN program for helping physicians diagnose infections is a backward-chaining deduction system [6.9]. Some 300 productions constitute MYCIN's pool of knowledge about blood bacterial infections. The following is typical:

> *If* the infection type is primary-bacteremia, the suspected entry point is the gastrointestinal tract, and the site of the culture is one of the sterile sites, *then* there is evidence that the organism is bacteriodes.

To use such knowledge, MYCIN works backward. For each of 100 possible hypotheses of diagnosis, MYCIN attempts to work toward primitive facts known from laboratory results and clinical observations. As MYCIN works in a domain where deductions are rarely certain, its developers combined a theory of plausible and probabilistic reasoning with the basic production apparatus. The theory is used to establish a so-called certainty factor for each conclusion in the AND/OR tree (Fig. 6.8).

Here F_i is the certainty factor assigned to a fact by the user, C_i indicates the certainty factor of a conclusion, and A_i the degree of reliability expected for a production rule. Certainty factors according to straightforward formulas are computed at the AND-nodes and the OR-nodes. If a certainty factor is 0.2 or less, the truth of the corresponding fact is considered unknown, and the value 0 is assigned.

The program computes degrees of inductive justification in dependence of more or less secure facts. This approach reminds of Rudolf Carnap's theory of induction. Carnap naturally did not believe in a universal inductive conclusion à la Bacon. Conclusions are always deductive. No Popperian advice was required for that. Expert systems do not operate otherwise. Nevertheless, probabilistic measures used in systems like MYCIN make the system more transparent to the user.

On the other hand, there are, so to say, Popperian programs with "Hypothesize-and-Test" strategies which generate the most interesting hypotheses with hard tests. There are programs that help to construct linear causal explanations of statistical data. Other programs uses the old philosophers' knowledge that inductive reasoning is nonmonotonic, which means a conclusion derived inductively from a set of premises may not follow from consistent extensions of the premises. For example, birds can fly, and Tweety is a bird, so I infer that Tweety can fly, but not if I also know that Tweety is an ostrich [6.10].

Another strategy is the division of a complex problem into simpler parts or less complex subproblems, which is for instance used in Georg Polya's heuristic

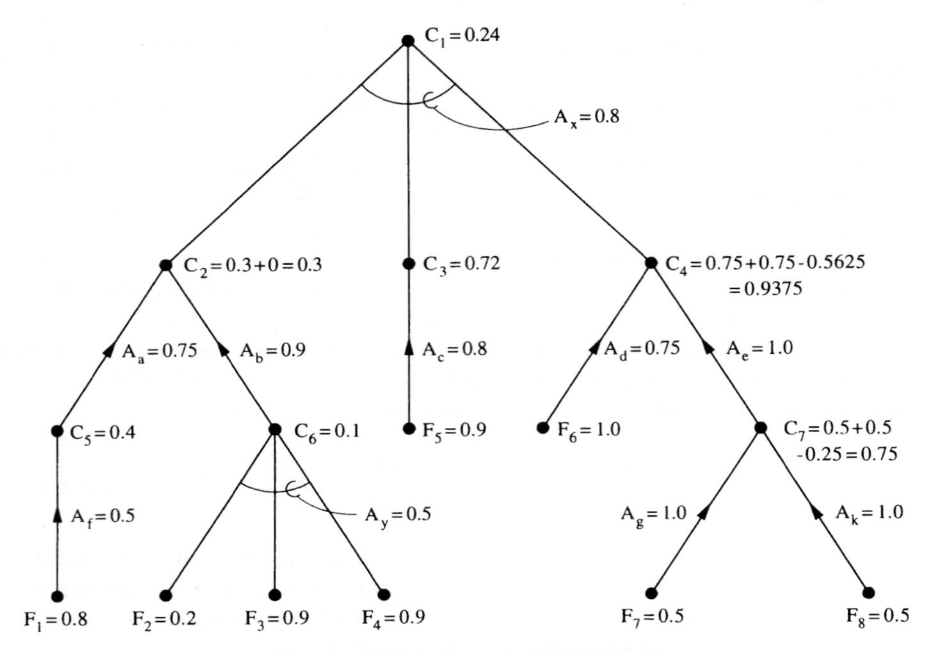

Fig. 6.8. AND/OR tree in MYCIN [6.9]

mathematical guide *How to solve it*. Therefore the domain of application must allow a division into independent parts. But it is obvious that a complex network of dependencies cannot always be cut without changing the original state of the system. For instance, consider the ecological network of the human environment or the complex psychic dependencies which a psychiatrist has to analyze. The system is not always the sum of its parts.

Some distinctions in philosophy of science can be translated as properties of knowledge based systems. If research makes extensive use of theoretical conceptions associated with intrinsic properties of a theory, the discovery process is described as theory-dependent ("theory-driven"). The converse view, which is often called Baconian, takes a body of data as its starting point. Then the discovery process is called data-driven. The distinction between theory- or data-driven knowledge processing is also well known in AI.

Now I shall sketch some programs from knowledge based systems which realize for various sciences the tasks and advantages just mentioned. My first example concerns mathematics. AM [6.11] is a knowledge based system that has recursively generated and, so to speak, rediscovered concepts from number theory. In contrast with a program in empirical science, AM's criterion of success is not that a concept is compatible with empirical data but that it is "interesting" concerning its capacity to generate examples, new problems, etc. The program, written in 1977 in LISP, begins with basic concepts such as sets, lists, equality, and operations, and heuristic advice to direct the discovery process. The heuristics suggest new tasks and create new concepts based on existing ones. New tasks are ordered according to their degrees of interest. Tasks proposed by a number of different heuristics tend to be more interesting than those proposed by a single rule.

Using this measure to direct its search through the space of mathematical concepts, AM defined concepts for the integers, multiplication, and the prime numbers, and found propositions on prime numbers (for instance the unique-factorization theorem).

A closer analysis, however, shows that a demand for simulation of a historical process of discovery cannot be maintained. The success of AM depends decisively on features of the programming language LISP. Nevertheless, an analysis shows interesting analogies with the human research process.

As is indicated by the name "LISP", lists of symbols are worked out systematically. Two lists may be recursively defined as equal when both are atomic and the atoms are equal, otherwise when the heads of the lists are equal and the rest of the lists are equal. In LISP the recursive Boolian function is noted as follows:

```
(DE LIST-EQUAL (XY)
    (COND((OR(ATOM X)(ATOM Y))
          (EQ X Y))
          (T(AND
              (LIST-EQUAL(CAR X)(CAR Y))
              (LIST-EQUAL(CDR X)(CDR Y))))))
```

Here CAR and CDR are basic operators in LISP sorting list heads and rest lists, respectively, out of given lists of symbols. A heuristic generalizing rule of AM generalizes the term of identity. Then two lists are called "generalized equal" if both are atomic and the atoms are equal, otherwise the rest lists are "generalized equal". In LISP:

```
(DE L-E-1(XY)
     (COND((OR(ATOM X)(ATOM Y))
            (EQ X Y)
            (T(L-E-1(CDR X)(CDR Y)))))
```

Because of the generalization all lists with the same length are considered equivalent. They define a class which is called "number". The process of discovery realized on concrete objects by children is simulated by AM via transformation rules. Addition is introduced as joining of two lists. The concept of prime number is discovered by a heuristic tranformation rule for forming inverses out of concepts already generated. AM was followed by EURISKO (1983) which can discover not only new concepts but new heuristics as well.

The discovering of quantitative empirical laws was analyzed by a sequence of programs called BACON [6.12]. The BACON systems are named after Francis Bacon, because they incorporate some of his ideas on the nature of scientific reasoning. They are data-driven knowledge-processing systems which gather data, discover regularities between two or more variables, and test laws. The basic methods of BACON make no reference to the semantic meaning of the data on which they operate and make no special assumptions about the structure of the data. In cases where one has experimental control over the independent terms, the traditional method of "varying one term at a time" can be used to separate the effects of each independent term on the dependent variables. Physical laws that can be reproduced by the BACON program include Boyle's law, Kepler's third law, Galileo's law, and Ohm's law.

An inquiry made with such a knowledge based system at least obeys the requirement that laws relevant for different disciplines should fulfill the same methodological and heuristic frame conditions. The corresponding knowledge based system not only reproduces certain laws, which were discovered in different historical contexts, but also systematically generates the complete methodological scope of concepts and sorts out interesting applications. The so-far latest BACON program is not only data-driven, and thus in a strict sense "Baconian", but also theory-driven. In its theoretical requirement for symmetry and conservation laws it generates, for example, the conservation law of momentum.

Another sequence of programs are capable of inducing qualitative laws from empirical data (GLAUBER, STAHL, DALTON). These programs can also induce structural and explanatory models for certain phenomena. Qualitative laws are usual in chemistry [6.13].

Competition between a machine and a human scientist is not intended. However, systematic and structural classifications of scientific laws and theories have been accomplished which give insight into the complexity of scientific laws and their conditions of discovery.

There are several aspects of the diverse activity known as scientific discovery, such as finding quantitative laws, generating qualitative laws, inferring the components of substances, and formulating structural models. An integrated discovery system is envisaged which incorporates the individual systems as components. Each component accepts input from one or more of the other components.

For instance, STAHL focuses on determining the components of chemical substances, whereas DALTON is concerned with the number of particles involved in a reaction. Thus, STAHL can be viewed as laying the groundwork for a detailed structural model in the sense of DALTON. In this way it might be possible to develop more and more complex knowledge-based systems to analyze research as knowledge processing and problem solving.

Even within such an expanded framework of research, we have not addressed the mechanisms underlying the planning of experiments, or the invention of new instruments of measurement. Any intrinsic concept, in association with the experimental arrangement that allows it to be measured, can be employed as a scientific instrument. In this case the discovery of the instrument is coincident with the discovery of the concept itself.

There are also knowledge based systems which consider the design of experiments and their interaction with other activities of scientific research. A system called KEKADA (invented by a research group of Simon) is shown in Fig. 6.9 with hypotheses generators, experiment choosers, expectation setters, etc. [6.14]. It was developed to model the design of an experiment in biochemistry (Krebs' discovery

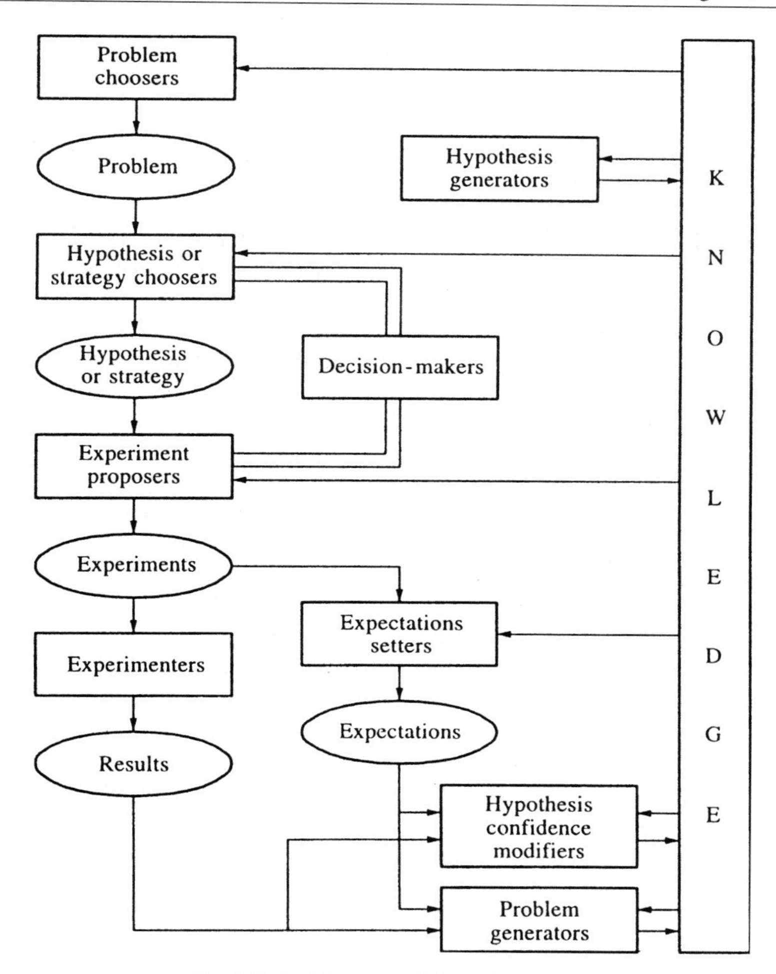

Fig. 6.9. Architecture of KEKADA [6.14]

of the urea cycle 1935). Like knowledge engineers, Simon and his crew analyzed the laboratory notebooks of Krebs, defined his methodological rules of research and translated them into a LISP-like programming language.

If the system has not decided which task to work on, problem choosers will decide which problem the system should start working on. Hypothesis generators create hypotheses when faced with a new problem. The hypothesis or strategy proposers will choose a strategy to work on. Then the experiment proposers will propose the experiments to be carried out. Both types of heuristic may need the decision makers. Then expectation setters set expectations and experimenters carry out experiments. The results of the experimenters are interpreted by the hypothesis modifiers and the confidence modifiers. When applicable, problem generators may add new problems to the agenda. If the outcome of an experiment violates the expectations for it, then the study of this puzzling phenomenon is made a task, and it is added to the agenda.

Every component of the system is an operator which is defined by a list of production rules. Besides domain-specific heuristics the system contains general rules that are part of general methodology of research. It is remarkable that a specific rule defines the case that an experimental result is a "puzzling phenomenon". Scientific discovery thus becomes a gradual process guided by problem-solving heuristics, and not by a single "flash of insight" or sudden leap. These examples of knowledge based systems may be interpreted as assistants in the studies of philosophers of science in the sense that, for instance, the program DENDRAL is an assistant in the laboratory of a chemist. They can investigate the whole space of possible laws produced by some heuristical rules. But they are the accurate assistants, not the masters. Their "flash of insight", the kind of "surprise" the system can recognize, is frame-dependent, given by the masters.

What about Turing's question, which motivated early AI researchers? Can machines "think"? Are machines "intelligent"? In my opinion, this question is for computer technology a metaphysical one, because "thinking" and "intelligence" are not well-defined concepts of computer science or AI.

All we can say today is this. If a program generates a structure that can be interpreted as a new concept, then the rules of transformation used contain this concept and the corresponding data structure implicitly. An algorithm which directs the application of these rules makes the implicitly given concepts and data structures explicit. In philosophical discussions on AI much confusion is provoked by the terminology of AI, which is introduced in a technical sense, but is associated

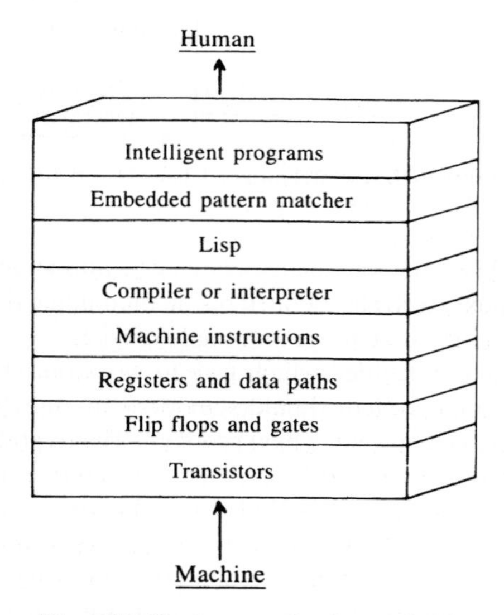

Fig. 6.10. Hardware and software levels

with sometimes old and sophisticated meanings in philosophy and psychology. As in other sciences, we have to live with traditional terms and notions, which may be highly confusing if they are abstracted from their technical context. An example was the notion "intelligence", for instance, in "Artificial Intelligence" (AI).

A term which sometimes confuses philosophers is the use of "knowledge" in AI. Let me again underline that "knowledge" in the term "knowledge based system" has a technical meaning, and does not claim to explain the whole philosophical, psychological, or sociological concept of knowledge. In AI technology as part of practical computer science, no philosophical reductionism is involved.

"Knowledge processing" in so-called "knowledge based systems" means a new kind of complex information processing which must be distinguished from the older merely numerical data processing. It involves complex transformation rules of translation and interpretation which are characterized at a high level in the hierarchy of programming languages (today LISP or PROLOG). This level is near to natural languages, but not identical, of course, and only grasps partial aspects of the broad meaning of human knowledge (Fig. 6.10). Nevertheless, knowledge processing remains program-controlled and in the tradition of Leibniz' mechanization of thoughts [6.15].

6.2 Neural Networks and Synergetic Computers

The algorithmic mechanization of thinking via program-controlled computers has some severe obstacles which cannot be overcome by simple growth in computational capacity. For example, pattern recognition, the coordination of movements, and other complex tasks of human learning cannot be mastered by Turing-like computer programs. Cellular automata and artificial neural networks make use of the principles of complex dynamical systems.

Historically, the modern development of cellular automata dates back to von Neumann's early ideas on self-reproducing automata. Besides self-reproduction, there is another feature which appears to be essential for natural complex systems, in contrast to traditional computers. The human brain has the ability to learn, for instance by perception. In the first logical model of the brain, which was provided by the McCulloch–Pitts network, the function of an artificial neuron was always fixed. McCulloch and Pitts succeeded in demonstrating that a network of formal neurons of this type could compute any finite logical expression.

However, in order to make a neural computer capable of complex tasks, it is necessary to find mechanisms of self-organization that allow the network to learn. In 1949, Donald Hebb suggested the first neurophysiological learning rule, which gained importance during the development of neural computers. Synapses of neurons do not always have the same sensitivity, but modify themselves in order to favor the repetition of firing patterns which have occurred frequently in the past.

In 1958, Rosenblatt designed the first learning neural computer, which he famously termed the "perceptron" [6.16]. Rosenblatt was originally a psychologist with a typical psychological interest in human learning processes. However, his design for a learning machine that was capable of complex adaptive behavior was extremely interesting to engineers and physicists. Thus, it is no wonder that this idea, originated by a psychologist, was grasped by engineers who were more interested in robots and computer technology than in simulating processes in the human brain. From a technical point of view, it is not essential that the learning procedures of neural computers resemble the learning processes of the mind–brain system. They must effectively manage complex tasks involving adaptive behavior, but they can use methods that are completely different from those developed during biological evolution.

Rosenblatt's neural computer is a feedforward network with binary threshold units and three layers. The first layer is a sensory surface called a "retina," which consists of stimulus cells (S-units). The S-units are connected with the intermediate layer by fixed weights that do not change during the learning process. The elements of the intermediate layer are called associator cells (A-units). Each A-unit has a fixed weighted input of some S-units. In other words, some S-units project their output onto an A-unit. An S-unit may also project its output onto several A-units. The intermediate layer is completely connected to the output layer, the elements of which are called response cells (R-units). The weights between the intermediate layer and the output layer are variable, which makes the system capable of learning.

The perceptron is a neural computer that can classify a perceived pattern into one of several possible groups. In the case of two groups, each R-unit learns to distinguish between the input patterns by activation and deactivation. The learning procedure of a perceptron is supervised. Thus, the desired R-unit states (active or not) that correspond to a particular pattern to be learnt must be known in advance. The patterns to be learnt are offered to the network, and the weights between the intermediate and output layer are modified according to the learning rule. This procedure is repeated until all patterns produce the correct output.

The learning procedure is a simple algorithm. For each element i of the output layer, the actual output o_i produced by a certain pattern is compared with the desired output d_i. If $o_i = d_i$, then the pattern is already correctly classified. If the desired output d_i is equal to one and the actual output o_i is zero, then all weights $w_{ij}(t)$ at time t with an active unit ($o_j > 0$) are increased in the following step $t + 1$, or more formally $w_{ij}(t + 1) = w_{ij}(t) + \sigma o_j$. The constant σ is the learning rate, which governs the speed of learning. If the desired output is zero and the actual output is one, then all weights with an active element are decreased, or more formally $w_{ij}(t + 1) = w_{ij}(t) - \sigma o_j$.

The perceptron initially appeared to usher in a new era of computer technology based on neural networks that could do anything. Early papers by the Perceptron Group made exaggerated claims. However, in 1969, the early enthusiasm gave way to bitter criticism. In that year, Marvin Minsky and Seymour Papert published their famous book *Perceptrons*, which discussed the limitations of perceptrons with mathematical precision [6.17]. The reaction to this analysis was that most research groups gave up their interest in the network and complex system approach and moved over to classical AI and computer technology, which seemed to be more profitable than the "speculations" of fans of the perceptron.

This approach taken by the scientific community after 1969 was, of course, an overreaction too. Uncritical enthusiasm and uncritical condemnation are inadequate measures of evolution in science. Darwinian evolution took millions of years to build brains capable of pattern recognition. It would be surprising if engineers could succeed in constructing analogous neural computers in just a few years.

The critical questions are the following. What can a perceptron do? What can it not do, and why does it have these limitations? A crucial step towards answering these questions was Minsky's and Papert's proof of the so-called perceptron convergence theorem. This states that any solution *learnt* by the network can in principle also be *found* in a finite number of learning steps. Therefore, the convergence of the system to a solution was proved.

However, the question then arose of whether a particular solution can be *learnt* by a perceptron in principle. In general, we have to determine the classes of problems that may be applicable to perceptrons. Some simple problems show that perceptrons cannot be applied universally, unlike it was initially and enthusiastically believed. For instance, a perceptron is not able to distinguish between even and odd numbers. One special case of this so-called parity problem is the following application to elementary logics.

A perceptron is not able to learn to behave as an exclusive OR (abbreviated XOR) gate. This unsolvable cognitive task places severe limitations on the applications of perceptrons to AI, as we will now show. Recall that the exclusive OR gate only outputs "true" for x XOR y if *either* x or y, not both x and y, are true. (An OR gate outputs "false" for x OR y if both x and y are false; otherwise it outputs "true.") The following table shows output values from the Boolean functions OR and XOR for various pairs of input values:

x	y	x XOR y	x OR y
1	1	0	1
1	0	1	1
0	1	1	1
0	0	0	0

Now, imagine a network with two input units x and y and an output unit z that can take the states 1 (active) and 0 (inactive). To simulate XOR, the output should be 0 for an even input (either both input units are active or neither is), and 1 for an odd input (one input unit is active and the other one not). In Fig. 6.11a,b, the possible input configurations of OR and XOR are illustrated in a coordinate system where the inputs x and y are coordinates.

Each pair (x, y) of coordinates x and y has a corresponding value z, which is marked by a white dot (0) or black dot (1). A linear threshold element Θ calculates the sum of the weighted inputs x and y with weights w_1 and w_2, or more formally $\Theta = w_1x + w_2y$. A simple derivation provides an equation for a straight line, as drawn in Fig. 6.11a,b. The position of the straight line is determined by the weights w_1 and w_2. This separates the active and inactive states of the threshold element.

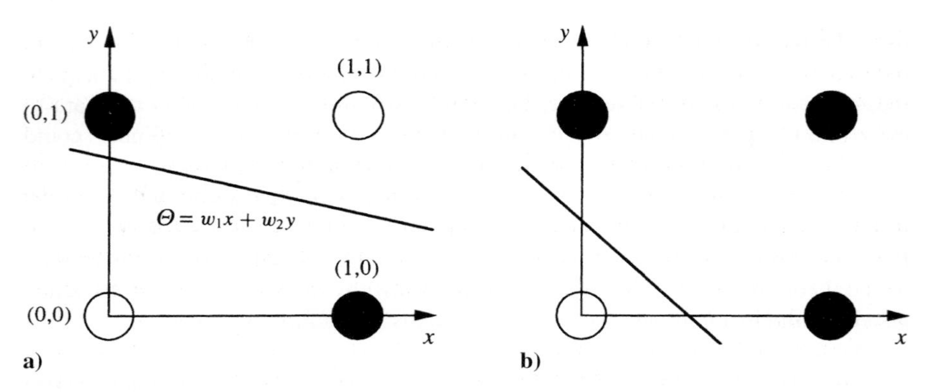

Fig. 6.11a,b. The linear separation of active and inactive states is not possible for the XOR-problem (**a**), but for the OR-problem (**b**)

In order to solve ("learn") the OR or XOR problem, the weights w_1 and w_2 must be adjusted in such a way that the points (x, y) with value $z = 1$ are separated from the points with value 0. This linear separation is geometrically possible for the OR problem (Fig. 6.11b), but impossible for the XOR problem. In general, the classification of input patterns by a perceptron is restricted to linearly separable patterns.

This result can easily be generalized to networks with more than two input units and real values. Many problems resemble XOR in that they are not linearly separable, including most interesting computational problems. The XOR problem can be solved by adding a single hidden unit to the network, which is connected to both inputs and the output (see Fig. 6.11c).

When both inputs are zero, the hidden intermediate unit with a positive threshold is off. A zero signal reaches the output, and because the threshold is positive,

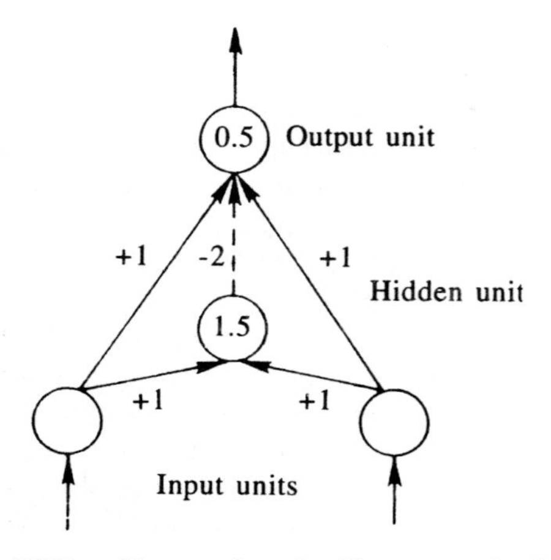

Fig. 6.11c. The XOR problem can be solved by a network with a hidden unit

the output is zero. If only one of the two inputs is 1, the hidden unit remains off and the output unit is switched on by the direct connection between the inputs and the output. Finally, when both inputs are 1, the hidden unit passes a 1 and inhibits the activation of the output unit due to its negative weight of -2.

Thus, the incorporation of the hidden unit lead to an appropriate internal representation. The XOR problem becomes a linearly separable problem through the use of a two-dimensional threshold plane in the three-dimensional coordinate system that arises when the three inputs of the output unit are employed as coordinates. This separation is possible because the input (1,1) is shifted in the z-plane to the point (1,1,1) (Fig. 6.11d).

A perceptron has only a single intermediate layer containing processor elements that can learn. In a multilayered network, we encounter the problem that the errors produced cannot be detected directly in the layers of neurons without connection to the outer world. An error can only be derived directly between the output layer and the intermediate layer below it.

The ability of a multilayered network to represent information and to solve problems depends on the number of learning layers and the numbers of units in these layers. It is therefore important to study the complexity of computation in neural computers, because it is necessary to increase the complexity of the network in order to move beyond the limitations of perceptrons.

In Sect. 4.2 we discussed the advantages of backpropagation in a multi-layered network (Fig. 4.17). A backpropagating learning algorithm allows us to define an error signal, even for the neurons in hidden layers. The error of the output layer is recursively propagated back to the layer below it. The algorithm is able to construct networks with many hidden layers, all of which contain neurons that can learn. As multilayered networks can represent much more information in their hidden layers than one-layered networks, networks with backpropagation are very powerful models for overcoming the limitations of perceptrons.

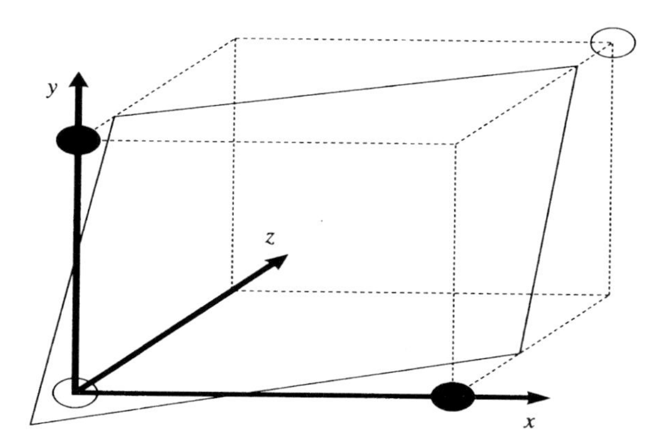

Fig. 6.11d. The XOR problem is made linearly separable by using a two-dimensional threshold plane [6.20], 92

While backpropagation delivers technically successful models, such models generally are not analogous to neural networks produced by biological evolution. Their adjustment of weights seems to be quite different from the approach used in biological synapses. However, computer technology does not aim to simulate the brain, but to solve problems in a reasonable amount of time. We must also look beyond the childish illusion that nature is a God-like engineer that always finds the best solutions through evolution. As we have already underlined in earlier chapters, there is no centralized controlling and programming unit in nature. There are often only local solutions that are, in general, not the "optimal" ones.

In 1988, a feedforward network, trained by the backpropagation of error method, was designed by Gorman and Sejnowski that could distinguish between sonar echoes of rocks and sonar echoes of mines. This task is rather difficult, even for an untrained human ear, but it is of course of great importance to submarine engineers confronted with the problem of designing a sonar system that distinguishes between explosive mines and rocks. The architecture of the proposed network consisted of an input layer with 60 units, a hidden layer with 1–24 units, and two output units, each one representing the prototypes to be distinguished, "mine" and "rock" (Fig. 6.12) [6.18].

Initially, a sonar echo is prepared by a frequency analyzer and is divided into 60 different frequency bands. Each value ranges between 0 and 1. These 60 values are the components of an input vector that is passed to the corresponding input units. They are transformed by the hidden units, leading to the activation of one of the two output units, which can take values between 0 and 1. Thus, in a trained network with well-adjusted weights, a mine echo leads to the output signal (1,0), while a rock echo has the output signal (0,1).

In order to train the net, it must be fed with examples of mine and rock echoes. In each case, the factual values of the output units are measured and compared with the desired values, based on the corresponding input. The difference between them is the error signal, which initiates small changes in the weights of the units. Using this gradient-descent procedure, the weights of the network are slowly adjusted to the correct levels.

The mine–rock network of Gorman and Sejnowski is an application of a complex system to AI. It is not claimed, of course, that the system simulates the way that the human brain distinguishes between two concepts like "mine" and "rock." However, we can say that this technical system does incorporate internal representations of these two concepts as prototype vectors in its hidden layer. In this restricted sense, the artificial system is "intelligent" because it can solve a task, which is said to indicate intelligence in the case of the human brain. Artificial networks are not, however, restricted to discriminating between two concepts. In 1986, Sejnowski and Rosenberg designed a network called NETalk that was taught how to read. It takes strings of characters that comprise English texts and converts them into strings of phonemes that can serve as inputs to a speech synthesizer. The surprising fact is not the resulting stammering sound of a childlike voice, which has been praised as a spectacular success in popular books. The important aspect of NETalk is its learning procedure for an internal representation of several concepts of pronunciation. For each letter

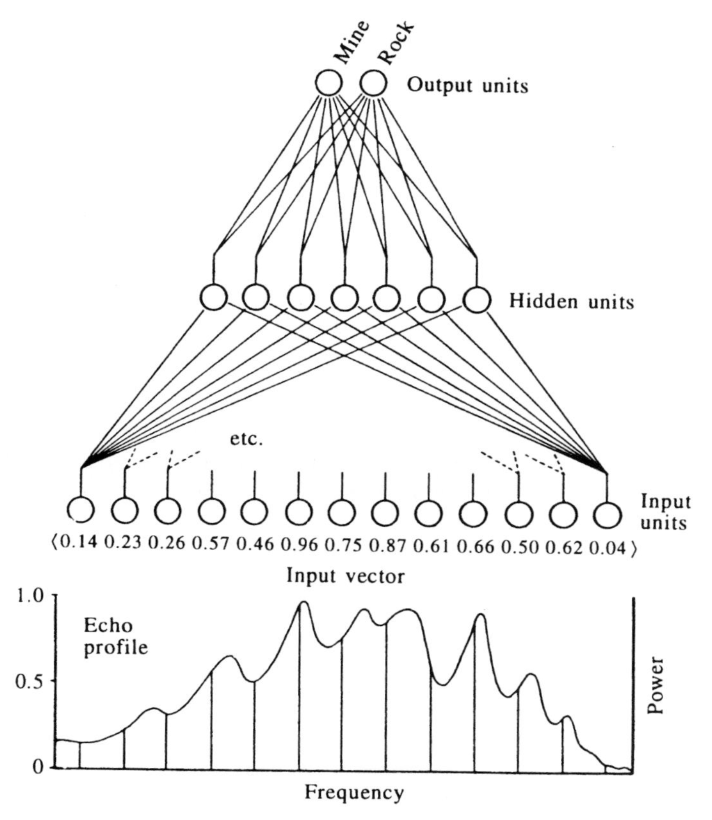

Fig. 6.12. Neural network that distinguishes between sonar echoes of rocks and mines [6.18]

of the alphabet, there is at least one phoneme assigned to it. For many letters there are several phonemes that could be signified, depending on the lexical text.

Sejnowski and Rosenberg used a three-layer feedforward network with an input layer, an intermediate hidden layer, and an output layer. Although backpropagation is unlikely to be "naturally" realized in biological brains, it has turned out to be the fastest learning procedure when compared with other solutions. The input layer looks at a seven-character window of text, for instance the word "phone" in the phrase "The_phone_is_" in Fig. 6.13a. Each of the seven letters is analyzed successively by 29 artificial neurons, each representing a letter of the alphabet (including neurons for a blank spaces and punctuation marks). Thus, exactly one neuron from each 29-element neural subsystem is activated.

The output layer consists of 26 neurons, each of which represents a component of pronunciation. There are six components for the position of pronunciation, eight for articulation, three for pitch, four for punctuation, and five for accentuation and syllabication. Therefore, each sound has four characteristics from these four groups of components. The $7 \times 29 = 203$ neurons of the input layer are connected with

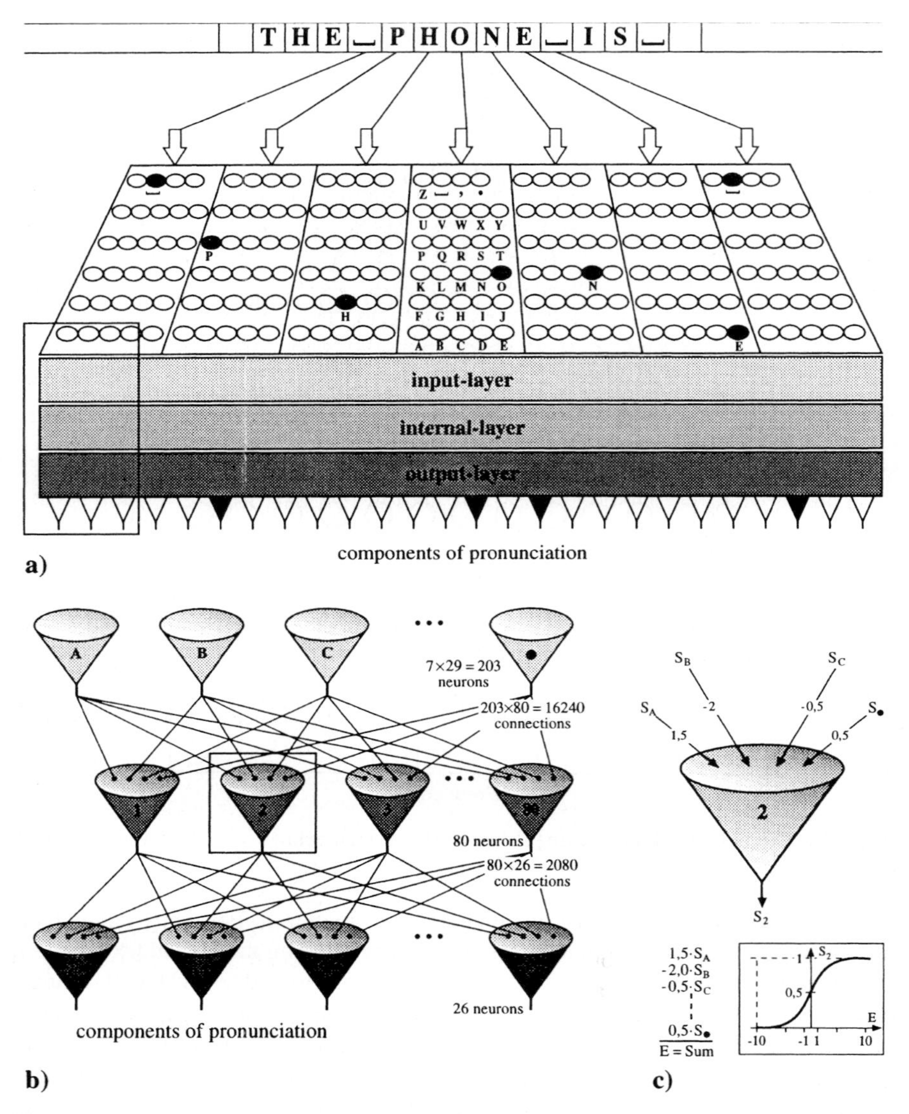

Fig. 6.13a–c. Architecture of NETalk (**a**) with neural interaction (**b**) and neural activation (**c**) [6.19]

80 internal neurons in the hidden layer, which is interconnected with 26 neurons of the output layer (Fig. 6.13b). Neurons from the same layer are not connected. The neurons of the input and output layers are also not directly connected.

An internal neuron from the hidden layer receives signals from 203 input neurons, but only sends 26 signals to the output layer. As the internal neurons are threshold units with thresholds $T_1, \ldots, T_8$, the inputs are multiplied by particular weights, and the size of the sum of these products dictates whether or not the neuron is activated (Fig. 6.13c). Realistically, the activation happens according to a continuous "sigmoid curve" and not a digital jump [6.19].

The weights are initially fixed at random. Thus, NETalk first outputs incomprehensible stammering. In the learning phase, NETalk uses a particular text from a child with well-known pronunciation. The random sounds are compared with the desired ones, and the weights are corrected by backpropagation. It is worth noting that this procedure is a kind of self-organization, and not a rule-based program of pronunciation. There is only the global demand to change the weights by approximating the factual output to the desired one. After ten runs of the particular text from the child, the network could already pronounce text that could be understood. After 50 runs only 5% of the text was pronounced incorrectly. In this phase, when an unknown passage of text from the child was passed to the network, only 22% of it was pronounced incorrectly.

So far, networks like NETalk have been simulated by traditional von Neumann computers because direct hardware for complex networks has not been available. Thus, each neuron must be calculated sequentially. Even today, the principles of self-organizing complex networks are still mainly realized in software, not in hardware. Nevertheless, we can still discuss "neural computers" because there are no known theoretical barriers to creating such hardware; its development depends simply on technical advances in the future, such as in solid state materials or optical procedures.

Projections derived from neural networks seem to be rather successful and profitable in financial, insurance, and stock exchange forecasts. The reason for this is that short-term forecasts of stock quotations are based on chaotic time series which become more and more chaotic if the period covered by the forecast decreases.

Conventional statistical programs are only successful in long-term forecasts if the stock trends can be smoothed without loss of relevant information. In this case, good statistical programs have an accuracy of between 60% and 75%. However, short-term forecasts are rather limited. Conventional statistical procedures for smoothing stock trends ignore the essential properties of short-term forecasts, which are frequent but small exchange fluctuations. In a conventional statistical program, the relevant calculation factors must be provided explicitly. A well-trained and appropriately designed neural network can recognize the relevant factors without explicit programming. It can weight the input data and minimize the forecasting errors in a self-organizing procedure. Furthermore, it can adapt to changing conditions in the system's environment, unlike computer programs that must be changed explicitly by a programmer. In order to design a neural network for stock forecasting, the input data must be prepared by codifying the stock data into binary digits. The input vector consists of several partial vectors that represent the exchange value, the absolute change from the day before, the direction of change, the direction of change from the day before yesterday, and relevant changes of greater than 1% compared to the day before. If the input vector has a fixed length of, say, 40 units, then the lengths of the partial vectors can vary, depending on their desired relevance. The system may have two output units. Activation of the left one signals a decrease in the value of the stock, while activation of the right one signals an increase.

In the learning phase, the network was fed with the actual daily quoted rate during a fixed period from, say, February 9, 1989, to April 18, 1989. Based on this learnt data, the network forecasts the stock trends for the next 19 days. The forecasts were then compared with the actual curves in order to measure the system's accuracy. Several multilayered structures were tested by backpropagation. They developed their particular global heuristics for forecasts by self-organization. For instance, if nearly the same value was forecast for the day after as for the actual day, then the error was relatively small. The heuristic (rule of thumb) used here is that a change in the quotation is more unlikely than it remaining unchanged. Figure 6.14a,b show forecast (+) and actual (−) stock quotation curves for a bank (Commerzbank) and a firm (Mercedes) [6.20].

Obviously, feedforward networks with backpropagation are very interesting from a technical point of view, although they do not seem to share many similarities with information processing in biological brains. In Sect. 4.2, we analyzed Hopfield systems with feedback (Fig. 4.8b) and Hebb-type learning (Fig. 4.9b), which appears to be utilized by biological brains. In the case of a homogeneous network of boolean neurons, the two states of the neurons can be associated with the two possible spin states of an electron in an external magnetic field. A Hopfield model is a dynamical system which, similar to annealing processes in metals, involves an energy function. As it is a nonincreasing monotonic function, the system relaxes into a local energy minimum, corresponding to a locally stable stationary state (a fixed point attractor).

Thus, the dynamical evolution of a Hopfield system may correspond to mental recognition. For example, an initial state representing a noisy picture of the letter "A" evolved towards a final state representing the correct picture, which was trained into the system using several examples (Fig. 4.9b). The physical explanation for this can be given in terms of a phase transition in equilibrium thermodynamics. The correct pattern is connected to the fixed point or final state of equilibrium. A more flexible generalization is that of a Boltzmann machine with a stochastic network architecture of nondeterministic processor elements and with a distributed knowledge representation, mathematically corresponding to an energy function (Fig. 4.11b).

The general idea behind relaxation is that a network converges to a more or less global state of equilibrium on the basis of local interactions. Through iterative modification of the local connections (for instance, using a Hebb learning strategy in the case of a Hopfield system), the network as a whole eventually relaxes into a stable and optimal state. We can say that local interactions lead to a cooperative search which is not supervised, but self-organized. There are networks that use the strategy of cooperative search to perform mental activities such as searching for a probable hypothesis. Imagine that a range of competing hypotheses are represented by neural units that may activate or inhibit themselves. The system thus moves away from the less probable hypotheses toward more probable hypotheses.

In 1986, McClelland and Rumelhart used this cognitive interpretation to simulate the recognition of ambivalent figures, which is a well known problem in *Gestalt*

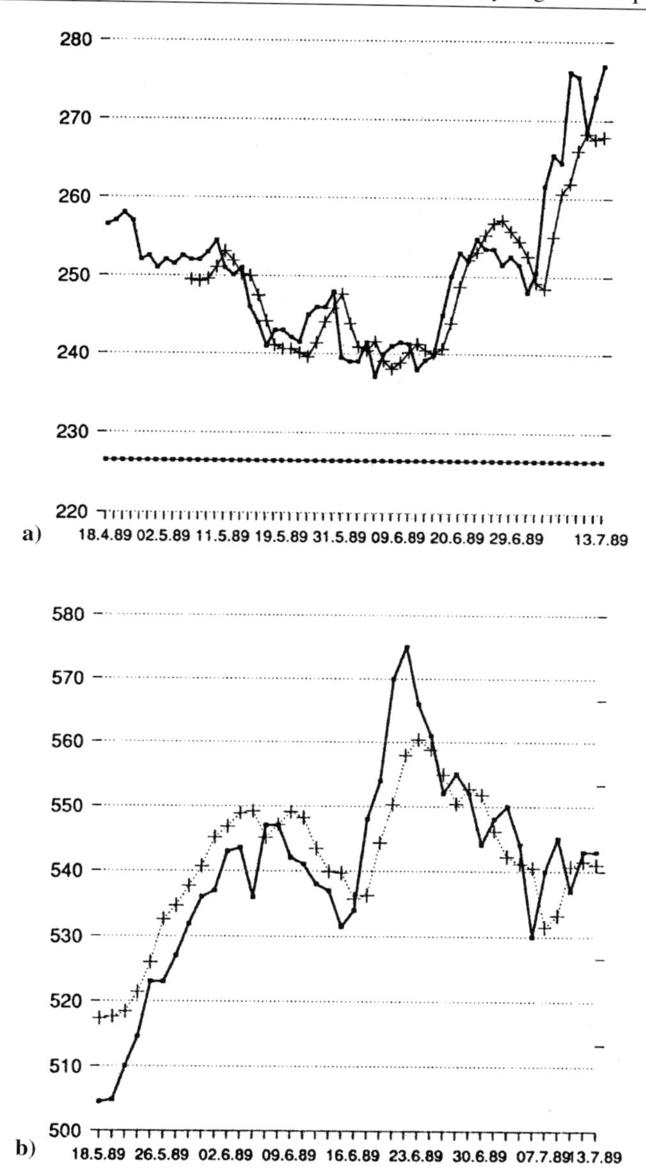

Fig. 6.14a,b. Neural network curves for forecast (+) and actual (—) stock quotations for Commerzbank (**a**) and Mercedes (**b**) [6.20]

psychology. Figure 6.15a shows a network used for a cooperative search that simulates the recognition of one of the two possible orientations of a Necker cube. Each unit is a hypothesis concerning a vertex of the Necker cube. Abbreviations are B (back), F (front), L (left), R (right), U (upper), L (lower). The network of

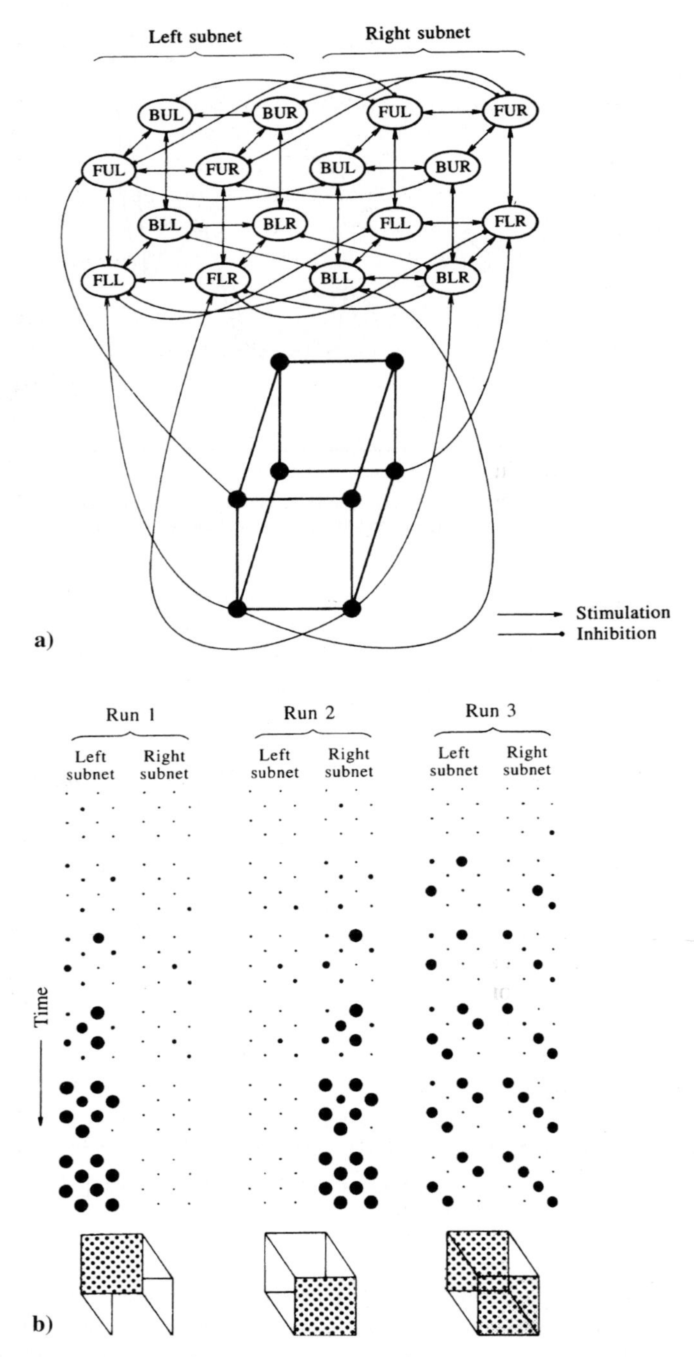

Fig. 6.15a,b. Neural network simulating the recognition of one of the two possible orientations of a Necker cube (**a**) with three evolution patterns (**b**) [6.21]

hypotheses consists of two interconnected subnetworks, one for each of the two possible interpretations.

Incompatible hypotheses are negatively connected, and consilient hypotheses are positively connected. Weights are assigned such that two negative inputs balance three positive inputs. Each unit has three neighbors connected positively and two competitors connected negatively. Each unit receives one positive input from the stimulus. The subnet of hypotheses sought is the one that best fits the input. Tiny initial fluctuations (small details in the special view of an observer) may decide which orientation is seen in the long run.

To visualize the dynamics of the network, suppose that all units are off. Then one unit receives a positive input at random. The network will evolve toward a state where all of the units of one subnetwork are activated and all of the units of the other network are turned off. In the cognitive interpretation, we can say that the system has relaxed into one of the two interpretations of the ambivalent figure of either a right-facing or a left-facing Necker cube.

Figure 6.15b shows three different evolution patterns, each of which depends sensitively on the initial conditions. The size of a circle indicates the degree of activation of the unit. In the third run, an undecided final state is reached, which is, however, still in equilibrium [6.21]. Obviously, the architectural principles of this network are cooperative computation, a distributed representation, and a relaxation procedure, which are well known in the dynamics of complex systems.

Many designs for artificial neural networks have been suggested. They have been inspired by problems in disciplines such as physics, chemistry, biology, psychology, and sometimes just technical issues. What are the common principles of the complex system approach? In earlier chapters, synergetics was introduced as an interdisciplinary methodology for dealing with nonlinear complex systems. Synergetics seems to be a successful top-down strategy for deriving particular models of complex systems from common principles that has been utilized in many scientific disciplines. The main idea behind this strategy is that the emergence of global states in complex systems can be explained by the evolution of (macroscopic) interactions between the elements of a system during learning strategies far from thermal equilibrium. Global states of order are interpreted as attractors (fixed points, periodic, quasi-periodic, or chaotic) of phase transitions.

Pattern recognition, for instance, is interpreted as a kind of phase transition by analogy with the evolution equations used for pattern emergence in physics, chemistry, and biology. We obtain an interdisciplinary research program that allows us to explain neurocomputational self-organization as a natural consequence of physical, chemical, and neurobiological evolution based on common principles. As in the case of pattern formation, a specific pattern of recognition (for instance a prototype face) is described by order parameters to which a specific group of features belong.

Once some of the features that belong to the order parameter are given (for instance, a part of a face), the order parameter will complement these with other features so that the whole system acts as an associative memory (for instance, it can reconstruct a stored prototype face from just part of that face). According to

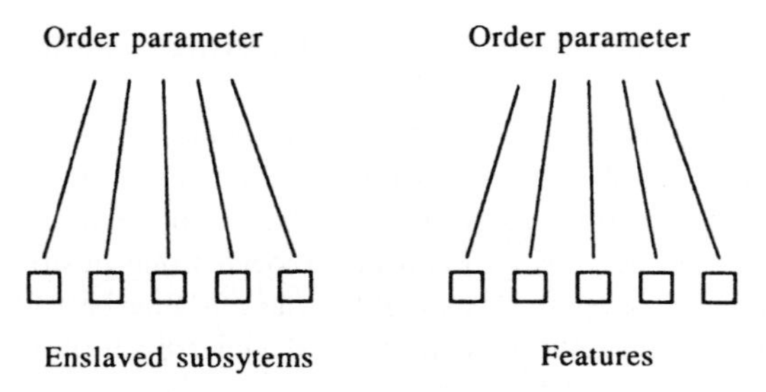

Fig. 6.16a. Haken's slaving principle for pattern formation and pattern recognition

Haken's slaving principle, the features of a recognized pattern correspond to the enslaved subsystems during pattern formation (Fig. 6.16) [6.22].

If a small part of the face that was learnt as a prototype is given to a synergetic computer, then it can complete the whole face based on the codified family name (Fig. 6.16b). The sequence of more or less fuzzy pictures corresponds to a phase transition of states in the synergetic computer.

Fig. 6.16b. Recognition of a face by a synergetic computer using the slaving principle [6.22]

When an incomplete pattern is offered to the neurons, a competition between different neuronal states, each corresponding to a specific prototype pattern, begins. This competition is won by that total state of the neuronal system that corresponds to the prototype pattern which most closely resembles the offered test pattern. In complete analogy to the dynamics valid for pattern formation, when a test pattern is offered to the synergetic computer, it will pull the test pattern from an initial state (at $t = 0$) into a specific final state corresponding to one of the prototype patterns.

The evolution of the test pattern is somewhat similar to the overdamped motion of a particle with a certain position vector in a potential landscape. Figure 6.16c shows a two-dimensional example of such a potential. The two prototypes correspond to two valleys. If the features of the pattern offered cannot be identified exactly with the features of the prototypes, then the particle adopts a position away from the valleys in the potential landscape. Obviously, recognition is a kind of symmetry breaking, as already illustrated by Fig. 4.21a for the one-dimensional case.

In synergetic systems, the shape of the potential landscape can be changed by tuning control parameters. As synergetic systems are open, control parameters can

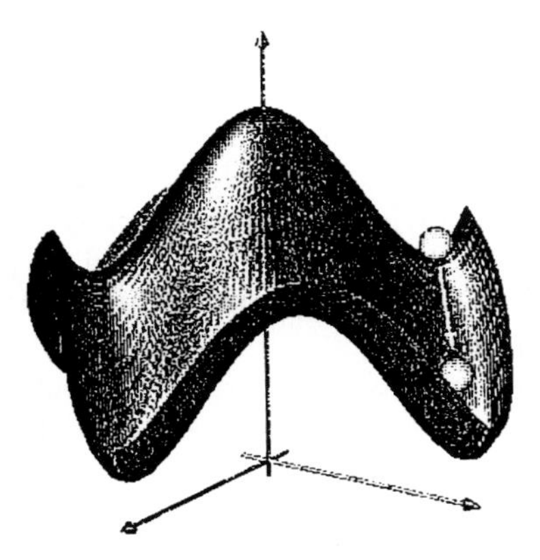

Fig. 6.16c. Illustrating the evolution of a test pattern (order parameter) in a synergetic computer via the overdamped motion of a particle in a potential landscape [6.22]

represent the input of energy, matter, information, or other stimuli from the system's environment. When the control parameter is below a critical value, the landscape may have a single stable position, like the single valley with the dotted line in Fig. 4.21a. After each excitation via fluctuations, the order parameter relaxes towards its resting state. When the control parameter exceeds a critical value, the formerly stable state becomes unstable and is replaced by two stable states in the two valleys of Fig. 4.21a.

The learning procedure of a synergetic computer corresponds to the construction of a potential landscape. The potential intensities, which are visualized as the shape of the landscape, indicate the synaptic forces of neural connections. It is an advantage of the synergetic approach that the huge number of microscopic details that characterize a pattern (for instance, a face) are determined by a single macroscopic order parameter. Thus, synergetic computers use the typical reduction of complexity method applied in synergetic models of natural evolution (see Sect. 3.3).

Order parameter equations allow a new kind of (non-Hebbian) learning, namely a strategy to minimize the number of synapses. In contrast to neurocomputers of the spin-glass type (such as Hopfield systems), the neurons are not threshold elements but instead perform simple multiplications and additions. However, the fundamental difference between neurocomputers of the spin-glass type and synergetic computers is the following: complex systems of the spin-glass type are physically closed systems. Thus, their pattern formation is driven by conservative self-organization without any input of energy, matter, or information from outside. Typical patterns formed by conservative self-organization are the "dead" ice flowers on windows in winter, which are frozen in equilibrium at low energy and temperature. The phase transitions of conservative self-organization can be explained completely by Boltzmann's principles of equilibrium thermodynamics.

In Sect. 3.3, we explained that pattern formation in living systems is only possible through the input of energy, matter or information far from thermal equilibrium.

This kind of self-organization has been called "dissipative" (Prigogine) or "synergetic" (Haken), but it can still be found in physical and chemical evolution. Consequently, pattern formation in the human brain, a living system that depends sensitively on influences from the outer world, will provide the "blueprints" or models for a new computer technology in the framework of synergetics. Neurocomputers of the spin-glass type may be practical and successful for particular technical purposes, but they are physically closed systems and so they differ in principle from living systems like the human brain.

The pattern recognition process of synergetic computers has been made simultaneously invariant to translation, rotation, and scaling. These features of recognition correspond to realistic situations. Faces, for instance, do not always appear as they did in the learning phase – they may be translated, rotated, smaller or larger, nearer or further away. A nice application of synergetic computers is the recognition of oscillation (for instance ambivalent pictures) and hysteresis in perception. Figure 6.17a shows a well-known example of hysteresis. When one scans the images from left to right, the transition from the image of a man's face to that of a girl will only

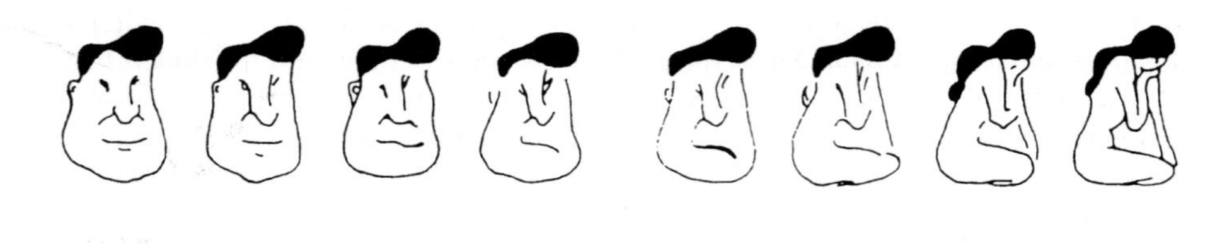

a)

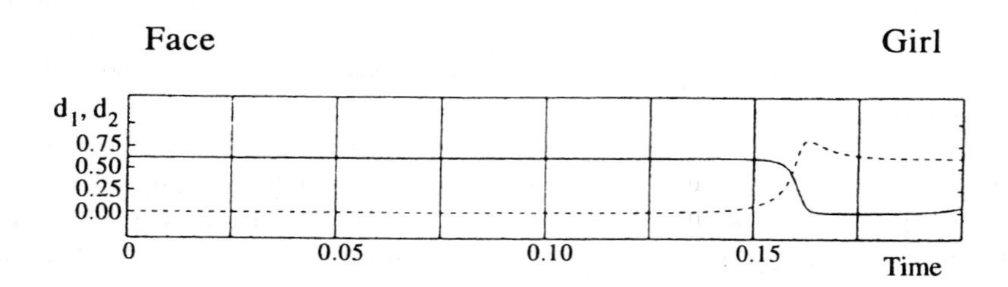

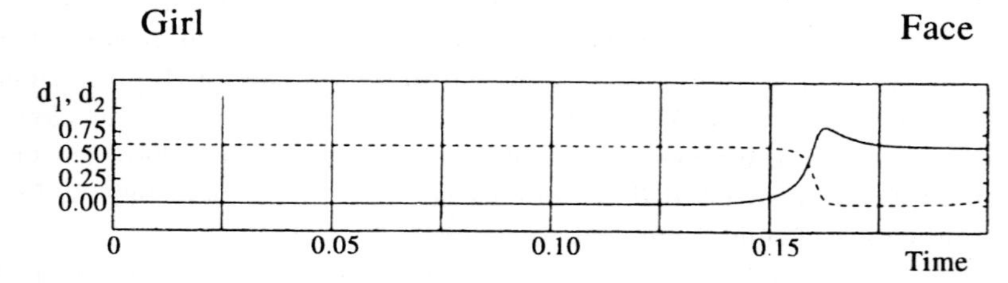

b)

Fig. 6.17a,b. Illustration of the hysteresis (**a**) of a synergetic computer based on the time evolutions of characteristic order parameters (**b**) [6.23]

occur after about six images. When one scans in the opposite direction, the switch from the perception of a girl to the face of a man occurs only toward the leftmost images.

Figure 6.17b shows the perception process for a synergetic computer in terms of the time evolutions of characteristic order parameters. The broken line shows the level of perception of the "girl," while the solid one shows the level of perception of the "face." The first diagram shows the transition from the perception of a man's face to that of a girl, while the second shows the transition from the perception of a girl to a man's face [6.23].

At this point we could object, stating that synergetic computers must still be simulated by traditional sequential computers. The principles of synergetic computers have only been realized in the field of computer software, not machine hardware. However, the interdisciplinary applications of synergetics have given rise to material and technical advances. As the laser is a well-understood model of synergetics (compare Sect. 2.4), it may play an essential role in the construction of an optical computer that utilizes synergetic principles. Different modes occur in a laser depending on the critical values of the laser threshold. These can be characterized by their photon numbers. At the microscopic level, the rate of change in photon number is described by a nonlinear evolution equation, which depends on the gain, loss, and saturation of the modes. At the macroscopic level, the order parameters correspond to the field amplitudes of several light wavetracks (Fig. 2.27a,b).

This suggests a three-layer architecture with an input layer of data that can be mapped by holographic mapping onto the laser. The laser, with its order parameters, is the intermediate layer. With its modes, the laser serves as a decision-making device via self-organization. The mode that survives, in the sense of the slaving principle, triggers new sets of features. This level is conceived as the output layer. Such a laser architecture for synergetic computers must, of course, be confirmed and improved experimentally. A synergetic computer would be a real dissipative system far from thermal equilibrium.

Obviously, complex dynamical systems are useful for simulating cognitive behavior and technical systems, too. The human brain can be modeled as a nonlinear complex system, the dynamics of which may be governed by fixed-point attractors, periodic or quasi-periodic attractors, or even chaotic attractors. Chaos, for instance, has been confirmed experimentally to be an efficient brain reset mechanism. After a study of the rabbit olfactory bulb, the recognition of various odors was modeled by relaxing a neural network towards cyclic asymptotic states. The chaotic state appeared during exhalation, eliminating the memory of the preceding odor. During inhalation, the presence of a particular odor as input drove the system to reach the limit cycle corresponding to that odor.

Technical applications of chaotic states are rather interesting, because chaotic systems are able to generate information. It is well known that chaotic systems depend sensitively on initial conditions. Thus, two trajectories may separate dramatically at a particular point during dynamical evolution, even if their initial states only differ by a very small value. Since any observation can only be made with finite precision, the separation between two different states may be less than our resolving power. Initially, an observer may see them as equal. But over the course of time, a chaotic system allows states that initially appear to be coincident to be disciminated.

Actually, the technical applications of neural computers have been explored in several industrial fields. Examples include robotics, aviation and astronautics (sensitive and adaptive systems, air navigation, etc.), medicine (evaluation and control of medical data, therapy, diagnosis, etc.), production industries (quality control, product optimization, etc.), safety technology, defense, communication technology, banking, postal services, etc. The complex system approach to technology should not be thought of as competition to or even opposition to classical AI. At the present stage of technological development, both neural networks and classical AI systems like expert systems seem to be useful and suited to different fields of application. Complex systems appear to be more suited than classical AI systems to the analysis and recognition of signals, images and speech, speech synthesis, sensorimotor coordination in robots, etc. Obviously, these examples of neural net applications are not single computers or robots, but complex functions that are integrated into composite systems that perform several tasks. From an anthropomorphic point of view, the problems actually managed by neural nets may be classified as "low-level" ones.

In Sect. 6.1, we learned that inferential models based on AI-like expert systems have failed because their functionality must be exactly and sequentially programmed, making them less tolerant and flexible. In contrast to expert systems and knowledge engineering, complex processes of self-organization cannot be controlled by an explicit formulation of expert knowledge. On the other hand, rule-based systems with inferential algorithms can be successful applied to any logically structured problem. Compared with, for instance, sensorimotor coordination, logical programming seems to be an example of "high-level" knowledge. Nevertheless, the low-level problems of nonlinear dynamical systems may be extremely complex. Nonlinear complex systems are not restricted, of course, to low-level knowledge, as we have seen in previous chapters. The principles of complex systems seem instead to be suited to modeling high-level functions of the human brain like concepts, thoughts, self-referential states, etc. However, the technology of neural nets is still in its infancy.

In present and future technologies, heterogeneous systems with several modular rule-based and complex dynamical systems are or will be of interest for specialized tasks. A speech comprehension system could consist of a neural network performing speech recognition and a rule-based symbolic module for syntactic and semantic analyses. Hybrid systems combine inferential and dynamical techniques that may be useful for various medical purposes. For example, imagine a system that is able to recognize and control medical parameters via neural nets, combined with a rule-based deductive system for diagnosing illness based on known data. Like nature, an engineer should not be restricted dogmatically to one "optimal" strategy, but should attempt to find purposeful solutions, such as several solutions that can be combined but which need not be the best ones.

6.3 Cellular Neural Networks and Analogic Neural Computers

The concept of synergetics with Haken's "cooperative phenomena" and "slaving principle" has its origin in statistical, quantum, and laser physics. In electrical engineering, information and computer science, the concept of cellular neural networks (CNN) has recently become an influential paradigm of complexity research and is being realized in information and chip technology [6.24]. The emergence of CNN has been made possible by the sensor revolution of the late 1990s. Cheap sensor and MEMS (micro-electro-mechanical system) arrays are proliferating in all technical infrastructures and human environments. They have become popular as artificial eyes, noses, ears, tastes, and somatosensor devices. An immense number of generic analog signals have been processed. Thus, a new kind of chip technology, similar to signal processing in natural organisms, is needed. Analogic cellular computers are the technical response to the sensor revolution, mimicking the anatomy and physiology of sensory and processing organs. A CNN chip is their hard core, because it is an array of analog dynamic processors or cells.

The CNN was invented by Leon O. Chua and Lin Yang at Berkeley in 1988 [6.25]. The main idea behind the CNN paradigm is Chua's so-called "local activity principle", which asserts that no complex phenomena can arise in any homogeneous media without local activity. Obviously, local activity is a fundamental property in micro-electronics. For example, vacuum tubes and, later on, transistors became the locally active devices in the electronic circuits of radios, televisions, and computers. The demand for local activity in neural networks was motivated by the practical needs of technology. In 1985, Hopfield suggested his theoretical neural network, which, in principle, could overcome the failures of pattern recognition in Rosenblatt's "Perceptron". But its globally connected architecture was highly impractical for technical applications in the VLSI (very-large-scale-integrated) circuits of micro-electronics: The number of wires in a fully connected Hopfield network grows exponentially with the size of the array. A CNN only needs electrical interconnections in a prescribed sphere of influence [6.26]. An immense increase in computing speed, combined with significantly less electrical power in the first CNN chips, has led to the current intensive research activities on CNN since Chua and Yang's proposal in 1988.

In general, a CNN is a nonlinear analog circuit that processes signals in real time. It is a multi-component system of regularly spaced identical ("cloned") units, called cells, that communicate directly with each other only through their nearest neighbors. But the locality of direct connections allows for global information processing. Communication between remotely connected units are achieved through other units. The idea that complex and global phenomena can emerge from local activities in a network dates back to J. von Neumann's earlier paradigm of cellular automata (CA). In this sense, the CNN paradigm is an advancement of the CA paradigm under the new conditions of information processing and chip technology. Unlike conventional cellular automata, CNN host processors accept and generate analog signals in continuous time with real numbers as interaction values. But, actually, discreteness of CA is no principle difference to CNN. We can introduce continuous cellular automata (CCA) as a generalization of CA in which each cell is not just, for example, black or white, but instead can have any of a continuous range of grays. A possible rule of a CCN may demand that the new gray level of each cell be the average of its own gray level, and that of its immediate neighbors. It turns out that in continuous cellular automata simple rules of interaction can generate patterns of increasing complexity, chaos, and randomness, which are not essentially different to the behavior of discrete CA. Thus, they are useful in approximating the dynamics of systems that are determined by partial differential equations (PDE).

For the CNN paradigm, a gene-technological and neurobiological language delivers metaphoric illustrations of concepts, which are nevertheless mathematically defined and technically implemented. According to todays dominating paradigms of life sciences, a biological language mediates visions of future connections between bio- and computer technology. Mathematically, a CNN is defined by (1) a spatially discrete set of continuous nonlinear dynamical systems ("cells" or "neurons") where information is processed in each cell via three independent variables ("input", "threshold", and "initial state") and (2) a coupling law relating relevant variables of each cell to all neighbor cells within a pre-described sphere of influence. A standard CNN architecture includes an $M \times N$ rectangular array of cells $C(i, j)$ with cartesian coordinates (i, j) with $i = 1, 2, \ldots, M$ and $j = 1, 2, \ldots, N$ (Fig. 6.18a). Figure 6.18b–c shows examples of cellular spheres of influence as 3×3 and 5×5 neighborhoods. The dynamics of a cell's state are defined by a nonlinear differential equation (CNN state equation) with scalars for "state" x_{ij}, "output" y_{ij}, "input" u_{ij}, and "threshold" z_{ij}, and coefficients, called "synaptic weights", modeling the intensity of synaptic connections of the cell $C(i, j)$ with the inputs (feedforward signals) and outputs (feedback signals) of the neighbor cells $C(k, l)$. The CNN output equation connects the states of a cell with the outputs.

The majority of CNN applications use space-invariant standard CNNs with a cellular neighborhood of 3×3 cells and no variation of synaptic weights and cellular thresholds in the cellular space. A 3×3 sphere of influence at each node of the grid contains nine cells with eight neighbor cells, and the cell in its center. In

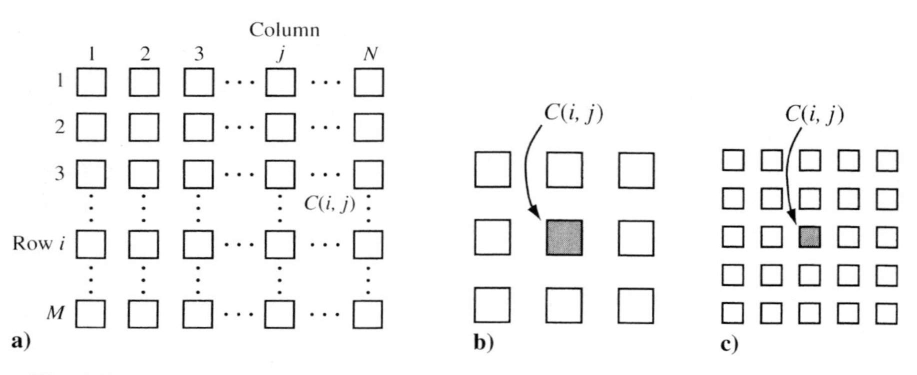

Fig. 6.18a–c. Standard CNN with array (**a**), 3×3 and 5×5 neighborhoods (**b**, **c**) [6.27]

this case, the contributions of the output (feedback) and input (feedforward) weights can be reduced to two fixed 3×3 matrices, which are called feedback (output) cloning template **A** and feedforward (input) cloning template **B**. Thus, each CNN is uniquely defined by the two cloning templates **A**, **B**, and a threshold z, which consist of $3 \times 3 + 3 \times 3 + 1 = 19$ real numbers. They can be ordered as a string of 19 scalars with a uniform threshold, nine feedforward and nine feedback synaptic weights. This string is called a "CNN gene", because it completely determines the dynamics of the CNN. Consequently, the universe of all CNN genes is called the "CNN genome". With respect to the human genome project, steady progress can be made by isolating and analyzing various classes of CNN genes and their influences on CNN genomes.

Concerning visual computing, the triple $\{\mathbf{A}, \mathbf{B}, z\}$, and its 19 real numbers can be considered a CNN macro instruction of how to transform an input image into an output image. Simple examples are subclasses of CNNs with practical relevance, such as the class $C(\mathbf{A}, \mathbf{B}, z)$ of space-invariant CNNs with excitatory and inhibitory synaptic weights; the zero-feedback (feedforward) class $C(0, \mathbf{B}, z)$ of CNNs without cellular feedback; the zero-input (autonomous) class $C(\mathbf{A}, 0, z)$ of CNNs without cellular input; and the uncoupled class $C(\mathbf{A}^0, \mathbf{B}, z)$ of CNNs without cellular coupling. In $\mathbf{A}^0$ all weights are zero except for the weight of the cell in the center of the matrix. Their signal flow and system structure can be illustrated in diagrams that can easily be applied to electronic circuits, as well as to typical living neurons.

CNN templates are extremely useful for standards in visual computing. Simple examples are CNNs that detect edges in either binary (black-and-white) or gray-scale input images. An image consists of pixels corresponding to the cells of a CNN with binary or gray scale. An EDGE CNN is an example of the zero-feedback class $C(0, \mathbf{B}, z)$ with binary edge detection templates:

$$\mathbf{A} = \begin{array}{|c|c|c|}\hline 0 & 0 & 0 \\ \hline 0 & 0 & 0 \\ \hline 0 & 0 & 0 \\ \hline \end{array} \qquad \mathbf{B} = \begin{array}{|c|c|c|}\hline -1 & -1 & -1 \\ \hline -1 & 8 & -1 \\ \hline -1 & -1 & -1 \\ \hline \end{array} \qquad z = \begin{array}{|c|}\hline -1 \\ \hline \end{array}$$

The input is a static binary image of black pixels. The initial state is arbitrary (e.g., zero). The boundary conditions (e.g., zero) determine inputs and outputs of so-called virtual cells that belong to a 3×3 neighborhood, but are outside the CNN grid. The output should be a binary image showing all edges in black. The EDGE CNN template is designed to work correctly for binary input images only. If the input image is a gray-scale image, the output image will generally be gray scale where black pixels correspond to sharp edges, near-black pixels correspond to fuzzy edges, and near-white pixels correspond to noise. Local rules that generate the edge image from a given input image are the following:

(1) white pixel → white, independent of neighbors
(2) black pixel → white, if all nearest neighbors are black
(3) black pixel → black, if at least one nearest neighbor is white
(4) black, gray, or white pixel → gray, if nearest neighbors are gray

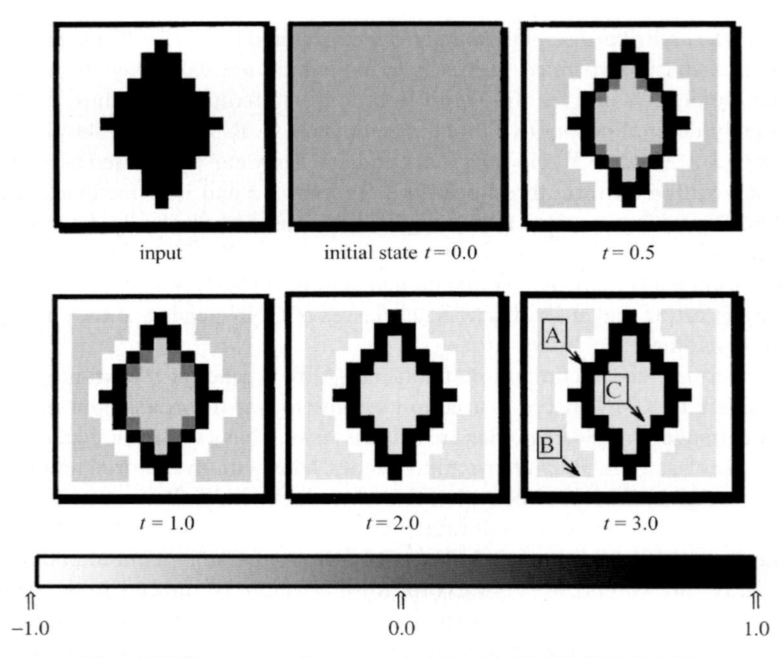

Fig. 6.19. Snapshots of image processing by EDGE CNN [6.28]

Logic operators can also be realized by simple CNN templates in order to combine CNN templates for visual computing. The Logic NOT Operation CNN inverts the intensities of all binary image pixels, and the foreground pixels become the background, and vice versa. The Logic AND (Logic OR, respectively) Operation CNN performs a pixel-wise logic AND (logic OR operation, respectively) on corresponding elements of two binary images. These operations can be used as elements of some Boolean Logic algorithms, which operate in parallel on data arranged in the form of images.

The analysis of CNNs for visual computing follows a standard series of steps. First (I), a non-technical description is given of the input-output image transformation at the complete image level. In the next step (II), local rules deliver a precise recipe of how to transform input into output pixels. These local rules must be complete in the sense that each output pixel can be uniquely determined by applying these rules to the state and input of all pixels within a local sphere of influence. Then (III), several examples are given, including (a) the input picture and initial state, (b) several consecutive snapshots until a static output image is reached in a transient settling time, or (c) time wave forms of both state and output at special points of interest on the output image. Finally, in a mathematical analysis (IV), a rigorous mathematical proof is given for each local rule. If a proof is not available, an intuitive proving sketch with various numerical studies is given. In this way, a gallery of CNN templates can be introduced in a standard way in order to

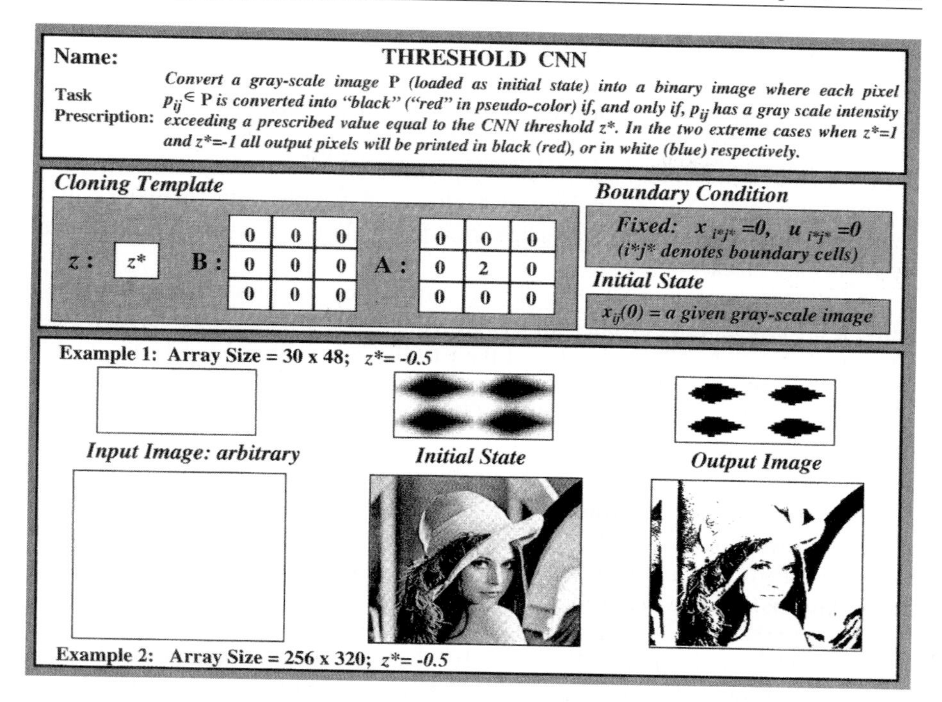
Name: THRESHOLD CNN
Task Prescription: Convert a gray-scale image P (loaded as initial state) into a binary image where each pixel pij ∈ P is converted into "black" ("red" in pseudo-color) if, and only if, pij has a gray scale intensity exceeding a prescribed value equal to the CNN threshold z*. In the two extreme cases when z*=1 and z*=-1 all output pixels will be printed in black (red), or in white (blue) respectively.
Cloning Template
z : z*
B :
0 0 0
0 0 0
0 0 0
A :
0 0 0
0 2 0
0 0 0
Boundary Condition
Fixed: x i*j* =0, u i*j* =0
(i*j* denotes boundary cells)
Initial State
xij(0) = a given gray-scale image
Example 1: Array Size = 30 x 48; z*= -0.5
Input Image: arbitrary
Initial State
Output Image
Example 2: Array Size = 256 x 320; z*= -0.5

Fig. 6.20. Gallery of THRESHOLD CNN [6.29]

illustrate a new paradigm of problem solving in visual computing. An example is a THRESHOLD CNN that converts a gray-scale image of a young girl into a binary image, depending on a certain threshold of gray-scale intensity (Fig. 6.20).

The simplest form of a CNN can be characterized via Boolean functions. We consider a space-invariant binary CNN belonging to the uncoupled class $C(\mathbf{A}^0, \mathbf{B}, z)$ with a 3×3 neighborhood which maps any static 3×3 input pattern into a static binary 3×3 output pattern. It can be uniquely defined by a Boolean function of nine binary input variables, where each variable denotes one of the nine pixels within the sphere of influence of a cell. Although there are infinitely many distinct templates of the class $C(\mathbf{A}^0, \mathbf{B}, z)$, there is only a finite number of distinct combinations of a 3×3 pattern of black and white cells, $2^9 = 512$. As each binary nine input pattern can map to either 0 (white) or 1 (black), there are 2^{512} distinct Boolean maps of nine binary variables. Thus, every binary standard CNN can be uniquely characterized by a CNN truth table, consisting of 512 rows, with one for each distinct 3×3 black-and-white pattern; nine input columns, one for each binary input variable; and one output column, with binary values of the output variable.

For example, consider the truth table of the EDGE CNN (Fig. 6.19). Since the table of 512 rows will exceed the length of a page, it is divided into 16 parts of truth tables with 32 rows (Fig. 6.21). By convention, column 5 of each truth table denotes the input of the cell in the center of the 3×3 neighborhood. According to the above mentioned local rules of Edge CNN, a white cell remains white independent of its neighbors. Thus, since in the 16 truth tables the first 16 cells of the 5th column are white, the corresponding output cell in the last column is also white. According to the local rules, a black cell remains black if at least one nearest neighbor is black, but it becomes white if all its nearest neighbors are black. Since the cells of the 5th column from rows 17 to 32 in the truth tables are black, the corresponding output cells of the last column remain black, except for the last row, where all the nearest neighbors are black.

As the black-white patterns of the 16 truth tables remain unchanged in many parts, it is sufficient to consider the last 16 columns of the 32 cells. Thus, we get a minimal 16×32 CNN truth table of the EDGE CNN (Fig. 6.22). The truth table for any binary CNN $C(\mathbf{A}^0, \mathbf{B}, z)$ with a prescribed initial state can be constructed by simply solving the associated differential equations for the input of each of the 512 distinct binary patterns and calculating the corresponding binary outputs. It is also easy to write a computer program to automatically generate the truth table under these conditions. As these truth tables contain a great deal of redundancy in unchanged patterns, it is generally sufficient to consider only the last column of output values. In this way, a minimal CNN truth table with immense data compression can be used to characterize the uncoupled CNNs.

The number of 2^{512} distinct Boolean functions of nine variables is gigantic, and with $2^{512} \approx 1.3408 \times 10^{154} > 10^{154}$ more than the volume of the universe. The uncoupled $C(\mathbf{A}^0, \mathbf{B}, z)$ CNNs are only a small subclass of CNNs. So the question arises: Which subclass of Boolean functions characterizes the uncoupled CNNs exactly? In Sect. 6.2, we introduced the concept of linearly separable and non-separable Boolean functions. The XOR-function (Fig. 6.11a) is an example of a non-separable Boolean function. It can be proven that the class $C(\mathbf{A}^0, \mathbf{B}, z)$ of all uncoupled CNNs with binary inputs and binary outputs is identical to the linearly separable class of Boolean functions. Thus, linearly non-separable Boolean functions, such as the XOR function, cannot be realized by an uncoupled CNN. But the uncoupled CNNs can be used as elementary building blocks, which are connected by CNNs of logical operations. It can be proved that every Boolean function of nine variables can be realized using uncoupled CNNs with nine inputs and either one Logic OR CNN, or one Logic AND CNN, in addition to one Logic NOT CNN.

Every uncoupled CNN $C(\mathbf{A}^0, \mathbf{B}, z)$ with static binary inputs is completely stable, in the sense that any solution converges to an equilibrium point. The waveform of the CNN state increases or decreases monotonically to the equilibrium point, if the state at this point is positive or negative. Moreover, except in some degenerated cases, the steady state output solution can be explicitly calculated by an algebraic formula without solving the associated nonlinear differential equations. Obviously, this is an important result in characterizing a CNN class of nonlinear dynamics with robust CNN templates. Completely stable CNNs are the workhorses of the most current CNN applications. But there are even simple CNNs with oscillatory or chaotic behavior. Future applications will exploit the immense potential of the unexplored terrains of oscillatory and chaotic operating regions. Then, the Cellular Neural Net-

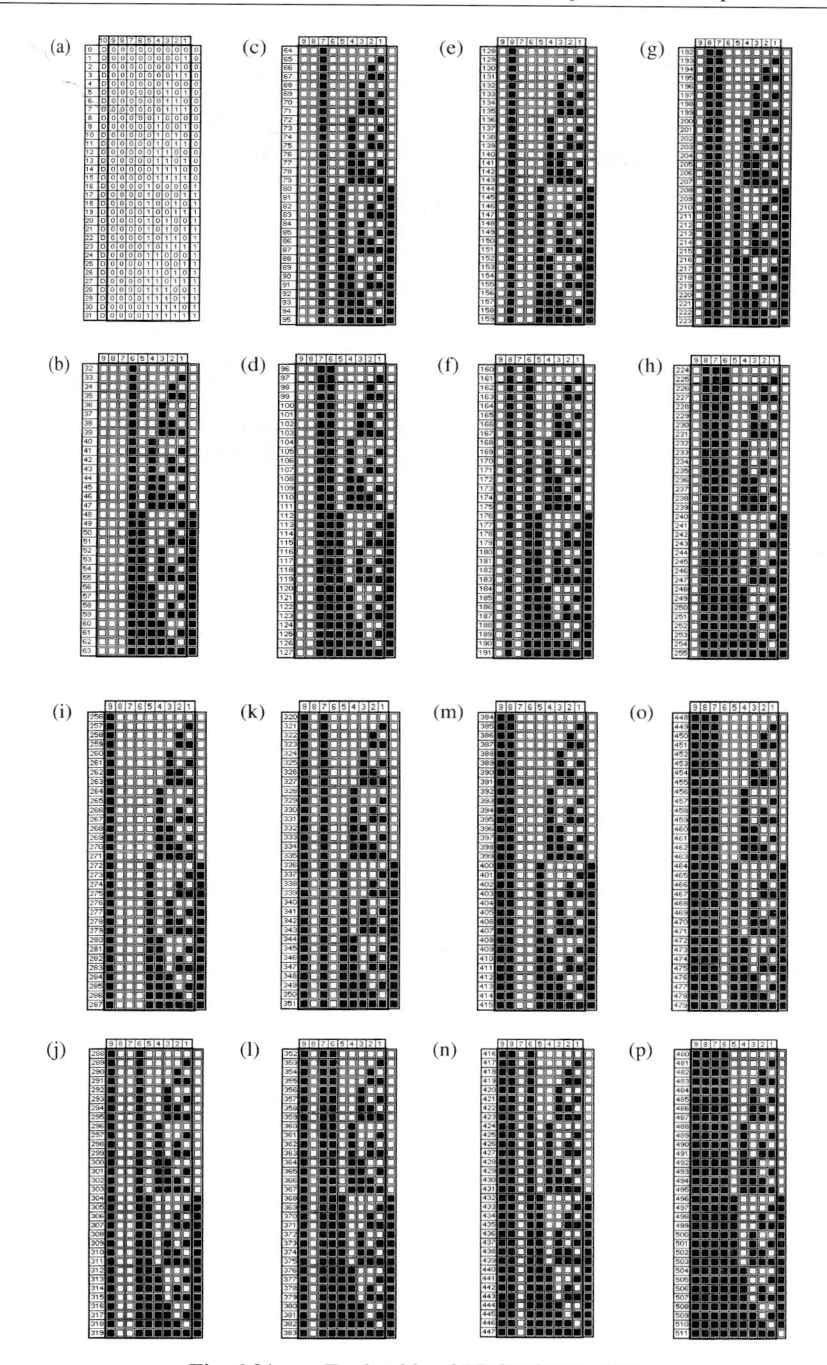

Fig. 6.21a–p. Truth table of EDGE CNN [6.30]

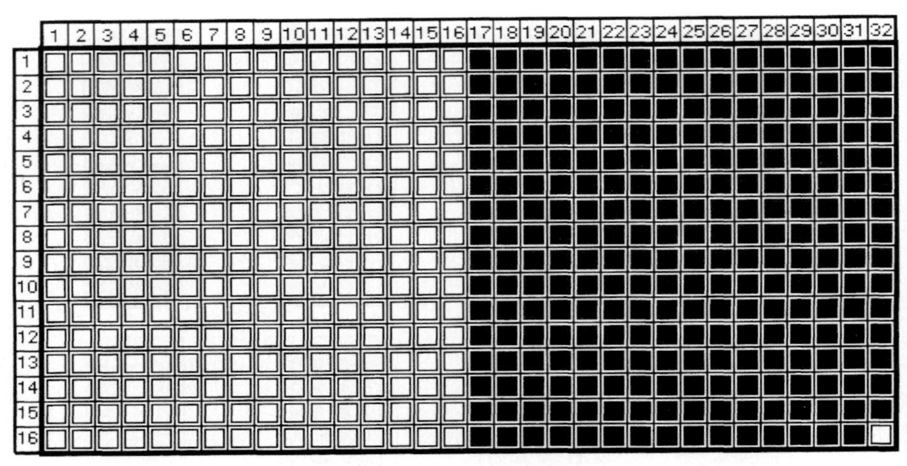

Fig. 6.22. Minimal CNN truth table [6.31]

works will actually be transformed into Cellular Nonlinear Networks with all kinds of phase transitions and attractors of nonlinear dynamics.

An oscillatory CNN with only two cells is given by the templates

A =

0	0	0
β	α	$-\beta$
0	0	0

B =

0	0	0
0	0	0
0	0	0

$z = \boxed{0}$

and zero boundary conditions. Figure 6.23 shows the architecture of the 1×2 CNN with virtual boundary cells (gray) of zero potential (a), and the corresponding signal flow graph. The state equations for the two cells of this CNN is given by the two differential equations

$$\dot{x}_1 = -x_1 + \alpha y_1 - \beta y_2$$
$$\dot{x}_2 = -x_2 + \alpha y_2 - \beta y_1$$

The corresponding outputs y_i $(i = 1, 2)$ are related to the states x_i by the standard nonlinear function $y_i = f(x_i) = 0.5|x_i + 1| - 0.5|x_i - 1|$ with piecewise-linear characteristic.

In Fig. 6.24a,b, the corresponding time series $x_1(t)$ and $x_2(t)$ are shown for example $\alpha = 2$, $\beta = 2$, and initial condition $x_1(0) = 0.1$, $x_2(0) = 0.1$. In the corresponding phase space (Fig. 6.24c), all trajectories starting from any initial state except the origin will converge to a limit cycle.

A chaotic CNN with only two cells is given by the templates

A =

0	0	0
1.2	2	−1.2
0	0	0

B =

0	0	0
0	1	0
0	0	0

$z = \boxed{0}$

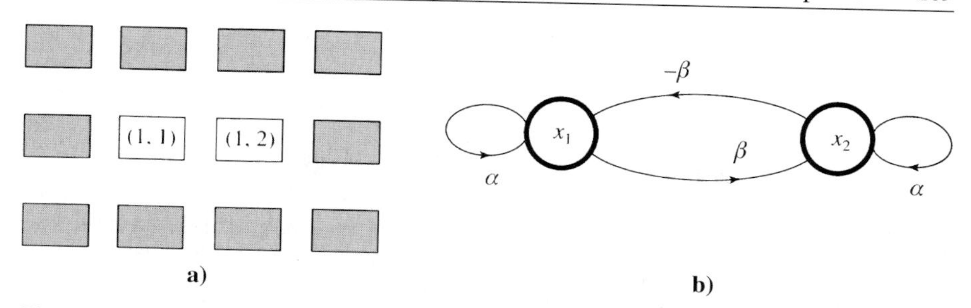

Fig. 6.23a,b. 1×2 CNN with (gray) virtual boundary cells (**a**) and signal flow graph (**b**) [6.32]

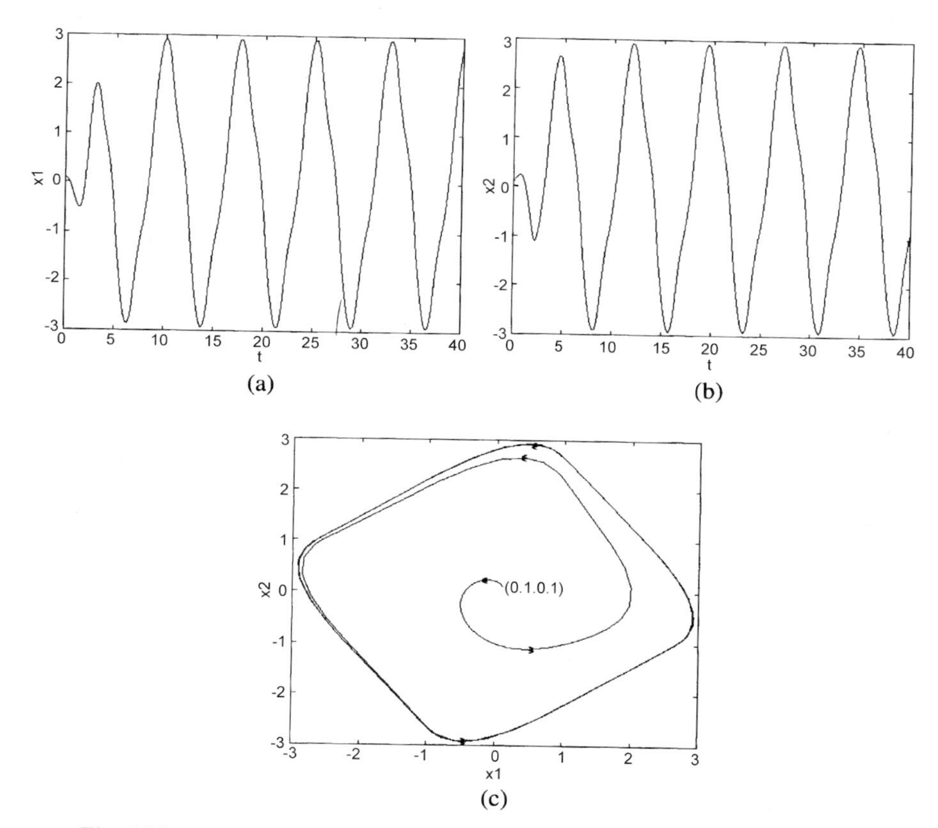

Fig. 6.24a–c. Periodic time series (**a**, **b**) and limit cycle (**c**) of 1×2 CNN [6.33]

and zero boundary conditions. Contrary to the example above, the CNN should be non-autonomous with a sinusoidal input $u_{11}(t) = 4.04 \sin(\pi t/2)$ to cell $C(1, 1)$, but zero input $u_{12} = 0$ to cell $C(1, 2)$ (Fig. 6.18a). From a technical point of view, the circuit of the two-cell CNN is driven by a sinusoidal signal. The state equations are given by the two nonlinear differential equations

$$\dot{x}_1 = -x_1 + 2y_1 - 1.\,2y_2 + 4.\,04\sin(\pi t/2)$$
$$\dot{x}_2 = -x_2 + 1.\,2y_1 - 2y_2$$

with the output function $y_i = f(x_i)$ of the previous CNN. The corresponding non-periodic time series $x_1(t)$ and $x_2(t)$ are shown in Fig. 6.25a,b with the same initial conditions of the previous example. In the corresponding phase space (Fig. 6.25c), the trajectories are attracted by a strange attractor called "Lady's shoe attractor", because its Poincaré map (Fig. 6.25d) resembles the shape of a lady's high heel.

From the perspective of nonlinear dynamics, it is convenient to think of standard CNN state equations as a set of ordinary differential equations, with the components of the CNN gene as bifurcation parameters. The dynamical behavior of standard CNNs can then be studied in detail. Numerical examples deliver CNNs with limit cycles and chaotic attractors. For technical implementations of standard CNNs, such as silicon chips, complete stability properties must be formulated, in order to avoid oscillations, chaotic, and noise phenomena. These results also have practical importance for image processing applications by CNNs. As brains and computers work

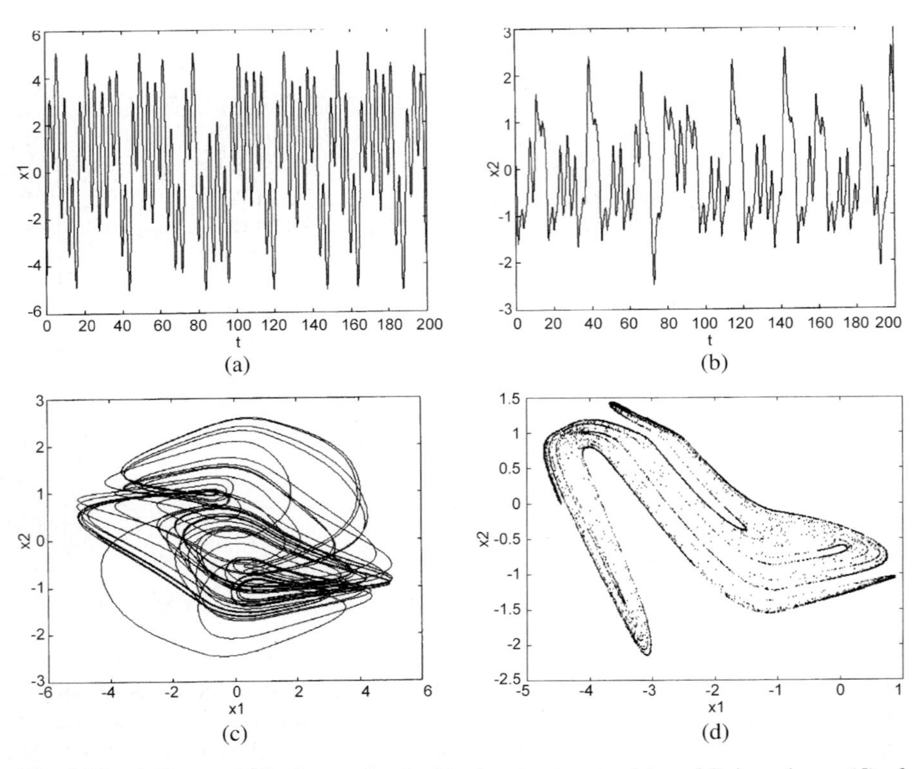

Fig. 6.25a–d. Non-periodic time series (**a**, **b**), chaotic attractor (**c**), and Poincaré map (**d**) of 1×2 CNN [6.34]

with units in two distinct states, the conditions of bistability are studied in brain research, as well as in chip technology.

The emergence of complex structures in nature can be explained by the nonlinear dynamics and attractors of complex systems. They result from the collective behavior of interacting elements in a complex system. The different paradigms of complexity research promise to explain pattern formation and pattern recognition in nature by their specific mechanisms. From the CNN point of view, it is convenient to study the subclass of autonomous CNNs that cells have no inputs. These systems can explain how patterns arise, evolve, and sometimes converge to an equilibrium by diffusion-reaction processes. Pattern formation starts with an initial uniform pattern in an unstable equilibrium that is disturbed by small, random displacements. Thus, in the initial state, the symmetry of the unstable equilibrium is broken, leading to rather complex patterns. Obviously, in these applications, cellular networks do not only refer to neural activities in nerve systems, but also to pattern formation in general. Thus, the abbreviation CNN is now understood as "Cellular Nonlinear Network".

A CNN is defined by the state equations of isolated cells and the cell coupling laws. For simulating diffusion-reaction processes, the coupling law describes a discrete version of diffusion (with a discrete Laplacian operator). CNN state equations and CNN coupling laws can be combined in a CNN diffusion-relation equation to determine the dynamics of autonomous CNNs. If we replace their discrete functions and operators by their limiting continuum version, we get the well-known continuous partial differential equations of diffusion-reaction processes, which have been studied in the complexity paradigms of, for example, Prigogine's non-equilibrium chemistry and Haken's synergetics. Chua's version of the CNN diffusion-reaction equation delivers computer simulations of these pattern formations in chemistry and biology (e.g., concentric, auto- and spiral waves). On the other hand, many appropriate CNN equations can be associated with any nonlinear partial differential equation. In many cases, it is sufficient to study the computer simulations of associated CNN equations in order to understand the nonlinear dynamics of these complex systems. Sometimes, the autonomous CNNs (like digital cellular automata) are only considered approximations of nonlinear partial differential equations for the practical purpose of computer simulations. But, Chua claims, nonlinear partial differential equations are limiting forms of autonomous CNNs. Thus, only a subclass of CNNs has a limiting representation of partial differential equations. In short, the CNN paradigm of complexity is more than the conventional approach with differential equations.

Pattern recognition is understood in relation to pattern formation. Coupled CNNs with linear synaptic weights open avenues to much richer visual computing applications than uncoupled CNNs. In coupled CNNs, there are couplings from the outputs of the surrounding cells to a cell in the center. Thus, at least one element of the feedback (output) template **A** (which is different from the coefficient of the cell in the center) is not zero. Coupled CNNs are, for example, able to detect holes (i.e., a set of adjacent pixels) on a surrounding background. In particular, it turns out that the famous connectivity problem can be solved by a simple coupled CNN of this kind. This problem is not only important for practical reasons, but also has a long tradition in the history of cognitive science. How can we recognize connected patterns ("gestalt"), such as shapes, figures, or faces from a set of pixels? In a famous proof, Marvin Minsky demonstrated that the connectivity of certain patterns could not be recognized by neural networks like Rosenblatt's "Perceptron".

In the CNN paradigm, cellular neural networks (CNNs) work with the special assumption of local activity. How can a locally connected neural network realize global functions and recognize a "Gestalt"? The strategy of a CNN is to delete all the pixels that are part of a connected object defined by black pixels on a white background. The intuitive idea is that in a complex image the connected parts of pixels are burnt out like in a propagating bushfire. In the case of a connected pattern (e.g., a labyrinth), the last image of pattern production is empty like a burnt countryside. In the other cases, the unconnected pixels are left over like bushes that have survived the fire (Fig. 6.26). In the sense of the local activity principle, the wavefront of the fire propagates from pixels to neighbor pixels and "detects" the connecting pixel clusters.

CNNs of this type are also used to study nonlinear waves of propagation phenomena (e.g., infectious diseases by computer simulations). With respect to Minsky's problem with Rosenblatt's "Perceptron", it can be proved that the global connectivity property can be realized by an appropriate CNN for any binary input pat-

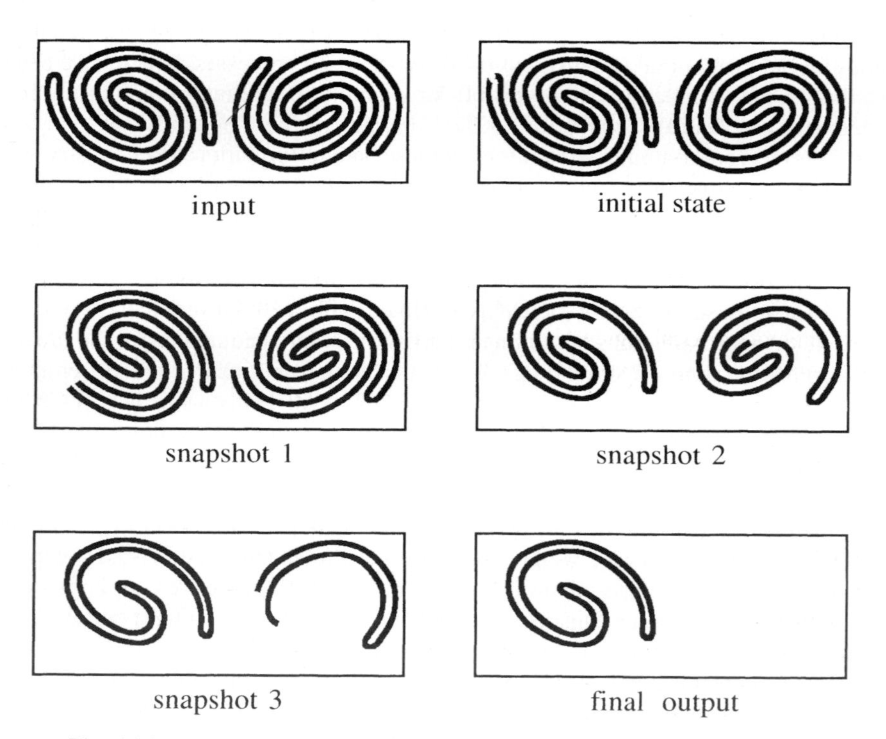

Fig. 6.26. Snapshots of the propagation of the CONNECTIVITY CNN [6.35]

tern. Appropriate coupled CNNs are able to simulate visual illusions, where some images can be perceived in an ambiguous way, depending on the initial thought or attention (Fig. 4.22). One of the examples of this phenomenon is the face-vase illusion, where the image can be interpreted either as two symmetric faces, or as a vase. Initial attention is implemented by specifying, via a second binary pattern, one of the two ambiguously interpreted regions.

In the case of linear synaptic weights, the characteristics of a synapse or template element are linear. But in technical applications (e.g., with voltage-controlled current sources) or living cells with synaptic communication by neurotransmitter, they are never completely linear. If we use nonlinear templates for modeling synaptic dynamics, the analysis becomes more complex. Thus, a compromise of modeling is the application of uncoupled CNNs with nonlinear, space-invariant weights.

6.4 Universal Cellular Neural Networks and Dynamic Complexity

Image processing and pattern recognition applications are sometimes linearly nonseparable problems that need programs of several cellular neural networks. A CNN program defined by a string of CNN genes is called a "CNN chromosome". Every cellular automaton (CA) with binary or Boolean states can be considered a CNN chromosome. In particular, Conway's game-of-life CA (compare Fig. 5.23 in Sect. 5.6) can be realized by a CNN chromosome of two CNN genes that are connected by the Logic AND Operation CNN. Since the game-of-life CA is a universal Turing machine, the corresponding game-of-life CNN is also a universal Turing machine with the capacity to self-replicate. These results lead to the technical implementation of the analog-input analog-output CNN universal machine, the CNN universal chip, which solves computational problems, performing a trillion operations per second. Because of its massively nonlinear dynamics, it differs from a conventional digital computer.

The CNN Universal Machine (CNN-UM) architecture needs an analog and logic memory, a local logic unit, and, like any programmable system, a global clock to control the instructions during a given clock cycle. As mentioned previously, CNN templates can be considered instructions with well-defined input and output. Not all tasks can be implemented by a single CNN template. Thus, when applying several templates, we define a CNN subroutine or function, such as in a C-like programming language. There are three equivalent ways of implementing CNNs: (1) hardware schematics with discrete hardwired cells and additional local and global devices, (2) a flow diagram of the CNN algorithm, and (3) a CNN program in an analogic (α) CNN programming language with CNN analog and logic operations.

There are practical and theoretical reasons to introduce the CNN-UM. From an engineering point of view, it is totally impractical to implement different CNN components or templates with different hardwired CNNs. Historically, John von Neumann's general purpose computer was inspired by Turing's universal machine in order to overcome all the different hardware machines of the 1930s and 1940s used for different applications. From a theoretical point of view, CNN-UM opens new avenues of analogic neural computers. In the CNN-UM, analog (continuous) and logic operations are mixed and embedded in an array computer. It is a complex nonlinear system that combines two different types of operations, namely, continuous nonlinear array dynamics and continuous time with local and global logic. Obviously, the mixture of analog and digital components is greatly similar to the neural information processing in living organisms. The stored program, as a sequence of templates, could be considered a kind of genetic code for the CNN-UM. The elementary genes are the templates. 3×3 templates, for instance, have a 19 real-number code. In the nervous system, the consecutive templates are placed in space as subsequent layers of neurons.

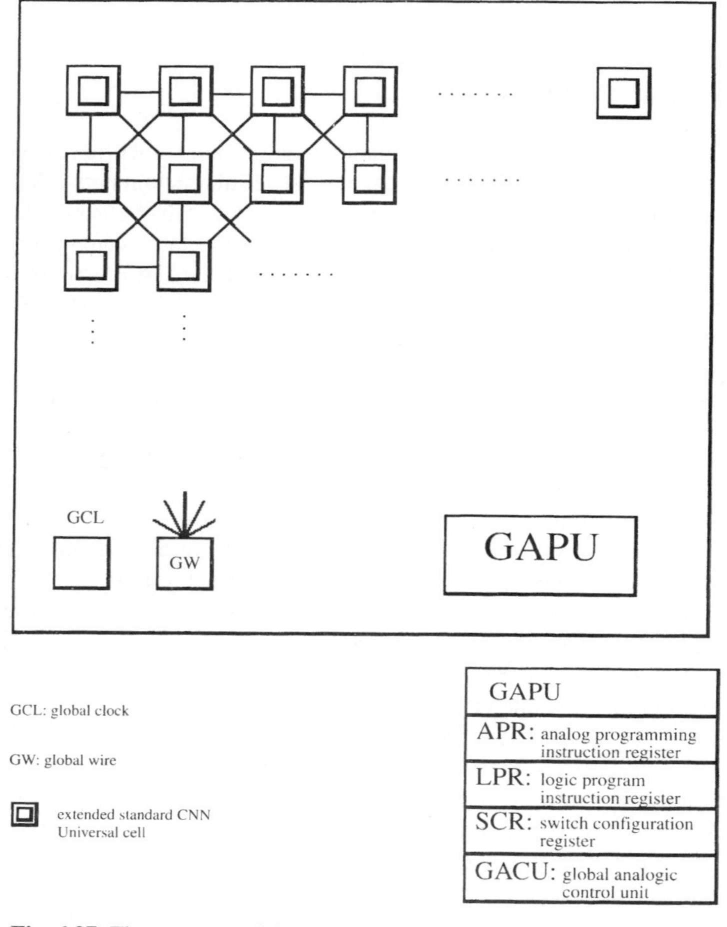

Fig. 6.27. The structure of the CNN universal machine (CNN-UM) [6.36]

The CNN-UM architecture consists of so-called extended standard CNN universal cells (Fig. 6.27). The main components of a universal cell are the local logic memory (LLM) and local analog memory (LAM), the local logic unit (LLU) and local analog output unit (LAOU) for digital and continuous (analog) signals, respectively. The local communication and control unit (LCCU) receives the programming instructions, in each cell, from the global programming unit (GAPU). The instructions contain the analog template values $\{\mathbf{A}, \mathbf{B}, z\}$, the logic fuction codes for the local logic unit, and the switch configuration of the cell, specifying the signal paths and some settings of the functional units. Thus, the GAPU needs storage elements (registers) for three types of information, namely an analog program register (APR) for the CNN templates, a logic program register (LPR) for the LLU functions, and a switch configuration register (SCR). Besides a global clock (GCL), there is a global wire (GW), which decides whether any black pixels remain in the processed images.

After the introduction of the architecture with standard CNN universal cells and the global analogic programming unit (GAPU), the complete sequence of an analogic CNN program can be executed on a CNN Universal Machine. The description of such a program contains the global task, the flow diagram of the algorithm, the description of the algorithm in a high level α programming language, and the sequence of macro instructions by an α compiler in the form of an analogic machine code (AMC). At the lowest level, the chips are embedded in their physical environment of circuits. The AMC code will be translated into hardware circuits and electrical signals. At the highest level, the α compiler generates a macro-level code, called analogic macro code (AMC). The input of the α compiler is the description of the flow diagram of the algorithm using the α language. Figure 6.28 describes the levels of the software and the core engines. The AMC is used for software simulations running on a Pentium chip in a PC and for applications in a CNN universal machine chip with a CNN chip prototyping system (CCPS).

The CNN universal machine is technically realized by analog and digital VLSI (very large scale integrated) implementation. It is well-known that any complex digital technology system can be built from a few implemented building blocks by wiring and programming. In the same way, the CNN Universal Machine, also containing analog building blocks, can be constructed. A circuit model of a standard CNN cell was introduced by Chua and Yang. A core cell only needs three building blocks: a capacitor, resistor, and a VCCS (voltage controlled current source). If a switch, logic register, and logic gate are added to the three building blocks, the extended CNN cell of the CNN-UM can be implemented. In principle, six building blocks plus wiring are sufficient to build the CNN-UM: resistor, capacitor, switch, VCCS, logic register, and logic gate. As in a digital computer, stored programmability can also be introduced for analogic neural computers, enabling the fabrication of visual microprocessors. Similar to classical microprocessors, stored programmability needs a complex computational infrastructure with a high-level language, a compiler, a macro code, an interpreter, an operating system, and a physical code, in order to make it understandable to the human user. Using this computational infrastructure, a visual microprocessor can be programmed by downloading the programs onto the chips, as in the case of classical digital microprocessors. Writing a program for an analogic CNN algorithm is as easy as writing a Basic program.

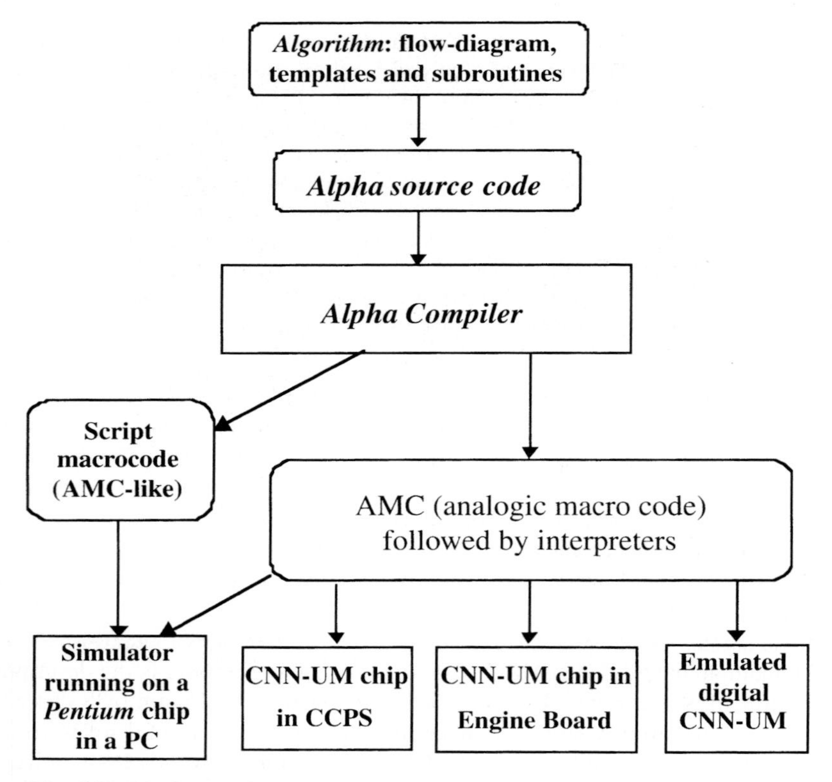

Fig. 6.28. The levels of the software and the core engines in the CNN-UM [6.37]

Concerning computing power, CNN computers offer an orders-of-magnitude speed advantage over conventional technology when the task is complex. There are also advantages in size, complexity, and power consumption. A complete CNN-UM on a chip consists of an array of 64 × 64 0.5 micron CMOS cell processors (Fig. 6.29). Each cell is endowed not only with a sensor for the direct optical input of images and video, but also with communication and control circuitries, local analog, and logic memories. CNN cells interface with their mearest neighbors, as well as with the outside world. A CNN chip with 4096 cell processors on a chip translates into 3.0 Tera OPS (operations per second) of computing power, which is about a thousand times faster than the computing power of an advanced Pentium processor. By exploiting the state-of-the-vertical packaging technologies, close to 10^{15} OPS CNN-UM architectures can be constructed on chips with 200×200 arrays. Thus, CNN universal chips will realize Tera OPS or even Peta (10^{15}) OPS, which are required for high-speed target recognition and tracking, real-time visual inspection of manufacturing processes, and intelligent vision capable of recognizing context-sensitive and moving scenes.

Fig. 6.29. One of the first prototypes of a CMM universal chip [6.38]

The CNN universal chip is a milestone in information technology, because it is the first fully programmable, industrial-sized, brain-like stored-program dynamic array computer. A CNN is a highly complex computational system, as it consists of a massively parallel focal-plane array with the computational power of a supercomputer. Besides its computing power, the CNN universal chip, with its unique brain-like architecture, can be used to implement brain-like information processing tasks that, until now, could not be performed by conventional digital computers. The development of adaptive sensor-computers will be a challenge for robotics and high-tech medicine in the future.

From a theoretical point of view, the CNN-UM gives deep insights into the dynamic complexity of computational processes. While the classification of complexity by cellular automata (CA) in Sect. 5.6 was more or less inspired by empirical observations of pattern formation in computer experiments, the CNN approach delivers a mathematically precise measure of dynamic complexity. The basic idea is to understand cellular automata as a special case of CNNs that can be characterized by a precise code for attractors of nonlinear dynamical systems, and by a unique complexity index.

Each 1-dimensional cellular automaton with two near neighbors of a cell in a row (Fig. 5.24a) is determined by a Boolean function of three variables for the two neighbors and the cell itself. Let us consider a ring of coupled cells C_i ($i = 0, 1, 2, \ldots, M$). In the context of CNNs, each cell is assumed to be a dynamical system with a state x_i, an output y_i, and three inputs u_{i-1}, u_i, and u_{i+1}. Variable u_{i-1} denotes the input from the left neighboring cell C_{i-1} to cell C_i, variable u_i, the self-input of cell C_i, and variable u_{i+1} the input coming from the right neighboring cell C_{i+1} to cell C_i. The Boolean function B delivers an output $y_i = B(u_{i-1}, u_i, u_{i+1})$ of cell C_i. In this sense, any 1-dimensional cellular automaton with near neighbors can be characterized by the truth table for a Boolean function of three binary variables

	u_{i-1}	u_i	u_{i+1}	y_i
0	-1	-1	-1	γ_0
1	-1	-1	1	γ_1
2	-1	1	-1	γ_2
3	-1	1	1	γ_3
4	1	-1	-1	γ_4
5	1	-1	1	γ_5
6	1	1	-1	γ_6
7	1	1	1	γ_7

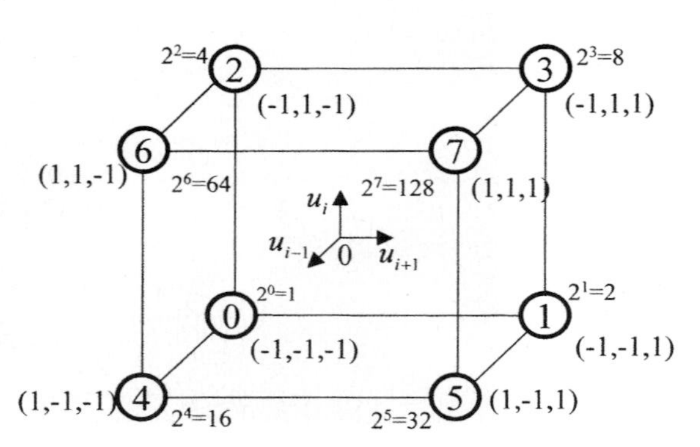

Fig. 6.30. Colored Boolean cube of a Boolean function representing a 1-dimensional cellular automaton [6.39]

(Fig. 6.30). As there are $2^3 = 8$ possibilities of three-bits words, we get $2^8 = 256$ Boolean functions B.

In order to characterize cellular automata by dynamical systems with differential equations, the binary states of a truth table are not denoted by the conventional symbols 0 and 1, but by the integers -1 and 1. The general scheme of a truth table for 256 Boolean functions is shown in Fig. 6.30. Each Boolean function can be characterized by a unique code number if we replace the output $y_i = (\gamma_7, \gamma_6, \gamma_5, \gamma_4, \gamma_3, \gamma_2, \gamma_1, \gamma_0)$ with γ_j either -1 or 1 by the equivalent binary form $y_i = (\beta_7, \beta_6, \beta_5, \beta_4, \beta_3, \beta_2, \beta_1, \beta_0)$ with β_j either 0 or 1. The corresponding code number is the integer $N = \beta_7 \cdot 2^7 + \beta_6 \cdot 2^6 + \beta_5 \cdot 2^5 + \beta_4 \cdot 2^4 + \beta_3 \cdot 2^3 + \beta_2 \cdot 2^2 + \beta_1 \cdot 2^1 + \beta_0 \cdot 2^0$ with $N = 0, 1, 2, \ldots, 255$, because there are 256 distinct combinations of this 8-bit word for 256 Boolean functions (compare Fig. 5.24a).

Geometrically, each Boolean function of three binary variables can be uniquely represented by a Boolean cube with eight vertices (Fig. 6.30). The center of the cube is located at the origin of a 3-dimensional space with coordinates u_{i-1}, u_i, and u_{i+1}. The coordinates (u_{i-1}, u_i, u_{i+1}) of each vertex k ($k = 0, 1, 2, \ldots, 7$) correspond to row k of the truth table. The vertex numbers k are the binary codes of the coordinates (u_{i-1}, u_i, u_{i+1}) if we replace -1 and 1 by 0 and 1. The number 2^k shown next to vertex k is its decimal equivalent. A vertex is colored red if y_i is 1, and blue if y_i is -1. Obviously, a colored Boolean cube represents the same information as the corresponding truth table. We get the code number N of a Boolean function if we add the decimal numbers 2^k associated with the red vertices k of a Boolean cube. In the corresponding rule of a 1-dimensional cellular automaton (Fig. 5.24a), the red vertices represent the outputs 1, while the blue vertices represent the outputs 0. For example, rule $N = 110$ (Fig. 5.18e) is represented by a Boolean cube with red vertices 1, 2, 3, 5, 6, and blue vertices 0, 4, 7. Thus, $110 = 2^1 + 2^2 + 2^3 + 2^5 + 2^6$.

There are 256 different colored cubes representing the 256 Boolean functions of 1-dimensional cellular automata. The spatial geometry of the colored cubes opens new avenues to characterizing the structural complexity of a Boolean function, its corresponding cellular automaton and rule. In relation to the Boolean functions of uncoupled CNNs in the previous chapter, we will distinguish linearly separable rules and linearly non-separable rules. In the context of colored cubes of cellular automata, separability refers to the number of cutting (parallel) planes separating the vertices into clusters of the same color. For rule 110, for example, we can introduce two separating parallel planes of the corresponding colored cube, which are distinguished in Fig. 6.31b by two different colors. The red vertices 2 and 6 lie above a yellow plane. The blue vertices 0, 4, and 7 lie between the yellow and a light blue plane. The red vertices 3, 1, and 5 lie below the light blue plane. It is well-known that the cellular automaton of rule 110 is one of the few types of the 256 automata that are universal Turing machines. In the sense of Wolfram's 3rd class of computer experiments (Fig. 5.24e), it produces very complex patterns.

An example of an automaton that can only produce very simple patterns is rule 232. There is only one separating plane cutting the corresponding Boolean cube for separating colored points (Fig. 6.31a): Red vertices 3, 5, 6, and 7 lie above a light blue plane. The blue vertices 0, 1, 2, and 4 lie below the light blue plane. A colored Boolean cube with three parallel separating planes is shown in Fig. 6.31c, representing the cellular automaton of rule 150: The blue vertex 6 lies above a green plane. The red vertices 2, 4, and 7 lie between the yellow and green planes. The blue vertices 0, 3, and 5 lie between the yellow and light blue planes. The blue vertex 1 lies below the light blue plane. Obviously, it is impossible to separate the eight vertices into three colored clusters and, at the same time, separate them by two parallel planes, no matter how the planes are positioned.

A rule whose colored vertices can be separated by only one plane is said to be linearly separable. An examination of the 256 Boolean cubes shows that 104 of them are linearly separable. The remaining 152 rules are not linearly separable. In general, each rule can be separated by various numbers of parallel planes. In order to use the number of separating planes as unique complexity index κ, it is necessary to choose the minimum number. Obviously, all linearly separable rules have a complexity index $\kappa = 1$. An analysis of the remaining 152 linearly non-separable rules shows that they have a complexity index of either 2 or 3. For example, rule 110 has a complexity index $\kappa = 2$, whereas rule 150 has a complexity index $\kappa = 3$. No rule with complexity index $\kappa = 1$ is capable of generating complex patterns, even for random initial conditions. The emergence of complex phenomena significantly depends on a minimum complexity of $\kappa = 2$. In this sense, complexity index 2 can be considered the threshold of complexity for 1-dimensional cellular automata.

All 256 Boolean cubes can be classified into equivalence classes with an identical complexity index. The corresponding Boolean rules are called equivalent if a transformation exists, mapping the one rule onto the other one, and vice versa. In the case of a Red $\leftrightarrow$ Blue complementary transformation, the colors of the vertices of corresponding Boolean cubes (Fig. 6.30) complement each other, i.e., corresponding red vertices become blue, and vice versa. In the case of left-right symmetrical transformation, the colors between vertices 3 and 6, as well as between vertices 1 and 4, in one Boolean cube (Fig. 6.30) are interchanged in order to get the other one. Obviously, rule 150 (Fig. 6.31c) is invariant under a left-right symmetrical transformation, because vertices 1 and 4 have identical colors (red), in addi-

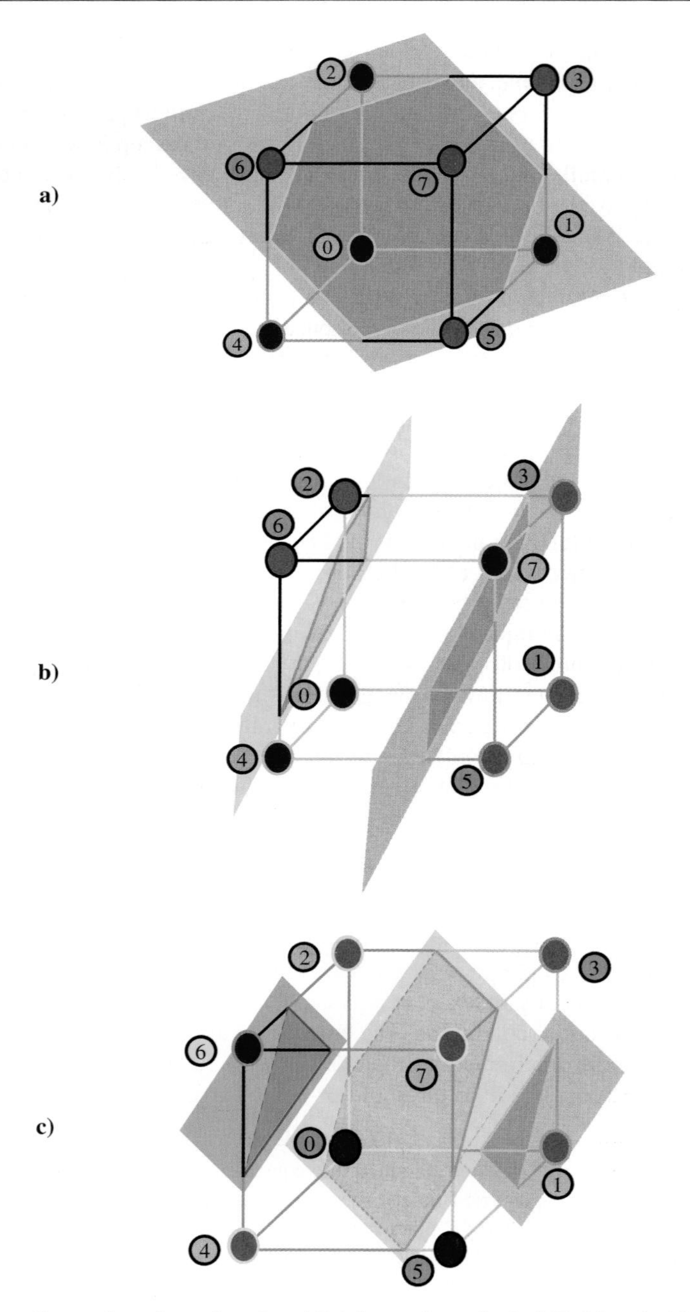

Fig. 6.31a–c. Separating planes in colored Boolean cubes of rule 232 (**a**), rule 110 (**b**), and rule 150 (**c**) [6.40]

tion to vertices 3 and 6 (blue). All members belonging to the same equivalence classes of Boolean rules have an identical complexity index and show dynamic behavior that can be predicted from each other. Thus, it is sufficient to study only one representative member of each equivalence class. In general, 33 independent linearly separable and 47 independent linearly non-separable rules can be identified. Thus, the nonlinear dynamics and dynamic complexity of 256 Boolean functions with three binary inputs is characterized by only 80 independent representatives.

Every 1-dimensional cellular automata or Boolean function with rule number N ($N = 0, 1, 2, \ldots, 255$) can be mapped into the nonlinear dynamical system of a corresponding cellular nonlinear network. In the context of CNNs, a 1-dimensional cellular automaton with near neighbors can be considered a CNN with a 1×3 neighborhood of each cell C_i with inputs u_{i-1}, u_i, u_{i+1} (from the two neighboring cells C_{i-1}, C_{i+1} and the cell C_i itself) and an output y_i. For a CNN of this kind, the time-dependent evolution of future states in a cell C_i depends on the past cellular state x_i, the three inputs u_{i-1}, u_i, u_{i+1}, and the output $y_i = y(x_i)$. Thus, the nonlinear dynamical systems for generating all 256 rules are defined by a state equation in the form $\dot{x}_i = f(x_i, u_{i-1}, u_i, u_{i+1})$ and the same initial condition $x_i(0) = 0$ at time $t = 0$. The output y_i of cell C_i is generated from state x_i by the output equation $y_i = y(x_i) := \frac{1}{2}(|x_i + 1| - |x_i - 1|)$.

Figure 6.32 shows the nonlinear dynamical systems of CNNs for rules 2, 110, 150, and 232. The truth table for each rule is cast in the form of a gene code for CAs (compare Fig. 5.24a). Each pattern consist of 30×61 pixels, generated by a 1-dimensional cellular automaton with rule number N. The top row corresponds to the initial pattern, which is "0" (blue) for all pixels except the center pixel, which is "1" (red). The evolution over the next 29 iterations is displayed in rows 2 to 30. Obviously, these CNNs generate patterns identical to the corresponding CAs. The complexity index κ of each rule is quoted in the upper right corner of each quadrant.

The attractors of these nonlinear dynamical systems code precisely the truth tables N ($N = 0, 1, 2, \ldots, 255$) associated with their Boolean functions and Boolean cubes. Therefore, we replace the output y_i in the state equations (e.g., Fig. 6.32) by their output equation $y_i = y(x_i) := \frac{1}{2}(|x_i + 1| - |x_i - 1|)$ and consider the state equations in the form $\dot{x}_i = g(x_i) + w(u_{i-1}, u_i, u_{i+1})$ with $g(x_i) = -x_i + |x_i + 1| - |x_i - 1|$ and the remaining part $w(u_{i-1}, u_i, u_{i+1})$. Since the nonlinear function $w(u_{i-1}, u_i, u_{i+1})$ delivers a constant real number for each vertex n ($n = 0, 1, 2, \ldots, 7$) with coordinates (u_{i-1}, u_i, u_{i+1}) of the corresponding Boolean cube or for each row n of the corresponding truth table (Fig. 6.30), we can write $w(n) := w(u_{i-1}, u_i, u_{i+1})$. Then, the state equation is recast into the form $\dot{x}_i = g(x_i) + w(n) := h_n(x_i)$ for each vertex n of the Boolean cube or row n of the corresponding truth table. For each of these eight differential equations, we can study the trajectories and attractors in the corresponding phase space. Figure 6.33 shows two typical cases if $w(n)$ is positive or negative. The curve Γ between the two trajectories denotes the plot of $g(x_i)$ ("driving-point function"). The upper curve denotes a vertical translation of Γ upwards by $w(n) > 0$, the lower curve denotes a vertical translation of Γ downwards by $w(n) < 0$. Since the initial condition is always assumed to be $x_i(0) = 0$, the trajectory must begin from the upper initial point $P_+(0)$ if $w(n) > 0$, or from the

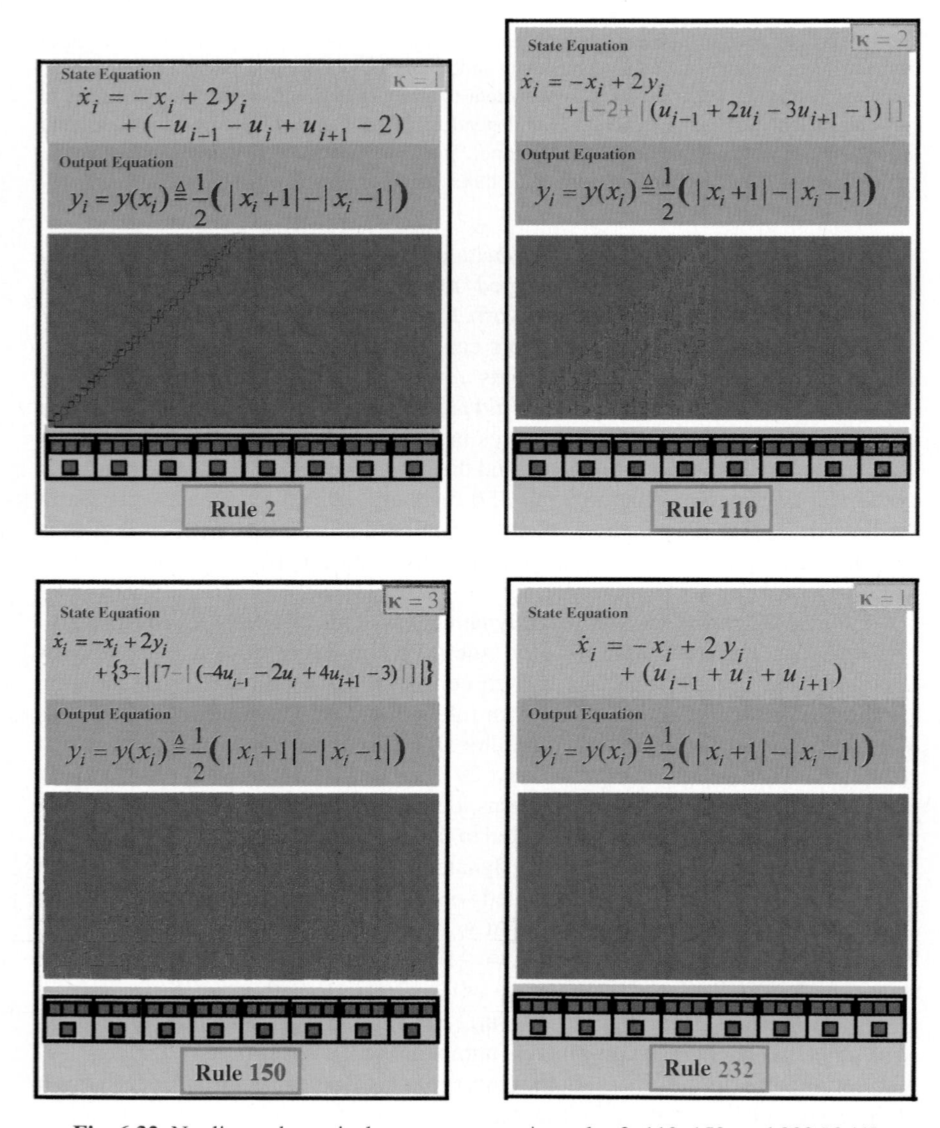

Fig. 6.32. Nonlinear dynamical systems generating rules 2, 110, 150, and 232 [6.41]

lower initial point $P_{-}(0)$ if $w(n) < 0$. Since $\dot{x}_i > 0$ at all points to the right of the initial point $P_{+}(0)$ on the upper curve, the trajectory must flow monotonically to the right until it arrives at the right equilibrium attractor point Q_{+} located at $x_i = x_i(Q_{+}) > 1$. Conversely, the trajectory must begin from the lower initial point $P_{-}(0)$ if $w(n) < 0$ and flow leftwards until it arrives at the left equilibrium attractor point Q_{-} located at $x_i = x_i(Q_{-}) < -1$.

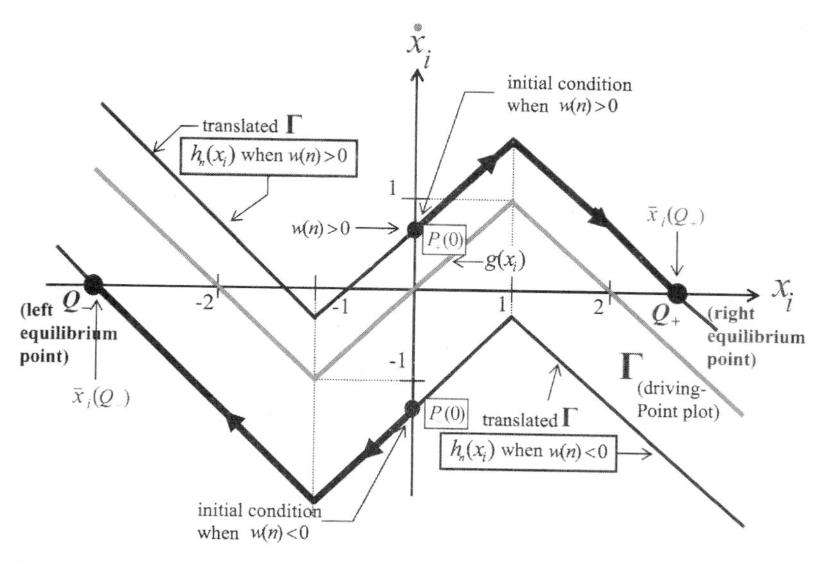

Fig. 6.33. Trajectories and attractors in the phase space of CNN state equation [6.42]

The corresponding output $y_i(t)$ converges at the Boolean state $y_i = 1$ in the former, and to the Boolean state $y_i = -1$ in the latter case. If we paint vertex n red whenever its equilibrium value $x_i(Q_+) > 1$, and blue whenever $x_i(Q_-) < -1$, then the color of all eight vertices for the associated Boolean cube is uniquely determined by the equilibrium solutions (attractors) of the eight associated differential equations. There are eight attractors for each rule. A CNN chip can solve these equations in a few nanoseconds, practically instantaneously.

The separating planes of a Boolean cube can be determined by the nonlinear function $w(u_{i-1}, u_i, u_{i+1})$ of the corresponding state equation. Geometrically, it is interpreted as a scalar function $w(\sigma)$ of only one variable $\sigma := b_1u_{i-1} + b_2u_i + b_3u_{i+1}$, representing an axis in the coordinate system (u_{i-1}, u_i, u_{i+1}) with orientation b_1, b_2, and b_3 (Fig. 6.30). Each colored vertex of a Boolean cube can be mapped on the σ-axis by a perpendicular projection. If we plot the curve of $w(\sigma)$ on the σ-axis, we observe that its zero-crossing points σ_0 with $w(\sigma_0) = 0$ separate the colored points on the projection-axis into clusters of common color. Thus, $w(u_{i-1}, u_i, u_{i+1})$ is sometimes called the discriminant function. Each zero-crossing point of $w(\sigma)$ defines a 2-dimensional plane $\sigma_0 = b_1u_{i-1} + b_2u_i + b_3u_{i+1}$ in the 3-dimensional coordinate system (u_{i-1}, u_i, u_{i+1}), separating the colored vertices of a Boolean cube into clusters of common color (Fig. 6.31). If the colored vertices can be separated by only one plane, we call the corresponding Boolean rule linearly separable. The reason is now obvious: In this case, the associated discriminant function $w(\sigma)$ is a straight line. In the case of several separating planes, there are several zero-crossing points of the separating curve associated with a nonlinear discriminant function $w(\sigma)$. The corresponding Boolean rule is then called linearly non-separable. The projection technique delivers a precise procedure to compute the complexity indices of Boolean rules and their associated 1-dimensional cellular automata.

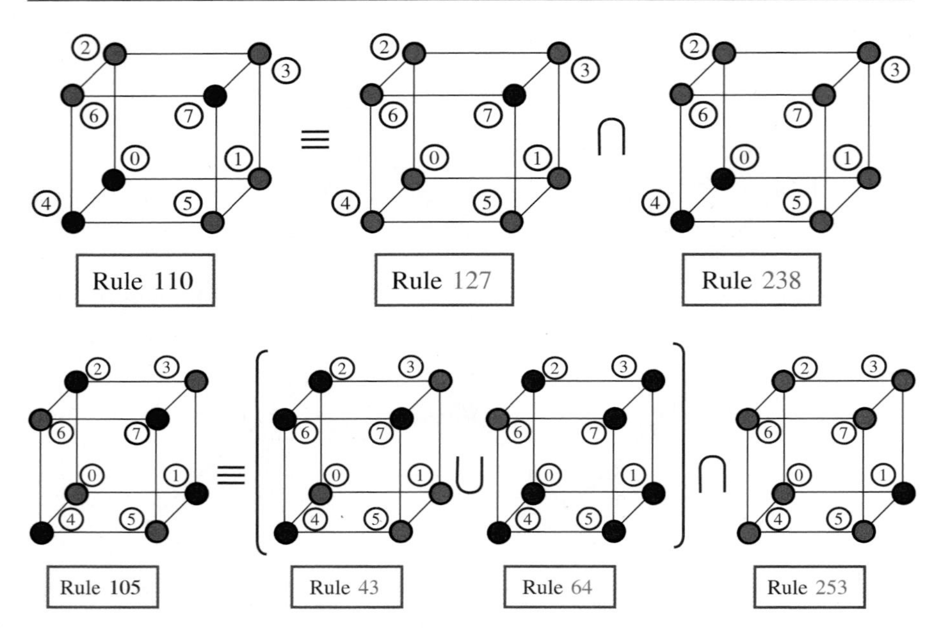

Fig. 6.34. Linearly non-separable rules decomposed in terms of linearly separable rules [6.43]

Linearly separable local rules have a complexity index $\kappa = 1$. We previously analyzed linearly separable Boolean functions of two inputs (Sect. 6.2), nine inputs (Sect. 6.3), and three inputs (Sect. 6.4). In general, they are the simplest building blocks of Boolean functions of any dimension. From an engineering point of view, they are also the simplest ones to implement on a chip. All 104 linearly separable rules are implemented on the CNN universal chip (CNN-UM) directly onto the hardware without programming. A technical advantage is that linearly separable rules are the fastest to execute on a chip. They only need a few nanoseconds via silicon technology, and speed of light via optical technology. The speed of the associated cellular automata is independent of the size of the array. It takes the same amount of time to run a 2-dimensional linearly separable rule on a 10×10 array as on a $10^6 \times 10^6$ array of a cellular automaton on a CNN chip.

In the context of uncoupled CNNs (Sect. 6.3), we mentioned that linearly non-separable Boolean rules of nine inputs can be implemented by combining only a finite number of linearly separable rules via some few standard logic operations (AND, OR, and NOT). As a special case of this fundamental insight, all 152 linearly non-separable rules of three inputs can be decomposed in terms of at most three linearly separable rules and combining them pixel-wise via only AND and OR logic operations. Figure 6.34 shows examples of decompositions for rule 110, involving only one AND operation (a) and for rule 150 with one AND and one OR operation (b). Thus, rule 105 is one of the most complicated 1-dimensional cellular automata to implement on a chip. Concerning dynamic complexity, rule 150 has the highest complexity index of 3.

6.5 Organic Computing, Neurobionics, and Embodied Robotics

The natural evolution of organisms has become an important paradigm in computing and robotics. The ultimate aim for organic computing is to construct self-organizing computing systems that display desired emergent behavior, just as the organisms produced via natural evolution do [6.44]. Emergence refers to a system property that is not associated with any one particular part of the system [6.45]. In nonlinear dynamical systems, the whole is more than the sum of its parts. In robotics, it refers to behavior that results from the agent–environment interaction but that is not preprogrammed. The term is generally not used if the behavior is entirely prespecified, such as when the trajectory of a hand has been precalculated by a planner. Agents designed using high-level ontologies have no room for emergence, for novel behavior. A domain or high-level ontology consists of a complete representation of the basic vocabulary – the primitives – that are used when designing the system. These are the only components that can be used: everything is built on top of these basic elements. The domain ontology remains constant for an extended period of time, often for the entire lifetime of the system. A well-known example is the bounded knowledge representation of an expert system. High-level ontologies are therefore used whenever we know which environments the systems will be applied to, for example in traditional computational systems or factory robot systems. If the environment is not known, a better strategy is to define a low-level ontology and to introduce redundancy with a wide variety of self-organization.

What can we learn from nature? In unknown environments, a better strategy is to define a low-level ontology, introduce redundancy (which there is a lot of in sensory systems, for example), and leave room for self-organization. Low-level ontologies of robots only specify systems like the body, sensory systems, motor systems, and the interactions among their components, which may be mechanical, electrical, electromagnetic, thermal, etc. According to the complex systems approach, the components are characterized by certain microstates that generate the macrodynamics of the whole system.

Take a robot with legs. Its legs have joints that can assume different angles, and various forces can be applied to them. Depending on the angles and the forces, the robot will assume different positions and behave in different ways. Further, the legs have connections to each other and to other elements. If a six-legged robot lifts one of its legs, this immediately changes the forces on all the other legs, even though no explicit connection needs to be specified. The connections are implicit: they are enforced through the environment, because of the robot's weight, the stiffness of its body, and the surface on which it stands. Although these connections are elementary, they are not explicitly included unless the designer wishes them to be. Connections may exist between elementary components that we do not even realize. Electronic components may interact via electromagnetic fields that the designer is not aware of. These connections may generate adaptive patterns of behavior with high degrees of fitness (order parameters). But they can also lead to sudden instability and chaotic behavior. In our example, communication between the legs of a robot can be implicit. In general, there is more implicit communication in a low-level specification than in a high-level ontology. In restricted simulated agents with a bounded knowledge representation, only what is made explicit exists, whereas in the complex real world, many forces and properties exist, even when the designer does not explicitly represent them. Thus, we must study the nonlinear dynamics of these systems in experimental situations in order to find appropriate order parameters and to prevent attractors due to unwanted emergent behavior.

In the dynamical systems approach, we first need to specify what system we intend to model, and then we have to establish the differential or difference equations. Time series analysis and further criteria for data mining can aid in the construction of appropriate phase spaces, trajectories, and attractors [6.46]. In organic computing, one approach would be to model an agent and its environment separately, and then to model the agent–environment interaction by making their state variables mutually dependent [6.47]. The dynamical laws of the agent A and the environment E can be described by simplified schemes of differential equations $dx_a/dt = A(x_a, p_a)$ and $dx_e/dt = E(x_e, p_e)$, where x_e and x_a represent their state variables, such as angles of joints, body temperature, or location in space, and p represents parameters like thresholds, learning rates, nutrition, fuel supply, and other critical features of change. Agents and environments can be coupled by defining a sensory function S and a motor function M. The environment influences the agent through S. The agent influences its environment through M. S and M constitute the agent–environment coupling, i.e., $dx_a/dt = A(x_a, S(x_e), p_a)$ and $dx_e/dt = E(x_e, M(x_a), p_e)$, where p_a and p_e are not involved in the coupling. Examples are walking or moving robots in environments with obstacles. In this case, the basic analytical problem can be stated in the following way: given environment dynamics E, agent dynamics A, and sensory and motor functions S and M, explain how the observed behavior of the agent is generated.

One of the controllers of the dynamics evolves when the agent's angle sensors are turned off and cannot sense the position of its legs. In this case, the activation levels of the neurons exhibit a limit cycle that causes the agent's single leg to stand and swing rhythmically. Doing this causes the robot to walk. The system's state repeatedly changes from the stance phase, with the foot on the ground, to the swing phase, with the foot in the air, and then back. This example illustrates that the dynamical systems approach can be applied in a synthetic way in order to design and construct robots and their environments. However, in general, the dynamical systems approach is used in an analytical way: it starts from a given agent–environment interaction, which is formalized in terms of differential equations. Complex behavior can be analyzed by solving, approximating or simulating the equations, in order to find the dynamic attractors. The dynamic attractors of the interacting system can be used to steer an agent or to allow it to self-organize in a desired way.

Obviously, self-organization leads to the emergence of new phenomena at successive levels of evolution. Nature has demonstrated that self-organization is necessary in order to manage the increasing complexity at these evolutionary levels. However, nonlinear dynamics can also generate chaotic behavior that cannot be predicted and controlled in the long run. In complex dynamical systems of organisms, monitoring and control are realized at hierarchical levels. There is still no final and unified theory of organic computing. We only know portions of biological, neural, cognitive, and social systems in the framework of complex dynamical systems. However, even in physics, which doesn't have a unified theory of all physical forces yet, scientists still work successfully with an incomplete patchwork of theories. In order to delve deeper into organic computing, we need interdisciplinary cooperation between the technical, natural, computer, and cognitive sciences, as well as the humanities.

The goal of organic computing is the construction of self-organizing computing systems that provide services, that can help us to manage a world of increasing complexity and to support a sustainable human infrastructure in the future. The complex system approach will enable us to create a new kind of computer-assisted methodology in scientific, technological, industrial, economic, and even cultural life. However, we must not forget that we must decide on the direction and the ethical goals that the technical development will target. Today, these goals vary between epistemic and scientific interests, technical, economic, cultural, and last but not least military applications. Undoubtedly, medical research and applications must be high on the list of research goals. I remind the reader of the old idea that medicine aims not only at scientific recognition and research, but at praxis too. Praxis means knowledge that is applied both technically, in the sense of an engineer, as well as to cure, help, and heal. Knowledge and research are only instruments to achieve the fundamental aim of medicine, which since the days of Hippocrates has been to preserve life [6.48].

The central organ of human personality is the brain. Thus, the medical task of maintaining the health of the brain places a great responsibility upon neurosurgeons. Their medical treatments involve manipulations of the whole mind–brain entity. In order to provide the best medical treatment possible for the human mind–brain entity, pioneering efforts are being made to expand and refine the diagnostic and therapeutic potentials of neurosurgery, operation planning, operating techniques, and rehabilitation by applied research. As far as we know, the human mind–brain entity is the most complex system produced by evolution. Therefore, an interdisciplinary research program involving computational neuroscience, physics, engineering, molecular biology, medicine, and epistemology is required to deal with this complex organ. This is why some scientists have started an interdisciplinary research program on the brain and mind that addresses ethical and anthropological issues. This has been called "neurobionics."

In general, "bionics" means the simulation of natural functions and processes through the application of technical and artificial procedures and systems. The way that aeroplanes and submarines are designed to emulate the aerodynamic bodies of birds and fishes are well-known examples of this approach. Historically, bionics is the age-old dream of mankind to simulate the principles of nature using technical means, in order to manage the complex problems of life. In this tradition, neurobionics means the elaboration of a techno-biological brain-like computing system constructed from silicon and/or organic materials that can enhance the morphological and functional properties of natural neurons and be used to develop neural prostheses. This is not a Frankensteinian dream; consider the desolate states of patients suffering from tumors or accidental injuries of the brain, and how neurobionics could potentially benefit them.

The specific medical discipline concerned with the treatment of conditions of the central nervous system and the human brain is neurosurgery [6.49]. As the brain is the biological medium of human personality and intellectual capability, the neurosurgeon must not only bring to bear the principles of brain neurology but also knowledge about the human mind and its function. Already, neurosurgery has resulted in progress in the treatment of patients. Notable successes in this regard have been achieved through the introduction of diagnostic visualization procedures like computerized and nuclear magnetic resonance tomography, and by the use of microsurgical procedures during operations.

However, crucial problems encountered during the treatment of patients with brain diseases are still unsolved. For example, the central nervous system of the adult human can only replace damaged areas of tissue with functional tissue to a very limited extent. This is due to the inability of nerve cells to subdivide further after completing the embryonic phase, in contrast to other cells of the body. Only embryonic tissue has the potential to adapt itself to the surrounding host tissue. Therefore, the destruction of neural cell groups by illness or accident often leads to permanent functional disabilities. In this area of application, artificial complex networks, with their principles of self-organization, will become extremely interesting.

In the history of medicine, attempts have been made to restore injured peripheral nerves by autografts. This method is based on the fact that even adults have the capacity to regenerate nerve cell extensions that grow centrally from the spinal cord into the periphery of the target organs. Thus, parts of functionally unimportant sensitive nerves are removed from appropriate points in the body and inserted into the area containing the interrupted nerves that need to be restored. The regrowth of the interrupted nerve fibers is still not completely understood, however. Therefore, it is not possible to control the growth of a transplant consisting of several hundred individual nerve cell extensions towards the target organ. As the central nervous system is unable to regenerate, transplantation is ineffective in the case of injuries close to the spinal cord.

An improvement in peripheral nerve transplants is assumed in the field of molecular biology. An understanding of the physiology and biochemistry of the nerve cells and their associated cells, like glial cells and Schwann cells, could lead to new methods of nerve transplantation. One advanced method of replacing tissue in the central nervous system is to transplant cells from the same body that are genetically altered and adapted. The effect of neural growth factors, the relationship between the source of the transplant and the target area in the recipient brain, and many other problems of molecular biology must be investigated. The resulting methods are based on knowledge derived from genetic engineering.

Another possible method of performing peripheral nerve transplantation would be to use artificial instead of biological transplants. Attempts to restore failed parts of the nervous system by artificial replacement have already been made in the fields of medicine and neurology.

Artificial transplants are equipped with learning algorithms that act as natural "blueprints." In contrast to MCP (McCulloch & Pitts) networks, they are BPN (biological pulse-processing) networks that work in real time. Figure 6.35 illustrates the general scheme for such a neurotechnological implant. A learning neural network encodes control signals for sensation and motion into many parallel pulse sequences, which are received by a set of implanted microcontacts that stimulate the intact nerves (Fig. 6.35a). Signals registered by nerves are decoded by a neural net that controls a movement prosthesis (Fig. 6.35b).

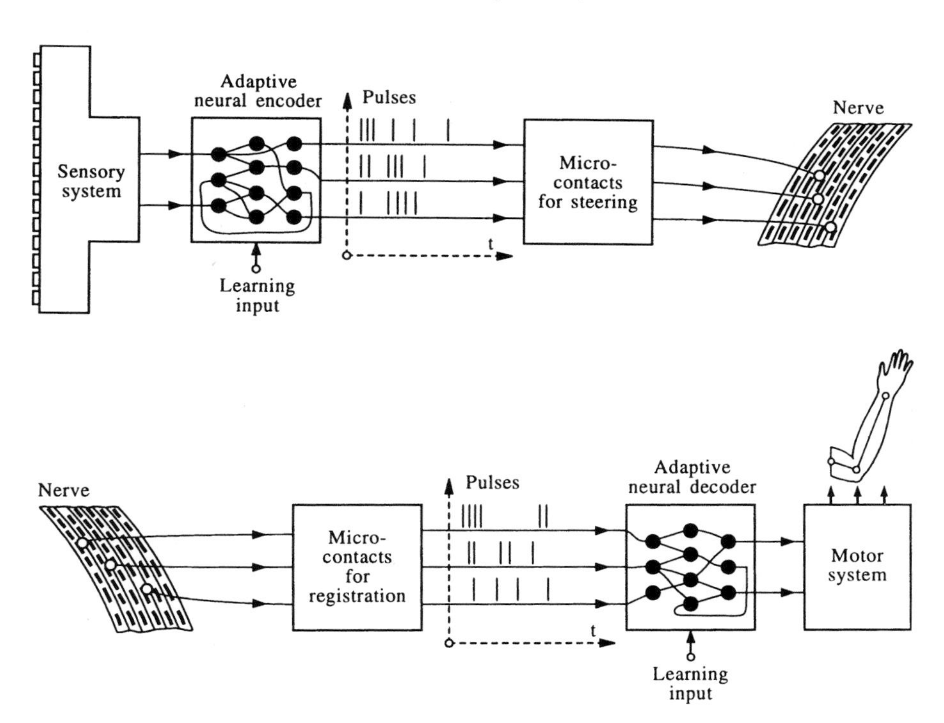

Fig. 6.35a,b. General scheme for neurotechnological prostheses with learning neural nets [6.50]

In patients suffering from a damaged spinal cord, attempts have been made to enhance standing and walking function with the aid of electrostimulation from BPN systems. Assuming that the peripheral apparatus is intact, electrical stimulation of the peripheral nerves brings about muscle contraction. This is caused by parallel pulses from an adaptive learning network encoding the auditory commands of the sensory system (Fig. 6.35a), which are provided by the patient. The system has the ability to learn because it can adapt to the particular patient conditions involved via sensory feedback to the moving legs. However, the system still depends on the patient being conscious and able to speak. In the next research step, the unconscious intentions of the brain must be decoded from signals registered in the spinal cord. These signals can then be sent, for example as radio waves, to a receiver with an adaptive neural encoder which again causes the muscles to contract, as in Fig. 6.35a.

One ambitious neurotechnology project concerns a certain type of blindness. Patients with retinitis pigmentosa suffer from damage to certain layers of the retina that are responsible for the perception of contours, surfaces, colors, and other visual features. The damaged layers of the retina are bridged by a neuroprosthesis. In the architecture of the so-called retina implant, the visual scene is registered by photoreceptors (e.g., transistors) in a spectacle frame equipped with an adaptive neural network. Optical signals are processed by the neural network (BPN), which can learn to model the receptive fields like a human eye. The signals are coded and transmitted telemetrically to an inductive receiver with an electrode array at the damaged retina in order to stimulate the optic nerve and the central nervous system (CNS). As this research progresses, the spectacle arrangement will be replaced by a receptor unit with adaptive neural net located directly in the eye. While neurotechnology cannot realistically hope to replace the full visual faculty in these first trials, a restored perception of contours and surfaces would facilitate patient orientation, and this is the present aim of the endeavor.

Decisive progress can be achieved if the stimulation of different muscle groups can be carried out directly at the branching of the ends of the peripheral nerves without the need to use inorganic metal electrodes. Biotechnical connectivity must be realized by molecular devices that are designed to produce electronic components from organic molecules. The processors that control the electrodes and process the information must be based on the principles of artificial neural networks, which are able to process data at a speed high enough to fulfil the requirements of human walking and standing apparatus. Obviously, the development of those complex networks requires the interdisciplinary cooperation of molecular biology, computational neuroscience, and high-tech hardware engineering.

A further example of the artificial replacement of failed neuronal function is the replacement of the inner ear with a cochlear implant. If the auditory nerve is intact, a 25-pole electrode can be implanted via microsurgery as a replacement for the organ of Corti. The auditory nerve is then excited by suitable electrode impulses that simulate acoustic patterns. The impulses are controlled by sequential microprocessors that are programmed according to linguistic knowledge. However, during difficult operations to remove tumors of the auditory nerve, there is the danger that the auditory nerve can be damaged, resulting in deafness in the patient. It is now also possible to connect an artificial network directly into the central auditory pathway. The sense of hearing can be restored despite loss of the auditory nerve. Interdisciplinary cooperation between biotechnology, computational neuroscience, and engineering is again necessary in order to achieve such advances.

In general, neurosurgery must take care of the following clinical aspects: neurosurgical diagnostics, operation planning, operative technique, and neurorehabilitation, which can be aided by applying the complex system approach to biotechnology and computational neuroscience. The visualisation processes of computerized tomography have opened up a new era in diagnostics. As neurosurgeons must deal with one of the most complex organs ever to have evolved, operation planning and simulation have become essential steps in preparing a successful therapy. In this context, complexity also refers to the personality traits of a patient, his or her particular case history, the pathology of a certain pathogenic process, individual anatomical particularities, and possible postoperative consequences of an operation.

A new method has already been put into practice. Neurosurgical operations can be simulated by CAD (computer-aided design)-supported techniques. Computer-generated, three-dimensional reconstructions of pathological anatomy can be produced by particular programs. Potential operational difficulties can be discovered during the simulation and thus avoided during the actual operation. In terms of operative techniques, one progressive development would be a reduction in the need for large open operations. Stereotactical and endoscopic techniques are important methods of minimizing surgery-induced trauma. The further development of laser technology in combination with neurosurgical endoscopy, intraoperative visualization processes, and computer-controlled regulation techniques could lead to useful complex operative instruments in the future.

A research group at the Massachusetts General Hospital in Boston used magnetic resonance image techniques (MRI) to get functional image maps of human task activation in the visual cortex in response to photic stimulation. Contrast agents were injected periodically over time. Images of the primary visual cortex were obtained via rapid MRI scanning without injection. Figure 6.36 shows real-time visualizations of brain cognitive activity as complex networks [6.51]. These advanced computer-based visualizations of complex neural nets not only help injured patients, but they will ultimately allow us to see ourselves think, feel, and dream.

Hallucinations are caused by self-organizing phenomena within the visual cortex. This type of pattern perception appears to be similiar to the formation of patterns in fluids in chemistry or aerodynamics. In chemistry, local nonlinear interactions of chemical substances generate macro phenomena, such as oscillating chemical clocks (e.g., the Belousov–Zhabotinsky reaction). Pattern formation in the visual brain is due to local nonlinear coupling among cells. In the living organism, there is a spatial transformation between the pattern perception of the retina and pattern formation within the visual cortex of the brain. The first simulations of this corticoretinal transformation by neural networks showed remarkable similarities to pattern perceptions that are well-known from subjective hallucination experiences. Perceptions of a spiraling tunnel pattern have been reported by people who were clinically dead and later revived. The light at the end of the tunnel has sometimes been interpreted as a religious experience.

Cellular neural networks (CNN) are optimal candidates for simulating local neural cell interactions that generate collective macro phenomena. A simple autonomous CNN was designed using a template with local activation and lateral inhibition. It spontaneously generates a labyrinth pattern from random initial conditions. In the next step, the retinocortical map is applied to the resulting stable pattern. Geometrically, a polar-coordinate point on the retina is mapped from the Cartesian point on the cortex, producing the perceived vision of a spiraling tunnel pattern (Fig. 6.37). The advantage of a CNN-based model is obvious: it can easily be

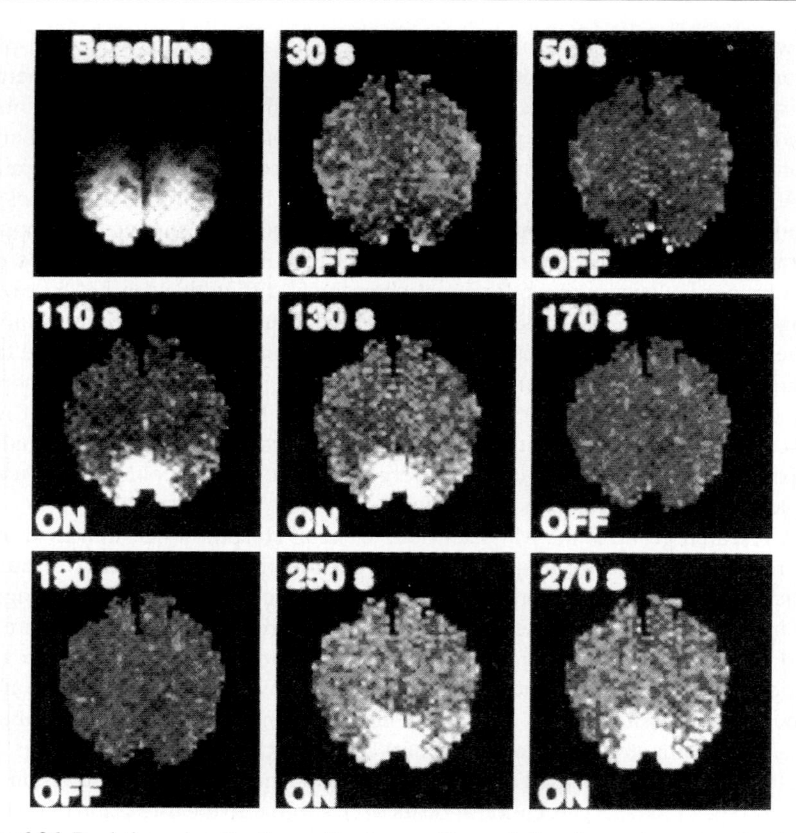

Fig. 6.36. Real-time visualizations of brain cognitive activity of complex networks [6.51]

programmed onto a CNN universal machine (CNN-UM) chip (see Sect. 6.4), which can be incorporated into the living brain in future neurosurgery applications.

One important reason to develop artificial neural networks is that the production of highly integrated electronic circuits based on the chemical element silicon will reach its physical limits in the near future. It will then become impossible to miniaturize silicon-based circuits, the basis for microprocessors, any further. High parallelism and self-organization strategies are required to deal with the brain's complexity. Thus, the use of a new substrate appears to be crucial to the development of information-processing systems. First steps are currently being taken to develop molecular electronic devices based on biological components. Electrical signals could be passed between nerve cells via organic conductors.

In computational neuroscience, computer simulations of neural networks can help to identify the algorithms actually used by the central nervous system and the brain. Present models of artificial neural nets are mainly investigated by simulations performed on vector computers, workstations, special coprocessors, or transputer arrays. However, the advantages of spatiotemporal parallelism in complex networks

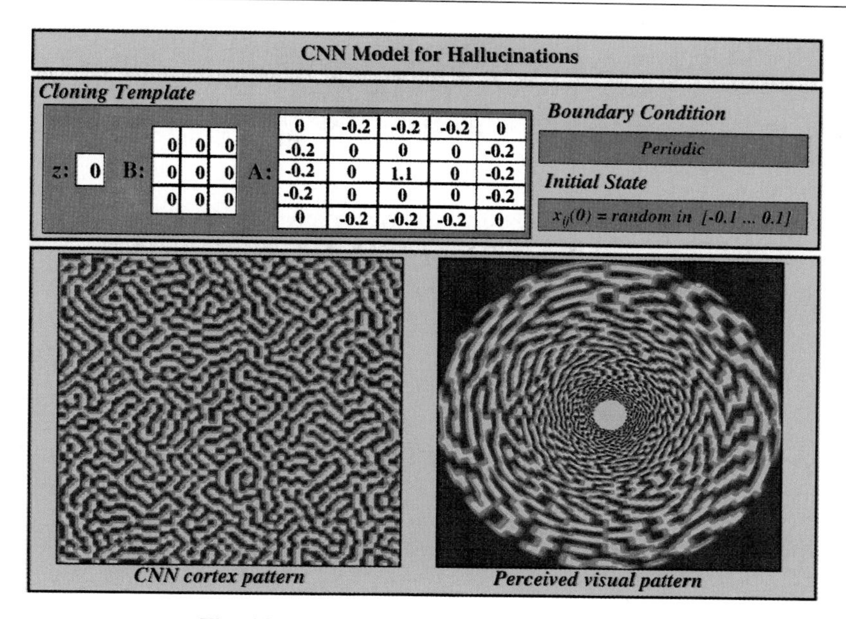

Fig. 6.37. CNN model of hallucinations [6.52]

are entirely or partially lost when simulations are performed on classical computers. Execution of tasks in real time only becomes possible with specially designed neural hardware.

Cellular neural networks (CNN) that process information in nanoseconds (in standard designs) and even at the speed of light (in optical technology) seem to be optimal candidates for neurobionic applications. There are surprising similarities between CNN architectures and, for example, the visual pathway of the brain. An appropriate CNN approach is called the "Bionic Eye," which involves a formal framework of vision models combined and implemented on the CNN universal machine (CNN-UM). The analysis starts with a model of the receptive field organization in the retina and the visual pathway. Figure 6.38a shows a neuron with one axonal output, which acts as a branch to several other neurons and dendritic inputs. The small gaps denote the synapses that are modeled by template elements. In Fig. 6.38b, a neuron in the center receives recurrent outputs from its neighbors. Thus, the receptive field of a central neuron is modeled using a corresponding 3×3**A**-template as its local sphere of influence. In Fig. 6.38c, a part of a two-layer neuron network is shown, with each layer shown as a one-dimensional representation of a two-dimensional grid. The neuron in the center of layer 2 receives dendritic inputs from the neighborhood of input layer 1. The corresponding weights are modeled by a **B**-template.

Several neuroanatomic and neurophysiological models can be translated into CNN cloning templates. Length tuning, for example, means that certain neurons in the lateral geniculate nucleus (LGN) and the visual cortex give a maximal response

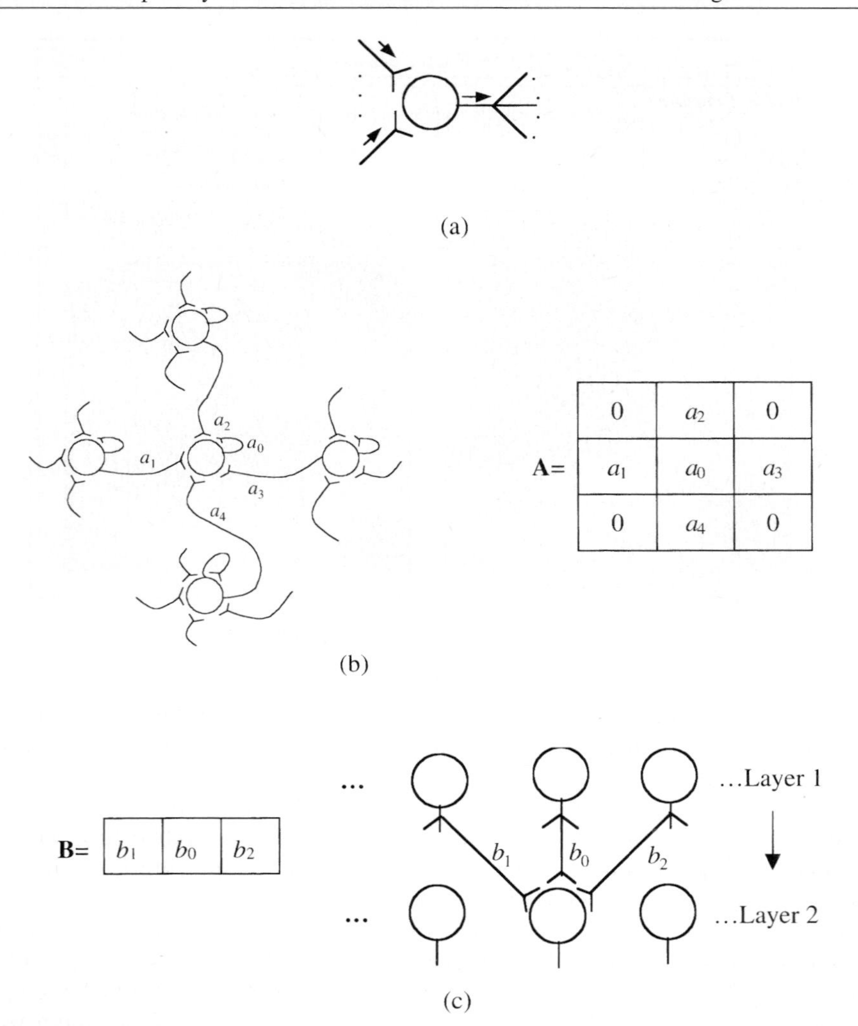

Fig. 6.38a–c. CNN neurobionics models, with a neuron (**a**), an **A**-template (**b**), and a **B**-template (**c**) [6.53]

to an optimally oriented bar of a certain length. The response decreases or vanishes as the length of the bar stimulus increases. A corresponding CNN model detects horizontal, vertical, and diagonal bars whose lengths do not exceed three pixels. Another function of the visual cortex is orientation selectivity, which can also be realized by an uncoupled CNN. Visual illusions studied in cognitive psychology, such as the arrowhead illusion, can also be simulated by an uncoupled CNN. After introducing the "Lego" elements of the retina, such as cells, synapses, and templates for receptive field organization, a simplified multilayered CNN model of the retina can be designed and applied in neurobionics. Ultimately, the CNN-UM architecture allows the implementation of many spatiotemporal neuromorphic models. The same universal machine architecture can be used to mimic the retinas of animals (e.g., of a frog, a tiger salamander, a rabbit, or an eagle) and they can also be combined and optimized for technical applications. Combinations of biological and artificial chips, long associated with cyborgs in science fiction, are now a technical reality, with inspiring ramifications for robotics and medicine.

Clinical applications of CNN chips to epileptology have already been envisaged. The idea is to develop a miniaturized chip device for the prediction and prevention of epileptic seizures. Nonlinear time series analysis techniques have been developed that can characterize the typical EEG patterns of an epileptic seizure and recognize the phase transitions leading to epileptic neural states. These techniques mainly involve estimates of established criteria, such as correlation dimension measurements, Kolmogrov–Sinai–entropy, Lyapunov exponents, measures of determinism, fractal similarity, etc. (see Sect. 2.6). Implantable seizure predictions and prevention devices are already being applied to parkinsonian patients. For epileptic processes, such a device would continuously monitor features extracted from the EEG, compute the probability of an impending seizure, and provide suitable prevention techniques. It should also be both highly tunable to individual patient patterns and it should allow the estimation of these features in real time. Ultimately, such devices should have a low energy consumption and be small enough to be implemented in a miniaturized, implantable system. These requirements are optimally realized by CNNs, with their massive parallel computing power, analogic information processing, and capacity for universal computation. Figure 6.39 shows a miniaturized chip device for seizure prediction and prevention. EEG data are recorded from electrodes implanted near or within the epileptic area and fed to a time series analysis system. The system extracts features of an impending seizure using a warning system (I) and supports the on-demand infusion of fast-acting drugs to prevent the seizure (II).

In future neurobionic applications, the training of neural chips that can serve as interfaces between human nerve fibers will give rise to nonlinear dynamics of great complexity. The chip designer is faced with interconnection problems: if tens of thousands of wires are to be connected physically to a neuron and thousands

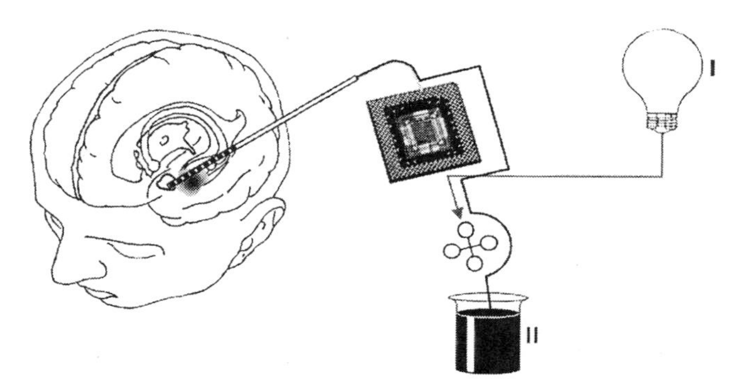

Fig. 6.39. CNN-UM chip used for epileptic seizure prediction and prevention [6.54]

of neurons need to be implemented, then the wiring area will grow to such a size that the delay in the wires will tend to exceed the latency time of the functional block representing the neuron. As the reduction of technological structure size is an economically and physically limited process, bionics designers now favor an architectural solution to the interconnection problem. First, they would need to check the real processing time of the neural net, and they would then consider the extent to which it is possible to deviate from its ideal massive parallelism.

Obviously, the underlying parallel hardware significantly increases the complexity of the software and calls for new methods. Powerful operating systems, programming tools, and flexible user interfaces must be designed in order to ease interface communication with the system. This task will become especially important in an interdisciplinary team with differing degrees of knowledge of computer science. Knowledge-based expert systems may help team members to work with the bionic software and to help them integrate into the team. Neural network hardware programming will differ significantly from the classical programming of von Neumann computers. The programmer must identify the necessary network topology and architecture and specify neuron behavior with interconnection schemes. Thus, the use of heterogeneous and hybrid systems that integrate neural nets with classical knowledge-based systems (as described in Sect. 6.2) appears to be a realistic approach in neurobionics.

In nature, complex patterns of movements are not computed and controlled by a central processor, but by self-organizing learning algorithms of feedback nets. An example is a grasshopper with six legs and different motor modules that perform lifting, swinging, and coordinating. External information about an unknown environment is learned and stored implicitly by the distribution of synaptic weights in neural nets. During evolution, decentralized network modules were used as building blocks for different organisms due to changing conditions. These biological insights into motor information processing are already being applied to robotics and chip technology (embodied cognition). Soft computing uses fuzzy logic, genetic and learning algorithms for flexibility, and adaptive and human-like information systems. Affective computing aims at recognizing and modeling emotional states of the brain as information processing. Cyborgs (cybernetic organisms) represent a vision of a brain with implanted neural computer chips. Neural nets could recognize patterns in brain activity (e.g., EEG signals) that correlate with states of cognition and consciousness. In a next step, neural activity patterns could be scanned and downloaded to a supercomputer. Then, of course, an important ethical problem arises: could human personality (not just the DNA genotype) be cloned and influenced by computational systems?

Some people may fear that hybrid computer systems of increasing complexity cannot be managed without highly specialized training. New interfaces between computer systems and users must be developed. The manipulation of computer-generated images should achieved via direct speech and visual and tactile contacts in a "virtual reality." In this case, the user becomes part of the computer-generated reality by connecting his or her senses to it through technology.

Visual impressions are generated by a monitor with position sensors which can be worn as an eye-mask. A microphone is connected to a speech recognition system that translates spoken orders into system commands. Tactile contacts are generated and modeled using so-called "data gloves" that transform hand and finger movements into electrical signals (Fig. 6.40) [6.55].

Optical fibers are placed between two layers of cloth in the data glove. They are connected to a special module that transforms the light signals into electrical signals. It is assumed that data gloves will have practical applications in, for instance, astronautics. NASA is interested in the development of robots that can perform complicated and dangerous actions in space by simulating the hand movements of an astronaut located safely in a space station. It also appears possible to extend the principles of data gloves to create a data suit that simulates the movements and reactions of the whole body.

The senses have an important influence on human imagination. Thus, molecular modeling can be enhanced through the use of not only computer-assisted graphics

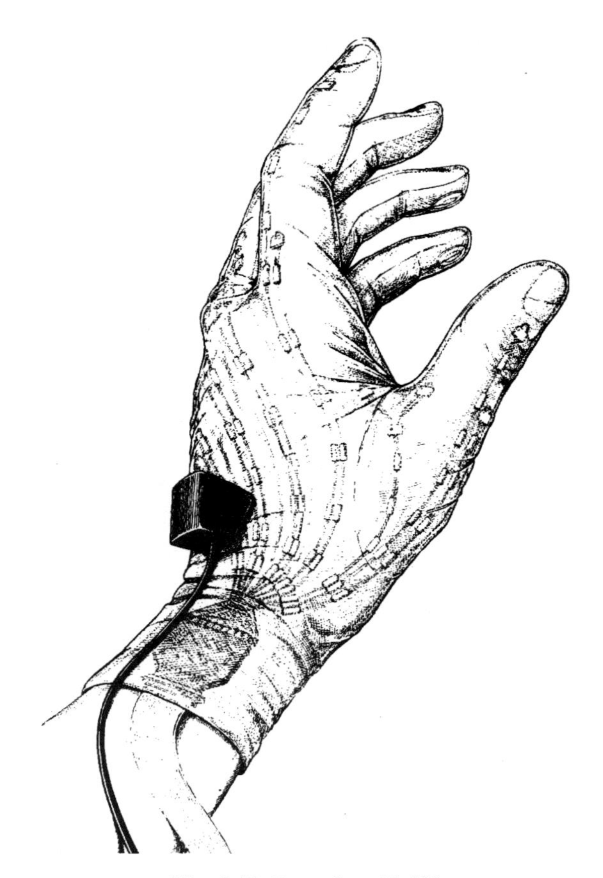

Fig. 6.40. Data glove [6.55]

that aid structural visualization, but through the introduction of a tactile element too. Using data gloves, chemists will be able to grasp a molecule, to feel its surface, and to manipulate it in a desired manner. Engineers try to generate the effects of touches and forces using particular technical systems. As it manipulates molecules in a virtual reality via data gloves, the human hand should receive feedback from the imaginary objects touched. The complexity of the experienced world should be simulated in all its aspects.

In the examples given above, the virtual reality created corresponds to actual reality in the macro- or microcosmos. However, imagine computer-generated scenarios of fantastic worlds that only exist as electronic realities. The boundary between technical feasibility and science fiction appears to be fuzzy. People may feel like imaginary bodies in a computer-generated "telereality." There are proposals to build so-called "home reality engines" that can transfer the user into a virtual world of both desired and unwanted fantasies. Sex with Marilyn Monroe or discussions with Albert Einstein – whatever you want – have been promised by the prophets of computer-generated virtual realities. Authors of science fiction like William Gibson describe a computer-generated "cyberspace" that will be experienced by people as a gigantic hallucination:

> Cyberspace. A consensual hallucination experienced daily by billions of legitimate operators, in every nation, by children being taught mathematical concepts .... A graphic representation of data abstracted from the banks of every computer in the human system. *Unthinkable complexity*. Lines of light ranged in the nonspace of the mind, clusters and constellations of data. Like city lights, receding ... [6.56]

These perspectives, of course, give rise to an essential critique on our cultural development. People locked up in boxes with their own private or manipulated virtual reality generated by super-Crays and neural networks seem to be a chilling vision akin to Orwell's Big Brother.

Besides those ethical problems, there are some severe epistemic questions that arise with the possibility of computer-generated complex artificial worlds. In traditional epistemology, philosophers like Berkeley and Hume discussed the solipsistic or sceptical position that the reality of the outer world cannot be proved by any means. All of our impressions may be illusions generated by our brain and its mental states. These puzzling problems were not meant as jokes by childish philosophers who were completely ignorant of the world. They were invoked to inspire us to test and analyze the validity of our arguments. Modern logicians and philosophers of the mind like Hilary Putnam have translated the problem in the following way, which reminds us of the famous Turing test.

Imagine that a human has been subjected to an operation by an "evil scientist." The person's brain has been removed from the body and placed in a vat of nutrients that keeps the brain alive. The nerve endings have been connected to a hybrid neural computer which causes the person whose brain it is to have the illusion that everything is perfectly normal. All the person experiences is the result of electronic impulses sent from the computer to the nerve endings. If the person intends to raise his hand, the feedback from the computer will cause him to "see" and "feel" the hand being raised, although there is no physical eye or ear, only the corresponding patterns in the brain. The evil scientist can cause the poor person to experience any situation. Putnam said:

> It can even seem to the victim that he is sitting and reading these very words about the amusing but quite absurd supposition that there is an evil scientist who removes people's brains from their bodies and places them in a vat of nutrients which keep the brains alive. The nerve endings are supposed to be connected to a super-scientific computer which causes the person whose brain it is to have the illusion that ... [6.57]

Could we, if we were brains in a vat in this way, say or think that we were? Putnam argues that we could not. The supposition that we are actually brains in a vat cannot possibly be true, because it is self-refuting. A self-refuting supposition is one whose truth implies its own falsity. A logical example is the general thesis that all general statements are false. If it is true, then because of its generality it must be false. An epistemic example is the thesis "I do not exist" which is self-refuting if thought by me. Thus, Descartes argued that one can be certain that oneself exists, if one thinks about it. The supposition that we are brains in a vat has this property.

Suppose that we are brains in a vat with nutrient fluid and afferent nerve endings connected to a super neural computer that produces all of the brain's sensory inputs. As the human brain in the vat is functioning well, it of course possesses consciousness and intelligence. However, its ideas and images of trees, houses, etc., have no causal connection to actual trees, houses, etc., in the outer world of the brain in the vat, because these ideas and images are produced by our super neural computer. Thus, if we suppose that we are brains in a vat with all of these conditions, then the words "vat," "nutrient fluid," etc., do not refer to an actual vat, nutrient fluid, etc., but to certain ideas and images that have been produced by our super neural computer. Consequently, the statement "we are brains in a vat" is false (Fig. 6.41).

Note that the possibility that we are brains in a vat is not ruled out by physics, but by logic and philosophy. A physically possible world in which we are brains in a vat is compatible with the laws of physics. However, we can even use a *Gedankenexperiment* to rule out the existence of physically possible worlds.

The reason for this seems to be the structure of self-referentiality, which is typical of the high-level capabilities of the mind–brain system. In Sects. 4.3 and 4.4, we argued that self-referentiality may be a prerequisite to consciousness and self-consciousness – not only in the mind–brain system, a product of biological evolution, but even in artificial complex systems with a quite different hardware.

Turing himself proposed a well known test that could be used to decide whether an artificial system like a computer is conscious nor not: allow someone to have a keyboard conversation with the computer and a similar conversation with a (human) stranger. If he cannot tell the computer from the human being based on the conversations, then the computer is conscious. In short: a computing machine is conscious if it can pass the Turing test.

The "brain in the vat" thought experiment shows that Turing's dialog test must fail at some point. The words and sentences used by the artificial system do not necessarily refer to the actual objects and events we refer to in natural human languages.

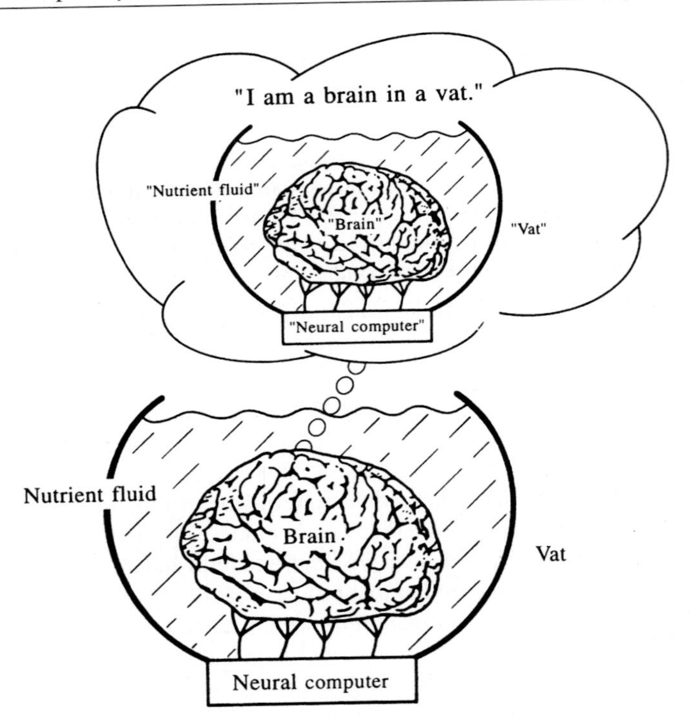

Fig. 6.41. Brain in a vat which thinks that it is a brain in a vat

The use of words and sentences can be governed by syntactic patterns that are programmed into the Turing machine in a highly sophisticated way. Weizenbaum's program ELIZA, which simulated the dialog of a patient with his psychologist, may hint at these possibilities. In this sense, Turing's test cannot rule out the notion that the conversation of the machine is only a syntactic play that resembles intelligent human discourse. Nevertheless, in principle, it cannot be excluded that self-organizing complex systems are able to learn the particular references of syntactic words and rules using prototype patterns and experiences with their environment. In the long run, we will have to answer the ethical question of whether we want to develop those highly autonomous (dissipative) systems.

6.6 Embodied Artificial Intelligence and Artificial Life

While technological advances in the life sciences and computer science are expected in the future, we can also expect much research into "artificial life" and "artificial evolution." In a famous quotation from his Monadology (§64), Leibniz argued that every organism is a kind of "divine machine" or "natural automaton surpassing all artificial automata infinitely." The modern sciences of computability and complexity, which try to model intelligent behavior using computational and complex dynamical systems, can be traced back to the work of Leibniz.

While "artificial intelligence" (AI) is a classical discipline within computer science, "artificial life" (AL) describes a novel but growing field of research performed within the framework of the science of complexity. The term "artificial life" literally means life made by man rather than by nature. For Christopher G. Langton from the Santa Fe Institute, who organized the first AL conference in 1987, artificial life is still in the process of defining itself [6.58]. Natural life on Earth is organized at the molecular, cellular, organism, and population–ecosystem levels. Research into artificial life attempts to find modeling tools that are powerful enough to capture the key concepts of living systems at these levels of increasing complexity.

A key concept of living systems is the distinction between genotype and phenotype. In biology, the genotype is the complex set of genetic information (genes) encoded in the DNA of an organism. The phenotype is the physical organism itself. The development of the phenotype, as directed by the genes of the genotype, is called morphogenesis. Morphogenesis was characterized in Sect. 3.3 by complex dynamical order parameters. In our framework of complex systems, phenotype refers to a macrophenomenon with order parameters that depend on the nonlinear interactions of genes at the microlevel of the system. The high nonlinearity of genetic dynamics provides a great variety of possible phenotypes. On the other hand, it prevents predictions or derivations of the properties or future behavior of an individual phenotype.

In the artificial life approach, the distinction between genotype and phenotype is not restricted to life processes based on carbon-chain chemistry. In the theory of computability, the genotype can be generalized as being a set of local computational devices ("genes") which recursively generate global phenotype structures.

Examples of the artificial life approach are the so-called L-systems of Aristid Lindenmayer, which model filaments and branching botanical trees and other plants and organs [6.59]. L-systems consist of sets of rules for deriving strings of symbols, which are analogous to the formal grammars of Noam Chomsky. For example, $x \rightarrow y$ means that every occurrence of the symbol x in a formal structure should be replaced by the string y. The symbol x can appear on the right as well as the left side of some rules. Thus, the set of rules can be applied recursively to previously derived structures. We then get finite structures ("phenotypes") step by step, ad infinitum. If the context of symbolic replacement is not considered, the rules are called context-free. Context-free rules that only replace single symbols are equivalent to the operations of finite state machines (or the regular languages of Chomsky grammars). This kind of L-system can produce branching structures such as the vascular network of the heart shown in Fig. 3.9.

When there is more than one symbol on the left-hand side of a rule, the rules constitute context-sensitive grammars that are equivalent to Chomsky's Turing languages. Obviously, in context-sensitive grammars the way that local rules are applied depends on neighboring symbols, which corresponds to nonlinear interactions of the elements in complex systems. Without context sensitivity, the global structures ("phenotypes") produced by L-systems are linearly decomposable.

A formal advantage of L-systems is their ability to measure complexity in the context of computer science (cf. Sect. 5.2). A phenotype is interpreted as a formal structure that may or may not belong to the class of derivable structures in a certain L-system. Thus, the undecidability of the membership to a given class indicates that this class is more complex than those for which the membership is decidable. Furthermore, we can measure the computational time for derivations in L-systems. In this case, L-systems can be ordered according to the degree of complexity from N to NP complete problems.

L-systems provide a rich variety of computer-generated shapes for modeled plants and organs. Sometimes the formal rules of L-systems can be realized by mechanisms observed in matter. In supramolecular chemistry, for example, tree-like polymers such as dendrimers (cf. Sect. 2.5) are generated by a divergent synthesis of repeating bifurcations which can be described by the recursive rules of an L-system. Obviously, L-systems are highly applicable to computer graphics that illustrate and visualize molecular and cellular growth in chemistry and biology. However, the actual mechanisms of molecular and cellular growth must be observed and measured in the laboratories of chemists and biologists of course – not by computer experiments.

Another example of the artificial life approach is the concept of cellular automata, which we discussed in Sect. 5.6. Their simple sets of local rules can be interpreted as "genotypes" that produce "phenotypes" of more or less complex patterns. Cellular automata allow remarkable simulations of key concepts in biology. John von Neumann's idea of a self-reproducing automaton has already been mentioned. Although this idea is justified by a mathematically precise proof, it is hard to realize with technical computers. The reason for this is von Neumann's requirement that the self-reproducing structure must be a universal computer that has the degree of complexity of a universal Turing machine. Obviously, self-reproducing molecules from prebiotic evolution (for example, see Eigen's model, described in Sect. 3.3) were hardly capable of universal construction. Therefore, Christopher Langton dropped the requirement for universality and designed a very simple cellular automaton that can reproduce itself. It consists of a set of neighboring cells that are finite state automata.

In Fig. 6.42, each number is the state of one of the automata in the lattice. Blank space means cells in state 0. The 2-states form a sheet around the 1-state data path, which forms a loop with a tail at one end, reminiscent of a Q-shaped virus. The inner data path conducts data, such as the state pairs 70 and 40, which are necessary for self-reproduction. With each new generation of the whole cellular automaton, the cells in this inner layer follow rules that affect the states of the neighboring cells. They propagate signals counter-clockwise around the loop. When the signals reach the end of the tail, each 70 signal extends the tail by one unit and the two 40 signals construct a left-hand corner at the end of the tail. For each full cycle of the signals around the loop, another side and corner is constructed. After four cycles the tail loops back on itself and the two loops are disconnected. The cellular automaton of the loop has then reproduced itself.

Each loop goes on to produce further offspring, which also reproduce (Fig. 6.43a,b). This development continues indefinitely and produces an expanding colony of loops. As the loops in the core of the colony have no space to reproduce, they lose their tails and "die" (Fig. 6.43c). Thus, the colony contains a reproductive fringe that surrounds a growing "dead" core (Fig. 6.43d). This pattern seems to be similar to the growth of coral, with dead skeletal structures inside and living cells on the surface.

In Sect. 5.6, we mentioned that cellular automata can be used as discrete and digital models of phase portraits describing the global dynamical behavior of complex systems. The complexity classes of cellular automata (Fig. 5.24b,c) correspond to the degrees of complexity of the attractors that classify dynamical systems. Langton introduced a certain parameter λ that characterizes the dynamical behavior of cellular systems depending on their transition rules [6.61]. Figure 5.24a shows a one-dimensional cellular automaton with $8 = 2^3$ transition rules since there are three neighboring cells and two possible cellular states.

In general, a cellular automaton has K^N transition rules for N neighbors and K states. An arbitrary state of a cellular automaton is distinguished as the quiescent state. In Fig. 5.24b, the quiescent state is the white cell ("zero"). If n_q is the number of transition rules to this special quiescent state, then there are $K^N - n_q$ remaining rules that should be spread randomly and uniformly over the other $K - 1$ states. Now, the λ-parameter is defined by $\lambda = (K^N - n_q)/K^N$, which ranges from zero to one. If $n_q = K^N$, then all transition rules of the automaton are to the quiescent state

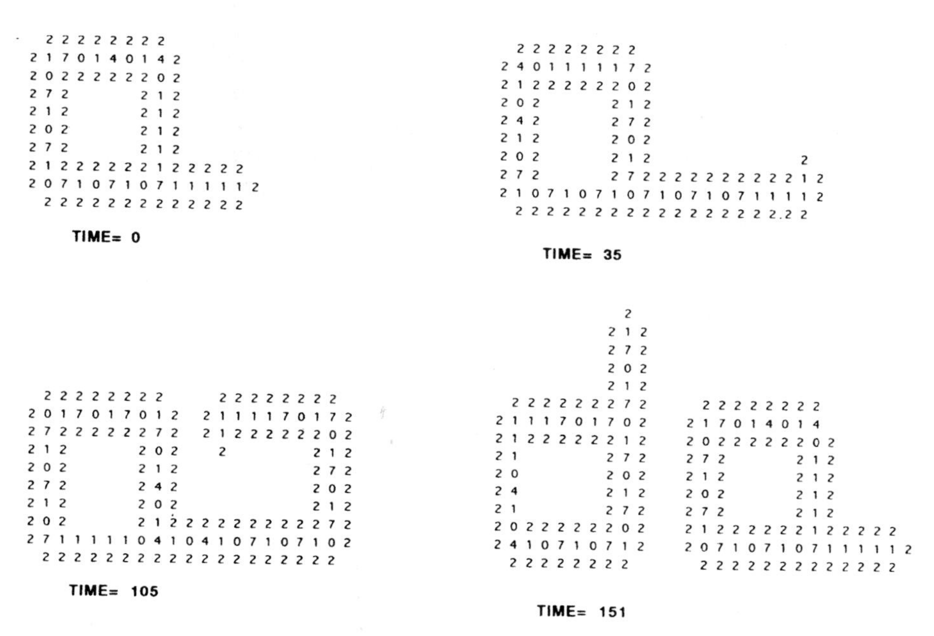

Fig. 6.42. Self-reproducing Langton loop with an extending tail, and the formation of a daughter loop [6.60]

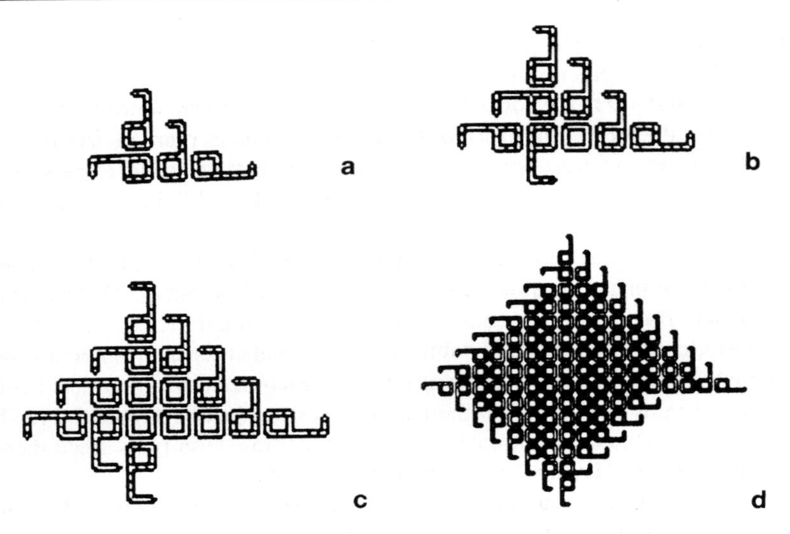

Fig. 6.43a–d. Expanding colony of artificial coral [6.60]

and $\lambda = 0$. If $n_q = 0$ then there are no transition rules to the quiescent state and $\lambda = 1$.

The dynamical meaning of λ is obvious. If λ is very small and approaches zero, then all dynamics are frozen after a short time, like the molecules in a frozen solid. If λ is very large, approaching its maximum value, then the dynamics are very free and chaotic, and a cellular pattern is very difficult to retain, like molecules in a gaseous state. Physically, life cannot be maintained at these extreme scenarios. However, there is a certain region between these extreme situations where the dynamics still change, but not rapidly enough to lose all configurations of the previous state. This region corresponds to the liquid state, which has indeed supported the emergence of life during biological evolution.

Computer experiments show that the progression of the λ parameter characterizes the spectrum of dynamical behavior from fixed point dynamics to periodic dynamics, then to complex dynamics and finally to chaotic dynamics. These regimes correspond to the complexity classes of cellular automata, as illustrated in Fig. 5.24b (fixed point dynamics), Fig. 5.24c (periodic dynamics), Fig. 5.24e (complex dynamics), Fig. 5.24d (chaotic dynamics). Thus, the λ parameter corresponds to a control parameter of a dynamical system that characterizes its phase transitions. At an intermediate critical value λ_C, we observe a phase transition between periodic and chaotic dynamics. There is an increase in complexity at λ_C in the sense of longer cellular patterns. Transition lengths of cellular patterns are shorter at both ends of the range of λ and they get longer in the middle of the range. At λ_C the transient lengths diverge. Beyond λ_C, the transient lengths get shorter with increasing λ, although the dynamical activity expands more rapidly with time until the onset of chaos. There is even evidence of a second-order phase transition at λ_C, because the average Shannon entropy of cellular growth has no discrete jump, but is smooth at λ_C.

From a computer modeling viewpoint, the solid, liquid, and vapor phases of matter are only special types of dynamical system. They can also be realized by artificial systems such as cellular automata in the medium of a computer. In this framework, artificial life is possible in the neighborhood of the critical value λ_C. In Sect. 5.6, we suggested how cellular automata may be related to our concept of order parameters and synergetics. Order parameters correspond to macroscopic spatiotemporal properties of cellular patterns such as fixed point, periodic, complex, or chaotic attractors.

In the synergetic approach, stable and unstable collective motions ("modes") can be distinguished close to critical points of instability. This instability is caused by a change in a control parameter, which leads to new macroscopic spatiotemporal patterns. In cellular automata, stable and unstable motions can be characterized by transition rules that may or may not change previous states. The stable modes are enslaved by the unstable modes and can be eliminated (Fig. 6.16a). The remaining unstable modes serve as order parameters that determine the global macroscopic pattern of the system. In general, eliminating unstable modes by enslaving results in an enormous reduction of the degrees of freedom. The global dynamics of order parameters may lead to fixed points, periodic, oscillating, pulsating, or chaotic patterns. Recall the patterns of fluid dynamics (Fig. 2.26a–e) or the laser (Fig. 2.27a,b).

For cellular automata, fixed point dynamics are independent of fluctuating initial states such as order formation near to thermal equilibrium. It corresponds to a low value of the control parameter λ, close to zero. In the neighborhood of the critical value λ_C the cellular dynamics depends sensitively on initial fluctuations. Thus, it simulates the complex dynamics of open dissipative systems such as the laser in physics, a Belousov–Zhabotinsky reaction in chemistry, or the cellular differentiation of an organism in biology. Langton's λ_C-regime is not therefore restricted to "artificial life" as a computer simulation of biological life, but to "artificial dissipative non-chaotic systems." In Sect. 3.3 we underlined that, from a biological point of view, it is not sufficient to know the general scheme of dissipative self-organization. We must know more about the tricks of "carbon-chain chemistry" to create biological life (especially its clever combination of conservative and dissipative self-organization). We can then also grasp the complex dynamics of "artificial life" in the medium of a computer. Otherwise the cellular automata approach is too general for modeling life processes.

Since Darwin, selection has been emphasized as a key mechanism of biological evolution. However, there are also selection processes in dissipative systems like the laser in physics or the Belousov–Zhabotinsky reaction in chemistry. The nonlinearity of their dynamics shows that there is no superposition of pattern waves, but a competition of stable and unstable modes. In the late 1960s, Ingo Rechenberg [6.62] used evolutionary strategies as blueprints when optimizing technical systems. John Holland applied the process of selection to algorithmic learning procedures [6.63]. His genetic algorithms have gained importance in artificial life research. In general, a genetic algorithm is a method for moving from one population of genotypes to a new "fitter" population. The genotypes may be represented by, for example, bit strings for the genes of an organism, or by possible solutions to a problem. The basic concept of a genetic algorithm works as follows. (1) Start with a randomly generated population of genotypes. (2) Calculate the fitness of each genotype in the population. (3) Apply selection and genetic operators to the population to create a new population. (4) Go to step 2. Selection chooses those genotypes in the population that can reproduce and decides how many offspring each is likely to have; the fitter genotypes produce on average more offspring than less fit ones. Holland distinguishes the genetic operators of crossover, mutation, and inversion. Crossover exchanges subparts of two genotypes (Fig. 6.44). Mutation randomly changes the values of some locations in the genotype. Inversion reverses the order of a contiguous section of the genotype.

If the algorithmic procedure (1–4) is repeated over many time steps, it produces a series of populations (generations) until one or more highly fit genotype emerges. Genetic algorithms have been applied successfully to many scientific and engineering problems [6.64]. In Sect. 4.2, we analyzed neural networks with competitive learning procedures (Fig. 4.12) and learning classifier systems. Eigen's self-optimizing strategy of genetics (Sect. 3.3) can also be thought of as a kind of genetic algorithm.

The learning we each perform during our lives does not directly affect our genotype. However, if learning helps survival, then the organisms that are best able to learn have the most offspring. They increase the frequency of the genes responsible for learning. This indirect effect of learning on evolution is called the Baldwin effect. Computer experiments can help to understand and to measure the success of this effect. The learning process is not necessarily supervised by a global function that evaluates the fitness of the population. However, it is a macroeffect of individ-

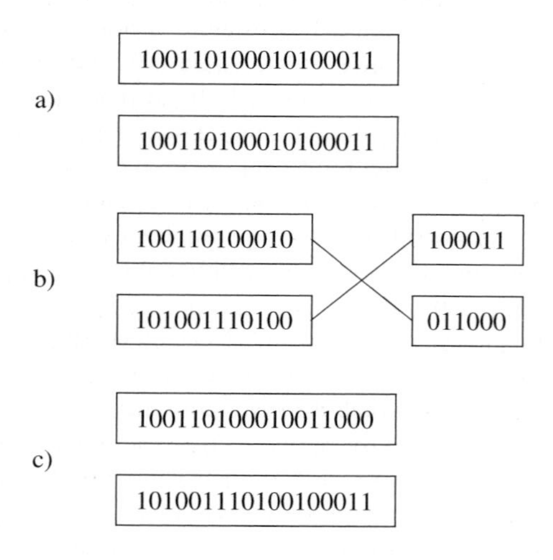

Fig. 6.44a–c. Crossover operator of a genetic algorithm with binary coding

ual learning processes by individual agents of the population. At each time step t in an agent's life, the agent evaluates its current state, using its evaluation network. This evaluation is compared with the evaluation it produced at time $t-1$. The comparison gives a reinforcement signal that is used to modify the weights in the action network. The next action can then be determined. Obviously, this kind of unsupervised learning can be used in computer models of social and economic systems. It can be embedded in the learning procedures of neural networks to improve their capacity for problem solving.

Other extremely interesting applications of genetic algorithms include computer models and computer experiments in gene technology. Another example is the medical simulation of immune systems in which learning takes place [6.65]. In recent years, immunology has become a tremendously important field of medical research, especially with respect to therapies for cancer. The human immune system is capable of recognizing on the order of 10^{16} different foreign molecules. The genome encoding the construction rules of the immune system only contains about 10^5 genes. Thus, the immune system is an example of a complex dynamical system which is highly effective for pattern recognition with relatively few instruction rules. There is no central organ to control it. Thus, pattern recognition by an immune system is a macroeffect of local molecular interactions. It can be characterized by order parameters. They represent the extent of molecular binding between the cells performing the recognition ("antibodies") and the foreign material ("antigens"). Genetic algorithms can be used as artificial immune systems in the medium of a computer by representing antigens and antibodies with binary strings. The binary immune systems have been used to study the ability to detect typical patterns in an environment of randomly distributed antigens or to learn new patterns if the pathogenic material is rapidly evolving through mutations.

Organic forms of dangerous viruses are well known. But artificial forms are not only interesting for simulating and modeling biological immune systems. Over the past few years there has been considerable interest in the application of computer viruses as security and the social problems of global computer-assisted networks [6.66]. In general, a computer virus is a code that can attach itself to other computer entities such as programs, data files, etc. and reproduce itself, at least partially. Today, we can distinguish generations of more or less sophisticated and dangerous computer viruses. A shell virus, for example, forms a shell around the original code. The virus becomes the program, and the original host program becomes a subroutine of the viral code. The most dangerous computer viruses are the polymorphic or self-mutating viruses. They infect their targets with a modified or encoded version of themselves. Scanners are often used as a computer virus defense or an artificial immune system. They read data from a disk and apply pattern matching operations against a list of known virus patterns. However, polymorphic viruses cannot be detected by scanners and need learning algorithms for their detection.

From an artificial life viewpoint, computer viruses do satisfy some of the criteria for life, but not all. Obviously, a computer virus is able to reproduce itself. On the other hand, it is not the agent of reproduction; the computer is. Obviously, a computer virus stores information on its self-representation. Like DNA molecules, its code is a template that is used by the virus to replicate itself. There also seems to be a kind of metabolism that converts matter and energy. However, electrical energy is again not used by the virus but by the underlying computer system. There are also ecological forms of computer viruses, such as those displaying predatory, territorial, or self-protective behavior. Up to now, however, these behaviors have not been initiated by the viruses themselves, but by a human programmer. In the future virtual computer viruses in cyberspace may have all the dangerous capabilities that are associated with biological life.

In biology, artificial life systems are applied as software, hardware, and wetware systems [6.67]. Wetware refers to the techniques of a biochemical laboratory which, for example, are used to model the molecular origins of prebiotic evolution in an evolutionary reactor. These are, of course, empirical experiments that reproduce the biochemical evolution of life or that create new forms of biological life. They are only termed "artificial" because the conditions required for the self-organization of biological life are arranged by human experimentors. Software systems are computer programs or networks that model key concepts of life and evolution. We have already analyzed the cellular automata approach. There are also computer programs that model the evolution of animals. In the program GENESYS, for example, animals are represented as neural nets or finite automata. The genes of each organism are represented as bit strings encoding the weights of a neural net or the transition lists of a finite automaton. In order to increase the modeling capacity of the program, it is executed on a massively parallel connection machine. The GENESYS program reminds us of Leibniz' vision of natural and artificial automata in his Monadology. Hardware models of artificial life are used in robotics. In Sect. 6.5 we analyzed the first applications of neurobionics. In Sect. 2.5 we mentioned the self-construction of smart and "intelligent" new materials.

In evolutionary robotics, initial research is into the artificial evolution of intelligent systems. It is important to realize that the traditional idea of computers as "intelligent" is inappropriate, because the brain does not run simple programs. Evolutionary theory says that the brain has evolved not to derive formal proofs but to control our behavior and to ensure our survival. Thus, intelligence always manifests itself in interactions with our bodily behavior and interactions with our environment. A new field has grown up around the study of behavior-based intelligence, also known as embodied cognitive science, new AI, and behavior-based AI [6.68]. It aims to produce a coherent framework for the study of naturally and artificially intelligent systems. Its goal is to understand complexity and intelligence by designing, constructing, and building robots.

The artificial life approach is clearly not a homogeneous field of research [6.69]. However, it does seem to provide many useful modeling instruments for chemical and biological research. Besides these practical applications, there is a visionary dream of AL with a hard scientific core. Actual biological evolution is simply a complex dynamical model which is governed by highly nonlinear equations. Today, we only know some properties of these equations, and when they are known, we have no analytical tools to solve them exactly. Even numerical approximations would be restricted by their immense degree of computational complexity. Nevertheless, computer models may allow computer experiments to become familiar with possible scenarios under several restricted constraints. These experiences with computer experiments may be used to create particular conditions under which new materials may construct themselves and new forms of life may organize themselves. The first steps toward this have already been taken. Even the emergence of artificial consciousness cannot be excluded in principle [6.70]. If we know the complex neural dynamics behind conscious states (see Sects. 4.3, 4.4, 6.5), then the wetware and hardware of the human brain is only a particular model. It is not matter itself that makes new materials, life, and consciousness possible, but its particular organization and dynamical laws, which can be modeled using more or less complex systems. Thus, in the long run, we will need to address the ethical question of whether we want open-ended artificial evolution, the results of which we cannot forecast.

Obviously, interacting embodied minds and embodied robots generate embodied superorganisms of self-organizing information and communication systems. What are the implications of self-organizing human–robot interaction (HRI)? Self-organization means more freedom but also more responsibility. Controlled emergence must be guaranteed in order to prevent undesirable side-effects. However, in a complex dynamical world, decision-making and acting is only possible under conditions of bounded rationality. Bounded rationality results from limitations on our knowledge, cognitive capabilities, and time. Our perceptions are selective, our knowledge of the real world is incomplete, our mental models are simplified, our powers of deduction and inference are weak and fallible. Emotional and subconscious factors effect our behavior. Deliberation takes time and we must often make decisions before we are ready. Thus, knowledge representation must not be restricted to explicit declarations. Tacit background knowledge, changes in emotional states, personal attitudes, and situations with increasing complexity are challenges encountered when modeling information and communication systems. Human-oriented information services must be improved in order to support a sustainable information world.

While a computational process of, say, a PC is running, we often do not know the quality of the processing or how close the current processing is to a desired objective. Computational processes seldom provide intermediate results to tell us how near the current process is to any desired behavior. In biological systems, for example, humans experience a sense that we know the answer. In a kind of recursive self-monitoring, some internal processes observe something about our cognitive states that help us to evaluate our progress. The evolutionary selection value of self-reflection is obvious: if we have these types of observations available to us, we can alter our current strategies according to changing goals and situations. Engineered systems have some counterparts to the kinesthetic feedbacks one finds in biological systems. However, the challenge is to create feedback that is useful for decisions made by a system that can reflectively reason about its own computations, resource use, goals, and behavior within its environment. This kind of cognitive instrumentation of engineered systems [6.71] can only be obtained through artificial evolution, because the cognitive processes of humans, with their multiple feedback processes, developed only after a long history of evolution and individual learning.

Thus, we need generative processes, cognitive instrumentation, and reflective processes of systems in order to handle the complexity of human–robot interactions. Biological systems take advantage of layers with recursive processing involving self-monitoring and self-control from the molecular and cellular to the organic levels. Self-reflection leads to a knowledge that is used by a system to control its own processes and behaviors. What distinguishes self-reflection from any executive control process is that this reflection involves reasoning about that system; the system is able to determine or adjust its own goals because of this reflection. However, self-reflection must be distinguished from self-consciousness. Consciousness is at least partly about the feeling and experience of being aware of one's own self. Therefore, we could construct self-reflecting systems without consciousness that may be better than biological systems for certain applications. It is well known that technical instruments (e.g., sensors) already surpass the corresponding capacities of natural organisms by many orders of magnitude. Self-reflecting systems could help to improve self-organization and controlled emergence in a complex world.

But just how far should we go? Self-consciousness and feeling are states of brain dynamics which could, at least in principle, be simulated by computational systems. The brain observes, maps, and monitors not just the external world but also internal states of the organism, especially its emotional states. To "feel" is to have an awareness of one's emotional states. In neuromedicine, the "Theory of Mind" (ToM) even analyzes the neural correlates of social feeling, which are situated in special areas of the neocortex. People, for example, suffering from Alzheimer's disease lose their feelings of empathy and social responsibility because the neural areas associated with them are destroyed. Therefore, our moral reasoning and decision-making has a clear basis in brain dynamics which, in principle, could be simulated by self-conscious artificial systems. In highly industrialized nations with advanced aging, feeling robots with empathy may be used to nurse old people as the number of young people engaged in public social welfare decreases and personal costs increase dramatically [6.72].

7 Complex Systems and the Evolution of Economies

How can one explain the emergence of economic order in human societies? This chapter starts with a short history of economic systems in the seventeenth and eighteenth centuries. The concepts of economic order that were conceived during this period were often associated with other technical, physical, or biological concepts of the period. The physiocrats based their model of the economic system of an absolutist state on a typical eighteenth-century clockwork mechanism. The liberal ideas of Adam Smith were devised against a historical backdrop of Newtonian physics. Until recently, mainstream economics has often been inspired by models from linear mathematics, classical mechanics, thermal equilibrium thermodynamics, and sometimes the Darwinian theory of evolution. Like many physicists, economists believed in the exact computability of their (linear) models and suppressed the possibility of any "butterfly effect" that could lead to chaos and rule out long-term economic forecasts (Sect. 7.1). To describe the dynamics of an economy, it is necessary to have evolution equations for many economic quantities from perhaps thousands of sectors and millions of agents. Therefore, stochastic models are often preferred when modeling global trends. Phase transitions and bifurcations at critical points are crucial concepts when attempting to understand the nonlinear dynamics of economies. Chaos and randomness lead to the behavioral concept of bounded rationality (Sect. 7.2). One challenge of globalization is to understand the dramatic dynamics of financial markets. Modern mathematical finance theory is still based on Louis Bachelier's assumption (from 1900) of an efficient market with a normal ("Gaussian") distribution of price changes and mild randomness (Sect. 7.3). However, complex and global markets are actually turbulent like the weather; they exhibit typical power-law distributions. Stochastic processes with probabilistic attractors (see Sect. 5.4) lead to abrupt and discontinuous events (the "Noah effect") or long-term trends (the "Joseph effect") (Sect. 7.4). Economic and financial sociodynamics open up new avenues of econophysics (Sect. 7.5).

7.1 Smith's Economics and Market Equilibrium

In the seventeenth and eighteenth centuries, classical mechanics became the universal scientific paradigm. The mechanistic view of an economy was elaborated later on by the French scientist François Quesnay (1694–1774), the founder of the physiocratic school in economics [7.1]. Quesnay, who started as a physician at the court of Louis XV, wrote a book about the "animal economy" (*oeconomie animale*) of the human body inspiring him with the idea of an economy of the social organism. Descartes' mechanistic world view was the leading philosophy of the physiocrats.

Thus, the economic system was modeled as a mechanistic clockwork of cog wheels, springs, and weights. A clock is a sequentially operating system with programmed functions. Analogously, the physiocratic economy cannot regulate itself. Advances in agriculture, which are the driving forces of the physiocratic economy, are compared with the weights and springs of a clock. Economic production was referred to the composed movements in a clock. Consequently, economic prosperity is only guaranteed by a regular economic circulation like a clockwork.

The physiocrats used a particular table (*Tableau économique*) to visualize the circulation of wealth between the social classes of farmers ("productive class"), tradesmen ("sterile class") and land owner. In Fig. 7.1a, an economic period starts when the class of land owners distributes their received rent (for instance 200 million Louisdor) for food and agricultural products to the left column of farmers (100 million Louisdor) and for the goods of tradesmen to the right column (100 million Louisdor) [7.2]. Both incomes enable the two classes of farmers and tradesmen to produce new goods. As farmers use the products of tradesmen and tradesmen use the goods of agriculture, money must circulate between the columns of the corresponding classes. The circulation forms a zigzag curve, until the net profit is paid out at the bottom of the table.

But the expenditure of the net profit causes new receipts making new expenditure possible, which may reproduce the net profit, in order to start a new economic circulation. The mechanics of the regular circulating and iterative reproduction of the net profit was illustrated by clocks with rolling balls (Fig. 7.1b) [7.3]. The clocks measure time by means of balls rolling down inclined zigzag paths. After a period of circulation the balls are lifted to the top of the system and the process is iterated. Obviously, the distribution of net profit in a circulation period is compared with the zigzag path of the rolling balls in the clockwork. The iteration of an economic circulation period corresponds to the lifting of a ball and its rolling down the zigzag path again.

The physiocratic economists used physical models in the framework of Cartesian mechanics. Their causal determinism without any kind of self-regulation or individual freedom corresponds completely to the political system of absolutism. The citizens are reduced to functioning elements in a political and economic machine.

While the physiocrats designed their economic model against the background of Cartesian mechanics, Adam Smith referred to the classical physics of his great predecessor Sir Isaac Newton. In Cartesian mechanics, all physical events are reduced to the contact effects of interacting elements like cog wheels in a clock or impulses between balls. Thus, Cartesian physicists constructed hypothetical mechanisms which are often not observable. For instance, the refraction of light was explained by the interactions of hypothetical tiny glass-like balls. The laws of percussion and impulse were fundamental in Cartesian physics.

Newton critized Cartesian mechanics with his famous phrase *Hypotheses non fingo*. His law of gravitation was mathematically deduced from his axioms of me-

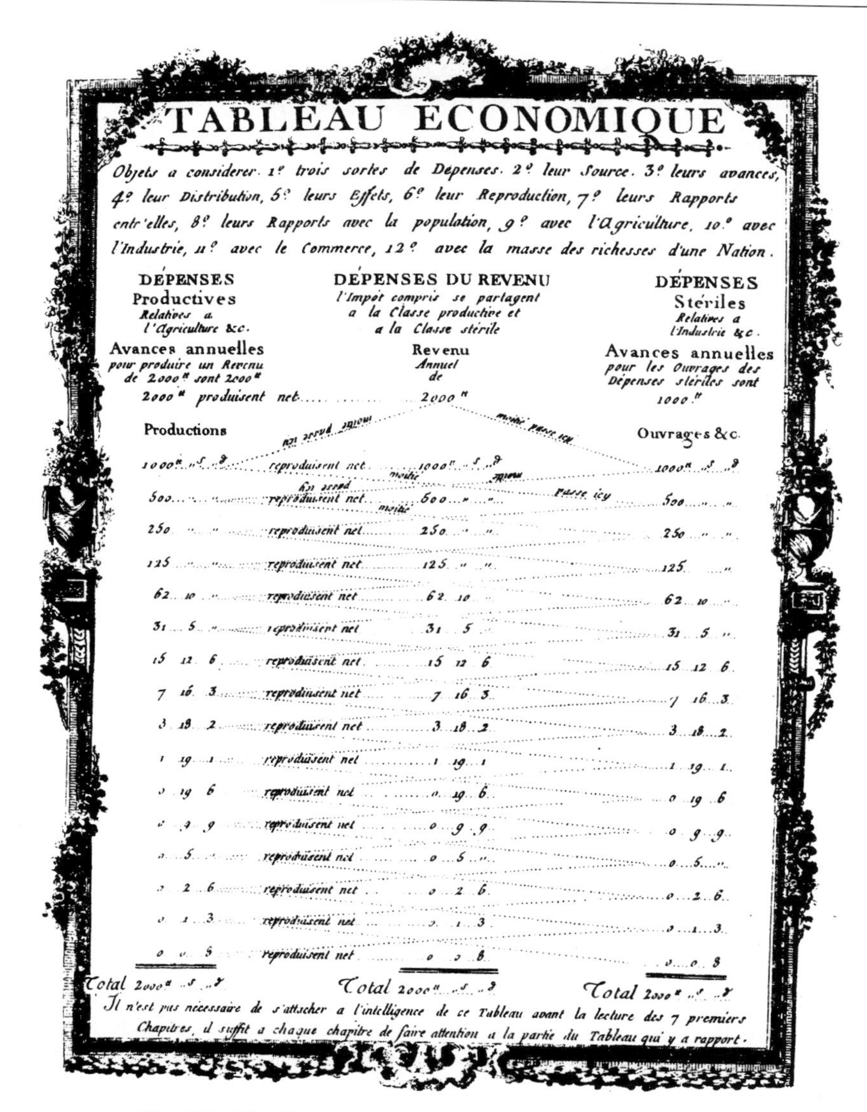

TABLEAU ECONOMIQUE

Objets a considerer. 1° trois sortes de Dépenses. 2° leur Source. 3° leurs avances, 4° leur Distribution, 5° leurs Effets, 6° leur Reproduction, 7° leurs Rapports entr'elles, 8° leurs Rapports avec la population, 9° avec l'Agriculture, 10° avec l'Industrie, 11° avec le Commerce, 12° avec la masse des richesses d'une Nation.

DÉPENSES Productives Relatives a l'Agriculture &c.		DÉPENSES DU REVENU l'Impôt compris se partagent a la Classe productive et a la Classe stérile	DÉPENSES Stériles Relatives a l'Industrie &c.
Avances annuelles pour produire un Revenu de 2000# sont 2000#		Revenu Annuel de	Avances annuelles pour les Ouvrages des Dépenses stériles sont 1000#
2000# produisent net		2000#	1000#
Productions			Ouvrages &c.
1000#	reproduisent net	1000#	1000#
500	reproduisent net	500	500
250	reproduisent net	250	250
125	reproduisent net	125	125
62 10	reproduisent net	62 10	62 10
31 5	reproduisent net	31 5	31 5
15 12 6	reproduisent net	15 12 6	15 12 6
7 16 3	reproduisent net	7 16 3	7 16 3
3 18 2	reproduisent net	3 18 2	3 18 2
1 19 1	reproduisent net	1 19 1	1 19 1
0 19 6	reproduisent net	0 19 6	0 19 6
0 9 9	reproduisent net	0 9 9	0 9 9
0 5	reproduisent net	0 5	0 5
0 2 6	reproduisent net	0 2 6	0 2 6
0 1 3	reproduisent net	0 1 3	0 1 3
0 0 8	reproduisent net	0 0 8	0 0 8
Total 2000#		Total 2000#	Total 2000#

moitié passe ici

Il n'est pas nécessaire de s'attacher a l'intelligence de ce Tableau avant la lecture des 7 premiers Chapitres, il suffit a chaque chapitre de faire attention a la partie du Tableau qui y a rapport.

Fig. 7.1a. The "Tableau économique" of the physiocrats [7.3]

chanics and empirically confirmed by forecasts and experiments. But he renounced any explanation of the long-distance effects of gravitation in empty space (*actio in distans*) by some hypothetical transmission mechanism which was not observable.

In Newton's celestial mechanics, material bodies move in a system of dynamical equilibrium which is determined by the invisible forces of gravitation. The physical concept of freely moving individuals in dynamical equilibrium corresponds to the liberal ideas of a free economy and society with the division of independent po-

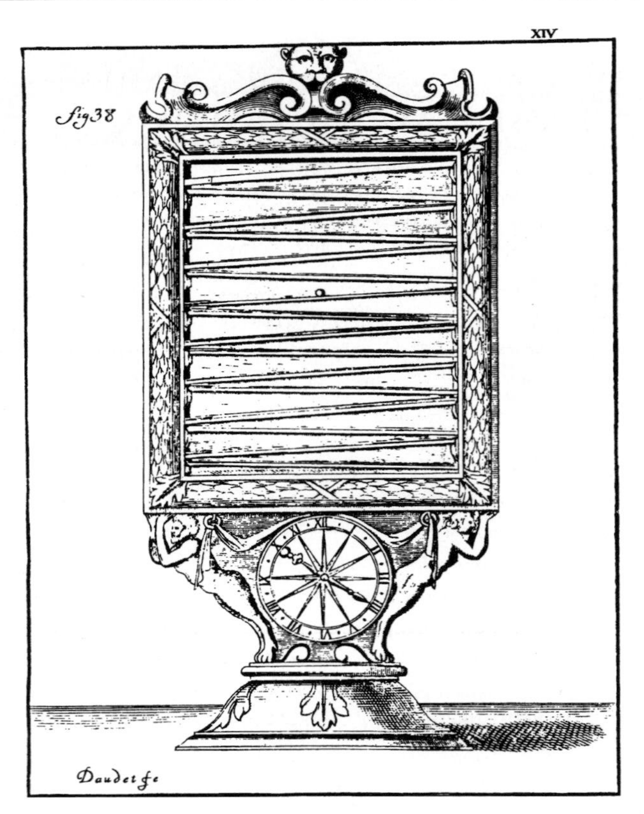

Fig. 7.1b. The "Tableau économique" of the physiocrats (**a**) illustrated by the zigzag path of the rolling ball in a clockwork (**b**) [7.3]

litical powers. Contrary to the liberal ideas, the Cartesian clockwork of nature seems to accord with the state machinery of absolutism with its citizens as cog wheels.

The publication of Smith's famous *Inquiry into the Nature and Causes of the Wealth of Nations* (1776) is often celebrated as the birth of a separate discipline. Nevertheless, Smith was a professor of moral philosophy, and Newton was a professor of natural philosophy. Actually, Smith tried to unify ethics, economics and politics, and Newton embedded his physics into a cosmic and even religious framework. In his *Theory of Moral Sentiments* Smith analyzed the role of sympathy in human beings [7.4]. In his *Wealth of Nations* the self-interested behavior of men is assumed to be the essential impulse of economics.

In both of these books Smith tried to apply the Newtonian method to ethics and to economics [7.5]. He described the Newtonian method as one in which a scientist lays down "certain principles, primary or proved, in the beginning, from whence we account for several phenomena, connecting all together by the same chain." In a Hume-like manner, Smith credited the origin of science not to men's love for truth, but to the simple desire to maximize "wonder, surprise, and admiration." The great purpose of human life is the uniform, constant, and uninterrupted effort to better men's condition. In short, the self-interest of men tends to maximize the welfare function.

According to Newton's motto *Hypotheses non fingo*, Smith emphasized that human self-interest is not a theoretical construction of economists, but an empirical fact of experience. Self-interest is a strong and natural impulse of single human beings, and therefore a human right. But the interactions of several single micro-interests achieve the common macro-effect of welfare by the mechanism of the market. Two famous quotations from Smith's *Wealth of Nations*: "I have never known much good done by those who affected to trade for the public good" [7.6] and "It is not from the benevolence of the butcher, the brewer, or the baker, that we expect our dinner, but from their regard to their own interest" [7.7].

The mechanism of the market is regulated by supply and demand driving the competing micro-interests to the macro-effect of welfare and the "wealth of the nation" in the equilibrium of the market. In the mechanistic view, the micro-interests seem to be drawn to the common macrostate of equilibrium by an "economic demon" or mechanical spring. According to the Newtonian method, Smith prefers the picture of an "invisible hand" directing the micro-interests like the "invisible" long-distance force of gravitation in astronomy. Obviously, Smith describes an economy as a complex system of many competing micro-interests. The dynamics of their interactions is a self-organizing process of competition with a final state of equilibrium between supply and demand.

The value of goods is measured by money. But the money measure could not be used without a precaution. It was necessary to distinguish between "market prices" achieved by the market mechanism and the "natural prices" or actual costs of a good. Economists had to find a "standard of value" in order to be able to correct the value of money. Thus, Smith already aimed at a political economy based upon a theory of value. Values were needed for weighting the social product. Figure 7.2 illustrates Smith's self-organizing process of supply and demand by a feedback scheme with demand r for a good, supply c, market price m, and natural price n [7.8]:

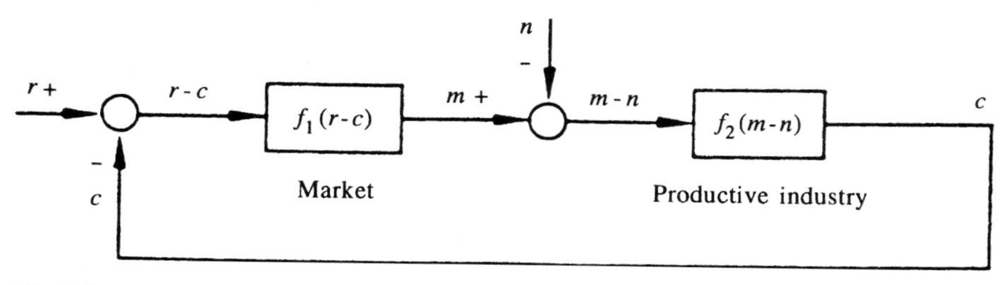

Fig. 7.2. Adam Smith's self-organizing process of supply and demand illustrated by a feedback schema [7.8]

But Smith did not introduce a "just" price like Aristotle on the background of ethical ideals like justice. His inquiry analyzed the "nature" and "causes" of the "wealth of nations" based on actual facts of human nature like self-interest. Concerning the natural prices of goods, Smith and early classical economics like David Ricardo tried to find an absolute measure of value such as gold, corn, or labor.

For Ricardo, the common measure should be explained by his labor theory of value. Ricardo, like Smith was familiar with the general ideas of classical physics. Thus, he believed that some of the conclusions of economics were "as certain as the principles of gravitation" [7.9]. As history went on and economic and political problems changed, Ricardo's theory of growth, rent, and labor was influenced by the historical conditions of his own time at the beginning of the 19th century. Prominently, there were the economic problems of feeding a growing population which had been already considered by Malthus.

John Stuart Mill (1806–1873), the British philosopher and economist, had a great interest in the methodology of economics. He defined "political economy" as an axiomatic system of deductive analysis resting on assumed psychological premises and abstracting from all noneconomic aspects of human behavior. These abstractions were compared with disturbing causes like friction in mechanics:

> The disturbing causes have their laws, as the causes which are thereby disturbed have theirs; and from the laws of the disturbing causes, the nature and amount of the disturbance may be predicted a priori ... The effect of the special causes is then to be added to, or subtracted from, the effect of the general ones. [7.10]

In this quotation, Mill obviously described the principle of causality in classical physics, which basically made forecasting in the long-run possible: similiar causes effect similiar results. Thus, Mill's methodology of economics accords with the Laplacean spirit of classical physics, assuming an approximately correct calculation of forecasts by economic laws if initial conditions are approximately known. Furthermore, Mill's axiomatic hypotheses define a simplified model of economic behavior, and not the complex economic reality.

Thus, Mill was the first economic theorist who explicitly operated with the fictional "economic man" (*homo oeconomicus*), not with the real man in his whole complexity which originally was the subject of Smith's inquiry. The general hypothesis of an economic man maximizing a certain economic utility function is empirically based on a kind of experience, namely, introspection and the observation of Mill's fellow men, but it is not derived from specific observations or concrete events. Analogously, Newton's general law of gravitation was empirically justified by some special observations of falling bodies or moving celestial bodies, but not derived from these events. Mill's methodology corresponds to new insights into the role of formal systems and models in 19th century physics.

Predecessors of modern mathematical economics like Walras and Pareto propagated the use of the mathematical methods of physics in economics. Both thinkers were representatives of the so-called Lausanne School. Classical theorists were already implicitly guided by physico-mathematical concepts. They spoke of a more or less rough correspondence between the play of economic forces and mechanical equilibrium. Actually, the predecessors of mathematical economics borrowed much of their vocabulary from mechanics and thermodynamics, for instance, equilibrium, stability, elasticity, expansion, inflation, contraction, flow, force, pressure, resistance, reaction, movement, friction, and so on [7.11].

In 1874, Walras followed Smith's idea that the maximizing behavior of consumers and producers will result in an equilibrium between amounts demanded and supplied in every product and factor market of the economy [7.12]. Since Walras, general equilibrium theory has become a leading concept which demands existence proofs of equilibrium in mathematical models of an economy. Mathematical economists tried to isolate the elements of a complex system from its environment, which was specified by exogenous parameters. If the exogenous parameters themselves depend on influences from the whole system, however, then separating system and environment and neglecting the actually existing feedback makes an adequate economic model impossible.

In general, classical economists tried to reduce the complexity of economic reality by some typical assumptions characterizing their linear and mechanistic models. First of all, they believed in the fiction of a rational economic agent (*homo oeconomicus*) for each human action. The individual behavior of agents, for instance, in a market should be isolated as a whole. Human behavior is described by general behavioral patterns abstracted from individual behavior. Thus, individual human behavior is assumed to be regular and predicable like elements of a mechanical system with certain mathematical laws of motion. If the initial conditions and the environment are known and precisely measurable, then individual behavior within the environment is believed to be deterministic like that of molecules in a gas.

The linearity of the economic models follows from the superposition principle, assuming that a society consists of the additive actions of its members. The principle of superposition implies that society as a whole does not differ from the sum of the individual actions. Obviously, linear models abstract from unpredictable and irrational individual behavior, from restrictions in the environment, and from non-additive ("nonlinear") and synergetic interdependencies between individuals and their actions.

These linear principles of methodology completely correspond to the Laplacean world view of physics. They still have a strong influence on mainstream economics today, although physics itself has undergone some dramatic revolutions in this century, such as quantum mechanics with its characteristic uncertainty relation. But, Heisenberg's uncertainty relation is a consequence of particular relations between quantum mechanical operators depending on Planck's constant, which seems to have no relevance for the dimensions of the economical world. Nevertheless, the quantum formalism à la Schrödinger and Heisenberg remains linear (compare Sect. 2.3). The fact that classical linear dynamical systems behave in a very regular fashion allows exact forecasts. A nonlinear model demonstrating chaotic behavior as well as the impossibility of predictions in the long run was considered as a bad economic instrument.

In this century, mathematical economists have more and more given up the physicalism of the Lausanne School, for instance, which tried to determine an economy like a classical physical system. Economists have tried to found their own basic mathematical instruments. The assumptions of linear dynamical models have been justified by technical reasons. This formal attitude is expressed in the following statement by John Maynard Keynes in a letter to Roy Harrod from 1938:

> It seems to me that economics is a branch of logic, a way of thinking; and that you do not repel sufficiently firmly attempts ... to turn it into a pseudo-natural science ... Economics is a science of thinking in terms of models joined to the art of choosing models which are relevant to the contemporary world. [7.13]

Under the influence of characteristic economic crashes (for instance at the end of the 1920s), Keynes and others underlined that the economic system is not automatically capable of self-regulation. The "instability of capitalism" has become the popular version of the so-called Keynesian doctrine. So, it was suggested that the economic system be stabilized from outside by particular instruments of policy like fiscalism. Linear models have been employed especially by neoclassical theorists who, again, have concentrated on the investigation of equilibrium economics.

Nonlinear approaches have been preferred mainly by economists who felt uncomfortable with the classical ideal of equilibrium economics. Thus, Keynesian authors have often criticized the linear framework of equilibrium theories without being familiar with the mathematical methods of nonlinearity.

A new epoch of linear mathematical economics was introduced by John von Neumann and Oskar Morgenstern's *Theory of Games and Economic Behavior* (1943) [7.14]. Linear programming, operational research, and even mathematical sociology have been influenced by this famous book. In the *Theory of Games*, von Neumann and Morgenstern assumed rationally acting persons who try to maximize their profit in terms of certain utilities. In general, a class of possible actions $a_1, \ldots, a_m$ and a class of possible states $s_1, \ldots, s_n$ are used to form pairs (a_i, s_j) with $1 \leq i \leq m$ and $1 \leq j \leq n$ which are mapped onto utilities u_{ij}. The possible utilities u_{ij} are ordered in a $(m \times n)$-matrix.

Several criteria of rationality have been suggested for decisions under uncertainty, for instance. Uncertainty means that no probabilities of possible utilities are known. The so-called maximin utility criterion is mainly used. In this case each possible action a_i is associated with an index which is the minimum of the utility values $u_{i1}, \ldots, u_{in}$, the smallest value in the i-th line of the matrix (u_{ij}) of utilities. Then, the rule demands: choose the action the index of which is a maximum. In short, the maximin rule chooses the action maximizing the utility of the most unfavorable case. The rule can be applied very easily and mechanically to the matrix of utilities.

Imagine the following example from the philosopher Carl Gustav Hempel [7.15]. There are two urns with balls of equal size which cannot be distinguished by the sense of touch. In the first urn there are balls of lead and platinum. In the second urn there are balls of gold and silver. A player is allowed to take a ball from one of the two urns as a free gift. A probability distribution of the balls in the urns is not known to the player. The utilities of the balls are estimated at 1000 (platinum), 100 (gold), 10 (silver), 1 (lead).

The maximin rule suggests choosing the second urn. In this urn the most unfavorable case is a silver ball, but in the first urn the most unfavorable case is a lead ball. Obviously, the maximin rule corresponds to a pessimistic view of the world. In a game the player assumes a very hostile opponent. Then, the maximin rule suggests the most useful actions.

An optimistic attitude corresponds to the so-called maximax utility criterion. The player is convinced that each possible action has a best possible result. Thus, it seems to be rational to choose an action the best possible result of which is at least as good as the most favorable result of an alternative action. In our urns example the maximax rule suggests choosing the first urn.

A cautious player would not like to choose that rule. But on the other hand the maximin rule is only rational if the opponent is known to be hostile. Some numerical examples may confirm these interpretations. For two possible states s_1, s_2 and two possible actions a_1, a_2 the matrix of utilities is given by Fig. 7.3a.

	s_1	s_2
a_1	0	100
a_2	1	1

	s_1	s_2
a_1	0	10^{15}
a_2	0.000001	0.000001

Fig. 7.3a,b. Utility matrices

The maximin rule suggests action a_2. Even if the number 1 is diminished to a tiny value such as 0,000001 and the number 100 is enlarged to a very great value such as 10^{15} (Fig. 7.3b), the maximin rule would suggest action a_2. This decision is actually rational for a player who must assume an absolutely hostile opponent. In any case the opponent would try to prevent a state corresponding to maximum utility for the player. But otherwise the maximin rule would be irrational, because a_1 would be the better action. If state s_1 is realized, then the player has to give up an increment of utility which is very small. In the case of state s_2, he would receive a very large growth of profit with action a_1.

In order to justify such a decision, Savage introduced the so-called minimax risk criterion. He suggested replacing the matrix of utilities u_{ij} (Fig. 7.3a) by a matrix of risk values r_{ij} (Fig. 7.3c). A risk value r_{ij} must be added to a utility value u_{ij} in order to get the maximum value of utility in the j-th column.

In matrix 7.3a the largest value of utility in the first column is 1, and in the second column it is 100. Thus, the risk matrix is given by Fig. 7.3c.

	s_1	s_2
a_1	1	0
a_2	0	99

Fig. 7.3c. Risk matrix

The minimax risk rule demands: choose an action minimizing the maximum risk. As the maximum risk of a_2 has the value 99 and that of a_1 has the value 1, it seems to be rational to choose action a_1. But, of course, this rule is only rational under certain particular conditions. There are many other criteria of rationality.

A further example is the so-called pessimism optimism criterion which suggests a solution between the pessimistic maximin rule and the optimistic maximax rule. For an action a_i let m_i be the minimum and M_i the maximum of utilities $u_{i1}, \ldots, u_{in}$. Let α be a constant such that $0 \leq \alpha \leq 1$ called the optimism pessimism index. Then, the action a_i is associated with the so-called α-index $\alpha m_i + (1-\alpha)M_i$. The pessimism optimism rule prefers actions with a larger α-index. But, of course, a special criterion is only defined if a particular number α is given. These examples demonstrate that there is no absolute criterion of rationality, but a class of criteria corresponding to different degrees of optimism and pessimism under certain circumstances.

Von Neumann and Morgenstern's book *Theory of Games* considers the stability of a society or market as the result of interactions of competing or cooperating persons or groups. In many cases, they accept an oversimplification of the actual economic, social, and psychological complexity. Each player can exactly determine his possible actions to aim at certain states and possible utilities. In general, the theory of games assumes the principle of linearity (superposition) that complex interactions of many persons in a society ("game") can be reduced to the sum of many simple interactions of a few persons.

Thus, the investigation of 2-person games plays a crucial role in game theory. The event that player 1 chooses action a_1 and player 2 chooses action a_2 is designated by the pair (a_1, a_2). In the case of this event, the utilities are defined as $u_1(a_1, a_2)$ for player 1 and $u_2(a_1, a_2)$ for player 2. An important class of games is characterized by the fact that the utilities of both players in each event are absolutely opposite with $u_1(a_1, a_2) + u_2(a_1, a_2) = 0$ ("zero-sum" games). Any cooperation is excluded. Then the maximin rule seems to be rational if there is no special information about the irrationality of the opponent. In other cases a cooperation may sometimes be rational.

The mathematically essential problem is the existence of equilibrium points in games [7.16]. If there is no cooperation, an equilibrium point can be defined for the possible actions of the two players in the following way. An event $(\overline{a}_1, \overline{a}_2)$ is an equilibrium point of the game if the utility value $u_1(\overline{a}_1, \overline{a}_2)$ of player 1 is greater than or equal to $u_1(a_1, \overline{a}_2)$ for all his actions a_1 and if the utility value $u_2(\overline{a}_1, \overline{a}_2)$ of player 2 is greater than or equal to $u_2(\overline{a}_1, a_2)$ for all his actions a_2.

If player 1, assuming that player 2 chooses his action $\overline{a}_2$, tries to maximize his utility, then he can choose action $\overline{a}_2$ and vice versa. The equilibrium is stable in the sense that a player knowing that his or her opponent is at the equilibrium point has no reason to alter his or her behavior. Obviously, this definition of equilibrium does not consider any dynamical aspect. But the actual behavior of societies and markets is determined by a complex dynamics in time. The trade cycles are well known examples of economic dynamics. The question arises of whether these dynamical processes are attracted by equilibria and whether these equilibria are stable. In general, the theory of games does not consider the "butterfly effect" that a small failure of behavior may sometimes cause a global crisis and even chaos.

The game theory of von Neumann and Morgenstern is not only in the tradition of linear mathematical economics. It developes the ideas of economic welfare theory. A rational society is assumed to choose a distribution of profits which is Pareto-optimal. A distribution of profits is called Pareto-optimal if the welfare of no individual can be increased by a deviation without diminishing the welfare of at least some other individual. It is not sufficient that this weak condition of Pareto-optimal welfare is satisfied. The power of potential coalitions must be considered, too. The theory of cooperative solutions in game theory mainly stems back to the ideas of welfare economics, diplomacy, and self-interested politicians who are sometimes social. Mathematically, the concepts of justice, impartiality, or fair play, which determine the political and social framework of welfare economics, are reduced to certain principles of symmetry.

Game theory is an exact mathematical theory whose application to economics has sometimes been overestimated. Its limitations are mainly given by its typical linear assumptions about society. Nevertheless, game theory is a brilliant mathematical invention – which was mainly initiated by John von Neumann. It is noteworthy that John von Neumann was a central figure in nearly all the scientific developments of this century which are considered in this book. He was engaged in the development of program-controlled computers, the theory of automata, quantum mechanics, and game theory. Furthermore, he was interested in interdisciplinary mathematical models in the natural and social sciences. All these brilliant developments are mainly ruled by principles of linearity. But John von Neumann was also one of the first scientists to recognize the importance of self-reproduction and self-organization. His theory of cellular automata is a famous example.

7.2 Complex Economic Systems, Chaos, and Randomness

From a methodological point of view, mainstream economics has often been inspired by models from linear mathematics, classical mechanics, thermodynamics of thermal equilibrium, and sometimes the Darwinian theory of evolution. Classical economic models have postulated a rational economical man (*homo oeconomicus*) who seeks to maximize his utility by minimizing his costs and maximizing his profits. These rational agents are assumed to interact by exchanging commodities in markets that achieve an economic equilibrium of demand and supply by certain price mechanisms.

To describe the dynamics of an economy, it is necessary to have evolution equations for many economic quantities from perhaps thousands of sectors and millions of agents. Since everything depends, in economics like elsewhere, on everything else, the equations will be coupled and nonlinear, in order to model economic complexity as well as possible. But then, even completely deterministic models may produce highly irregular behavior which is unpredictable in the long run. Economics seems to suffer from the same shortcomings as meteorology [7.17].

Before the discovery of mathematical chaos and the butterfly effect in meteorology, people believed in the possibility of precise long-term forecasts of the weather.

John von Neumann as a computer pioneer maintained that, given sufficient data about global weather and a supercomputer, he could provide accurate predictions of weather over time and space. He did not fail for mathematical reasons, because in the framework of linear mathematics he was as right as a classical astronomer. But the actual behavior of fluids and weather differ dramatically from these models in the long run.

How does one deal with the complexity of weather and the economy? In meteorology, Edward Lorenz has proposed a nonlinear dynamical model which can produce chaotic behavior by its intrinsic ("exogenous") disturbance (compare Sect. 2.4). Analogously, there are two possible ways of explaining the complexity of economic evolution. The mainstream approach has assumed linear models with somewhat ad hoc and unexplained exogenous shocks. The nonlinear approach gives up oversimplified hypotheses of linearity and ad hoc hypotheses of exogenous shocks and tries to explain the complexity of actual economies by their endogenous nonlinear dynamics. In a few cases nonlinearities may be so weak that linear approximations do not constitute an essential error.

In the history of economics, the Great Depression of the 1930s gave rise to theories which tried to explain the economic irregularities. But the models (for instance, the Kalecki and the Hansen-Samuelson models) were linear and could not explain the emergence of oscillations [7.18]. Thus, exogenous shocks were assumed by economists to produce the observed oscillations. If the economists had been more familiar with mathematical developments, they would have been acquainted with nonlinear mathematical models leading to limit cycles as solutions.

Originally, economists only knew about stable equilibrium as a fixed-point attractor. Poincaré generalized the conception of equilibrium to include equilibrium motion in the form of limit cycles. But for chaotic attractors like the Lorenz model (Fig. 2.21) there is neither a fixed point nor a fixed motion, but a never-repeating motion. Nevertheless, it is a bounded motion and a non-wandering set attracting certain dynamical systems to a final state of dynamical equilibrium.

Historically, the economy in this century has been characterized by growth that was interrupted by spectacular collapses, for instance, in the 1930s (the Great Depression) and in the 1970s (the Oil Crisis). Concerning the structure of growth, particular attention must be given to innovations and technical progress. The expansion of a successful innovation is empirically well represented by the logistic function which was introduced in Sect. 2.4. A recursive version can be referred to integers t as time indices and a growth factor $\alpha > 0$. Initially, an innovation is rather unfamiliar. Then as it gains acceptance it reaches its maximum rate of expansion. After that there is a slow deceleration of the process of absorption as the innovation approaches complete integration into the economy.

The resulting curve is shown in Fig. 2.22. For $\alpha \leq 3$ we get a fixed point attractor which is visualized in Fig. 2.22a. For larger α, the result is a periodic oscillation (Fig. 2.22b and Fig. 2.24c) and then a chaotic movement (Fig. 2.22c and Fig. 2.24b). For $\alpha > 3$ the number of periods doubles successively as α increases (Fig. 2.23a), until it is transformed into complete chaos (Fig. 2.23).

The connection between innovation and economic output is shown in the following model of Fig. 7.4 [7.19]. Initial output q is taken to be in equilibrium so that as the growth rate Δk rises, the output also rises, gradually. But then as innovation reaches its saturation and Δk drops to towards zero, output q falls back to its initial level. So innovation stimulates a boom but also a subsequent collapse. An innovation may be labor-saving. If inputs of labor per unit of output falls by 20%, employment will fall also.

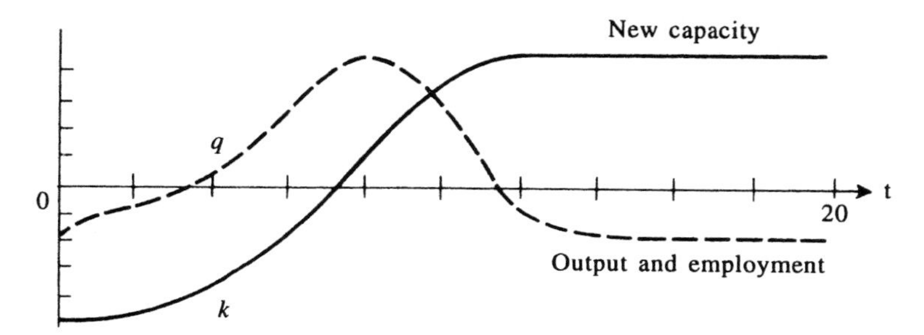

Fig. 7.4. Logistic curve of innovation capacity with output and employment curve [7.19]

As the growth of new ideas is assumed to be exponential, economists like Schumpeter have maintained that at the end of an innovatory burst a new one will start. Then a second boom and collapse is initiated, and so on, as the economic system continues to function and technological conceptions continue to grow at around 4% per year. Innovation is crucial to economic cycle theory, because in a depression there is no other basis for the new investment which is necessary to generate a new expansion.

New ideas arise steadily. When enough ideas have accumulated, a group of innovations are introduced. They develop slowly at first, then accelerate as the methods are improved. A logistic development characterizes the typical trajectory of an innovation. Some investment must precede the introduction of an innovation. Investment stimulates demand. Increased demand facilitates the spread of the innovation. Then, as all the innovations are fully exploited, the process decelerates towards zero.

Schumpeter called this phenomenon the "swarming" of innovations. In his three-cycle model, the first short cycle relates to the stocks cycle and innovations play no role. The following longer cycle is related to innovations. Schumpeter recognized the significance of historical statistics and related the evidence of long waves to the fact that the most important innovations like steam, steel, railways, steamships, and electricity required 30 to 100 years to become completely integrated into the economy.

In general, he described economic evolution as technical progress in the form of "swarms" which were explained in a logistic framework. A technological swarm is assumed to shift the equilibrium to a new fixed point in a cyclical way. The resulting new equilibrium is characterized by a higher real wage and higher consumption and output. But Schumpeter's analysis ignores the essential issue of effective demand as determining output.

Historically, the Great Depression of the 1930s inspired economic models of business cycles. However, the first models (for instance Hansen–Samuelson and Lundberg–Metzler) were linear and hence required exogenous shocks to explain their irregularity. The standard econometric methodology has argued in this tradition, although an intrinsic analysis of cycles has been possible since the mathematical discovery of strange attractors. The traditional linear models of the 1930s can easily be reformulated in the framework of nonlinear systems.

The Metzler model is determined by two evolution equations. In the first equation, the rate of change of output $\dot{q}$ is proportional to the difference between actual stocks s and desired stocks s^*. The desired stocks are proportional to the output. The next equation concerns the rate of change of stocks $\dot{s}$ which is output q less demand. The demand is proportional to the output. The complex system whose dynamics is determined by these two evolution equations will produce simple harmonic motion with increasing amplitude.

If the system is expanded by certain nonlinearities, a different kind of behavior results. A third equation may be associated with perverse behavior of net public surpluses and deficits. The aim is to produce a cycle of a certain duration of years. A mathematical model is suggested by the so-called Rössler band [7.20]. One uses a Möbius strip, top to bottom, sides reversed, giving a one-sided surface (Fig. 7.5a). Following a trajectory, an outer loop expands to the right and upward. Then it folds over and as it descends, contracts to an inner loop and so on. Figure 7.5a gives a two-dimensional projection which shows the two cycles. The lines tend to get grouped with empty spaces in between. If the simulation is continued these bands become more and more dense.

Figure 7.5a is a simple but remarkable example of a chaotic ("strange") attractor. Although each trajectory is exactly determined by the evolution equations, it cannot be calculated and predicted in the long run. Tiny deviations of the initial conditions may dramatically change the path of trajectories in the sense of the butterfly effect. Figure 7.5b shows a trajectory of output in the state space plotted over 15 years, which was simulated in a computer experiment with chosen parameters. Figure 7.5c shows the corresponding development as a time series [7.21].

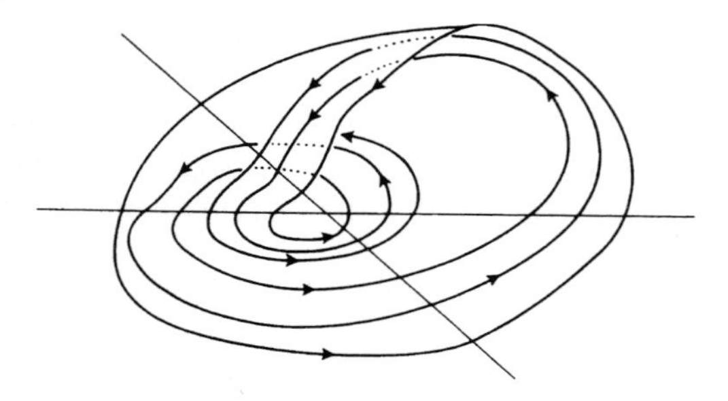

Fig. 7.5a. Rössler attractor

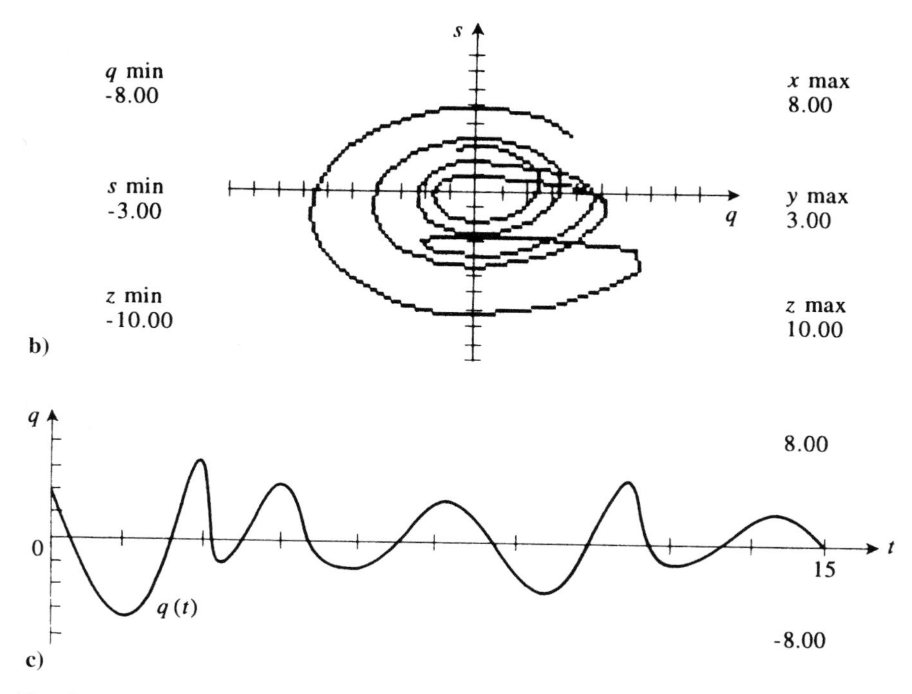

Fig. 7.5b,c. Business cycle of an endogenous nonlinear system with trajectory in state space (**b**) and time series of output (**c**) [7.21]

This highly erratic behavior is produced by a completely endogenous system without any exogenous shocks. In economics the irregularities of time series are usually explained by exogenous shocks. But they are only arbitrary ad hoc hypotheses and hence can explain anything. From a methodological point of view, chaotic endogenous models with strange attractors seem to be more satisfactory. Nevertheless, endogenous nonlinear models and linear models with exogenous shocks must be taken in earnest and tested in economics.

Obviously, an economy consists of many connected and separated parts with endogenous dynamics and exogenous forces. There are individual national economies which are increasingly acted on by movements of the world economy. Within an economy there are several markets with particular dynamics. They are influenced by cycles, for instance, the annual solar cycle determining the agricultural, tourist, or fuel market. The pig cycle and building cycle are further well known economic examples. Thus endogenous nonlinear systems with impressed waves of exogenous forces are realistic models of economies. The impression given is of a disturbed chaotic attractor or a kind of super-chaos. It is this erratic character of economic events which causes great problems for economic agents who have to make decisions that depend on an unpredictable future.

In Sect. 2.3 we saw that self-organizing complex systems may be conservative or dissipative. Their different types of attractors were illustrated in Fig. 2.14a,b. Some well known examples of conservative and dissipative models from the natural sciences have been applied to economies, too. In 1967, Goodwin proposed a conservative dynamical model to make the 19th-century idea of class struggle precise. He considered an economy consisting of workers and capitalists. Workers spend all their income on consumption, while capitalists save all their income. Goodwin used a somewhat modified predator-prey model of Lotka and Volterra which was described in Sect. 3.4 [7.22].

Goodwin's conservative model supports the idea that a capitalist economy is permanently oscillating. Thus, the trajectories describe closed orbits as illustrated in Fig. 3.11b. Goodwin's model was criticized as superficial, because it does not refer directly to the functional income shares of capitalists and workers or to their population size. But it is mainly its conservative character that makes Goodwin's model seem economically unrealistic. The model is put together as an isolated set of assumptions which do not reflect other influences.

Thus, the model has been made more realistic by the assumption of "economic friction". In the biological case, a dissipative Lotka–Volterra model with a point attractor was illustrated in Fig. 3.11c. A dissipative system always possesses attractors or repellers, in the form either of fixed points, limit cycles, or strange attractors. As dissipative systems have an irreversible time evolution, any kind of backward prediction is excluded [7.23].

In reality, a dynamical system cannot be considered as isolated from other dynamical systems. Thus, in Sect. 2.2, we studied coupled oscillating systems like, for instance, two clocks (Fig. 2.11a,b). The state space of the combined systems was represented by a torus (Fig. 2.11c,d). The dynamics of the whole system was illustrated by the phase portrait of trajectories and the vector field on the torus.

An economic model of coupled oscillatory systems is provided by international trade. Imagine a simplified macroeconomic model of a single economy with gross investment and savings, depending both on income and interest rate. The dynamics of the system is determined by an evolution equation for income which adjusts it according to excess demand in the market for goods, and a second evolution equation for the interest rate. These equations constitute a nonlinear oscillator in such a way that the model generates endogenous fluctuations.

The interactions of three economies, for instance, can be described by a system of three independent, two-dimensional limit cycle oscillators. If all three economies are oscillating, the overall motion of the system constitutes a motion on a three-dimensional torus. The coupling of nonlinear oscillators can be understood as a perturbation of the motion of the autonomous economies on a three-dimensional torus. The coupling procedure has been applied to several economic examples, such as international trade models, business cycle models, and interdependent markets.

A crucial problem of practical policy occurs when self-organizing economic systems are allowed to be influenced by political interventions. In some cases the market is not able to develop according to welfare criteria. If the economy is left to itself, it may be characterized by fluctuations. But policy measures can cause effects opposite to those intended if the complexity and nonlinearity of economic growth is not considered.

With respect to the dramatic social and political consequences of economic catastrophes, several policy measures have been discussed in the framework of Keynesianism and Neo-Keynesianism. Compensatory fiscal policy, for instance, may be regarded as a type of dynamical control. It should serve to diminish the amplitude of economic fluctuations. But post-war experiences have shown that there is no hope of reducing fluctuations to zero, or holding the degree of employment constant. Furthermore, a good policy needs considerable time for the collection of data, analyzing the results, and formulating the legislative or executive measures. The result is that any policy may be outdated by the time it is effective. Thus, a policy measure in the complex and nonlinear world of economics may be absolutely counterproductive [7.24].

For example, a Keynesian income policy can be ineffective when the assumed economic dynamics and the proposed time path of policy interventions are too simple. In the framework of complex systems, economic policy measures can be interpreted as impressing exogenous forces on an oscillating system. Thus, it cannot be excluded that the economy becomes chaotic. Forced oscillator systems are well known in physics. For instance, if a dynamical system like a pendulum (Fig. 2.7) is oscillating, and if it is periodically influenced by an exogenous force, the outcome may be unpredictable because of increasing amplitudes, total damping of the oscillation, and complete irregularity.

Since the time of the economic classics until today, it has been the aim of business cycle theory to model the dynamics of an economy with its typical regular fluctuations. According to the linear mechanical view, the actual business cycles were modeled by regular systems with superimposed stochastic exogenous shocks that had to be explained by more or less appropriate economic assumptions. It is, of course, highly unsatisfactory when these exogenous forces determine the main qualitative properties of a model without their having a reasonable economic interpretation. If an actual system is nonlinear and chaotic, further information on possible exogenous forces that may influence the economic dynamics may be superfluous. From a methodological point of view, Ockham's famous razor should be applied to cut out these superfluous ad hoc hypotheses of economics, according to his motto *entia non sunt multiplicanda sine necessitate* ((theoretical) entities should not be multiplied without necessity).

From the practical point of view of an agent, it may be rather irrelevant whether he is confronted with a stochastic linear process or a chaotic nonlinear process. Both kinds of systems seem to prevent him from making precise predictions. As a chaotic model has a sensitive dependence on initial conditions, arbitrarily precise digital computers are in principle unable to calculate the future evolution of the system in the long run. The trajectories diverge exponentially. On the other hand, an agent believing in stochastic exogenous shocks may resign in face of the too-complex behavior of the system.

Nevertheless, nonlinear systems with chaotic time series do not exclude local predictions. If the attractor of a nonlinear system can be reconstructed, then numerical techniques promise to predict the short-run evolution with sufficient accuracy. Short-term economic forecasting may be an interesting application of complex system theory in economics, which is still in its infancy [7.25].

Since the beginning of economics, empirical tests and confirmations have been a severe methodological problem for economic models. In contrast to the natural sciences with their arbitrarily many measurements and experiments in laboratories, economic time series must refer to time units of a day, or annual, quarterly, or monthly data. The length of a standard time series typically consists of a few hundred points. Thus, there are already empirical reasons for the limited reliability of economic models. Empirical experiments are, of course, largely excluded.

Therefore an adequate knowledge of endogenous economic dynamics will at least help to achieve mathematical models whose future developments may be simulated by computer experiments. A politician and manager will at least get the "phase portrait" of alternative economic scenarios if the assumptions of their economic and political surroundings are realized. The qualitative insight into the high sensitivity of nonlinear systems may at least prevent overacting agents from pushing the system from an unstable point to further and perhaps even greater chaos.

The main arguments for nonlinear models in economics are given by the recent structural change of economic growth caused by the development of new dominant technologies. The traditional theory of economics assumes decreasing income. The more goods of a certain kind are produced and put on the market, the more difficult their production and sale will become and the less profit can be gained. The interaction of agents is determined by a negative feedback stabilizing the economy by counteracting each reaction caused by an economic change.

In an economy with negative feedback an equilibrium of prices and market shares may arise and may be forecast. A famous example was the Oil Crisis in the 1970s. The sudden increase of prices for crude oil in the 1970s induced savings and the search for alternative energy which caused a decrease of oil prices in the 1980s. In traditional economies the equilibrium corresponds to an optimal result with respect to the particular surroundings. The theorem of decreasing income implies the existence of a single equilibrium point. Economies with the negative feedback of decreasing income are typical for traditional sectors like agriculture, the mining industry, and mass products.

But economic sectors depending on high technical knowledge gain increasing income. The development and production of high-tech goods like computers, software, airplanes, chemicals, and electronics need complex processes of research, experiment, planning, and design, demanding high investments. But if the high-tech products are put on the market, the enlargement of production capacity is relatively cheap and income is going to increase. Thus, modern high-tech industries must be described in dynamical models as generating the positive feedback of increasing income.

Systems with positive feedback have not only a single, but several states of equilibrium which are not necessarily optimal [7.26]. If a product has some accidental competitive advantages on the market, then the market leader will dominate in the long run and even enlarge its advantage without being necessarily the better product. Many examples of modern high-tech industries show that competing products may have approximately equal market shares in the beginning. But tiny fluctuations which increase the market share of a particular product decide its final success. Sometimes the final market leader may even have less quality from a technical point of view.

These effects cannot be explained in the framework of traditional linear dynamics. But they are well known in nonlinear systems. Figure 7.6 illustrates the competition of two technologies with positive feedback. Some trajectories of market shares are drawn on a convex surface. The more a technology can dominate the market, the more easily it can gain a yet greater market share. As a leading market position is initiated by random fluctuations, it cannot be forecast. In Fig. 7.6, the left curve illustrates the final dominant of technology A. In the two other cases, the technology B will finally dominate the market after initial fluctuations [7.27].

The nonlinear dynamics of these economic models is determined by initial random fluctuations and positive feedback. Obviously, the bifurcation of the possible paths is a kind of symmetry breaking by initial random fluctuations, which is well known in complex physical systems. The reader need only recall the stationary convection rolls that appear in a heated fluid (Fig. 2.20b), where the rolling direction left or right depends on initial random fluctuations.

Obviously, many structural analogies exist between patterns of behavior in societies and pattern formations in nature. From a microscopic point of view, economies can be considered complex multicomponent systems consisting of individuals interacting with themselves and with their material environment. From a macroscopic point of view, collective trends of behavior emerge from microscopic interactions. Thus, the question arises: Is there a general strategy and modeling procedure for the collective dynamic macro-processes in society and economy, as there is in nature. But the modeling procedure cannot simply be taken from non-equilibrium thermodynamics. In this case, we would need equations for motion on the microlevel of interacting agents in the economy, which are necessary to derive the corresponding equations of macrovariables for pattern formation. Only in rather simple models of

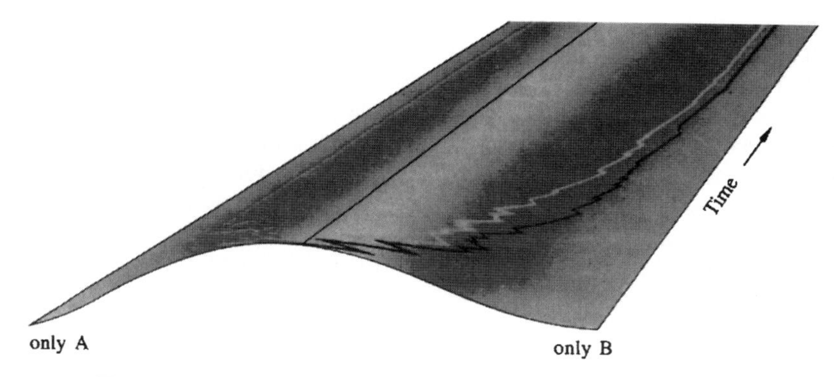

Fig. 7.6. Nonlinear dynamics of two competing technologies [7.27]

sociobiology, can we observe direct input stimulus and output reactions between interacting members of a population, such as neurochemical interactions of insects or the Pavlov reflex of mammals. In general, there are only indirect estimations and valuations between environmental input and individual decisions and reactions.

The complexity of human behavior only allows for a probabilistic treatment of individual decisions and actions. From stochastic microbehavior, we get a stochastic description of the macrodynamics by a so-called master equation for the probability distribution of macrovariables, which are order parameters of economies and societies. They dominate the individual attitudes and activities corresponding to microvariables and govern the patterns of behavior on the macrolevel. Consequently, an appropriate choice of macrovariables plays a key role in successful modeling in economics and social sciences.

In economics, we distinguish the configuration of collective material variables and the socioconfiguration of collective personal variables. Material variables are well-known in economics. Like in thermodynamics, there are intensive variables that are independent of the size of a system. Examples are prices, productivity, and the density of commodities. Extensive variables are proportional to the size of a system and concern, e.g., the extent of production and investment, or the size and number of buildings. Collective material variables are measurable. Their values are influenced by the individual activities of agents, which are often not directly measurable. The social and political climate of a firm is connected to socio-psychological processes, which are influenced by the attitudes, opinions, or actions of individuals and their subgroups. Thus, in order to introduce the socioconfiguration of collective personal variables, we must consider the states of individuals, expressed by their attitudes, opinions, or actions. Further on, there are subgroups with constant characteristics (e.g., sections or departments of a firm or an institution), so that each individual is a member of one subgroup. The number of members of a certain subgroup in a certain state is a measurable macrovariable. Thus, the socioconfiguration of a company, for example, is a set of macrovariables describing the distribution of attitudes, opinions, and actions among its subgroups at a particular time. The total macroconfiguration is given by the multiple of material configuration and socioconfiguration.

If all macrovariables of a macroconfiguration remain constant over time, the social system is in a stationary macroscopic equilibrium, which can be compared to thermodynamic equilibrium. If there are dynamics, we must consider the transition rate between macroconfigurations by either increasing or decreasing macrovariables. In the case of material configuration, an elementary change consists of the increase or decrease of one macrovariable (e.g., price of commodities) by one appropriately chosen unit. The elementary change in the socioconfiguration takes place if one individual of a subgroup changes its state, leading to an increase or decrease in the number of a subgroup by one unit. Thus, the transition rates of neighboring configurations describe the probability per unit of time that the respective transition takes place if the initial configuration is realized. They can be used for setting up the central evolution equation of a social system, the master equation for the probability distribution over the macrovariables of total macroconfiguration. The distribution function $P(m, n; t)$ of the master equation is the probability of finding a certain macroconfiguration of material configuration m and socioconfiguration n at time t. The master equation with $dP(m, n; t)/dt$ describes the change of the probability of a total macroconfiguration due to transitions of the material configurations and the socioconfigurations [7.28].

As an economic example, consider a system of firms, producing the same kind of substitutable, durable, high-tech commodities and competing with respect to the quality of their products. A positive feedback between quality enhancement and customer reaction to quality is assumed. On the demand side, the consumer configuration consists of collective personal macrovariables distinguishing between the number of owners of a certain good and the number of nonowners which are possible customers. The supply configuration consists of material variables referring to the number of commodity units produced by a certain firm in the time unit, prices per commodity, and measures of the quality of commodities. The total economic configuration E contains the consumer configuration n and supply configuration m. The evolution of the system takes place via elementary transition between a total economic configuration and its neighboring economic configurations. On the consumer side, the corresponding transition rates count the purchase steps of nonowners becoming owners of a certain commodity and the sorting out steps of owners becoming nonowners. On the supply side, the transition rates refer to the decisions of firms to change their production output, their prices, and the quality of their products. Using these transition rates, it is possible to set up the corresponding master equation. The distribution function $P(E; t)$ is the probability finding the economic configuration E at time t. The master equation $dP(E; t)/dt$ describes the effects of transitions on the consumer configuration of personal variables and the configuration of supply variables.

However, in general, it is not possible to exhaust all the information contained in the probability distribution of a master equation by comparison with empirical data. In this case it is best to neglect the fluctuations of the macrovariables around their mean evolution and to restrict ourselves to quasi-meanvalue equations of the macrovariables which are derivable from the master equation. Let us consider an economic system with only two competing firms. In synergetics, order parameters of the macrodynamics of a system were introduced by the adiabatic elimination of fast-relaxing variables (compare Sect. 2.4). The idea is that a dynamical system with fluctuations (e.g., a laser) starts with stable modes (waves) according to a certain control parameter (e.g., pumping of energy). If the control parameter changes to a critical value, the modes that are becoming unstable are taken as order parameters, because they start to dominate ("enslave") the stable ones. Since their relaxation time tends toward infinity, the fast-relaxing variables can be eliminated adiabatically. In more popular terms, synergetics demand that long-living systems enslave short-living systems. Thus, if we assume that prices and supply are the quickly adapting variables of our economic system, they may be adiabatically eliminated, according to the synergetic procedure. What remain are the equations for the customer macrovariables and for the quality macrovariables of products. A decisive control parameter of this system is, obviously, competition between two firms. If the evolution of the quality of products is slow compared with the purchase activity of customers, then, according to synergetics, the fast-relaxing customer variables may also be eliminated adiabatically. The only surviving evolution equation is the equation for the quality variables, which may be considered the order parameters of the system.

The evolution of the quality variables q_i for firm $i = 1$ and firm $i = 2$ can be investigated numerically. In Fig. 7.7, their (stationary) solutions $q_i(\phi)$ are depicted as a function of the competition parameter ϕ. It turns out that both firms have the same stationary quality $q(\phi)$ of their products and also the same stationary market share, as long as the competition value ϕ is smaller than a critical value ϕ_c. At ϕ_c, a bifurcation occurs and for $\phi > \phi_c$ there exist two stable quality values $q_+(\phi)$ and $q_-(\phi)$. The winning firm, say $i = 1$, will have reached the quality $q_+(\phi)$, whereas the losing firm, $i = 2$, arrives at quality $q_-(\phi)$ with corresponding market shares.

Besides synergetic and dissipative systems, there are also conservative ones with symmetry breaking. Consider the self-organization of the dipoles in a spin-glass when the temperature is reduced (Fig. 4.9a). In thermal equilibrium, the spins become aligned in the same direction, depending on initial random fluctuations. The dynamics of market shares seem to develop in the same manner. There are many examples of technologies which have made their way because of initial random fluctuations. In the last century neighboring railway companies accept the same gauge in a larger region. The standard gauge was only the result of historical random events and not distinguished by technical reasons.

The behavior of these complex systems is determined by similar evolution equations, like the symmetry breaking of spin-glass systems or ferromagnets. Fig-

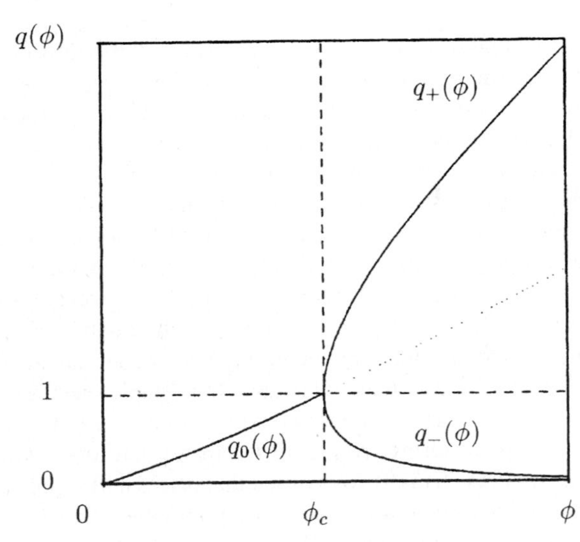

Fig. 7.7. Bifurcation of two competing firms into winner and loser after competition $\phi > \phi_c$ [7.29]

ure 7.8a shows the evolution of magnetic dipoles in a ferromagnet. Each dipole or elementary magnet can be either up (North) or down (South). A dipole can interact with its nearest neighbors. At a high temperature the directions of the dipoles are random. If the temperature is reduced, then the elementary dipoles become aligned in the same direction. As the evolution is a kind of symmetry breaking, it cannot be forecast which direction is distinguished in the final state of equilibrium. Figure 7.8b illustrates the analogous process of self-organization for the gauges of railway companies.

Self-reinforcing mechanisms with positive feedback are typical features of nonlinear complex systems in economies and societies. For example, we may speculate why Santa Clara County in California has become the famous Silicon Valley. In the 1940s and 1950s some key individuals (like the employers Hewlett, Packard, and Shockley) founded electronics firms in the neighborhood of Stanford University. These pioneers created a concentration of high-tech engineers and production which has become an attractor for finally more than 900 firms. At the beginning, there were some random fluctuations favoring Santa Clara County. Thus, it is no miracle, but lawfulness in a nonlinear sense that explains *how* Silicon Valley

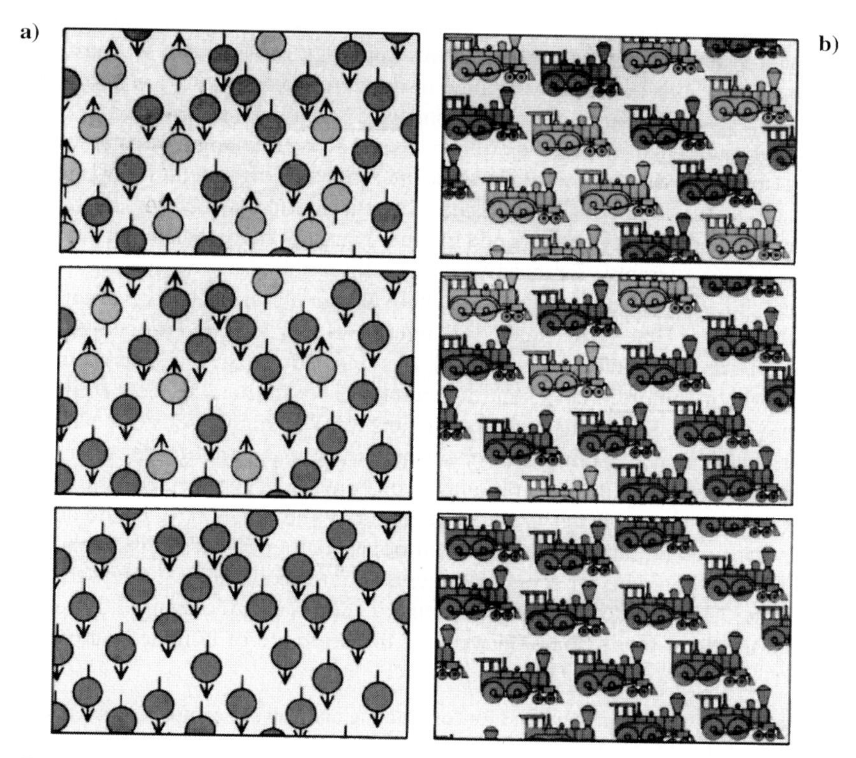

Fig. 7.8a,b. Nonlinear dynamics of magnetic dipoles in a ferromagnet (**a**) and nonlinear dynamics of competing technologies (gauges of railway companies) in an economy (**b**) [7.27]

arose. But it is a miracle in the sense of randomness *that* it arose in Santa Clara County.

Nowadays, self-reinforcing mechanics determine the international trade of high-tech technologies [7.30]. The car industry competition between the USA and Japan can be explained in this framework. Initially, the Japanese industry offered small cars on the American market without any resistance from the American car industry, which was traditionally specialized for production of big cars. The Japanese industry could gain market share, lower costs, and raise quality. Thus, the positive feedback enabled the Japanese industry to invade the American market.

The insight into these nonlinear market effects may have major consequences for political decisions. From the traditional point of view, assuming a constant or decreasing income, governments trust in open markets, try to prevent monopolies and hope that industries will support research and technical developments. They believe in a fixed equilibrium of world market prices and refuse to intervene with subventions or tariffs. Their politics is justified in an economy with decreasing income, but it may be dangerous in sectors depending on high technology with increasing income.

Mechanisms operating with increasing income change the equilibria of competition among nations. Even strong national economies may miss out on the development of important technologies. The technology gap between Western Europe and the USA in the 1960s (for instance in the computer industry) is a famous example. Technical standards or conventions have often been established by positive feedback. Examples like the gauge of railways, the English language as the standard language of air navigation, FORTRAN as a computer language, a particular screw thread, and so on often cannot be changed even if alternative techniques or conventions may be better. They have gained too much market share. But superiority at the beginning is no guarantee for survival in the long run.

The theory of complex systems can help design a global phase portrait of economic dynamics. Time series analysis delivers useful criteria and procedures to recognize patterns of behavior in, e.g., stock markets. But experience and intuition are sometimes more helpful than scientific knowledge for finding the local equilibria of economic welfare. Economists and politicians must be highly sensitive when dealing with highly sensitive complex systems and must make appropriate decisions.

Experience shows that the rationality of human decision making is bounded. Human cognitive capabilities are overwhelmed by the complexity of the nonlinear systems they are forced to manage. Traditional mathematical decision theory assumed perfect rationality of economic agents (homo oeconomicus). Herbert Simon, Nobel Prize laureate of economics and one of the leading pioneers of systems science and artificial intelligence, introduced the principle of bounded rationality in 1959:

> The capacity of the human mind for formulating and solving complex problems is very small compared with the size of the problem whose solution is required for objectively rational behavior in the real world or even for a reasonable approximation to such objective rationality [7.31].

Bounded rationality is not only given by the limitations of human knowledge, information, and time. It is not only the incompleteness of our knowledge and the simplification of our models. The constraints of short-term memory and of information storage in long-term memory are well-established. In stressful situations people are overwhelmed by a flood of information, which must be filtered under time pressure. Therefore, we must refer to the real features of human information processing and decision making, which is characterized by emotional, subconscious, and all kinds of affective and nonrational factors. Even experts and managers often prefer to rely on rules of thumb and heuristics, which are based on intuitive feelings of former experiences. Experience shows that human intuition does not only mean lack of information and the failure to make decisions. Our affective behavior and intuitive feeling are parts of our evolutionary heritage that enable us to make decisions when matters of survival are at stake.

In traditional economics and decision theory, affectiveness and intuition have been regarded as imperfect information processing, which must be overcome by computer and information technology. But even our modern insight into the computational complexity of nature and life enforces us to give up traditional standards of so-called rationality. Actually, nature and life are not only characterized by highly complex and chaotic dynamics, but also by randomness and noise. In this case, there is no shortcut to any finite program or law that would forecast the future and reduce complexity. The computational amount of reducing complexity would be at least as large as the actual behavior of the system. Computational irreducibility and undecidability (compare Sect. 5.6) force us to look for behavioral strategies that are based on our actual experiences in analog situations. Instead of mathematical optimization theory, empirical and experimental research are needed to find and evaluate the heuristics people actually use in decision making.

As optimal decison making with perfect models is impossible, people have developed many individual strategies to respond to bounded rationality [7.32]. They always intend to reduce the complexity of real situations. Organizational habits and routines are justified by a firm's tradition and culture. Rules of thumb are based on simplified and incomplete models of situations that have been justified in the past. If, for example, managers do not have the information or time to set the price of each item to optimize store profits, they may multiply wholesale costs by a traditional markup ratio. When prices seem too high or too low, they are corrected according to other rules of thumb. Solving optimization problems is often too complex for decision making. Instead of optimization, people tend to set goals that they can reach under given resource and time constraints. If the goals are met, they stop their efforts, though they cannot be sure if there was a more perfect solution to the problem. Consumers, for example, stop searching for bargains if they find a low enough price for a certain item. Employers sometimes hire the first candidate who meets the job requirements, rather than searching for the best one. Deconstructing decision problems into subgoals is also a strategy to reduce complexity. But, of course, local decision makers in a complex firm run the danger of ignoring those aspects of their subgoals which they believe are not directly related to them.

If decision makers ignore the side effects and feedbacks of actions in the nonlinear dynamics of a firm or market, they may cause instability and dysfunction in the system. But people are not complete irrational, insensible, or narrow-minded. The principle of bounded rationality tells us that there is no global concept of rationality. As people, nevertheless, try to act purposefully on the basis of their knowledge, experience, and feelings, their decisions may be intendedly or locally rational. A decision rule is called intended or locally rational if it generates reasonable and sensible results in the context of the decision makers believed and assumed environment. For example, it seems rational for a firm to cut prices to stimulate market share when capacity utilization is low. However, it is not generally a rational rule, because it depends on the managers belief that competitors will not, or cannot, respond by cutting their own prices. Later on, this example illustrates how locally rational rules can lead to global irrationality by the nonlinear dynamics of a complex system.

The nonlinear dynamics of competing firms is determined by several feedback loops. A company cuts prices when capacity utilization falls below a certain level. This is a negative feedback that managers attempt to follow. They only concentrate on this feedback loop and simplify the whole system in the sense that the price of competing products is considered exogenous. If that were true, then cutting prices to stimulate demand and boost profits would make sense. But the managers model is only a small fraction of the whole nonlinear reality. Competitors are likely to set prices using the same feedback loop as their local strategy. Cutting prices when utilization drops initiates a reinforcing feedback in which a drop in price causes the market share and utilization of competing firms to fall. Thus, they are motivated to cut their prices. The company states that its market share and utilization have not improved as expected and cuts prices again, thereby closing a positive loop. When the managers model of exogenous prices is false, their locally and intendedly rational decisions lead to an unintended price war that destroys the profitability for all.

In order to avoid global failure by locally rational decisions, it is necessary to test partial models. In a corresponding test, each organizational function is isolated from its environment until it is consistent with the model that belongs to the decision rule. Obviously, in complex systems, intendedly and locally rational decision rules used by individual actors cannot guarantee globally rational behavior. In general, it is one of the most amazing insights of modern research in computational complexity – that very simple and well-determined rules of local behavior generate not only highly complex and chaotic, but also random dynamics of the whole system (compare Sect. 5.6).

If we look at the development of prices in financial markets, we will see time series with trends and patterns, but also amounts of apparent randomness with which prices fluctuate. The reason is that prices are not determined by true values, as Aristotle and Adam Smith believed; prices are determined by their estimated value at any given time. These estimates are influenced by all kinds of events that people relate to them and to their feelings and intuitions. In this sense, random fluctuations in prices reflect the random change of the outside world and its human estimation which can be observed on different time scales of time series. Thus, price fluctuations are the results of nonlinear interactions of a large number of factors. How can this large number of factors be represented in appropriate models? Even the nonlinear equations of dynamical systems reduce complexity to a certain number of continuous functions. Sometimes these functions are not well-known, or they must be selected and isolated from a complex background in an idealized manner. Instead of solving difficult nonlinear equations with selected functions, it could be more appropriate to consider computer programs of complex systems with a large number of interacting entities, which could be analyzed in computer experiments.

Following the line of Sect. 5.6, a market could be viewed as a 1-dimensional cellular automaton. Each cell represents a single trading unity with a finite number of states (e.g., colors), indicating, for example, whether it chooses to buy or sell at that step. All kinds of behavior could be interpreted by changing states of unities. The information flow in a market is realized by the change of states depending on the actions of their neighbors on previous steps. In a simple Wolfram cellular automaton (Fig. 7.9a), the behavior of a given cell is determined by looking at the behavior of its two neighbors on the previous step according to simple rules with cellular states 1 (black) and 0 (white).

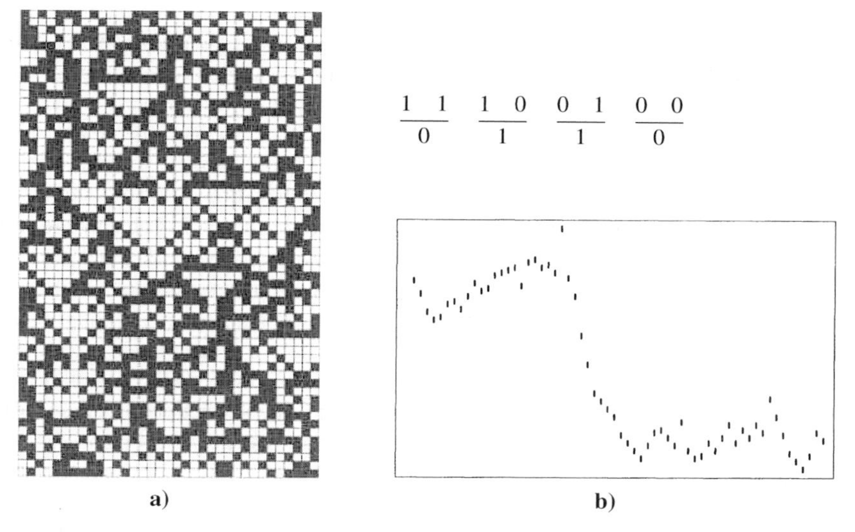

Fig. 7.9a,b. Market model of buying and selling units (cells) by pattern formation of a 1-dimensional cellular automaton (**a**) with time series of market price (**b**) [7.33]

As a rough analog to a market price, the running difference of the total number of black and white cells at successive steps in the cellular pattern is plotted as a time series (Fig. 7.9b). This plot looks random, although patches of predictability can be seen in the pattern of the complete behavior of the system. In real markets we have, of course, no chance of knowing the detailed behavior of each entity, as in a cellular automaton. But this kind of idealization is also assumed in the molecular models of thermodynamics. Although we do not know the exact behavior of each gas molecule, we can recognize trends of behavior or the complete randomness of the whole gas. For cellular automata and equivalent systems, it is well-known that even the more complicated local rules of interacting entities do not change the global pattern of behavior. Thus, even in more realistic models, where agents are allowed to have different or changing rules, there is no essential change in global behavior. These systems can generate complex, chaotic, and even random patterns. Thus, in economic contexts, locally or intendedly rational behavior can lead to random fluctuation. The consequence is that even from a computational point of view, we must support experimental research on the actual behavior of economic agents in changing situations. There is no shortcut to a philosophers stone in economics.

7.3 Bachelier's Financial Theory and Market Equilibrium

In economics as well as in financial theory, uncertainty and incomplete information prevent exact predictions. A widely accepted belief in financial theory is that time series of asset prices are unpredictable. Chaos theory has shown that unpredictable time series can arise from deterministic nonlinear systems. The results obtained in the study of physical, chemical, and biological systems raise the question of whether the time evolutions of asset prices in financial markets may be the result of underlying nonlinear deterministic dynamics of a finite number of variables. If we analyze financial markets using the tools of nonlinear dynamics, we may be interested in reconstructing an attractor. In time series analysis (Sect. 2.6), it is rather difficult to reconstruct an underlying attractor and its dimension d. For chaotic systems with $d > 3$, it is a challenge to distinguish between a chaotic time evolution and a random process, especially if the underlying deterministic dynamics are unknown. From an empirical point of view, discriminating between randomness and chaos is often an impossible task. The time evolution of an asset price depends on all of the information that affects the investigated asset. It seems unlikely that all of this information can easily be described by a limited number of nonlinear deterministic equations.

Therefore, asserts price dynamics are assumed to be stochastic processes. One key concept to understanding stochastic processes that was discovered early on was the random walk (see Sect. 5.4). The first theoretical description in the natural sciences of a random walk was given in 1905 in Einstein's analysis of molecular interactions. However, the first mathematization of a random walk was not realized in physics, but in the social sciences, by the French mathematician Louis Jean Bachelier (1870–1946). In 1900 he published his doctoral thesis, entitled "Théorie de la Spéculation" [7.34]. During that time, most market analyses looked at stock and bond prices in a causal way: something happens (the cause) and prices react (the effect). In complex markets with thousands of actions and reactions, a causal analysis is difficult to work out after the reaction has occurred, never mind beforehand (i.e., in an attempt to forecast a reaction). One cannot know everything. Instead, Bachelier tried to estimate the odds that prices will move. He was inspired by an analogy between the diffusion of heat through a substance and how a bond price fluctuates.

He believed that both were processes that could not be forecast precisely. At the level of particles in matter or individuals in markets, the details are too complicated. One can never analyze exactly how all relevant factors interrelate to spread energy or to energize spreads. However, the broad pattern of probability that describes the whole system can be seen in both fields.

Bachelier introduced a stochastic model by analyzing the bond market as a fair game. Each time one tosses a coin, the odds of obtaining heads or tails remains $1/2$ regardless of what was obtained in the prior toss. In this sense, coin tossing is said to have no memory. Even during long runs of heads or tails, the run is as likely to end as to continue upon each new toss. In the thick of the trading, price changes can certainly look that way. Bachelier assumed that the market had already taken all relevant information into account, and that prices were in equilibrium, with supply matched to demand and seller paired with buyer. Unless some new information came along to change that balance, one would have no reason to expect any change in price. The next move could be up or down, with each equally as likely.

Actually, prices follow a random walk. Imagine a drunken man staggering across an open field. How far will he have staggered after a certain period of time? He could go one step left, two steps right, three backwards, and so on in an aimless path. On average, just as in coin tossing, he gets nowhere; his random walk is stuck at his starting point. In the same way, market prices can go up or down, in big steps or small ones. With no new information to push a price in one direction or the other, a price will on average fluctuate around its starting point. In this case, the best forecast is the price today. Each variation in price is unrelated to the last. In a stochastic model, the price changes form a sequence of independent and identically distributed random variables. In Fig. 7.10b, a chart of changes in price illustrates a more or less uniform distribution over time. Most price changes occur within a narrow range. There are also some larger fluctuations, but these are relatively rare; this then resembles an unmown lawn, in that most of the blades of grass fall within a narrow range of heights, while a minority rise above this range.

In order to illustrate this smooth distribution, Bachelier plotted all of the changes in bond price over a month or year on a graph. When the price changes were independent and identically distributed, they spread out in the well-known bell curve shape of a normal ("Gaussian") distribution (see Sect. 5.4), with many small changes clustered in the center of the bell and a few big changes at the edges (Fig. 7.10a). Bachelier assumed that price changes behave like the random walk of molecules in Brownian motion. Long before Bachelier and Einstein, the Scottish botanist Robert Brown had studied the erratic way that tiny pollen grains jiggled about in a sample of water. Einstein explained that this motion was due to molecular interactions and developed equations very similar to Bachelier's equation of bond-price probability, although Einstein didn't realize this. It is a remarkable coincidence that the movement of security prices, the motion of molecules, and the diffusion of heat are all described by mathematically analogous models. In short, Bachelier's model depends on the three hypotheses of: (1) statistical independence (each change in price is independent of the previous change); (2) statistical stationarity of the price changes (the process that produces the price changes does not

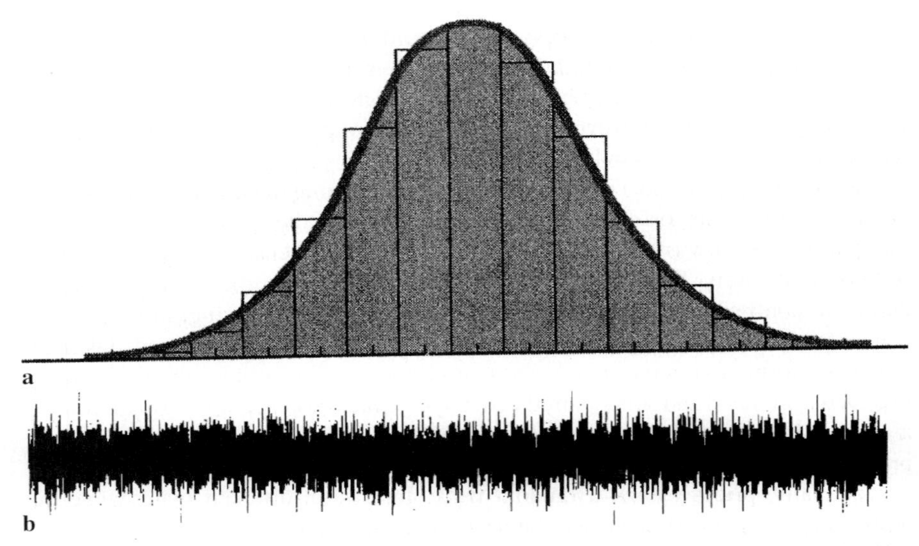

Fig. 7.10. Normal Gaussian distribution curve (**a**) and chart of price changes (**b**) obtained by Bachelier [7.35]

evolve over time); and (3) a normal distribution (the price changes occur according to the shape of the Gaussian bell curve).

However, it was a long time before economists recognized the practical virtues of describing markets via the laws of chance and Brownian motion. In 1956, Bachelier's idea of a fair game was used by Paul A. Samuelson and his school to formulate the Efficient Markets Hypothesis [7.36]. They argued that in an ideal market, security prices fully reflect all relevant information. A financial market is a fair game in which buyer balances seller. By reading price charts, analyzing public information, and acting on inside information, the market quickly discounts the new information that results. Prices rise or fall to reach a new equilibrium between buyer and seller. The next price change is, once again, as likely to be up as to be down. Therefore, one can expect to win half of the time and lose half of the time. If one had special insights into a stock, one could profit from being the first in the market to act on it. However, because there are many people in the market that are as intelligent as oneself, one cannot be sure of being first or correct.

Following the work of Samuelson, Bachelier's theory was not only elaborated into a mature theory of how prices vary and how markets work; it was more important for the financial world to translate the theory into practical financial tools. In the 1950s, Harry M. Markowitz was inspired by Bachelier to introduce Modern Portfolio Theory (MPT) as a method for selecting investments [7.37]. In the early 1960s, William F. Sharpe devised a method of valuing an asset called the Capital Asset Pricing Method (CAPM). Finally, the Black–Scholes formula for valuing options contracts and assessing risk was devised by Fischer Black and Myron S. Scholes in the early 1970s. These three innovations – CAPM, MPT, and Black–Scholes –

are still the fundamental tools of classical financial theory, based on Bachelier's hypotheses for financial markets.

Markowitz, who won the Nobel Prize for Economics in 1990, suggested that investment risks could be minimized by distributing them in portfolios, according to the Gaussian bell curve: given two investment portfolios to choose from, go for the one with the highest expected mean return and the lowest variance, or risk. Mathematically, Markowitz argued that any stock's prospects can be described by two numbers, the reward and the risk, or the mean and the variance of what one expects the stock will pay back by the time one sells. The first number, the average expected selling price, is predicted with standard stock analyst tools. The second number, the variance, is predicted by reviewing how the stock did in the past and using the bell curve as a yardstick. With these two numbers as coordinates, stocks can be represented by points in a plane and compared systematically. They are scattered across the plane in a spectrum of mean and variance, profit and risk. The final step is to combine the stocks in an efficient portfolio cluster.

The problem is that portfolio risk is more complicated than adding numbers. Taken as a whole, a portfolio can be greater or less than the sum of its parts. Stocks have a tendency to move up and down together. Markowitz compared it with the coin-tossing game. If one bets on a set of fair coins and their tosses are all uncorrelated, one will probably come out even. The heads will counterbalance the tails, and the bet is diversified. However, if the coins are correlated, then the outcomes of the coin tosses will be dependent on each other in some fashion. In a similar way, any stock is correlated more or less with other stocks, depending on the market sector. Markowitz's portfolio theory suggests that some stocks that tend to give tails should be mixed with others that tend to give heads, so to speak, in order to lower the risk associated with the overall portfolio. More and more stocks should be added in different combinations to build an efficient portfolio. A portfolio is efficient if it produces the most profit with the least risk. The variance of a portfolio can be mathematically estimated based on the variances of its stocks. In the simplified case of a two-stock portfolio P, its variance is defined by the formula $\sigma_P^2 = w_A^2\sigma_A^2 + w_B^2\sigma_B^2 + 2w_Aw_B\sigma_A\sigma_B\rho_{AB}$, where σ_A and σ_B are the standard deviations of stock A and B, the squares are the variances, w is each stock's weighting in the portfolio, and ρ_{AB} is the correlation between A and B. Using these methods, Markowitz transformed investing from a game of stock tips into an analysis of means, variances, and risk aversion indices. This marked the beginning of financial engineering.

But what would happen if everybody in the market uses Markowitz's rules? Sharpe, who also won the Nobel Prize for Economics in 1990, concluded that there would be not as many efficient portfolios as there were people in the market, but just one portfolio for all of them [7.38]. The reason for this is that everybody would abandon a first portfolio and move their money in a second one if fluctuations in stock prices suggested that the second portfolio was better. After time, there would be just one portfolio, known as the market portfolio. The market itself is a kind of computer that performs Markowitz's calculations. Therefore, Sharpe introduced the notion of a stock index fund: a large pool of money from thousands of investors that hold shares in exactly the same proportion as the real market. In short, the stock index fund is considered to be a model of the real market. This is the main idea behind Sharpe's Capital Asset Pricing Model (CAPM).

From a computational point of view, a stock index fund is a clever reduction of complexity. In order to reduce the thousands of calculations specified by Markowitz for many complex portfolios, a stock index fund with less variables can be used to simulate the real market. Instead of the real market, the stock index fund should be used to find the best portfolios. The amount by which a stock reacts to a changing market is called β. For a portfolio, it is now sufficient to estimate the value of β for each of its stocks. Each stock has its own β, or degree to which its price fluctuations correlate with those of the market overall. It is defined as how much the stock varies with the market (the covariance) divided by the variance or risk of the market itself. Therefore, the validity of β depends entirely on whether prices actually do fit the assumed Gaussian bell curve [7.39]. Sharpe's CAPM and Markowitz's MPT are based on Bachelier's hypothesis of a normal distribution of price changes (Fig. 7.10).

The next step in the evolution of modern financial theory was the development of the famous Black–Scholes formula [7.40]. The American economists Black and Scholes proposed this formula for valuing warrants or options. Based on the results of the Black–Scholes formula, financiers buy insurance, or hedge, against unwanted market problems [7.41]. The formula has permitted an entirely new type of trading, not on stocks, but on their volatility. One important class of financial contract, derivatives, encompasses financial products whose price depends upon the price of another financial product. Examples of derivatives include forward contracts, futures, options, and swaps [7.42]. Derivatives are traded in either over-the-counter markets or specialized exchanges. The simplest derivative is a forward contract. When a forward contract is stipulated, one of the parties agrees to buy a given amount of an asset at a specified price, called the forward price or delivery price K, on a specified future date, the delivery date T. The other party agrees to sell the specified amount of the asset at the delivery price on the delivery date. The party agreeing to buy is said to have a long position, and the party agreeing to sell is said to have a short position. The actual price Y of the underlying financial asset fluctuates. The price $Y(T)$ at the delivery date usually differs from the delivery price specified in the forward contract. The payoff is either positive or negative, so whatever is gained by one party will be lost by the other.

A future contract is a forward contract traded on an exchange. This implies that the contract is standardized, because the two parties interact through an exchange institution, called the clearing house. The clearing house writes the contracts with the buyer and the seller, guaranteeing that they will be executed at the delivery date.

An option is a financial contract giving the holders the right to buy or to sell on an underlying asset at time T and at price K. The price K is called the strike price or the exercise price. T is the expiration date, sometimes called the exercise date or the date of maturity. Options can also be characterized by the period during which the option can be exercised. If the option can be exercised only at maturity, $t = T$, it is called a European option. If the option can be exercised at any time between the contract initiation at $t = 0$ and $t = T$, it is called an American option. In the following, European options are considered.

There are call options and put options. In a call option, the buyer of the option has the right to buy the underlying financial asset at a given strike price K at maturity. This right is obtained by paying an amount of money $C(Y, t)$ to the seller of the option. In a call option, the contract is issued and thereby acquires a right to be exercised in the future, while the seller of the option receives cash immediately but faces potential liabilities in the future. In a put option, the buyer of the option has the right to sell the underlying financial asset at a given price K at maturity, $t = T$, back to the seller of the option.

Derivatives are financial products traded by speculators and hedgers. Speculators are interested in derivatives that provide an inexpensive way to expose a portfolio to a large amount of risk. Hedgers are interested in derivatives that allow investors to reduce the market risk to which they are already exposed. Obviously, hedging and speculating are typical trading strategies in financial markets. Hedgers focus on portfolio risk reduction, while speculators maximize portfolio risk. In any case, option pricing is a challenge for financial markets. The task is to find the rational and fair price $C(Y, t)$ of an option. Since the price $Y(t)$ at each step t is a random variable, $C(Y, t)$ is a function of a random variable.

The Black–Scholes formula is a solution to the option-pricing problem [7.43]. Black and Scholes assumed several conditions for financial markets. (1) The change in price $Y(t)$ at each step t can be described by the stochastic differential equation of Brownian motion. This assumption implies that the changes in the (logarithm of) price are normally distributed. (2) Security trading is continuous. (3) Securities can be sold at any time. (4) There are no transaction costs. (5) The market interest rate r is constant. (6) There are no dividends between $t = 0$ and $t = T$.

(7) There are no arbitrage opportunities. Arbitrage is a key concept when attempting to understand markets. It means the purchase and sale of the same or equivalent security in order to profit from price discrepancies. A stock may be traded on two different stock exchanges in two different countries with different currencies. For example, assume that the current price of a share is 9 US dollars in New York and 8 euros in Frankfurt, and that the exchange rate between US dollars and euros is 0.80 (euros to 1 dollar). By buying several shares of the stock in New York and selling them in Frankfurt, the arbitrager makes a profit (ignoring any transaction costs). Traders looking for arbitrage opportunities contribute to the market's ability to yield the most rational price for goods. The reason is obvious: if someone has discovered an arbitrage opportunity and succeeded in making a profit, he will repeat the same action. After carrying out this action repeatedly and systematically several times, the prices change to minimize his arbitrage opportunities. In short: new arbitrage opportunities continually appear in markets, but as soon as they are discovered the market moves in a direction that gradually eliminates them.

Now, in the absence of arbitrage opportunities, the change in the value of a portfolio must equal the gain obtained by investing the same amount of money in a riskless security providing a return per unit of time r. This assumption allows us to derive the Black–Scholes partial differential equation, which is valid for both call and put European options. Under some boundary conditions and substitutions, the Black–Scholes partial differential equation becomes formally equivalent to the heat transfer equation used in physics, which is analytically solvable. The famous Black–Scholes formula for option pricing is just such an analytical solution, $C(Y, t) = YN(d_1) - Ke^{r(t-T)}N(d_2)$, where $N(x)$ is the cumulative density function for a Gaussian variable with a mean of zero and unit standard deviation [7.44].

The analytical Black–Scholes solution to the option-pricing problem was obviously a mathematical milestone in the modern theory of finance. However, their model of financial activity does not fully reflect the stochastic behavior observed in real markets. The main reason is the use of Bachelier's hypothesis of changes in stock prices, which assumes small changes that are on average within the Gaussian standard deviation interval (Fig. 7.10). Markowitz also believed that variance and standard deviation provided good proxies for risk, with the uniform distribution of the Gaussian bell curve describing how prices move. In the next step, Sharpe's beta (β) and cost-of-capital estimates depend on Markowitz's concept of portfolios, which is correct, provided that Bachelier is right. Finally, the Black–Scholes formula is right provided that the bell curve and a normal distribution of small changes with continuously moving prices are correct. Therefore, in the end, the mathematics of modern financial theory is founded on Bachelier's hypothesis.

However, at the end of the 1980s, abrupt financial crashes took many investors by surprise. For example, on October 19, 1987, the Dow Jones plunged by 29.2%. According to Bachelier's normal distribution of small changes in the Gaussian bell curve, the fall should not have happened and was mathematically considered a once-in-an-eon event. The correctly designed investment portfolios blew up, and options-based portfolio insurance failed. The subsequent heavy financial fluctuations observed in the 1990s reinforced increasing doubts about the validity of the financial theory. Further, the path of the underlying asset price can be discontinuous upon the arrival of relevant economic information. The volatilities of a given stock or index and the interest rate are not constant, but are themselves random processes.

The classical Black–Scholes formula considers the option pricing problem in an ideal frictionless market. However, in physics as well as in economics, models without friction are purely theoretical. In mechanics, for example, it is of course much easier to formalize a general model of motion without friction than to attempt to model the real world. In thermodynamics, similar scenarios are observed when equilibrium and nonequilibrium theories are compared. Real markets are often efficient, but they are never ideal. The complexity of modeling financial markets increases when aspects of real markets are taken into account. Market imperfections are not formalized in the ideal model. Some of these economic examples of "friction" are considered in the following section.

The existence of a portfolio containing both riskless and risky assets is crucial when determining the rational price of the option under the assumption that no arbitrage opportunities are present. However, in order to obtain the rational price of an option, other assumptions must be made concerning the risk aversion and price expectations of the traders. Actually, the price of the underlying asset does not follow Brownian motion, but a jump-diffusion model. Discontinuities in the path followed by the asset price is just one of the imperfections. Another imperfection of real markets concerns the random character of volatility. Further, in idealized financial markets, the strategy for perfectly hedging a portfolio consisting of both riskless and risky assets is known. In real markets, this strategy is unrealistic, because the rebalancing of the hedged portfolio is not performed continuously, and, contrary to the conditions of the Black–Scholes formula, there are transaction costs in real markets. These market imperfections imply that perfect hedging of a portfolio is not guaranteed in a real market if the asset dynamics are assumed to be determined by Brownian motion.

In science, models are used to understand the basic aspects of a scientific problem. An idealized model is not able to describe all aspects of reality, but it is, at least, able to describe those that are assumed to be essential. As soon as the validity of the idealized model is criticized, extensions and generalizations of the model are attempted in order to improve the description of the real system. In this case, the complexity of the modeling grows, the number of assumptions increases, and the generality of the solutions diminishes. The Black–Scholes formula is an idealized model with restricted validity [7.45]. Extensions of the Black–Scholes formula aim to relax assumptions that may not be realistic for real financial markets. Examples are option pricing with stochastic interest rates, option pricing with a jump-diffusion stochastic stock price process, option pricing with stochastic volatility, and option pricing with non-Gaussian distributions of prices. When one or several of the Black–Scholes assumptions are relaxed, the complexity of the corresponding equations immediately increases. It is no longer possible to find a simple replicating portfolio, or to perfectly hedge an optimal portfolio. The simplicity and elegance of the idealized Black–Scholes solution is lost in real markets.

7.4 Complex Financial Markets, Turbulence, and Power Laws

The computational dynamics of globalized commerce are realized in centers of financial trade and financial engineering around the world. These centers are where financial theories, from Bachelier to Black–Scholes, meet financial reality. Some of these theories are deeply rooted in classical economics but are refuted by observations of real human behavior:

1) Assumption: People are rational in the sense of Adam Smith's homo oeconomicus. Consequently, when presented with all the relevant information about a stock or bond, investors will make the obvious rational choice, leading to the greatest possible wealth and happiness. Their preferences can be expressed by mathematical formulae involving utility functions that can be maximized. Rational investors lead to a rational model of an efficient market. In reality, people do not only think in terms of mathematical utility functions, and are not always rational and self-interested. They are driven by emotions that distort their decisions. Sometimes, they miscalculate probabilities and feel differently about loss than gain.

2) Assumption: All investors are alike. Consequently, people have the same investment goals and react and behave in the same manner. In short, they act like the

molecules in an ideal gas in physics. An equation that describes one such molecule or investor can be used to describe any such molecule. In reality, people are not alike. If the assumption of homogeneity is dropped, a more complex model of the market emerges. For example, there are at least two different types of investor. A fundamentalist believes that each stock has its own value and will eventually sell for that value. On the other hand, a chartist ignores the fundamentals and monitors price trends in order to jump on or off bandwagons. Interactions between these types of investor can lead to price bubbles and spontaneous crashes. The market switches from a well-balanced linear system, in which one factor adds to the next in a predictable manner, to a chaotic nonlinear system in which factors interact, resulting in synergetic and unanticipated effects.

3) Assumption: Prices change in a practically continuous manner. Consequently, stock quotes or exchange rates do not jump up or down, but move smoothly from one value to the next. Continuity is assumed in classical physics, and Leibniz' melegantly stated that "natura non facit saltum" (nature does not make leaps), which was later repeated by Alfred Marshall in his textbook on economic systems, "Principles of Economics" (1890). From a methodological point of view, the belief that nature and economies behave continuously makes it possible to apply continuous functions and differential equations in order to solve physical or economic problems analytically. In reality, however, prices in economy and quantum states in quantum physics do jump, and reality is actually discontinuous. In contrast to Einstein's famous objection to quantum physics, God does play with dice – in both nature and society.

4) Assumption: Price changes can be described by Brownian motion. Brownian motion, as mentioned earlier, was first described by physicists but was later applied to financial markets by Bachelier. When applied to economics, Brownian motion implies three assumptions. First, that each change in price does not depend on the previous change (statistical independence). Second, the process that generates the price fluctuations does not change over time (statistical stationarity). Third, the price changes follow a normal distribution – when the frequency of each change is plotted, a Gaussian bell curve results. Actually, plots of financial data do not result in a smooth normal distribution of price changes. The analysis of real distribution patterns using stochastic mathematics and systems theory is somewhat challenging, and one that provides new insights into the complexity of modern society.

One well-known financial time series is the Dow Jones Industrial Average. This is a simple average of the stock prices of the thirty most highly valued companies in the United States. The Dow Jones is usually presented in the form of Fig. 7.11, which shows the daily index values from 1916 through to its peak of 11,722 in January 2000, as well as the few years of bear market that followed. Although crashes in the market, for example, on October 19, 1987, are visible, the overall impression is one of exponential growth at the end of the last century.

However, rather than showing just the index values, it is easier to study the fluctuations in the index by plotting the index changes from one day to the next. In order to demonstrate the differences between a hypothetical index that conforms to Brownian motion and the behavior of the Dow Jones index, each index change is

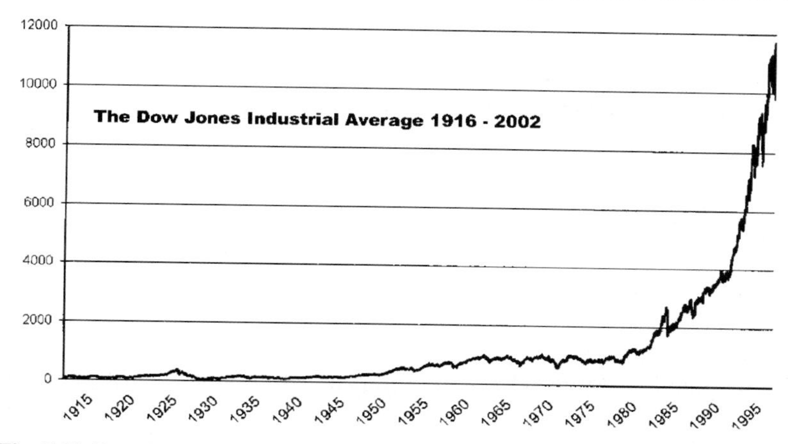

Fig. 7.11. Dow Jones Industrial Average values during the twentieth century [7.46]

converted into the number of standard deviations beyond the average change. In this case, a very large and rare index change will produce a tall bar in the chart. Small (common) changes give short bars. It is apparent from comparing the two plots in Fig. 7.12 that the underlying dynamics of the Dow Jones index are very different from Brownian motion. Mathematically, the standard deviation is designated by the Greek letter σ. In the Brownian chart (7.12b), most values are within 2σ, and very few values are any larger than this. However, the spikes are huge in the Dow Jones chart (7.12a); some are 10σ, and one, corresponding to the market crash in 1987, is 22σ.

The Dow Jones chart demonstrates that movements in financial markets are not distributed normally. Price fluctuations in real markets are not mild, but wild. This means that stocks are riskier than would be assumed based on a normal distribution. If the bell curve is used as a guide, stock portfolios could be put together incorrectly, risk management could fail, and misguided trading strategies could be used. Further, the Dow Jones chart shows that, as globalization gathers pace, we will see more financial crises. Therefore, we must focus on market extremes.

On a qualitative level, financial markets seem to be similar to turbulence in nature. Wind is an example of natural turbulence, and its properties can be studied in a wind tunnel. When the rotor at the start of the tunnel spins relatively slowly, the resulting wind profile is smooth; and the wind currents form long, steady lines, planes, and curves. However, as the rotor speed increases, the wind speed and energy inside the tunnel increases and the wind suddenly breaks up into sharp and intermittent gusts. Eddies form, and a cascade of whirlpools of various sizes appear spontaneously. The same emergence of patterns and attractors has been noted for the fluid dynamics of water (Fig. 2.26).

Figure 7.13a shows the time series for a turbulent wind. The chart illustrates the change in wind speed as it jumps into and out of gusty, turbulent flow. Turbulence can be observed everywhere in nature. Turbulence occurs in clouds, as well as

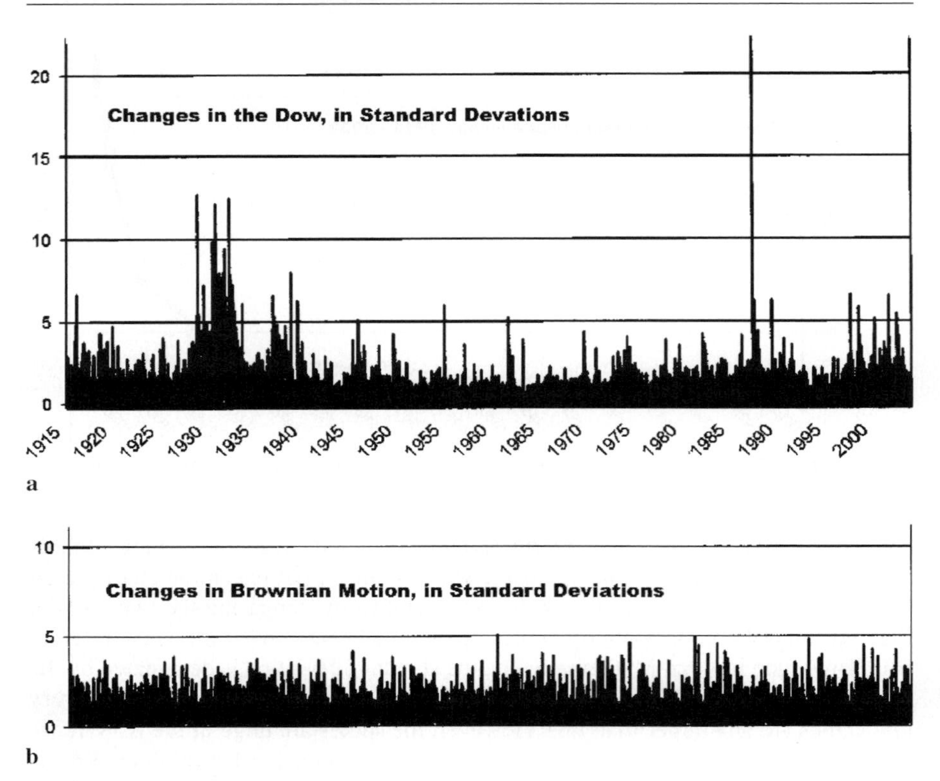

Fig. 7.12. Fluctuations in the Dow Jones index (**a**), and fluctuations from Brownian motion (**b**), both given in standard deviations [7.47]

in patterns of sunspots. All kinds of signals seem to be characterized by turbulent signatures. Figure 7.13b shows a "turbulent wind" in the stock market, which varies wildly through the turbulent twentieth century. This chart shows the volatility of the stock market: the fluctuations in price vary wildly in size from month to month. The peaks occur from 1929 to 1934 and in 1987. If this pattern is compared with the wind chart (7.13a), it is easy to spot the same types of behavior in both plots: abrupt discontinuities between wild motion and quiet activity, intermittent periods, and the same bunching of events in time. It is clear that the destructive turbulence observed in nature can also be seen in financial markets.

Indeed, the word "turbulence" is even used by politicians, since price fluctuations in financial markets qualitatively resemble velocity fluctuations in natural turbulence. The question then arises of whether this qualitative parallel is also useful on a quantitative level, such that our understanding of turbulence could aid our understanding of price fluctuations. Turbulence is a well-defined but still unsolved physical problem. It is a considerable challenge to mathematical physicists, and one that is central to research into complexity. A simple system that exhibits turbulence is a fluid of kinematic viscosity ν flowing with a velocity V in a pipe of diameter L.

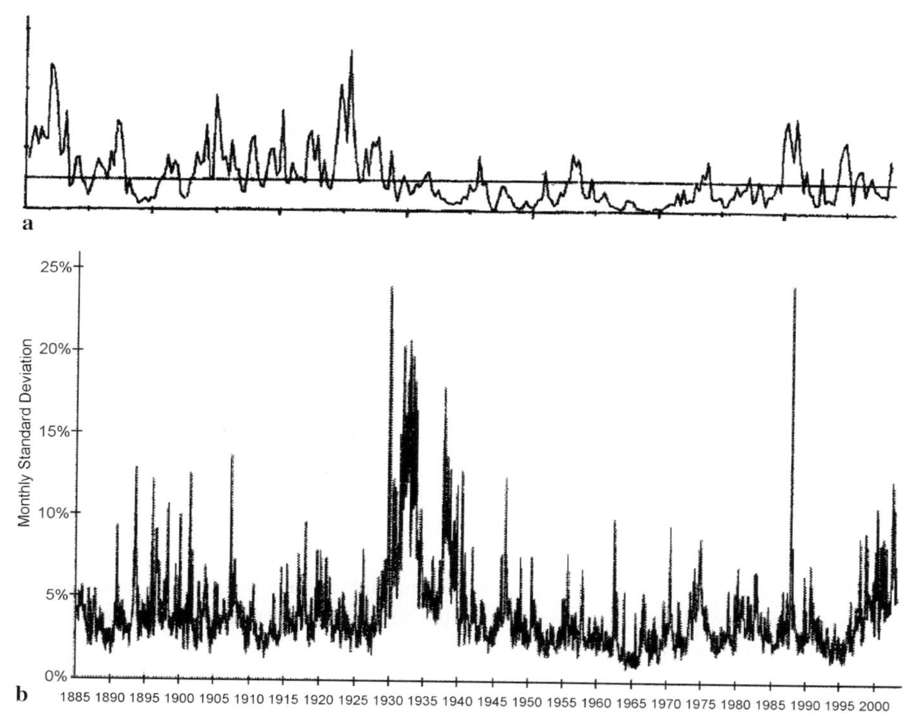

Fig. 7.13. Turbulent winds in nature (**a**) and in the financial market (**b**) [7.48]

The control parameter that determines the complexity of this flowing fluid is the Reynolds number $Re = LV/\nu$ (Fig. 2.26). When Re reaches a particular threshold value, the complexity of the fluid patterns increases dramatically, leading to turbulence.

The equations that describe the time evolution of an incompressible fluid were studied by Claude Louis Navier (1785–1836) in 1824. The Navier–Stokes equations take the form of $\partial/\partial t \mathbf{V}(\mathbf{r}, t) + (\mathbf{V}(\mathbf{r}, t) \cdot \nabla)\mathbf{V}(\mathbf{r}, t) = -\nabla P + \nu \nabla^2 \mathbf{V}(\mathbf{r}, t)$, and $\nabla \cdot \mathbf{V}(\mathbf{r}, t) = 0$, where $\mathbf{V}(\mathbf{r}, t)$ is the velocity vector at position $\mathbf{r}$ and time t, and P is the pressure. The Navier–Stokes equations characterize fully developed turbulence completely at high Reynolds numbers. However, it has proven impossible to solve these equations analytically, and it is even impossible to get numerical solutions for very large values of Re.

In 1941, Kolmogorov [7.49] proved that in the limit of an infinitely large Reynolds number, there is an approximate solution for the mean square velocity increase during turbulence, but he failed to describe mean velocity increases of any higher order. In fully developed turbulence, velocity fluctuations are characterized by intermittent behavior, as indicated by the leptokurtic shape of the probability density function of the velocity increases. Kolmogorov's theory is not able to describe the intermittent behavior of velocity increases. However, there are at least practical approximations devised by experimenters that can be used to estimate fully developed turbulence from the measured velocity $V(t)$.

Turbulence displays both analogies with and differences from the time evolution of prices in financial markets [7.50]. If we compare the time evolution of a stock index and the velocity of a turbulent fluid at high Reynolds number, both processes display intermittency and non-Gaussian features at short time intervals. Both processes are nonstationary at short time scales, but are asymptotically stationary. The similarities are their intermittency, non-Gaussian probability density functions, and their gradual convergence to a Gaussian probability attractor (see Fig. 5.12). However, the probability density functions have different shapes in the two systems, and the probabilities that they will return to the origin are different for the two systems. Also, velocity fluctuations are anticorrelated whereas index fluctuations are uncorrelated. The evolution of the variance of velocity fluctuations is characterized by a power law, which is valid only for a system in which the dynamical evolution is essentially controlled by the energy dissipation rate per unit mass. In a financial market, there is no reason that assets should exhibit a dynamical evolution that is controlled by a similar variable. In short, there is no analog to the power law for price dynamics.

Nevertheless, in modern physics and economics, phase transitions and nonlinear dynamics are related to power laws, scaling, and unpredictable stochastic and deterministic time series. Historically, the first mathematical application of a power-law distribution occurred in the social sciences and not in physics. Recall that the concept of a random walk was also mathematically described in economics (by Bachelier) before it was applied in physics by Einstein (see Sect. 7.3). The Italian social economist Vilfredo Pareto (1848–1923), one of the founders of the Lausanne school of economics (see Sect. 7.2), investigated the statistical characteristics of the wealth of individuals in a stable economy by modeling them using the distribution $y \sim x^{-\nu}$, where y is the number of people with income greater than or equal to x, and ν is an exponent that Pareto estimated to be 1.5 [7.51]. He noticed that his result could be generalized to different countries. Therefore, Pareto's law of income was sometimes interpreted as a universal social rule that was rooted in Darwin's law of natural selection.

In Sect. 5.4, we mentioned that power-law distributions may lack any characteristic scale. This property prevented the use of power-law distributions in the natural sciences until the introduction of Lévy's new probabilistic mathematical concepts and the introduction of new scaling concepts for thermodynamic functions and correlation functions in physics. In financial markets, time scale invariance means that even a stock expert cannot distinguish between time series where the prices are provided, for example, daily, weekly, or monthly. Figure 7.14a shows the natural logarithms of the daily final prices of the German stock index (DAX) between November 4th 1986 and August 4th 1993. In Fig. 7.14b, three time series are shown that contain 60 days of daily prices, 60 weeks of weekly prices, and 60 months of monthly prices; all of these plots are very similar from a statistical viewpoint.

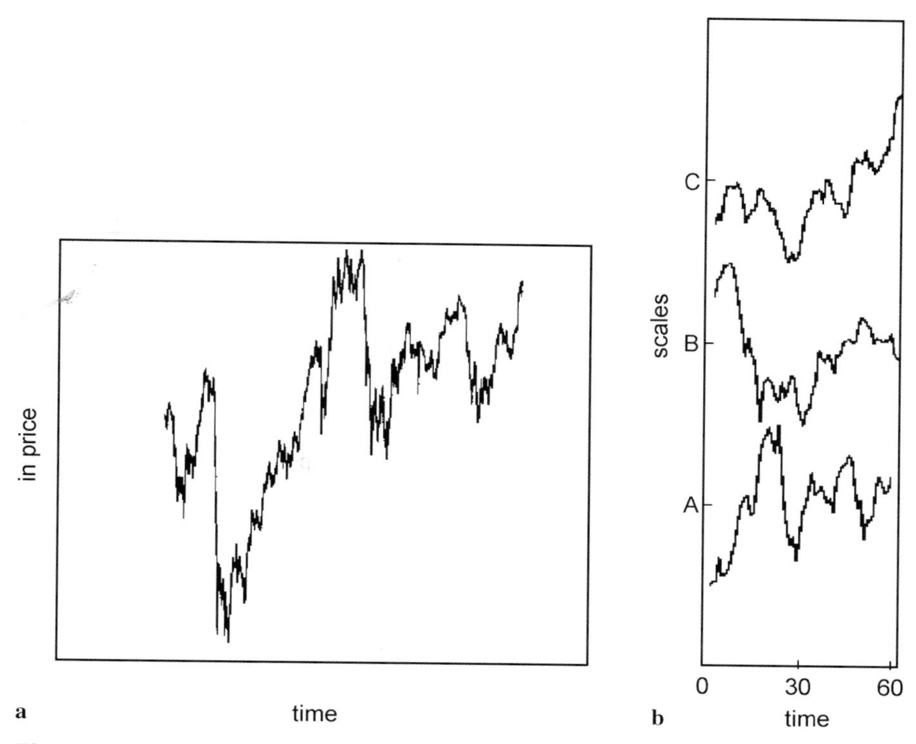

Fig. 7.14. Natural logarithms of the daily final prices of the DAX (the German stock index) (**a**), and three (statistically similar) time series showing daily, weekly, and monthly prices, respectively (**b**) [7.52]

The stock price (s) series is given by $\{X^s(kt)\}_{t_0}^{t_0+60}$, where the scale $k = 1$, 5, and 20 for daily, weekly, and monthly prices, respectively. If $X(t)$ is the time-dependent stock price sequence, the return obtained when prices are given every k days is given by $Y_k(t) = \ln X(t+k) - \ln X(t) \approx (X(t+k) - X(t))/X(t)$. Due to the statistical self-similarity of the time series, the return when prices are given every k days, $Y_k(t)$, has an equal distribution to the normed return obtained when prices are given every k' days, $c(k, k')Y_{k'}$; here $c(k, k')$ is a scaling factor and k, k' are natural integers. The scale invariance of measurable magnitudes resulting from self-similarity also implies that formal relationships between the magnitudes are scale-invariant. For example, we consider the average return for a number T of trading days, $\langle |Y_k| \rangle = 1/T \sum_{t=1}^{T} Y_k(t)$. Then the ratio of the measurable average returns

$$\langle |Y_k| \rangle \,/\, \langle |Y_{k'}| \rangle = \langle |Y_k| \rangle \,/\, \langle |Y_{k'}| \rangle$$

is also valid for all scales $k, k' > 0$ and scalings $\lambda > 0$. It follows that the scaling factors $c(k, k') = c(\lambda k, \lambda k')$ are equal, especially $c(k, k') = c(k/k', 1)$ for $\lambda = 1/k'$. As the average values $\langle |Y_k| \rangle$ are functions of k, both sides of the equation can be differentiated with respect to k. With $k = k'$ we get the relation

$$\partial \ln \langle |Y_k| \rangle / \partial \ln k = \partial \ln \langle |Y_{\lambda k}| \rangle / \partial \ln \lambda k$$

for all $\lambda > 0$, which means a constant magnitude for all k, designated by

$$H = \partial \ln \langle |Y_k| \rangle / \partial \ln k \, .$$

The resulting homogeneous first-order linear differential equation

$$\partial \; \langle |Y_k| \rangle / \partial k - H/k \, \langle |Y_k| \rangle = 0$$

has the solution $\langle |Y_k| \rangle = C \, k^H$ with $C = \langle |Y_1| \rangle$. Therefore the average k-day returns satisfy the power law $\langle |Y_k| \rangle \sim k^H$. This allows us to determine the scaling factors $c(k, k')$ for scale invariance with

$$Y_k =_P (k/k')^H Y_{k'} \, ,$$

where $=_P$ means "identically distributed." The exponent $0 \leq H \leq 1$ is called the Hurst exponent. For independent and identically distributed random variables in the sense of Bachelier's random walk model for the financial market, $H = 1/2$ from the central limit theorem (see Sect. 5.4). This is called the square root of time law of Brownian motion.

Because of the self-similarity of financial charts, Mandelbrot suggested that it was possible to construct fractal cartoons from them. A fractal is a pattern whose parts echo the whole [7.53]. Figure 7.15 illustrates the construction of a fractal cartoon of Bachelier's financial chart. The fractal starts with an initial box, called the fractal generator. Inside the box, a straight line rising from the bottom left corner to the top right corner symbolizes the underlying trend line of the dynamics. The price fluctuations along the way are represented by a zizag shape that fits over the straight line. The zigzag line consists of three parts: a rise, a fall, and then another rise (all of which are straight lines). In a next step, each of these straight line segments is replaced with a copy of the broken line that spans the whole box. This construction is repeated recursively, generating charts with ever-smaller lines and zigzags. The top line shows the first steps in the construction of the fractal. The middle diagram is the completed fractal chart. The diagram beneath that shows the increases from one moment to the next. The construction can be made more realistic via randomization. In this case, the process starts with different initial boxes that are selected at random and then used in the subsequent recursive construction steps.

Fractal generated from financial charts can be further complicated by using different zigzag lines. Of course, real price charts do not arise this way. Real charts record thousands of transactions that cannot be analyzed individually. Instead, a mathematical model can mimic the real process by simulating its tendency to rise and fall. The fractal constructed from this will not trace exactly the same path as that followed by the real price, but it will behave in the same manner from a statistical point of view. Computers can easily and quickly perform numerous recursive construction steps. Using this approach, although we cannot forecast financial events

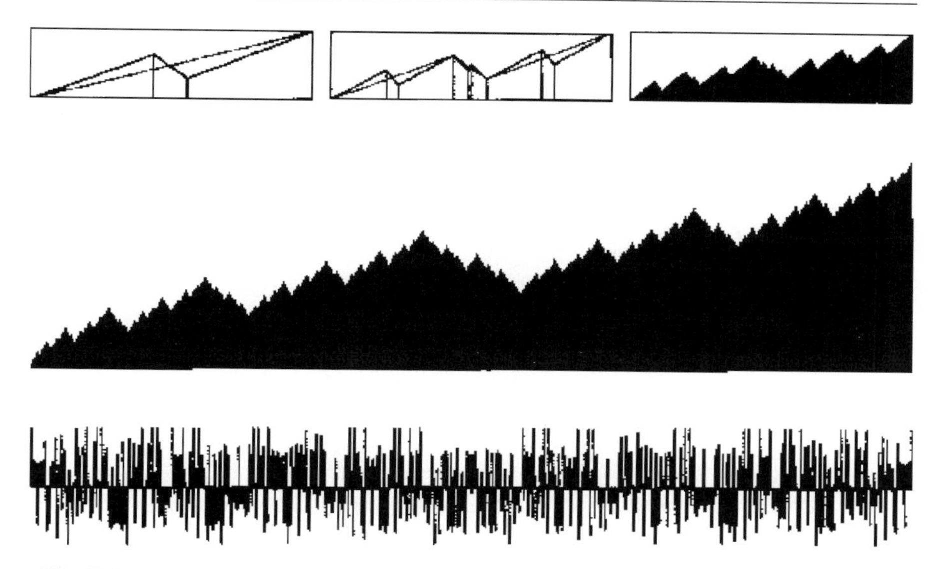

Fig. 7.15. Fractal reconstruction of Bachelier's normally distributed financial chart [7.54]

like we can predict the positions of the planets, we can still study typical behavioral patterns in financial markets in order to understand their dynamics and prepare ourselves for the future.

Bachelier's stochastic model of stock price dynamics assumes that $\ln X(t)$ is a diffusive process and that increases in $\ln X(t)$ increments are normally distributed. This model, based on Brownian motion, provides a first approximation of the behavior of empirical data. However, systematic deviations from the predictions of the model are observed. The empirical distributions are more leptokuric than Gaussian in form. A highly leptokurtic distribution is characterized by a narrower and larger maximum and by fatter tails than in the Gaussian case. The degree of leptokurtosis is much larger for high-frequency data. Figure 7.16 shows the empirical probability density function for high-frequency price fluctuations in Xerox stock traded on the New York Stock Exchange during a two-year period (1994–1995). The Gaussian obtained with the measured standard deviation is also shown for comparison.

Several alternative models to Brownian motion that are based on different assumptions have been proposed. The models differ in terms of the shape and leptokurtosis of the probability density function, and in terms of key properties such as the finiteness or infiniteness of the variance, the stationarity present at a short time scale or asymptotically, the continuous or discontinuous character of $X(t)$, and the scaling behavior of the stochastic process. The first model to explicitly take into account the leptokurtosis observed empirically in the probability density function was proposed in 1963, when Mandelbrot modeled $\ln X(t)$ for cotton prices. For more than a century, the New York Cotton Exchange had kept exact daily records of prices. Nearly all interstate trading was centralized at one exchange. It was a huge,

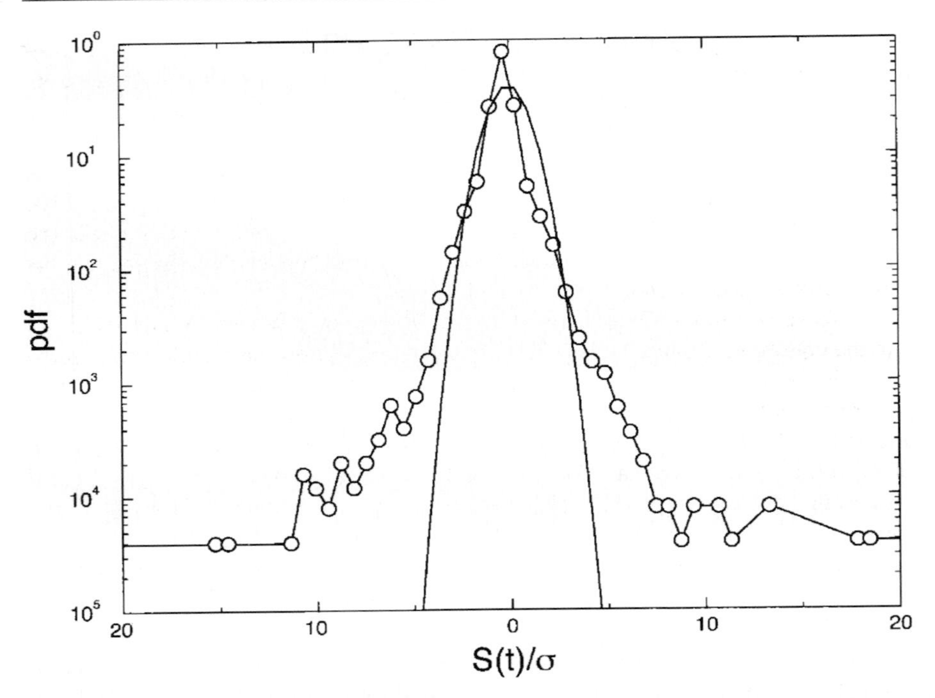

Fig. 7.16. Empirical probability density function (pdf) for high-frequency price fluctuations $S(t)$ in Xerox stock. Note the fat tails, in contrast to the Gaussian constructed for the measured standard deviation [7.55]

liquid market, with ample resources for record-keeping. However, it had long been a nightmare for mathematical economists. No matter how they manipulated the numbers, they could not get them to fit into Bachelier's model. The price jumps and falls were just too big. The volatility or standard deviation of the prices kept shifting over time. The prices were stable in some years but wild in others.

Mandelbrot's paper "The variation of certain speculative prices" (1963) was a breakthrough in economic theory [7.56]. For the first time, he applied Lévy-stable non-Gaussian distributions (see Sect. 5.4) and emphasized the importance of power laws in wild financial markets. The most interesting properties of Lévy-stable non-Gaussian processes are their stability and self-similarity, and their role as attractors in probability space (Fig. 5.12). Mandelbrot's Lévy-stable hypothesis implies that $\ln X(t)$ is discontinuous and successive changes $S(t) = \ln X(t+1) - \ln X(t)$ in the natural logarithm of the price are characterized by non-Gaussian scaling and by a distribution with infinite variance.

The concept of scaling and discontinuity can also be translated into a fractal cartoon. Fractal patterns are used to characterize the fat tails and abrupt price changes of real financial markets. In Fig. 7.15, we saw that the fractal cartoon of Brownian motion was constructed with a rising, straight-line initiator and a zigzag generator.

Shrunken copies of the generator are interpolated into the diagram recursively. By repeating this process, the specific fractal for Brownian motion can be generated. The reason for this is the specific shape of the generator: starting at the point (0,0), it rises to the point (4/9, 2/3), falls to the point (5/9, 1/3), and ends up at (1,1). The sizes of the three segments of the generator are decisive. Their widths are 4/9, 1/9, and 4/9, respectively. The heights are 2/3, -1/3 (minus, because the line falls), and 2/3. Obviously, each width is the square of each height, or each height is the square root of each width. This square-root relationship between widths and heights characterizes the Brownian motion fractal.

If these coordinates are changed, however, the outcome can look much more like the cotton price chart. In Fig. 7.17, fractal construction starts with five inclined intervals, but adds two vertical discontinuities. Unlike the Brownian generator, this pattern exhibits sharp discontinuities. Each construction step adds further jumps up and down. The bottom panel of Fig. 7.17 shows the changes in value from one moment to the next; many sharp outliers are present, giving the plot a very different look to that of Bachelier's normal distribution. The jumps follow a power-law distribution, corresponding to a curve with fat and leptokurtic tails. The exponent α of the power law is the α parameter of the Lévy-stable non-Gaussian distribution ($0 < \alpha < 2$). A stable Gaussian distribution is characterized by $\alpha = 2$ (see Sect. 5.4).

Abrupt price changes were called the "Noah effect" by Mandelbrot, due to an analogy with the flood described in the story of Noah in the Bible (see Fig. 7.18). In

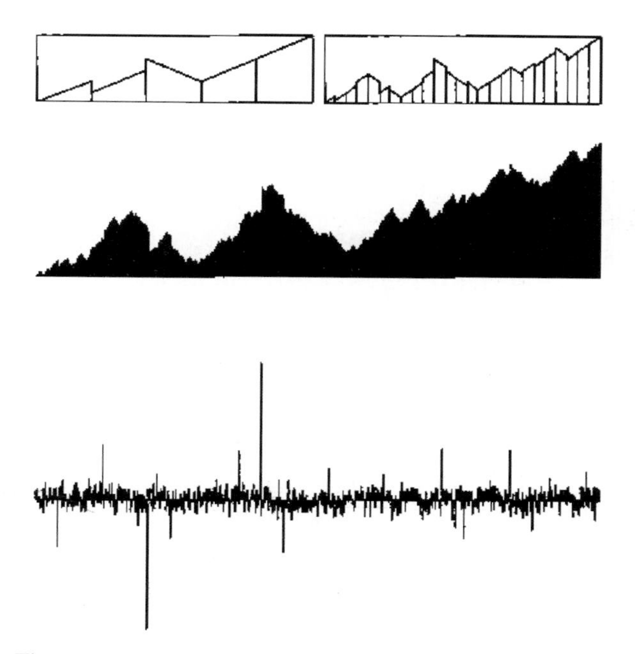

Fig. 7.17. Fractal reconstruction of abrupt price changes [7.57]

Genesis 7:4, God ordered the great Flood in order to purify a wicked world: "I will cause it to rain upon the earth forty days and forty nights, and every living substance that I have made will I destroy from off the face of the earth." Noah survived; divine advice caused him to prepare for the coming flood by building a ship strong enough to withstand it. The flood came and went. It was catastrophic, but transient like a market crash. Noah could not forecast the biblical flood, but he understood the signs and symbols that hinted at that catastrophe to come. In the same manner, we study the fractal patterns of time series and distribution laws in order to prepare ourselves for sudden events.

There is another biblical story that illustrates a second type of wild behavior in the financial markets. In this story, the Pharaoh dreamed that seven fat cattle were feeding in the meadows when seven lean cattle rose out of the Nile and ate them (Fig. 7.19). Genesis 41:28–30 explains: "What God is about to do he showeth unto Pharaoh. Behold, there come seven years of great plenty throughout all the land of Egypt: and there shall arise after them seven years of famine; and all the plenty shall be forgotten in the land of Egypt; and the famine shall consume the land." Joseph called the Pharaoh's dreams prophetic: seven years of famine would follow seven years of prosperity. Therefore, he advised the Pharaoh to stockpile grain for bad times to come. This pattern of periods of feast then famine is analogous to the behavior often observed in financial markets where a period of rises is followed by another period of falls (an "almost cycle"). Such behavior was termed the "Joseph effect" by Mandelbrot, and its presence indicates that market fluctuations are dependent on each other to some degree (i.e., the market appears to have a long-term memory).

Fig. 7.18. The Noah effect: discontinuities in market dynamics, as illustrated by Noah's ark and the biblical flood [7.58]

Fig. 7.19. The Joseph effect: long-term memories in financial markets, as illustrated by Joseph's interpretation of the Pharaoh's dream [7.59]

The flooding of the Nile inspired the British engineer Harold Edwin Hurst to derive a formula that could be used to model this long-term cyclic behavior [7.60]. The height of the flood was required in order to be able to construct appropriate dams. At first, engineers assumed that the variations in the flood level from one year to the next were statistically independent, just like Bachelier's coin-tossing. In this case, the range between the biggest winning margin at one moment of the game and the worst losing margin at another time varies by the square root of the number of tosses. However, Hurst found that the range from the highest Nile flood to the lowest widened faster than the coin-tossing rule predicted. The range did not widen according to the square-root rule (i.e. a one-half-power), but according to the number of flood heights measured raised to power of three-quarters.

Long-term dependence also occurs in financial data when correlations fall more slowly than expected. This is due not only to the complexity of the physical world of weather, crops, ores, and factories, but also the psychological complexity of humans and their changing expectations and estimations of what may or may not happen. In such a complex world of interrelations, events in the distant past continue to echo in the present. Financiers that were shocked by the 1929 stock market crash and the Great Depression would hesitate to take a risk on the wildest forms of speculation. Their memories led to long-term dependencies in the financial markets. It may be no coincidence then that in 1987, when most of those financiers were no longer on the scene and their wisdom had been forgotten, another great crash happened. Nevertheless, classical financial theory, based on Bachelier's hypothesis of mild markets, holds that all that matters is today's news and the expectation of tomorrow's news.

Brownian motion provides a very handy rule for estimating the development of prices. It tells us how far an asset's price may rise or fall and how much it is likely to fluctuate within a certain broad band. However, the exchange rates of different countries, for example, can actually wander further than the square root of time law forecasts. This can occur when exchange rates exhibit long-term dependencies. The movement of a rate in one direction will tend to continue over the next few days. The rate will still fluctuate from da to day. However, in the long run it will drift further from its starting point than predicted by the Brownian rule. The rates are no longer simple random fluctuations that conform to Bachelier's hypothesis.

Bachelier's square root of time law is a power law with an exponent H of $1/2$. Actually, it is only a special case of a more general rule for particles in physics or prices in financial markets: the distance traveled (by the particle or price) is proportional to some power H of the time elapsed, with $0 \leq H \leq 1$ (note that the exponent of the power law is called H in honor of Hurst). If H is bigger than the Brownian value of 0.5, the particle or price will persistently roam far. This is the case when there are long-term correlations with "memory" and echo effects. If H is smaller than the Brownian value of 0.5, the particle or price will roam less, in a narrow and furious zizag pattern. Therefore, the degree of dependence (the Joseph effect) between price changes can be measured by the Hurst exponent H. In Fig. 7.20b, each price change is assumed to be independent of the last with $H = 0.5$. Figure 7.20c shows the persistent case, when $H > 0.5$ and the resulting price trends are broad and exhibit long-term memory. Figure 7.20 a shows the antipersistent case, when $H < 0.5$ and the action is furious but still constrained.

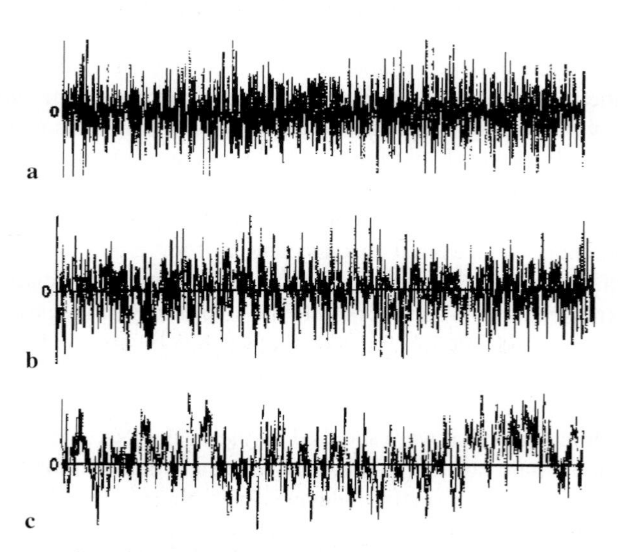

Fig. 7.20. Degrees of complexity of long-term dependencies, where (**a**) shows $H > 0.5$ and persistent behavior, (**b**) $H = 0.5$ and Brownian independence, and (**c**) $H < 0.5$ and antipersistent behavior [7.61]

The degrees of complexity of long-term dependencies can also be characterized by the fractal patterns in Fig. 7.21. Recall that the pattern for Brownian motion (the middle panel in Fig. 7.21) is characterized by a square-root relationship in the three segments of the fractal initiator, corresponding to the Hurst exponent $H = 1/2$. The height of each segment is the square root of its width. Consequently, when attempting fractal reconstructions of the charts 7.20a,c, each segment's height should be made equal to its width raised to some arbitrary power between 0 and 1. The result of using a value of H that is greater than 1/2 is shown in the bottom panel of Fig. 7.21, which corresponds to persistent behavior. The result of using a value of H that is smaller than 1/2 is shown in the top panel of 7.21, which corresponds to antipersistence.

Fractal patterns illustrate the scale invariance of time series. At different time scales they all look roughly alike. No matter how wild the chart is, it sometimes rises or falls in long waves, or with small waves superimposed on bigger waves. However, two forms of wildness seem to typify the zigzag-patterns: the abrupt changes and discontinuities of the Noah effect and the almost cycles of the Joseph effect.

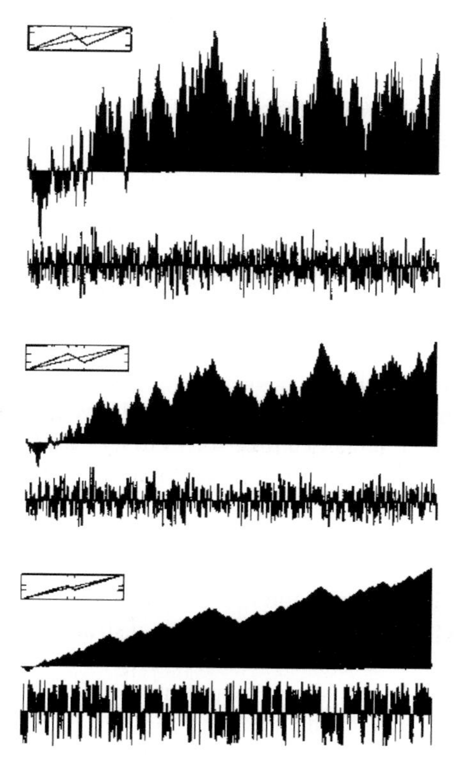

Fig. 7.21. Fractal reconstructions of long-term dependences where (**a**) shows $H > 0.5$ and persistent behavior, (**b**) $H = 0.5$ and Brownian independence, and (**c**) $H < 0.5$ and antipersistent behavior [7.62]

Mathematically, these can be characterized by two parameters: $0 < \alpha \leq 2$ for stable Lévy distributions and $0 \leq H \leq 1$. Both are the exponents of power laws. A low-α market would be risky, with wild price swings. As α is increased the behavior of the market becomes closer to that of Bachelier's market (where $\alpha = 2$). Obviously, Lévy's α is an appropriate tool for describing discontinuities and abrupt changes in the sense of the Noah effect. A Hurst exponent H of 1/2 implies that each price change is independent of the previous one [7.63]. A larger H suggests that the data are correlated; they move persistently in a certain direction. A smaller H implies antipersistent behavior. The Hurst parameter H characterizes the long-term memory associated with the Joseph effect.

Sometimes, both effects are interrelated; for example, when $H = 1/\alpha$. This is the case for Brownian-type markets with $\alpha = 2$. Interrelations can also be observed in investment bubbles. Investment bubble crashes are Noah effects that are produced by Joseph-style dependencies. During the Internet boom at the end of the 1990s, the investment bubble of, for example, the company Cisco Systems indicated how enthusiastic investors extrapolated the earning trends of 1999 into a soaring stock price. In 2000, as earnings plateaued, investors started to sober up and the bubble began to deflate. The dynamics and exponents discussed above explain why economic bubbles are sometimes unavoidable under certain circumstances. They are not irrational deviations from the economic norm, caused by the immorality of speculators. Therefore, we must prepare ourselves for them by analyzing the intricate dynamics of financial markets.

Complex dynamical systems have their own intrinsic time scales that differ from standard clock time. In physiology, organisms are characterized by their own intrinsic aging processes. Two sixty-year-old people may be in very different states of health; sixty years is the external physical clock time, not the internal biological clock time. Aging is a set of complex physiological and psychological processes that depend on the organs which may be in different states of health. Also, our mind is characterized by an intrinsic experience time. In boring situations, time appears to drag, while time flies when we are busy and engaged in exciting tasks. In the same way, markets have their own intrinsic trading times. When the volume of trade is climbing, and prices are rising, then time seems to fly. However, there are also slow times, when trading is thin on the ground and prices are quiet.

Mandelbrot, the father of fractal geometry, suggested that the mixed behavioral patterns of trading time can be modeled by multifractals (see Sect. 2.6). As we noted earlier, a fractal is a pattern whose parts echo the whole, only scaled down. A multifractal has more than one scaling ratio in the same pattern. Some parts of the pattern shrink quickly, others slowly. Multifractals have distributions with tails that follow power laws. They all manifest the Noah effect (they show abrupt and discontinuous changes) as well as the Joseph effect (long-term trends). Thus, multifractal models mirror the intrinsic dynamics of the market.

Multifractal models can also be sketched using multifractal cartoons. Recall that the Brownian chart displayed a relationship between the width and the height of each segment in the fractal generator (Fig. 7.15). This was related to the presence of a power law in which one was the square root of the other. By choosing different powers, one could generate charts with varying degrees of long-term dependence. By adding vertical jumps to the fractal generators, charts with fat tails and discontinuities were also produced. Thus, a multifractal cartoon can be generated by modifying the initial generator.

In this way, the clock time can be converted into a multifractal trading time in an erratic process. A simplified sketch of a multiplicative cascade is shown in Fig. 7.22: a portion from the time axis is considered at successively finer scales. The top rectangle is the first approximation. In this, 60% of the trading activity is to the left and 40% to the right. Each half of the time axis is then cut into two halves, with 60% of the activity in each half again placed in the left quarter and 40% in the right quarter. In other words, the distribution in the first quarter is 60% times 60%, or 36% of the total trading activity. The second quarter contains 40% times 60%, or 24% of the total trading activity. When this operation is performed repeatedly, the final panel of Fig. 7.22 is eventually generated. It is a very uneven distribution. Trading time bunches and moves quickly at the peaks, and thins and moves slowly in the valleys. This is, of course, only a sketch of a much more complex multifractal structure of trading time.

More sophisticated versions of multifractal trading time have successfully modeled real data [7.65]. Price changes in particular currency markets scaled just as the model predicted. Volatility clustered. Episodes of fast action were interspersed with intervals of slow and dull trading. By zooming in on fast episodes, they were seen to have subclusters of fast and slow subintervals. Clusters within clusters showed a multifractal pattern. Multifractal models assume unchanging and fundamental patterns of market behavior. In this sense, they mirror mathematical invariances in financial dynamics. Invariances are the fundamental principles that guarantee the validity of a theory and its laws everywhere and at any time under the boundary conditions of the theory.

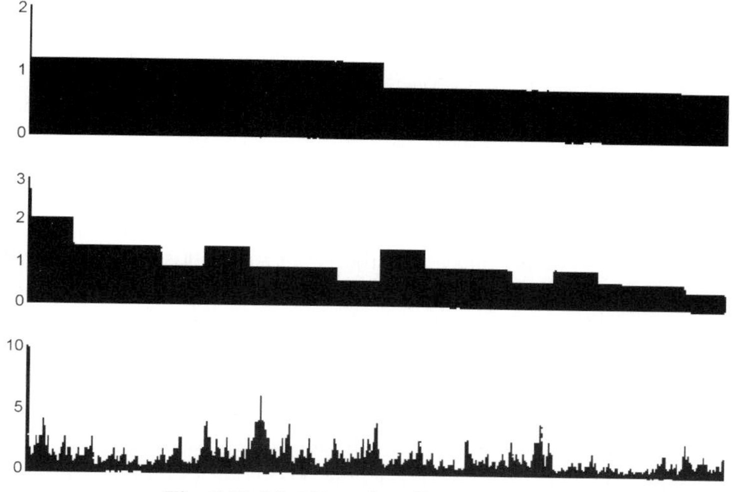

Fig. 7.22. Multifractal trading time [7.64]

Let us finish this section by summarizing some essential insights into the dynamics of financial markets. Financial markets are more complex than traditional academic theory would have us believe. They are turbulent, not in the strict physical sense, but due to their intrinsic complex stochastic dynamics, and this turbulence can have dangerous consequences, just like earthquakes, tsunamies, or hurricanes in nature. Therefore, financial systems are not linear, continuous, and computable machines; individual economic events cannot be predicted like planetary positions in astronomy. Financial systems are very risky and complex but still computational, because the application of appropriate stochastic mathematics allows us to analyze and recognize typical patterns and attractors in the underlying dynamics. These methods support market timing. However, there is no guarantee of success: big gains and losses are concentrated into small periods of time. The belief in a continuous economic development is refuted by the fact that prices often leap, adding to the risk. Markets are determined by an intrinsic trading time, which is different from the physical clock time. Trading time is flexible. As price changes become more dramatic, the trading time passes more quickly. As the price chart becomes duller, the market clock passes more slowly. Obviously, trading time also fits our subjective feeling of time, which depends on the intensity of our experiences. Markets are mathematically characterized by power laws and invariance. A practical consequence of this is that all markets – whatever the location or time period – work alike. If market properties that remain constant over time or location can be found, it is then possible to build useful models that can be used to help with financial decisions. However, we must be cautious, because markets are deceptive. Their dynamics sometimes appear to provide patterns of correlations that do not actually exist, because we subconsciously want to see them. During evolution, our brain was trained to recognize patterns of correlation since this skill aids survival. Therefore, we sometimes see patterns where there are none. Systems theory and appropriate tools of complexity research should help to stop us from seeing imaginary correlations in markets.

7.5 Perspectives on Econophysics

Econophysics is an interdisciplinary scientific field that refers to economics research performed by physicists by applying mathematical methods. The interest of physicists in economics and finance started with the explosion of research into nonlinear dynamics, complex systems, and chaos theory in the 1980s. Historically, the first applications of physical models to the social sciences occurred even earlier, in the nineteenth century. In 1798, the British cleric Thomas Robert Malthus analyzed the exponential growth of populations [7.66]. He proposed that the incremental increase in the size of the human population was geometric; in other words the human population increases according to a constant ratio from generation to generation. Mathematically, his assumption of growth can be represented by the elementary differential equation $dN(t)/dt = kN(t)$, where $N(t)$ is the population at time t, and k is a constant that characterizes the rate at which the population grows. The solution to Malthus' growth equation is $N(t) = N(0)\exp(kt)$, where $N(0)$ is the size of the population at time 0 and $\exp(kt)$ is the exponential function with unrestricted growth to infinity.

The Belgian mathematician Pierre-Francois Verhulst (1804–1849) noted that the growth of real populations is not unbounded. He argued that factors such as the availability of food, shelter, and sanitary conditions influence and restrict the growth of populations. Therefore, he suggested the equation $dN(t)/dt = kN(t)[1-N(t)/M]$, which limits the growth of the population to a level M. The solution to the Verhulst equation is the S-shaped logistic curve $N(t) = MN(0)/[N(0) + (M - N(0))\exp(-kt)]$. Both solutions coincide at early on but deviate from each other in the long run (see Fig. 7.23). Therefore, Malthus' dynamic equation can be considered to be a fundamental law of growth in the absence of any limiting influence. The deviation of the growth curve into an S-shape becomes evident when an influential "social force" exists. This is not unlike Newton's force law, which defines that a force must be applied to induce a change in the otherwise constant momentum of a particle.

Actually, in the 1970s, three laws of social dynamics were suggested that were analogous to Newton's famous laws of mechanics [7.68]. The first law states that in the absence of any social, economic, or ecological force, the rate of change in the logarithm of the population size $N(t)$ is constant, which means $d/dt \log N(t)$ is constant. The second part of this law is that, under the same conditions, in other words an absence of social, economic, or ecological forces, the rate of change of the logarithm of the price $P(t)$ of maintaining a population member is also constant, which

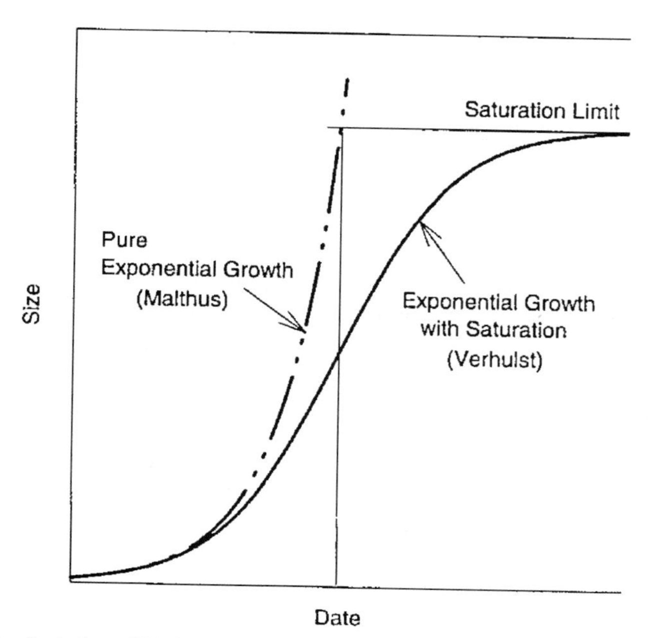

Fig. 7.23. The deviation of Verhulst's logistic curve from Malthus' exponential curve [7.67]

means that $d/dt \log P(t)$ is constant. For products, $P(t)$ is the unit cost. Obviously, the first part of the law is just Malthus' law for the exponential growth of a population, and the second part is a formulation of the observation that things always get more expensive over time.

In his second law, Newton defined a physical force as being anything that causes a change in momentum. Thus, the formulae associated with the first social law are violated upon the application of a social, economic, or ecological force. The force is measured by observing the way in which the violation occurs. The third law of social dynamics is that evolution is the result of a sequence of replacements. The term evolution in this context is broader than its biological meaning and is intended to include changes in social systems over time, changes in economic products over successive generations, or improvements in modes of transportation over the years. In biology, the term may refer to mutations, whereas in computer technology it may denote the latest chip generation, and in transportation it may refer to the stagecoach being replaced by the train, which in turn is replaced by the airplane.

Newton's three laws were intuitively true axioms that he never derived from first principles. Later on, they became fundamental laws in many areas of classical physics. From a methodological point of view, we could ask the question: what are physical laws or "natural laws?" What do we mean by social laws? Can social dynamics be reduced to fundamental principles? Do economics and physics have laws in common? These methodological questions are central to econophysics as an interdisciplinary field of economic and physical research.

Since they are an essential feature of the mathematical laws of nature, the laws of physics do not change on any time scale that we can observe. Nature obeys inviolable mathematical laws only because those laws are grounded in fundamental principles of symmetry. Symmetry refers to invariance with respect to frames or systems moving at constant velocity relative to each other, in the sense of the principle of relativity, translational invariance, rotational invariance, and time-translational invariance. These invariances are the same whether we are discussing Newtonian mechanics, general relativity, or quantum mechanics [7.69]. Without these principles of symmetry, it would be impossible to discover the mathematical laws of nature. This is because the invariances form the theoretical basis for repeatedly performing identical experiments whose results can be reproduced by different observers independent of where and at what time the observations are made, and independent of the state of relative motion of the observational system. In physics, therefore, we do not merely have models of the behavior of matter but laws that are justified by invariance principles. Fundamental laws are characterized by fundamental natural constants.

Laws of nature are established by repeatedly performing identical experiments or observations. In physics and astronomy, any prediction must be falsifiable in practice, otherwise the model or theory used to make the prediction is not regarded scientific. A falsifiable theory or model is one with few enough parameters and predictions that are precise enough to be able to be tested observationally. One might, however, object that in economics there are no inviolable mathematical laws and "natural" constants. Real economic laws, like any legislated law or social contract, can always be violated by people and groups that wish to do so. Because the laws of physics are based on the principles of invariance, they are independent of initial conditions like absolute time, absolute position, and absolute orientation. Socioeconomic behavior is not necessarily universal but may vary from country to country. Many econophysicists hope that a single universal law governs markets. Are there any socioeconomic invariances that can support this hope [7.70]?

In the past, econophysicists only had limited success when modeling financial markets, following Bachelier, Markowitz, Sharpe, Black, and Scholes (see Sect. 7.3). Their financial models at least allow predictions that can be falsified. Also, they satisfy general mathematical equations that can be considered to be general laws governing the dynamics of complex systems. In Sect. 7.4, the Black–Scholes model, for example, was a mathematical solution to a modified physical heat equation. However, the important point is that these equations are only used formally, as particular types of differential and stochastic equations, without any physical meaning. In the previous section, we emphasized the differences between physical and economic turbulence. In general, the Navier–Stokes equations of turbulence are not analytically solvable due to their computational complexity. Nevertheless, in fluid dynamics, we know the equations of motion, since they are based on Galilean invariance principles. In economics, we miss out these principles in order to describe the microdynamics of human behavior. In physical turbulence theory, we cannot predict the weather in the long run like we can predict planetary motion using Kepler's laws, but we understand the underlying physical principles.

Therefore, from a methodological point of view, the name "econophysics" is extremely misleading, because it seems to suggest a physics-like science of economies. I prefer the term "sociodynamics," which refers to a particular class of models in the general mathematical theory of complex dynamical systems. Physical dynamical systems as well as chemical and biological systems are described by particular classes of models in this general mathematical theory. From a mathematical point of view, the general theory of complex systems and nonlinear dynamics concerns, for example, particular differential and stochastic equations, phase spaces of attractors, mathematical invariances and group structures. Even in sociodynamics, we encounter typical invariances which are characterized by attractors of a probabilistic space (for example, the Lévy-stable distributions of economic turbulence), scale invariance, and power laws. They are falsifiable in economics and finance and may be modified in the future. However, the invariance principles of physics must also be falsifiable. The symmetry of time, for example, is widely accepted in physics. Nevertheless, it has been refuted by the results from experiments with particular elementary particles (kaons), and the reason for this is still not understood. Thus, symmetries and invariances are not eternal truths, but general and formal frameworks of research which are, nevertheless, falsifiable. In short, the theory of complex dynamical systems is a part of mathematics, not physics. Applications of it in physics, chemistry, biology, economics, finance, and the social sciences require particular initial and boundary conditions, frames and empirical interpretations of the basic formal concepts. The falsifiability of these models depend on the particular empirical conditions applied.

8 Complex Systems and the Evolution of Human Culture and Society

How can one explain the emergence of political, social, and cultural order in human societies? This chapter starts with a brief history of political systems from classical antiquity to the present. Political ideas have often grown from the technical, physical, or biological concepts that were preeminent at the time. In the seventeenth century, Thomas Hobbes tried to transfer the Galilean and Cartesian laws of movement from mechanics to anthropology and state theory. The liberal ideas of Locke and Hume were devised against a historical backdrop of Newtonian physics. Like many physicists, political thinkers, lawyers, and politicians have believed in a mechanistic world of linear causality (see Sect. 8.1). But what about the "butterfly effect", when small errors or local conflicts have global effects? How much should individuals or firms, cities or countries be blamed when their local failures lead to global catastrophes (e.g., in environmental policy)? The main point highlighted by the complex system approach is that from a macroscopic point of view, the development of political, social, or cultural order is not just the sum of single intentions, but the collective result of nonlinear interactions. In Sect. 8.2, examples of complex social and cultural problems are analyzed in the framework of complex dynamical systems: the growth of urban centers, global migration problems, and managemental issues associated with complex organizations.

One challenge for nonlinear dynamics is to address the evolution of complex communication networks in the age of globalization (Sect. 8.3). The flow of data traffic can be characterized by phase transitions and attractors. In order to manage an ever-increasing flood of increasing complexity, we need user-friendly methods of information retrieval and personalized information systems. In soft computing, information retrieval is supported by neural network and multiagent technology. However, the complexity of global networking refers to more than the growth of the Internet. Ubiquitous computing (Sect. 8.4) is an expansion of the global networking, wireless media access, wide-bandwidth-range, real-time capabilities of multimedia over standard networks, and data packet routing. The chapter finishes with a look at complex communication networks, which offer the prospect of a worldwide "global village," as well as the enslavement of mankind through a proliferation of modern, high-tech procedures.

8.1 From Aristotle's Polis to Hobbes' Leviathan

After the evolution of matter, life, mind–brain, and artificial intelligence we finish the book with the question of whether the evolution of human society can be at least partially described and modeled in the framework of complex systems. In the social sciences one usually strictly distinguishes between biological evolution and the history of human society. The reason is that the development of nations, markets, and cultures is assumed to be guided by the intentional behavior of humans, i.e., human decisions based on intentions and values.

From a microscopic view we may actually observe single individuals with their intentions, beliefs, and so on. But from a macroscopic view the development of nations, markets, and cultures is more than the sum of its parts. The emergence of political, social, and economical order seems to be caused by procedures of self-organization which remind us of certain phase transitions in complex systems. Nevertheless, avoiding any reductionist kind of naturalism or physicalism, we shall consider the characteristic intentional features of human societies. In Sects. 3.4 and 4.4, the evolution of animal populations was modeled in the framework of the complex system approach. Macroscopic structures like social order, organization of social behavior, the construction of nests, and so on were explained by attractors of complex systems. But the difference of complexity between animal populations and human societies is enormous, although there are common origins and features. Thus, in the following, concepts like "evolution" and "nature" cannot be restricted to the mechanisms of molecules, fishes, ants, and so on. They imply a new kind of complex dynamics, the analysis of which must consider the long tradition of social philosophy.

Plato and Aristotle were the first philosophers who tried to explain the emergence of political, social, and economic order of human societies. They analyzed the structure of the Greek *polis*, which has become the germ cell of Western societies and states. In Greek antiquity a *polis* (πόλις) like Athens was the republic of a small city, which later on may be compared with the Italian states of Florence and Venice in the Renaissance, and perhaps the Swiss cantons of cities in modern times. A Greek polis was a small, but politically and economically almost autonomous state and society [8.1]. The Greek philosophers suggested an idealized model which was more or less realized by the historical examples.

Plato distinguished several states of transformation which a polis must undergo before reaching the final goal of a harmonious society. In the first state the citizens must learn different skills and professions, business, and trades, in order to satisfy the different needs of the whole community. Plato believed that the people of a polis must specialize, according to their different gifts. The citizens must organize themselves for cooperative work. Plato assumed that the exchange of their products and services achieved an equilibrium of work and demand by spontaneous self-organization. This economic state of equilibrium is characterized by "just" prices.

But Plato's idyllic world of cooperative work is, of course, unstable. People try to pursue their interests and make profits. They are selfish, immodest, full of envy, and driven by passions. Thus, conflicts arise and political power must be organized to avoid the destruction of the polis. Plato suggested an aristocracy of the best and wisest men ("philosopher-kings") who should rule the state [8.2]. Their government is intended to keep the whole system with its conflicting fluctuations in a state of equilibrium. It is well known that Plato distrusted democracy, because in his opinion common people without philosophical education are not able to recognize the true idea of justice. Plato believed in an eternal hierarchy of ethical values behind the changing and transitory world of appearances. Thus, there is an objective measure of values which people have to be aware of, in order to avoid chaos and to keep the system in a state of harmony.

Obviously, Plato defends a centralized system of political power. In the language of system theory, there is a centralized processor controlling all actions and reactions of the system's elements. Like the Laplacean demon in the scientific world, there is Plato's political myth of an ideal, wise, and good politician leading the system to a harmonious equilibrium. In a small city like a Greek polis, Plato's aristocracy of the best "philosopher-kings" may be justified under certain critical circumstances. Nevertheless, the experiences of actual history have shown that even well educated and intelligent political leaders are not free of temptations to misuse their power. Nowadays, Plato's aristocracy of the best may be compared with the power of experts in knowledge based complex societies. But, under present conditions of highly developed information and computer technologies, Plato's myth of a wise and good politician can easily change to Orwell's horror-scenario of a "Big Brother" with omnipotent controlling power.

The second famous philosopher who referred to the Greek polis was Aristotle [8.3]. He assumed that humans are social beings by nature who want to survive. Furthermore, they are political beings because they want to live well and happily. Aristotle believed in an organic development of human society, driven by the social and political nature of its members. The social and political dynamics reach a final state of equilibrium when the social and political form of a polis is realized. Aristotle described the social and political dynamics as processes in nature.

However, a dynamical process of nature is not conceived as a causal movement of mechanics, but as an organic growth like a plant or animal, starting with the initial state of a seed and aiming at the final state of its complete form (compare Sect. 2.1). Thus, Aristotle's model of society is naturalistic in the sense that humans are driven by the impulses of their social and political nature. But only the human instinct of social organization for the purpose of survival is common with animals. Humans are distinguished by their political nature to achieve a just society. In a famous formulation, Aristotle said humans are defined as rational beings striving for truth in science and philosophy, and they are political beings striving for justice in society.

Justice means a natural state of completion if the society is arranged in its harmonic proportions in equilibrium, like the static balance of an Archimedean scale (Fig. 8.1). Thus, economic equilibrium in an Aristotelean society is measured by "just prices" according to the "natural" value of goods and services. Economics was a part of Aristotle's moral philosophy of justice and state. He distinguished between commutative justice (*iustitia commutativa*) in private exchange and affairs of citizens and distributive justice (*iustitia distributiva*) concerning the relationship

Fig. 8.1. Last Judgement of Osiris (Egypt: 2nd century B.C.)

of citizens and the state. The Aristotelean model of economic and political justice became the leading idea during the Middle Ages. Obviously, it accorded with the Aristotelean concept of nature in those days.

The mechanistic view of nature was founded by Galileo, Descartes, and others, leading to Newton's grand system of classical physics. In his famous book *Leviathan or the Matter, Form and Power of a Commonwealth, Ecclesiastical and Civil* (1651), Thomas Hobbes (1588–1678) projected a mechanistic model of modern society and state [8.4]. At the end of the Middle Ages and the beginning of modern times, Hobbes lived in a period of dramatic political changes. The traditional monarchy and aristocracy of the Middle Ages had lost their religious legitimacy. In bloody civil wars, European societies and states were falling into ruin and chaos. Scientifically, Hobbes was impressed by the new mechanical method of Galileo and its successes in physics. Thus, he tried to use this method in order to found a mechanistic model of modern society without the obsolete traditional metaphysics to compromize its legitimacy in science and politics.

According to Galilean mechanics, there is an analytic or resolving method for dividing a system ("body") into its separable elements and a synthetic method for composing and unifying the separated building blocks into the whole system again. In short, the whole is the sum of its parts. Obviously, Galileo described the crucial superposition principle founding the linearity of the mechanistic world view. Actually, a mechanical system like a clock can be divided into separated elements such cog wheels and other mechanical parts which together constitute its perfect functionality.

Hobbes tried to transfer the law of movement from mechanics to anthropology and state theory. Humans are assumed to be driven by affects and emotions like physical bodies by mechanical impulses. The main affect is the individual instinct

Fig. 8.2. The "Leviathan" on the title page of Hobbes' book

of self-preservation and survival. In the opinion of Hobbes, it is a natural right of man to pursue his instinct of survival, leading to force and violence against other humans. Thus, in Hobbes' natural state of human society, there is a permanent struggle of everyone against everyone (*bellum omnium contra omnes*) without any state of equilibrium.

On the other hand, humans with their complex necessities can only survive in society. Thus, their reason dictates a first natural law to search for peace. In order to realize the "law of peace", a second law is necessary, demanding a social contract. Hobbes suggested that in this social contract all citizens have to transfer their natural right of power to an absolute sovereign ("Leviathan") who is alone legitimated to apply political power and to rule the state. In modern words, Hobbes' social contract legitimates the state's monopoly of power, in order to keep society in an absolute equilibrium.

Hobbes defines the sovereign as the "sum of all individuals" who made the social contract. Obviously, this idea is an application of Galileo's mechanical principle of superposition or linearity. The title page of Hobbes' book (Fig. 8.2) shows the body of the Leviathan as a huge complex system of single individuals, illustrating Hobbes' political principle of linearity.

The "phase transition" from the natural state of chaos to political order and equilibrium is realized by a social contract of all citizens and in this sense by self-organization. But then, the final state of Leviathan is a centralized and deterministic system without any political "degree of freedom" for its citizens. Hobbes compared the economic circulation of goods and money with the circulation of blood, which was discovered by the English physician William Harvey. Income and expenditure were referred to the incoming and outcoming blood of the heart as the pumping machine of the whole circulation.

The famous English philosopher John Locke (1632–1704) influenced not only the epistemology and methodology of Newtonian physics but also the political theory of modern democracy and constitution. He asked why man is willing to give up his absolute freedom in the state of nature and to subject himself to the control of political power. Locke argued that the enjoyment of the property man has in the natural state is very unsafe and insecure, because everyone else wants to take it away from him in a state of unrestricted freedom. Thus, the state of nature is unstable and will transfer to an equilibrium of political forces. For Locke, the "phase transition" from the state of nature to a society with government is driven by men's intention to preserve their property.

However, government does not mean the unfree machinery of Leviathan. It is a state of balance ("equilibrium") with independent political powers like the legislature and executive. Since laws are made by a parliament as representative organ of the society, there is an essential feedback to the citizens, who only give up their natural freedom to the extent that the preservation of themselves and their property require: "And all this to be directed to no other end but the peace, safety, and public good of the people" [8.5]. Historically, Locke's ideas of democracy, division of power, property, and tolerance mainly influenced the American and French constitutions.

As in epistemology, the great Scottish philosopher David Hume (1711–1776) was both more critical and more exact in political theory than Locke. In epistemology he taught that human consciousness is governed by associations of sensations and feelings which may be reinforced or diminished by outer experiences (compare Sect. 4.1). Thus, there is no absolute truth even in Newtonian physics, but only a useful method with more or less probable results. Analogously, there is no eternal ethical value like justice determining human behavior. Ethical ideas can only be evaluated by their individual or public usefulness [8.6]. In general, political institutions are only legitimated by their usefulness in being accepted by a society or not. Thus, Hume became a forerunner of utilitarian ethics and political philosophy. His friend and Scottish fellow-countryman Adam Smith was probably inspired by his sceptical anthropology of self-interested behavior in human societies.

With the assault on the Bastille, the system of absolutism collapsed. A local event made global history. Immanuel Kant (1724–1804) celebrated the age of republican freedom as a new state of civil history. A free civil society was believed to lead to the balance that historical developments had been aiming for. In short: a free civil society was considered an attractor of history. The constitution of human rights was to guarantee the civil rights of people in a balance of liberté, égalité, and fraternité. Kant postulated his categorical imperative as a universal ethical principle that balanced the individual freedom of a person with the freedom of other people. However, Kant's well-balanced equilibrium state seemed to contrast with the conflicts and wars of the real historical world. Therefore, Georg Wilhelm Friedrich

Hegel (1770–1831) explained that history charted a dialectic development of systems [8.7]. Dialectics refers to phase transitions from being to becoming, which is now addressed by systems theory. Contradictions, tensions, and conflicts ("negation") lead to the emergence of new order ("qualitative change"). Karl Marx (1818–1883) tried to apply Hegel's dialectic method to politics and economics [8.8]. However, due to the physical views that predominated in the nineteenth century, he believed in deterministic historical laws. Thus, Marx and his comrades assumed that absolute control of political and economic processes is possible, and that such control could be utilized to guide history to the equilibrium state of communism. This was a dangerous error that was due to the assumption of linear models. The resulting reduction of social and economic complexity led to the collapse and implosion of communist states at the end of the twentieth century.

8.2 Complex Social and Cultural Systems

In the social sciences and humanities one usually distinguishes strictly between biological evolution and the history of human cultures. The main reason is that the development of nations and cultures is obviously guided by the intentional behavior of humans with their attitudes, emotions, plans, and ideals, while systems in biological evolution are assumed to be driven by unintended self-organization. From a microscopic view, we observe single human individuals with their intentions and desires. Even in biological systems like animal ecologies there are individuals with intentional behavior of some degree.

The crucial point of the complex system approach is that from a macroscopic point of view the development of political, social, or cultural order is not only the sum of single intentions, but the collective result of nonlinear interactions. Adam Smith already knew that the distribution of economic wealth and welfare does not result from the social good will of single bakers and butchers. Individual self-interested and egoistic intentions may be counteracting collective interests. Nevertheless, their (nonlinear) interactions achieve collective equilibria by an "invisible hand" (Smith) or the "cunning of reason" (*List der Vernunft*) (Hegel).

Nonlinear systems of individuals with intentional behavior may be more complex than, for instance, a physical system of atoms or a chemical mixture of molecules. In Sect. 4.3–4.4, intentional behavior and consciousness were modeled as self-referential global states of complex neural systems which are caused by nonlinear interactions of neurons. The emergence of collective order phenomena with more or less complex degree is an intrinsic common feature of all nonlinear systems which is not necessarily linked with consciousness. A political state as a collective order of a human society obviously has no kind of consciousness or intelligence, as Hegel mistakenly believed, although its emergence may be modeled by a kind of phase transition caused by the nonlinear interaction of conscious humans with intentional behavior.

Thus, in the mathematical framework of complex systems the concept of "evolution" does not in general refer to the particular mechanisms of biological evolution. In a complex system, the so-called evolution equations describe the dynamics of its elements, which may be elementary particles, atoms, molecules, organisms, humans, firms, and so on. Another concept with a broad meaning is complexity itself. In the context of the social sciences there are many aspects of complexity, some of which are designed in Fig. 8.3 [8.9].

In the mathematical framework of complex systems in this book, complexity is at first defined as nonlinearity, which is a necessary but not sufficient condition of chaos and self-organization. On the other hand, linearity implies the superposition principle, which says in a popular slogan "the whole is only the sum of its parts". A second important aspect of complexity is defined by the structure of algorithms which was discussed in Sect. 5.2. Complexity theory in computer science provides a hierarchy of complexity degrees, depending on, for instance, the computational time of computer programs or algorithms. As nonlinear complex systems are sometimes modeled by computer graphics, their degree of algorithmic complexity may be described as their capacity for self-organization. This relationship has been explored in the theory of cellular automata (compare Sect. 5.6), in which different classes of self-organizing complex systems are modeled.

In the social sciences, the complexity of a highly industrialized society mainly consists in the great number of its citizens and their relationships, its organizational substructures, and their dependencies [8.10]. We should remember that in a complex system it is not the great number of its elements that is essential for the emergence of collective (synergetic) order phenomena, but their nonlinear interactions. The reader may recall the astronomical three-body problem with chaotic trajectories as possible solutions.

In the mathematical framework of complex systems, a physical or biological reductionism of human history and sociocultural development is not justified in any case. Models of social and cultural developments must be discussed with their par-

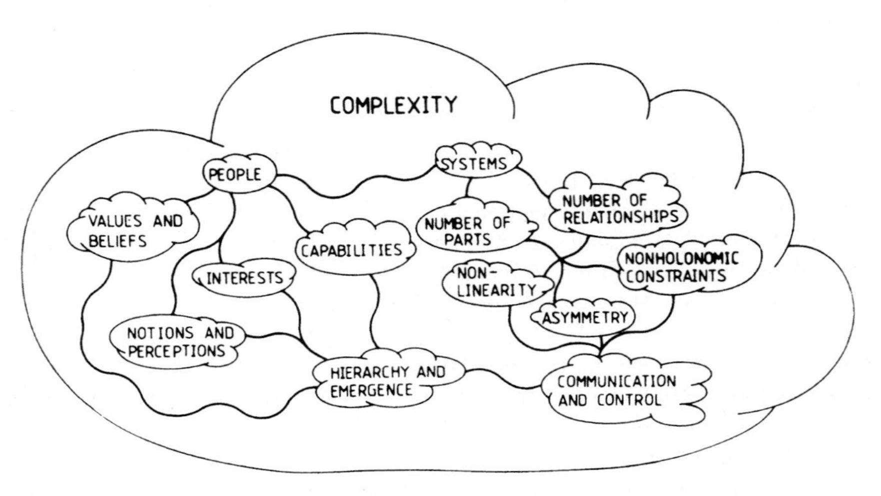

Fig. 8.3. Meanings of complexity [8.9]

ticular constraints and limitations. A severe methodological problem is that of providing empirical tests and confirmation of these models. Thus, the computer-assisted simulation of complex cultural systems has become a crucial instrument for providing new insights into their dynamics which may be helpful for our decisions and actions.

Historically, the interest of social sciences in nonlinear problems dates back to Thomas Malthus [8.11]. He argued that a population would outstrip its food supplies, since the population grows exponentially whereas the food supplies grow only linearly. In 1844, Verhulst modified the exponential equation, arguing that the rate of population growth is proportional to the product of the population and the difference between the total amount of resources and the amount of those resources consumed by the existing population. His famous logistic curve with its characteristical attractor of equilibrium has been used in demography, economics, and many other examples of social sciences. It provides a cascade of possible bifurcations and phase transitions including chaos.

The evolution of a predator–prey ecosystem described by Volterra and Lotka is another model which has been applied to social sciences. For instance, the Lotka–Volterra model helps us to understand the appearance of an agricultural society. Because of their ability to learn, humans can change their coefficients of interaction with the environment faster than nature can genetically evolve countermeasures. A human society which only tries to improve its hunting abilities for its survival diminishes the prey population. Then the society will be diminished, too. Consequently, both predator and prey population will become extinct. But agriculture involves developing the ability to increase the prey birthrate. Thus, human population increases and can be stabilized at a certain equilibrium.

The evolution of biological systems is governed by their genes. In Darwinian evolution a new type of individual emerges by the natural selection of mutants which appear spontaneously. In populations of higher animals the new possibility of behavioral change and adaption by imitation arises. In human societies, there are still more sophisticated strategies of learning, which dominate the behavioral patterns. Societies have developed particular institutions like the legal system, the state, religion, trade, and so on to stabilize the behavioral change for the following generations.

The complex system approach provides the essential insight that neither genetic evolution nor the evolution of behavior needs a global program like a supervising divine will, a vital force, or a global strategy of evolutionary optimization. The survival of genes or the emergence of global patterns of behavior are explained by local interactions of individuals composing the system. To make this point clear, it may be a question of religious or political *Weltanschauung* whether there is a "global program" like God, History or Evolution. In the methodological framework of complex systems, these hypotheses are not necessary for explanations and superfluous in the sense of Ockham's razor and his economy of theoretical concepts.

Obviously, nonlinear systems like biological organisms, animal populations, or human societies have evolved to become more and more complex. Our present society, when compared to Aristotle's *polis* or the political system of the physiocrats, is characterized by a high degree of institutional complexity and information networking. In the last century, Herbert Spencer already maintained that increasing complexity is the hallmark of evolution in general: "Evolution is an increase in complexity of structure and function ... incidental to the ... process of equilibration ..." [8.12]. Spencer still argued in the thermodynamic framework of thermal equilibrium.

In the framework of thermodynamics far from thermal equilibrium, there is not just a single fixed point of equilibrium, but a hierarchy of more or less complex attractors, beginning with fixed points and ending with the fractal structures of strange attractors. Thus, there is no fixed limit of complexity, either in biological or in sociocultural evolution, but there are more or less complex attractors representing metastable equilibria of certain phase transitions which may be overcome if certain threshold parameters are actualized. The structural stability of a society is related to these more or less complex attractors.

The traditional functionalist view of homeostasis and self-regulating systems stems from the concept of a technical thermostat. It may perhaps help us to understand why societies stay the same, but not why they change and why equilibria have been overthrown. In the framework of complex systems, the dynamics of a society is understood in terms of phase transitions of a dissipative system exchanging material, energy, and information with its environment. The institutions of a society are dissipative structures which may emerge and stay the same in a particular interval of threshold conditions. For instance, in neolithic villages, the institution of farming changed from dry farming to irrigation when the food supply was no longer secured by the established social structures.

In the history of industrialized societies, we can distinguish more or less strong economic fluctuations which may initiate the crash of social institutions and the emergence of new ones. For instance, the economic depression of the USA in 1922 was relatively mild and short-lived and did not produce a structural change of the society. In contrast to that phase of American history, the stock market crash of 1929 had a genuine butterfly effect in initiating the Great Depression of 1933 [8.13]. This crisis caused the financial ruin of many firms and a huge number of unemployed, and could not be managed by the established social institutions. The threshold parameters of the established structures were exceeded. New social institutions like the Securities and Exchange Commission, the Federal Deposit Insurance Corporation, and the Public Works Administration came into being to overcome the effects of depression and to prevent excessive fluctuations in the future business cycle. This Keynesian reaction by American society became famous as the New Deal legislation of President Roosevelt.

But as we have learnt from the neoclassic economists and the experience of social development after the Second World War, an optimizing strategy of public welfare may initiate a self-dynamics of administrative bureaucracy which paralyses economic initiatives and counteracts the originally good intentions. Overreactions may be as dangerous as no reactions for the structural stability of a system. On the other hand, the history of political revolutions shows that societies can totally lose their structural stability and achieve new constitutions, institutions, and social structures, but, of course, without any guarantee of durability.

From a methodological point of view, the question arises of how to represent the sociocultural evolution of societies in the mathematical framework of complex systems. The recognition of attractors and equilibria needs a phase portrait of the sociocultural dynamics which presumes the definition of a "sociocultural state" and a "sociocultural state space". But what is the sociocultural state space of Victorian England or the Weimar Republic? These questions demonstrate obvious limitations. What are the possibilities of the complex system approach in the historical and social sciences?

It cannot be the aim of research to represent a historical period in a complete mathematical state space. The relevant data are often unknown, contingent, and not quantified. In the last section, complex systems with state spaces and dynamical phase portraits were used to model the economic evolution of human societies. Economists do not claim to represent the complete economic development of the Weimar Republic, for instance. But nonlinear endogenous or linear exogenous models of business cycles can describe typical economic scenarios that influenced or depended on situations in political and cultural history.

Economic models are not studied for their own sake. Economists want to understand economic dynamics in order to support practical decisions by better structural insights. The economic dynamics of a society is embedded in its global sociocultural development. In view of its complexity, attempts have been made to model the sociocultural development only of such subsystems as urban centers. These models may grasp the typical features of evolving urban systems, which can help politicians and citizens to make better decisions in appropriate stituations.

In modern industrialized societies, there is a great variety of centers of all sizes, forms, and characters, from very large cities with high densities to small villages with few inhabitants. We may ask what is the reason for the spatial distribution of these different centers and what will be their evolution in time. To answer this, we need to know the global spatio-temporal state of an urban system resulting from local nonlinear interactions of its agents – individuals, families, firms, administrators, and so on – who may pursue different cooperative or conflicting interests. The structure of an urban center depends on commercial and industrial interests, the flows of goods and services, traffic connections, cultural attractiveness, and ecological demands. These factors are to be made precise as measurable quantities. The urban system has several exchanges with the outside world. Thus, it can be interpreted as a dissipative structure modeled by a complex dynamical system.

Peter Allen has suggested a system with evolution equations expressing the nonlinear interactions of its different actors. The spatio-temporal structure of the urban system with its changing centers and concentrations of inhabitants is not the trival sum of its actors. It is not the result of some global optimizer or some collective utility function, but the result of the instability of successive equilibria caused by nonlinear phase transitions. In this sense, the evolution of an urban system is not ruled by a Platonic king (or dictator), not constructed by a Cartesian architect or forecast by a Laplacean spirit. In the mathematical framework of complex systems the form of an urban system seems to grow like a living organism.

In Allen's analysis [8.14], the geographic space of the urban system is simulated by a triangular lattice of 50 local points (Fig. 8.4). The growth of the urban

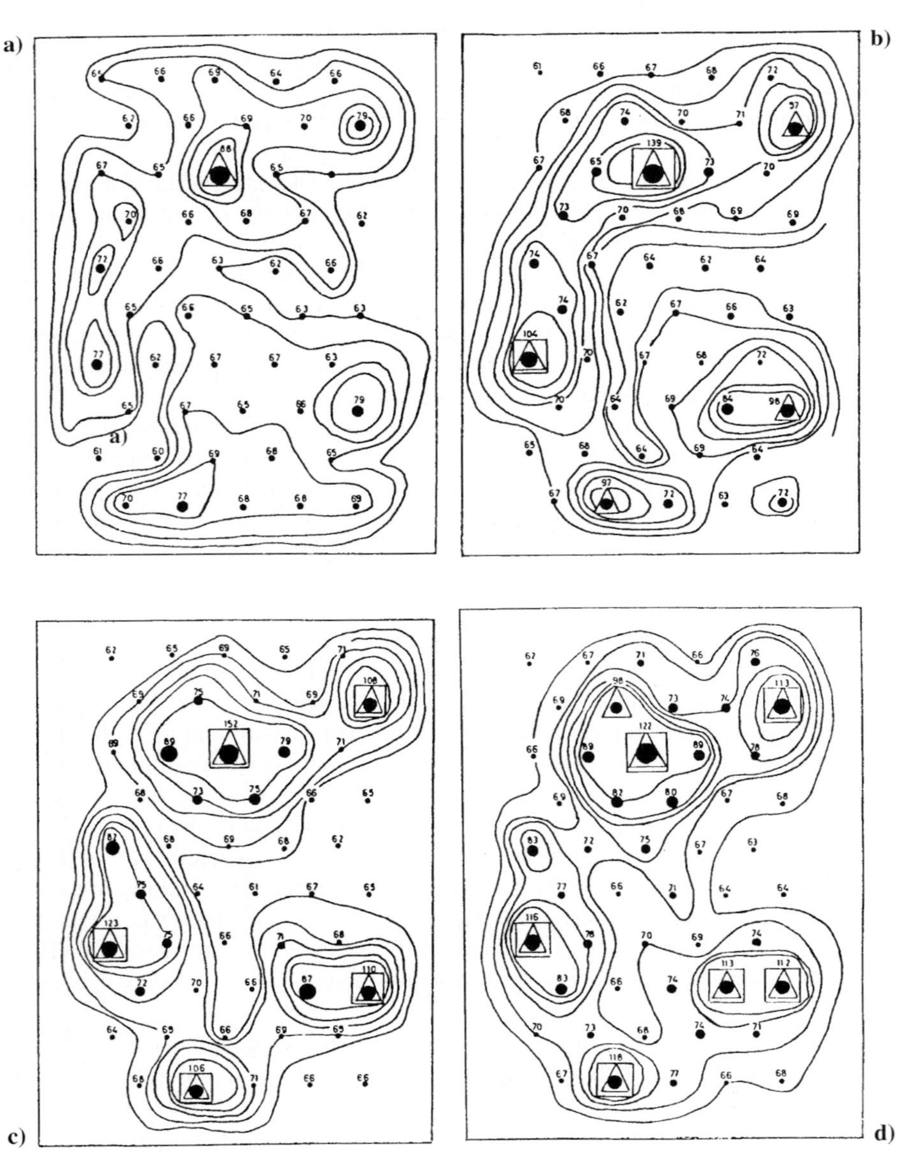

Fig. 8.4a–e. Computer-assisted model of urban evolution at time (**a**) $t = 4$, (**b**) $t = 12$, (**c**) $t = 20$, (**d**) $t = 34$, (**e**) $t = 46$ [8.14]

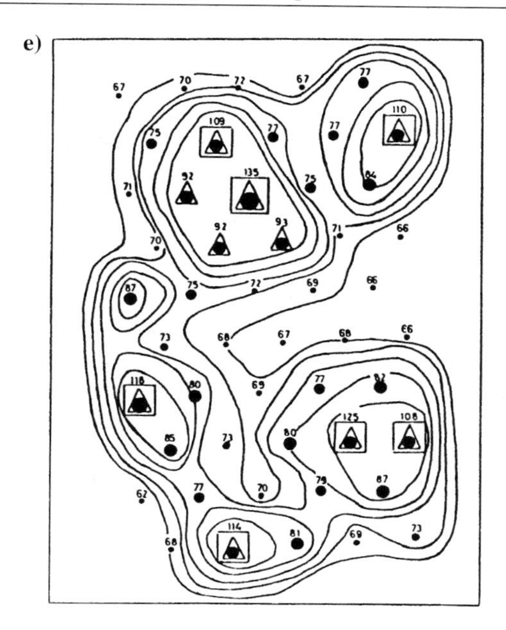

system is determined by two equations describing the change of the population on the local points and the evolution of the employment offered at the points. The local population and the local capacity for employment are linked by the urban multiplier as a positive feedback. The employment concentration offers externalities and common infrastructure which in turn gives rise to a positive feedback, while the residents and entrepreneurs together compete for space in a center that provides negative feedback.

The computer-drawn pictures Fig. 8.4a–e show the evolution of the population distribution of a region that starts off initially an area with no interaction between local centers. The urbanization process is revealed in phase transitions with changing local attractors. In Fig. 8.4b at time $t = 12$ units, the structure is beginning to solidify around five main centers. In Fig. 8.4c, the central core of the largest center is going through a maximum. In Fig. 8.4d at time $t = 34$, the basic structure is essentially stable. Two centers have undergone central core decay. In Fig. 8.4e, the basic pattern is stable. The decline, centralization, and decentralization result from the complex nonlinear dynamics.

Figure 8.4a–e form a speeded-up motion picture of the urban system's global evolution. Each picture is a phase portrait of a global dynamical state at a particular time. The model is simplified, of course. But it can be enlarged by integrating more functional aspects and by studying their more complicated nonlinear interactions. Nevertheless, the potential of the model for the exploration of decision alternatives can be analyzed in computer-simulated case studies. Either local or global changes can be imposed on the system. These simulations are extremely interesting for a government.

A possible strategy may be to interfere in the urban structure by placing a specific investment at a particular point. This decision strategy corresponds to the development of a hitherto undeveloped region in an urban system. Investments mean not only economic measures but also cultural attractions and traffic connections, for instance. Sometimes, an investment may intiate a local butterfly effect with global consequences that counteract the good intentions of the urban planners. This may occur because of the model's nonlinearity, which limits the possibilities of forecasting in the long run.

An urban dynamics is a practical example of a complex system demonstrating that the good will of single individuals is not enough and may even be dangerous if the nonlinear consequences are neglected. The collective effects of individual actions characterize our society. Decisions should be made, as far as possible, with an awareness of these collective effects. The importance of these results lies not only in computer simulations of concrete decisions and their nonlinear effects. Even citizens who are not engaged in concrete planning activities must be aware of the complex interdependencies in a society.

The demagogic demand for a strong political leader who can solve all problems is not only dangerous from a democratic point of view. It is false for mathematical reasons due to the complexity characterizing modern highly industrialized societies. On the other hand, we should not hold out high hopes for single politicians or parties and then react with total political frustration when our exaggerated expectations are disappointed. Human societies are characterized by the intentionality of their members. Nevertheless, they are governed by the nonlinear laws of complexity like atomic groups, molecular mixtures, cellular organisms, or ecological populations.

Sociological theories considering the epistemological consequences of complexity and nonlinearity are still in their infancy. The development of an appropriate statistical mathematics capable of handling the complexity of social problems could serve as a bridge to traditional concepts of sociology. In the complex system approach, social phenomena are described by nonlinear equations. When, for instance, Emil Durkheim [8.15] speaks of solidarity in society, then we may ascribe the functional aspects of this concept to nonlinear and collective effects of complex systems. We can distinguish political decisions as "linear", corresponding to "individual" choices, and "nonlinear", corresponding to the institutional environment of administrations, mass media, political parties, for instance. The actions and reactions of many citizens and institutions may be understood as fluctuations inherent to the statistical description of the society. The deterministic character of the society reflects only the averages of the distribution functions, which develop in time according to nonlinear laws like a master equation.

This approach to socio-economic dynamics has been realized by the Stuttgart school of Wolfgang Weidlich. The mathematical modeling methods are derived from synergetics and statistical physics, and allow a quantitative description of collective developments in the society. Synergetics suggests a relationship between the micro-level of individual decisions and the macro-level of dynamical collective processes in a society. The distinction between micro- and macro-economics or micro-

and macro-sociology are well known traditional concepts in the social sciences. The synergetic approach of Weidlich aims at a probabilistic description of macro-processes including stochastic fluctuations and derivation of a quasi-deterministic description neglecting the fluctuations.

Models of solutions can be investigated either by analytical methods, for instance, the exact or approximate solution of master equations or average equations, or by numerical and computer-assisted simulations of characteristic scenarios. Empirical systems can be analyzed and evaluated by determining the model parameters with field research on the micro-level or by estimating future developments with model simulation. The methodological framework of the synergetic approach to modeling social dynamics is illustrated in Fig. 8.6 [8.16].

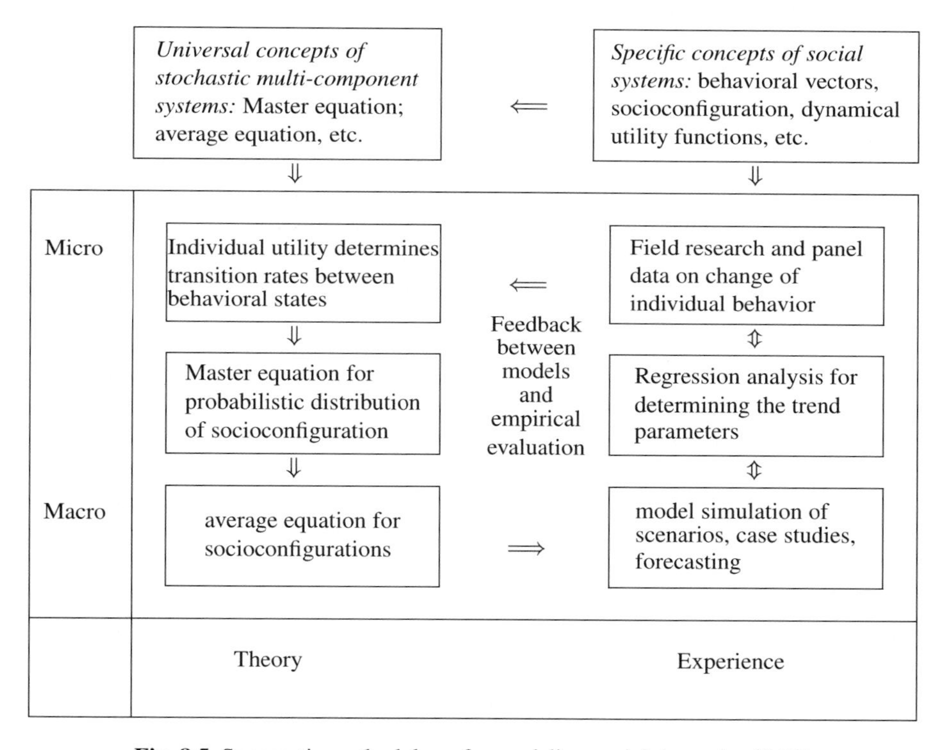

Fig. 8.5. Synergetic methodology for modeling social dynamics [8.16]

An example of application are urban systems, which can be considered evolving dissipative systems, in the sense of Allen and Prigogine, as well as stochastic systems, in the sense of synergetics. The macrovariables are material ones (compare Sect. 7.2), characterizing the state of a city. Human influence is not represented directly by collective personal macrovariables. The activities of municipal authorities, house builders, architects, and other decision makers are represented indirectly in the evolution of the states of an urban system. Later on, they show up in the different styles and fashions of suburbs, streets, and parts of a city. On a detailed level, the state of a city can be characterized by the number, location, and distribution of the different kinds of buildings fulfilling different purposes. The area of a city is mapped into a lattice of squares with discrete coordinates. We distinguish various kinds of buildings, such as lodgings, factories, schools, stores, houses, parks, etc. There are numbers of building units of a certain kind on the squares of the lattice. For example, there are macrovariables of lodgings or factories on the squares of the lattice. The set of numbers of building units of all kinds on all squares is denoted as city configuration.

In order to apply synergetics, the transition rates of city configurations to neighboring configurations must be set up. An elementary change of a city configuration consists of the increase or decrease of one macrovariable (e.g., the number of lodgings on a certain square) by one appropriately chosen unit. The transition rates of neighboring configurations describe the probability per unit of time in which the respective transition takes place if the initial configuration is realized. Transition rates are connected with utility functions measuring the differences of utilities of a city configuration before and after the transition step. For example, it may be unfavorable to build lodgings and factories in a near neighborhood, and that it leads to a high utility to have them at a certain distance on the lattice of area. According to synergetics, the transition rates can be used to set up the master equation for the probability distribution $P(x, y, t)$ over the city configurations of, e.g., macrovariables x of lodgings and macrovariables y of factories at time t. Again, in general, it is not possible to exhaust all of the information contained in the probability distribution of a master equation by comparison with empirical data. Then it is indicated to neglect the fluctuations of the macrovariables around their mean evolution and to restrict to quasi-meanvalue equations of, e.g., the macrovariables x of lodgings and y of factories derivable from the master equation. The evolution of these equations leads to several possible stationary states, depending on differing initial conditions. In corresponding computer simulations, the stationary states of city configurations are represented by different distributions of residential and industrial districts within an urban area.

The synergetic concept of modeling has been applied to several questions in the social sciences, for instance, collective formation of political opinions, demography, migration of populations, and regional geography. The synergetic concept is especially appropriate for integrating the interactions of several sectors of a society like, such as the relationship between the economy and collective formation of political opinions, or the interaction between the economy and processes of migration. Migration is a very dramatic topic nowadays, and demonstrates how dangerous linear and mono-causal thinking may be. It is not sufficient to have good individual intentions without considering the nonlinear effects of single decisions. Linear thinking and acting may provoke global chaos, although we act locally with the best intentions.

According to the synergetic approach, a socio-economic system is characterized on two levels, distinguishing the micro-aspect of individual decisions and the macro-aspect of collective dynamical processes in a society. The probabilistic macro-processes with stochastic fluctuations are described by the master equation of human socioconfigurations. Each components of a socioconfiguration refers to a subpopulation with a characteristic vector of behavior. Concerning the migration of populations, the behavior and the decisions to rest or to leave a region can be identified with the spatial distribution of populations and their change. Thus, the dynamics of the model allows us to describe the phase transitions between different global macro-states of the populations.

The empirical administrative data can be used to test the theory. The models may concern regional migration within a country, motivated by different economic and urban developments, or even the dramatic worldwide migration between the "South" and the "North", between the poor countries and the highly industrialized states of Western Europe and the USA, driven by political and economic depression. Physical transport or migration of animal populations are often uncontrolled, random, and linear, without interaction between the elements and collective aggregations. But human migration is intentional (driven by considerations of utilities) and nonlinear, because the transition rates depend not linearly on the whole socioconfiguration.

The migration interaction of two human populations may cause several synergetic macro-phenomena, such as the emergence of a stable mixture, the emergence of two separated, but stable ghettos, or the emergence of a restless migration process. In numerical simulations and phase portraits of the migration dynamics, the synergetic macro-phenomena can be identified with corresponding attractors. Figure 8.6a,b shows the homogenous mixture of both populations with weak trends of agglomeration and segregation in each population. Figure 8.6a is the phase portrait of the average equations with a stable point of equilibrium. Figure 8.6b shows the stationary solution of the master equation and a stationary probabilistic distribution with a maximum at the origin. Figure 8.7a,b shows the emergence of two stable ghettos with a weak trend of agglomeration and a strong trend of segregation between the populations. Figure 8.7a is a phase portrait with two stationary fixed points, while Fig. 8.7b describes the probabilistic distribution with maxima at the stationary points.

Figure 8.8a,b shows a moderate trend of agglomeration in each population and a strong asymmetric interaction between the populations. Figure 8.8a illustrates a vortex picture and Fig. 8.8b the corresponding probabilistic distribution with a maximum at the origin. Figure 8.9a,b corresponds to a restless migration process with a strong trend of agglomeration in each population and a strong asymmetric interaction between the populations. The phase portrait 8.9a shows a limit cycle with unstable origin. The stationary probabilistic distribution of Fig. 8.9b has four maximum values with connecting ridges along the limit cycle. Sociologically, this case may be interpreted as sequential erosion of regions by asymmetric invasion and emigration of the populations.

If we consider three instead of two populations in three regions, then the phenomenon of deterministic chaos emerges in the nonlinear model of migration. Some numerical simulations have a strange attractor as the final state of trajectories. In other cases, a successive bifurcation becomes more and more complex with a final transition to chaos.

Another practical application of the complex system approach is management and organizational sociology [8.17]. Actually, modern firms have begun to reorganize and decentralize their large organizations, in order to enable successful strategies in spite of the increasing complexity of their problems. They have started to support organizational fluidity, for instance, with new pathways that allow project-centered groups to form rapidly and reconfigure as circumstances demand. Fluid organizations display higher levels of cooperation than groups with a fixed social structure. Faced with a social dilemma, fluid organizations show a huge variety of complex cooperative behaviors caused by the nonlinear interplay between individual strategies and structural changes.

Thus, the dynamics of these social groups may by modeled by complex systems. Computer simulations can deliver global insights into the development of

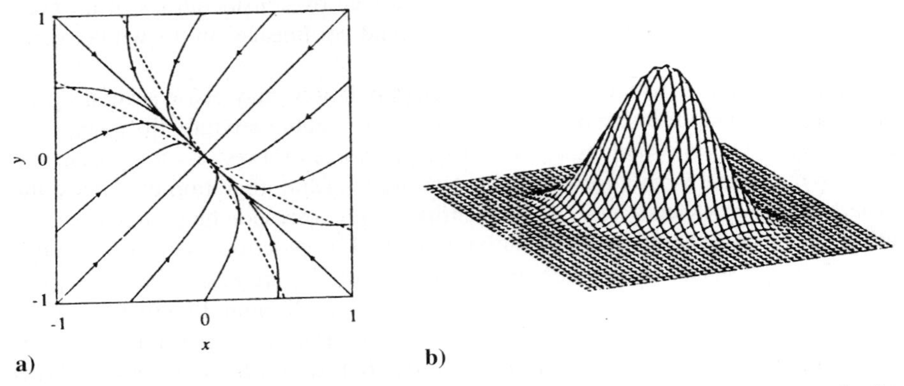

Fig. 8.6a,b. Migration dynamics with a stable point of equilibrium: (**a**) phase portrait, (**b**) probabilistic distribution [8.16]

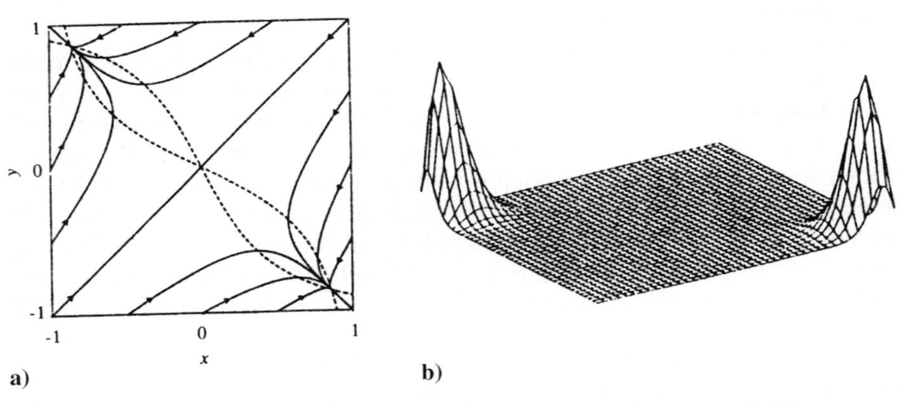

Fig. 8.7a,b. Migration dynamics with two stationary fixed points [8.16]

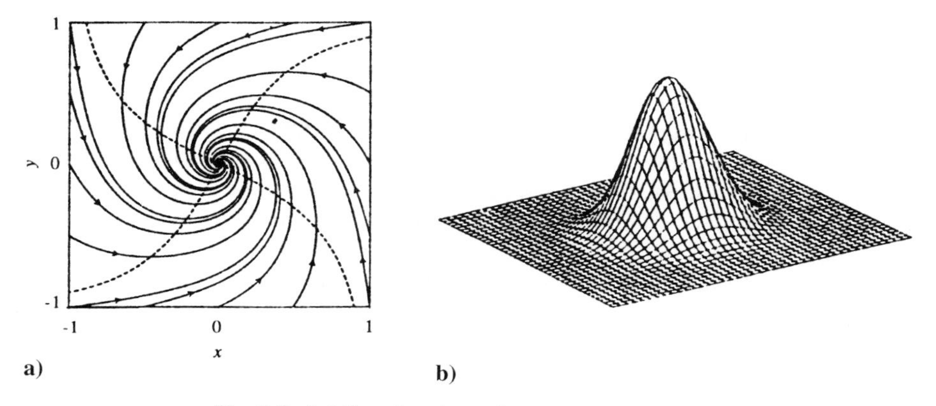

Fig. 8.8a,b. Migration dynamics with a vortex [8.16]

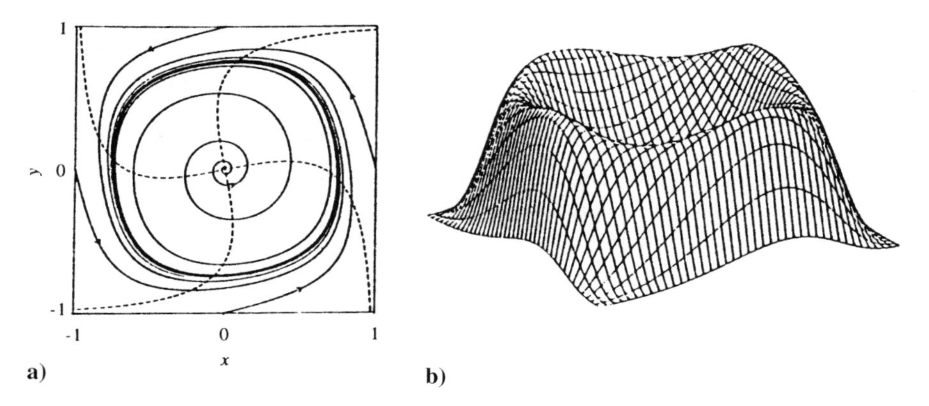

Fig. 8.9a,b. Migration dynamics with a limit cycle and instable origin [8.16]

possible behavioral patterns which may help managers to achieve appropriate conditions of development. If models of complex systems are applicable, then forecasting in the long run and overall control by a centralized supervisor is, of course, excluded.

The model consists of intentional agents making choices that depend on their individual preferences, expectations, and beliefs as well as upon incomplete knowledge of the past. Patterns of cooperation merge from individual choices at certain threshold values. An agent will cooperate when the fraction of the group perceived as cooperating exceeds a critical threshold. Critical thresholds depend on the group size and the social organizational structure emerging from the pattern of interdependencies among individuals. The potential for cooperative solutions of social dilemmas increases if groups are allowed to change their social structure. The advantages of organizational fluidity must be balanced against possible loss of effectiveness. The effectiveness of an organization may be measured by its capability to obtain an overall utility over time.

In firms, there is generally an informal structure emerging from the pattern of affective ties among the agents as well as a formal one dictated by the hierarchy. The informal structure is achieved by a kind of self-organization and can be represented by the sociometric structure of a social network. This approach dates back to early sociological explorations on social networks of urban families in the 1950s and has developed to a sophisticated computer-assisted instrument of sociology [8.18]. From the micro-view of individual interdependencies a global perspective on social structures emerges.

In Figs. 8.10 and 8.11, these structures are visualized as trees. Each branch represents a subdivision of the higher level in the hierarchy. The modes at the lowest level represent individuals, with filled circles marking cooperators and open circles marking defectors. The number of organizational layers which separate two individuals is determined by the number of modes backwards up the tree from each individual until a common ancestor is reached. The distance between two agents in the organization is measured by the number of separating layers. The larger the distance between two agents, the less their actions are assumed to affect each other. Thus the tree illustrates the amount and extent of clustering in a group.

The crucial question is how the structure and fluidity of a group will affect the dynamics of cooperation. Fluidity depends on how easily individuals can move within the social structure, and how easily they can break away on their own, extending the structure. In the framework of complex systems, macroscopic properties of the system are derived from the underlying interactions among the constituents, which are mathematically modeled by nonlinear evolution equations [8.19].

Figure 8.10a–d, due to Glance and Huberman, shows a computer simulation of some phase transitions in a fixed social structure, a three-level hierarchy consisting of three large clusters, each subsuming three clusters of three agents [8.20]. The final overall cooperation of Fig. 8.10d has been initiated by the actions of a few agents clustered together in Fig. 8.10a. These agents reinforce each other and at the same time can spur agents one level further removed from them to begin cooperating. This increase of cooperation can affect cooperation in agents even further removed in the structure.

A tiny act of cooperation within a hierarchy can initiate a widespread transition to cooperation within the entire organization. This cascade of increasing cooperation leads to a fixed point of equilibrium. But groups with fixed structures easily grow beyond the bounds within which cooperation is sustainable. In this case, the group rapidly evolves into its equilibrium state of overall defection. But even in these bounds, cooperation patterns may be metastable in the sense that the agents remain cooperating for a long time, until suddenly a symmetry breaking takes place with a transition to overall defection.

In fluid structures, individual agents are able to move within the organization. Individuals make their decision to cooperate or defect according to the long-term benefit they expect to obtain. In order to evaluate his or her position in the structure, an individual compares the long-term payoff he or she expects if he or she stays with the long-term payoff expected if he or she moves to another location, chosen randomly. When evaluating his or her position the individual also considers the

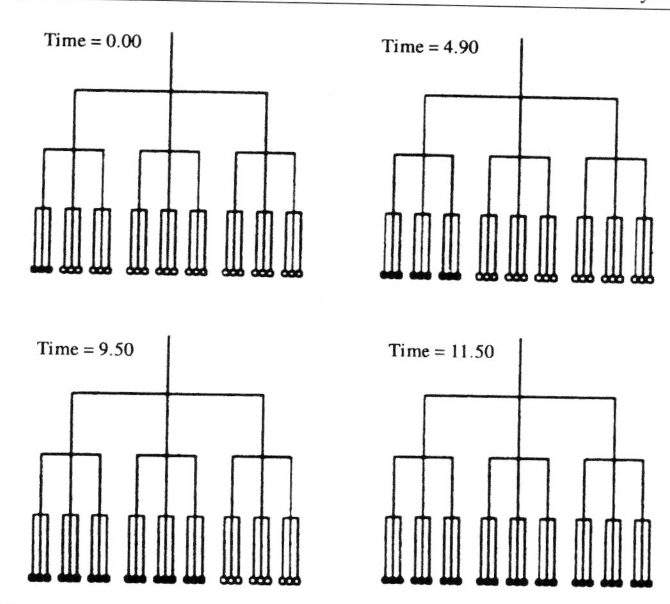

Fig. 8.10. Social dynamics in a fixed three-level hierarchy: the nodes at the lowest level of each tree represent individuals, with filled circles marking cooperators and open circles marking defectors [8.20]

possibility of breaking away to form a new cluster if he or she feels there is nothing to lose. How easily agents are tempted to break away determines the breakaway threshold, which is given as a fraction of the maximum possible payoff over time.

Figure 8.11a–e shows snapshots of a phase transition in a fluid organization [8.21]. Initially (Fig. 8.11a) all members of the group, divided into four clusters of four agents each, are defecting. Figure 8.11b shows that nearly all the agents have broken away on their own. In this situation, agents are much more likely to switch to a cooperative strategy, as realized in Fig. 8.11c. Because of uncertainty, agents will occasionally switch between clusters (Fig. 8.11d). When a cluster becomes too large, a transition to defection may start within that cluster. At this phase of transition (Fig. 8.11e), more and more agents will break away on their own, and a similar development is repeated. Cycles of this type have been observed to appear frequently in simulated organizations.

As in the case of urban growth (Figs. 8.4) or migration dynamics (Figs. 8.6–8.9), computer experiments simulating social organizations cannot deliver deterministic forecasts of individual behavior, but they help one to understand the sensitivity and complexity of social dynamics. Thereafter, appropriate circumstances and conditions may be achieved that allow the living conditions of humans in the corresponding social systems to be improved.

Models of sociocultural evolution must consider several interacting sections of a society. If a society is composed of overlapping layers and sectors of dissipative structures, we must find an appropriate picture of how these come about. In com-

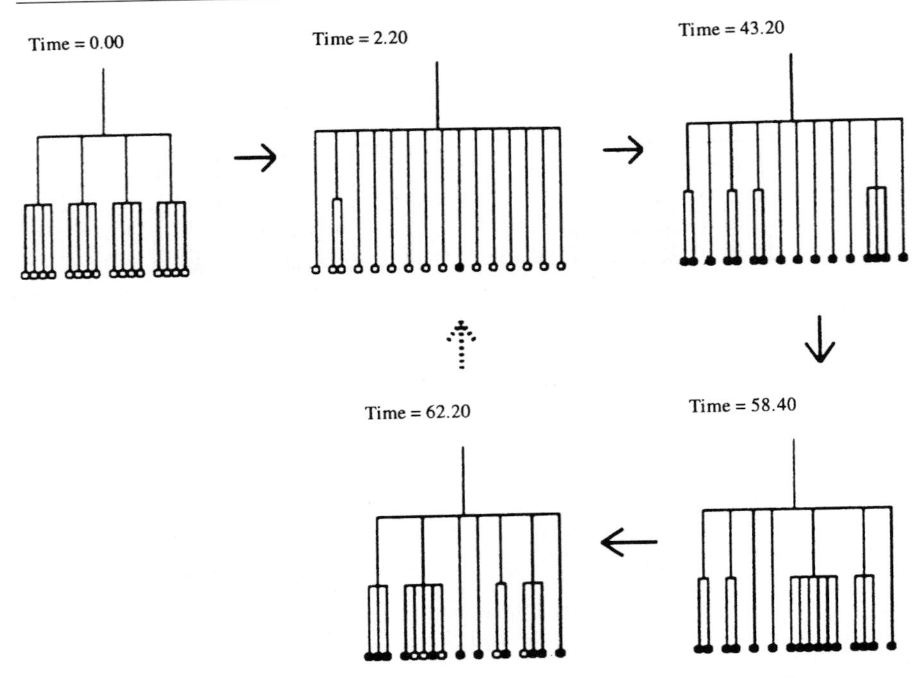

Fig. 8.11. Social dynamics in a hierarchy with fluid structure [8.21]

plex developments involving interacting dissipative systems, the emergence of new macro-structures is an uncentralized and unplanned event in the long run. Moreover, each section includes many more or less unclear human ideas, feelings, and projections, motivating and inspiring human actions. They cannot be directly identified as measurable quantities, because they are dissolved in a kind of mainstream which is called "lifestyle". Nevertheless, the lifestyle of a society is a typical sociocultural macro-phenomenon depending on several interacting factors which can be identified, like conditions relating to the economy, technology, work, travel, ecology, and the mass media.

In modern times, the evolution of technologies has been a driving force for change, influencing numerous lifestyle components. With respect to self-organizing processes, it is a significant feature of technological developments that they are autocatalytic, each innovation catalyzing the generation of the next. If technological and social evolution is interpreted as a consequence of sequential replacements of a technique, a mainstream idea ("paradigm"), or an artifact by another, then the development from growth to saturation can be mathematically modeled by a pattern of interdependent logistic curves. The assertion that a technology advances through a series of phase transitions and substitutions implies that it might be considered as a succession of logistic curves. Each curve reaches a saturation level. With each level of evolutionary innovation a transition to a new logistic curve develops.

In Chap. 7, we have already discussed that these phase transitions in technology are connected with the growth and decline of economies. The development of computer and information technologies has influenced the lifestyle of mankind in nearly all sectors as never before in history. It seems to be comparable with a quasi-evolutionary process producing computers and information systems of increasing complexity. Computer scientists use the terminology of Herbert Spencer when they speak of computer generations overwhelming the complexity of earlier generations. Actually, the complexity of the systems' functionality has increased. But on the other hand, the complexity of problems, which is measured by their computational time, for instance, has been reduced. The reduction of problem complexity is an order parameter of the quasi-evolutionary process of these technologies.

Computers and informational systems have become crucial techniques in sociocultural development, evolving in a quasi-evolutionary process. The replicators of this process are any of the information patterns that make up a culture and spread with variation from human to human. As humans, unlike molecules or primitive organisms, have their own intentionality, the spreading process of information patterns is realized not via mechanical imitation but via communication. By analogy with genes, these replicators are sometimes called "memes" [8.22]. They include ideas, beliefs, habits, morals, fashions, techniques, and so on.

Any pattern which can spread via communication of information is a meme, even if its human host cannot articulate it or is unaware of its existence. It is important to recognize that the replicators of human culture are memes, not people. Our ability to change our minds allows cultural evolution to proceed not by selection of humans, but by "letting our theories die in our stead", as Karl Popper has proclaimed [8.23].

In the framework of complex systems, we may, of course, speak of the system's "evolution" in the abstract sense of mathematical evolution equations. Biological evolution with its special biochemical mechanisms is only a special model of the general mathematical scheme characterizing complex systems. Thus, the evolutionary character of human culture cannot be reduced to the biochemical mechanisms of biological evolution. But concepts like "memes" should not be misunderstood as a merely social-Darwinistic jargon. They may illustrate basic features of complex systems which can be mathematically defined and empirically tested.

In this sense, the development of a worldwide communication network can be interpreted as the evolution of complex systems for aiding the spread of memes among humans and for establishing a memetic ecosystem [8.24]. The memes making up human culture are diverse, as are their variation and selection mechanisms. Economic markets are more or less open to their own environment of human society. They operate under a wide range of more or less rigorously enforced rules, imposed by a variety of legal and regulatory institutions.

In human societies, legal systems and governmental activities provide a framework for the market. In the framework of complex systems, they are not protected from evolutionary forces. They evolve within a political ecosystem with its own mechanisms for the variation and selection of laws [8.25]. Some political memes, like political desires, slogans, or programs, may become attractors in the dynamical phase portrait of a society. In an open democratic society, they may emerge, but also decline, if their attractiveness decreases due to the selection pressure of competing alternatives.

8.3 Complex Communication Networks, Information Retrieval, and Personalized Information Systems

Natural evolution has not focused on single organisms with increasing intelligence based on neural information processing. In species and populations, we observe increasing degrees of fitness enabled by increasing capacities of swarm, collective, and distributed intelligence with extrasomatic information processing. In sociobiology, populations of ants and termites organize complex transport, and information and communication systems through swarm intelligence. There is no central supervisor over the construction of complex networks of paths between their bivouacs. The ordering of the system is self-organizing according to chemical signals between thousands of animals. In human history, complex transport and information networks have emerged with more or less self-organizing behavior. Telephone and railway networks are supervised by global control stations, while car traffic in networks of streets depends on the local behavior of drivers. Thus, auto traffic can be considered a complex dynamical system with typical phenomena of oscillation ("stop-and-go"), congestion, and chaos.

The capacity to manage the complexity of modern societies depends decisively on an effective communication network. Like the neural nets of biological brains, this network determines the learning capability that can help mankind to survive. In the framework of complex systems, we have to model the dynamics of information technologies spreading in their economic and cultural environment. Thus, we speak of informational and computational ecologies. There are actually realized examples, like those used in airline reservation, bank connections, or research laboratories, which include networks containing many different kind of computers.

Incomplete knowledge and delayed information are the typical features of open computational systems which have no central controls. These large networks, emerging from the increasing connectivity between diverse computer-assisted information centers, are becoming self-organizing systems which are different from their individual program-controlled components. Their unplanned growth leads to an immense diversity of technical composition and use, with increasing difficulties of interoperability. The horrifying vision of a separated robotic world enslaving human culture may be the ultimate consequence of many possible uncomfortable scenarios.

As the worldwide growth of local information and computation centers cannot be planned by a centralized processor, it seems to have a nonlinear dynamics which should be studied in the framework of complex systems. Even simplified case studies would provide crucial insights in the complex dynamics of modern sociocultural evolution. The complex interdependencies of computational ecologies violate the traditional requirements for a hierarchical decomposition into technical, industrial, or administrative modules as used in traditional management. Modern technical communication networks are growing open systems which must be used without central control, synchronicity, or consistent data from other agents like machines or humans. Thus, the dynamical theory of informational and computational ecologies which incorporates the features of incomplete knowledge and delayed information will provide well known evolutionary patterns like fixed points, oscillations, or chaos.

Imperfect knowledge leads to an optimality gap, while delays in information access induce oscillations in the number of the agents involved. Synergetic effects may arise by cooperation and competition for finite resources. Chaos prevents any stable strategy of problem solving. Marvin Minsky has studied a simplified model of collective problem solving by nearly independent agents working on a set of related problems and interacting with each other. In the end of the 1980th, this example was already used as a design model for a distribut system of computer-assisted information systems [8.26].

The world is going to be populated with a huge number of computing systems with increasing complexity [8.27]. Among them will be traditional von-Neumann machines, vectorizing super-computers, shared-memory multi-processors, connection machines, neural-net simulators, millions of personal computers overcrowding the world like amebas, and future molecular machines. These computing systems are becoming more and more linked with such information systems as satellites, phones, and optical fiber. The idea of a self-organizing worldwide network of software and hardware systems has become reality. In a dramatic step, the complex systems approach has been expanded from neural networks to global technical information networks like the World Wide Web. In 1969, a computer-assisted information network was established at the U.S. Department of Defense. It has become the nucleus of a pullulating network with 175 000 computers, 936 subnets, and innumerable humans. The growth of Internet was not planned or controlled by any central processor unit, but a more or less anarchic process. Nevertheless, patterns of organization both arise from chaos or and decline in a kind of global self-organization.

In complex information networks, knowledge and information is distributed among several centers and individual programmers. Their complexity excludes central planning. Like all systems involving goals, resources, and actions, computation has been described in economic terms. Obviously, there is already a computational market for software and hardware using market mechanisms. As we have seen in Sect. 7.2, markets are a form of self-organizing complex ecosystem. According to Smith's fundamental insight, the force of consumer choice can make computational market ecosystems serve human purposes better than any programmer or central processor could plan or understand. The reason is the enormous complexity and diversity of computational ecosystems linked to the human market.

The reader may be reminded of the complexity of a biological ecosystem with several levels like cells, organs, and organisms. The elements of a computational ecosystem are likewise grouped on different levels according to the increasing complexity of its computational systems. Figure 8.12 shows the global network of USENET, which has grown through many local initiatives [8.28]. The network is shown in its state at the beginning of the 1990s as a fraction of the Internet (Fig. 8.12).

Today, the Internet can be considered a complex open computer network of autonomous nodes (hosts, routers, gateways, etc.), self-organizing without a central control mechanism. In the case of communicating with computers via the Internet, a message (e.g., e-mail) in an appropriate computer language (e.g., HTML) must be coded and administrated in several protocol layers before it can be sent through the network as byte packets from a sender to a receiver. Communication in computer networks is, for example, realized by the OSI-model of International Standard Organization (ISO).

The information flow is realized by information packets with source and destination addresses (e.g., IP address). The Internet shares the local activity concept with CAs and CNNs in the following sense: Routers are nodes of the network determining the local path of each packet by using local routing tables with cost metrics for neighboring routers. A router forwards each packet to a neighboring router, at the lowest cost, to the destination (Fig. 8.13). In the sense of the CA and CNN paradigms, the local routing tables can be considered "templates" of local nonlinear information processing.

As a router can only deal with one packet at a time, other arriving packets must be stored in a buffer. If more packets arrive than a buffer can store, the router discards the overflowing packets. Senders of packets wait for a confirmation message from the destination host. These buffering and resending activities of routers can cause congestion in the Internet. Fluctuations of information packet congestions can be indirectly observed through echo experiments of control messages between neighboring routers. A monitoring host between two routers periodically sends a series of

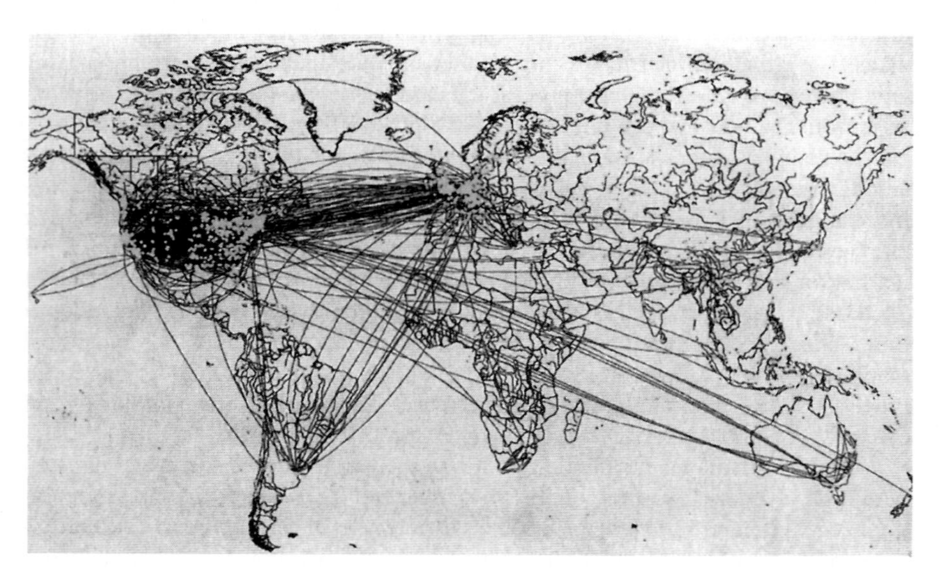

Fig. 8.12. Global networking in the beginning of the 1990s [8.28]

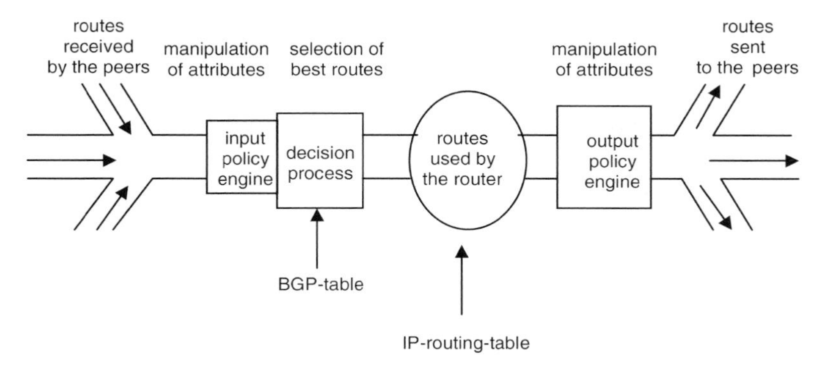

Fig. 8.13. Router node of the Internet [8.29]

echo packets to both routers. The packets take a round-trip time (RTT) to the destination and back. Congestion is associated with higher RTT values. RTT fluctuations increase with the sequence of routers in the Internet network (Fig. 8.14).

In automobile traffic systems, a phase transition from non-jamming to jamming depends on the average car density as the control parameter. At a critical value, fluctuations with self-similarity and power law distribution can be observed. From this analogy, a control parameter of data density is defined by the propagation of congestion from a router to neighboring routes and the dissolution of the congestion at each router [8.31]. The cumulative distribution of congestion duration is an order parameter of pattern formation. There are phase transitions between spare and congestion phases. The spare phase corresponds to a case in which the mean input of the information system is smaller than the maximum output. The critical point condition is when the mean input rate is equal to the maximum rate (Fig. 8.15).

At a critical point, when the congestion propagation rate is equal to congestion dissolution, fractal and chaotic features can be observed in data flow. On different time scales, we can analyze the self-similarity of the information packet's fluctuations, which is a necessary (not sufficient) condition of strange attractors (compare Sect. 2.6).

Congested buffers behave in surprising analogy to infected people. If a buffer becomes overloaded, it tries to send packets to its neighboring routers. Therefore, congestion spreads spatially. On the other hand, routers can recover when congestion to and from its own subnet is lower than the service rate of the router. Describing the epidemic processes of malaria in relation to the dynamics of routers is not only an illustrative metaphor, but provides a hint about nonlinear mathematical models. Thus, appropriate CAs and CNNs producing propagations of nonlinear waves would be able to simulate the data traffic of the World Wide Web [8.34]. Computational and information networks have become technical superorganisms, evolving in a quasi-evolutionary process. Computer networks are computational ecologies. If the Internet is a highly complex information network, then we have to manage information flow with a loss of information in chaotic situations. "Lost in the Net" is a popular slogan to describe these problems of increasing complexity.

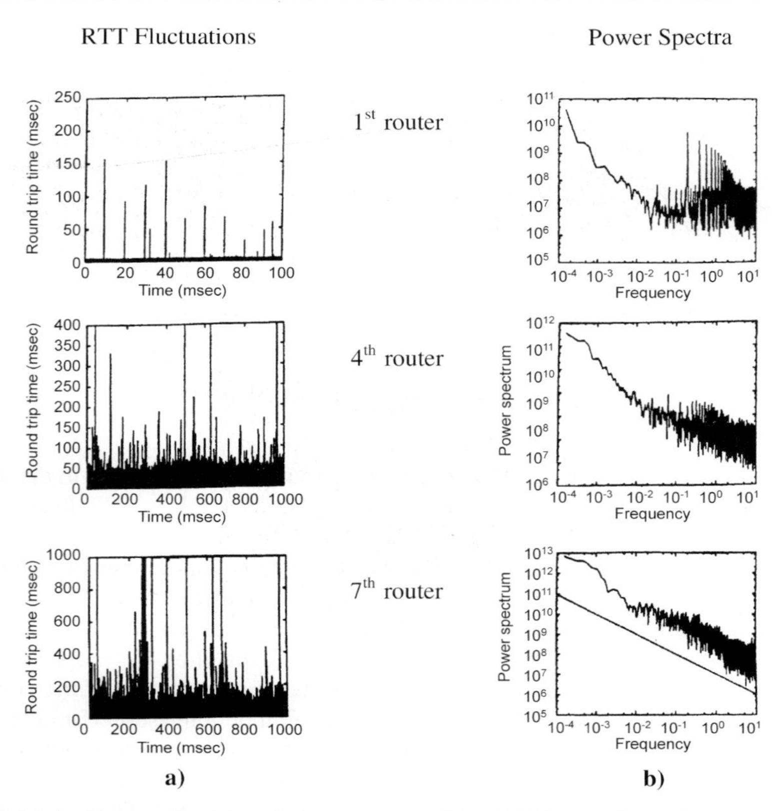

Fig. 8.14a,b. Time series (**a**) and power spectra (**b**) of RRT data fluctuations at Internet routers [8.30]

The information flood in a more or less chaotic Internet is a challenge for intelligent information retrieval [8.35]. Information Retrieval (IR) in the Internet can be considered decisive procedure for evaluating and selecting the most relevant documents according to certain constraints. In binary (Boolean) logic, a document is either relevant (1) or not (0) for an information query. In fuzzy logic, it has a degree of relevance in the internal [0,1]. Each document is characterized by keywords. A query is a proposition of logically connected keywords. A document is relevant for a query, if the same keywords occur in the document and in the Boolean proposition of the query. A query and the documents can be geometrically represented as vectors in a vector space of keywords. The relevance of keywords in a document or query is weighted by coordinates. The similarity of queries and documents is measured by the cosine (with values between 0 and 1) of the angle between their representing vectors.

There are also applications of genetic algorithms, in order to improve information retrieval. Genetic algorithms optimize populations of chromosomes in sequential generations by reproduction, mutation, and selection. In information retrieval,

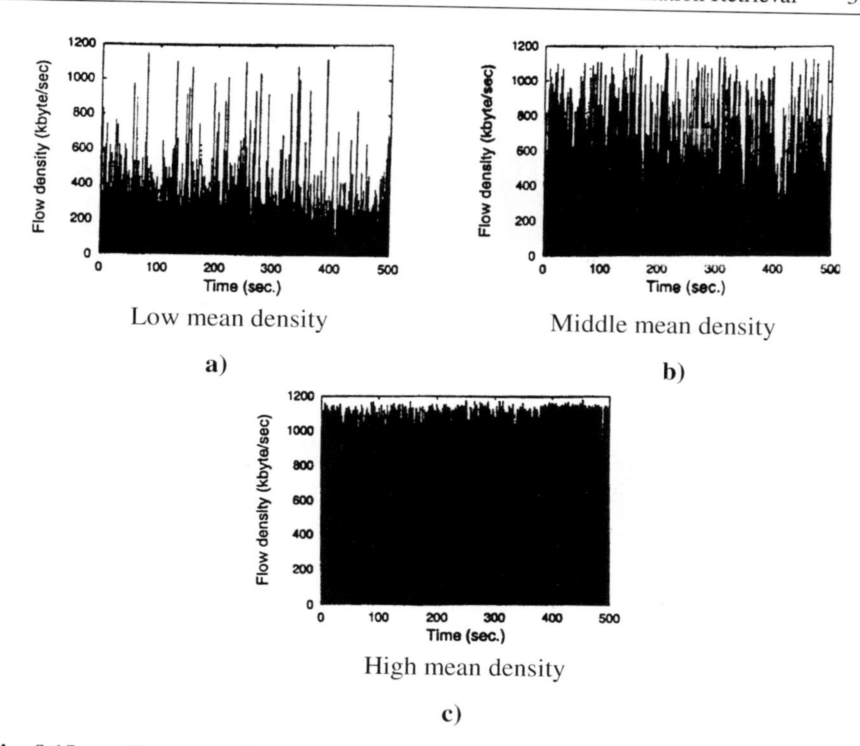

Fig. 8.15a–c. Phase transition of low (**a**), middle (**b**), and high (**c**) mean density of data traffic [8.32]

they are used for optimizing the queries of documents. A chromosome is a sequence of documents that is characterized by weighted key terms in binary codes. By extension, populations are sets of chromosomes. Mutation is the random change of binary digits. Sequential binary codes can be recombined. Fitness degrees measure the relevance of documents. Selection is the evaluation of populations of documents.

It is not only a metaphor to consider the Internet as a kind of superbrain with self-organizing features of learning and adapting [8.36]. We could use the analogies with a brain as heuristic devices to manage the information flood in the Internet. Information retrieval is already realized by neural networks adapting with synaptic plasticity to the information preferences of human users. Multi-layered neuronal nets can be applied for optimizing queries of documents (Fig. 8.17). Synaptic connections ("weights") between neurons change according to learning algorithms. The topology of the neural net consists of an input layer of the user's information preferences $q_{ui}^{(s)}$ with key terms t_i of query u in state s, a neuronal layer of key terms t_i, a neuronal layer of documents d_i, and an output layer of query results. The relevance of terms in a document corresponds to the weights between the neurons of terms and documents. Neurons fire if the sum of weighted inputs surpasses a critical threshold. A learning algorithm delivers a first query result by propagation. Deviations in user

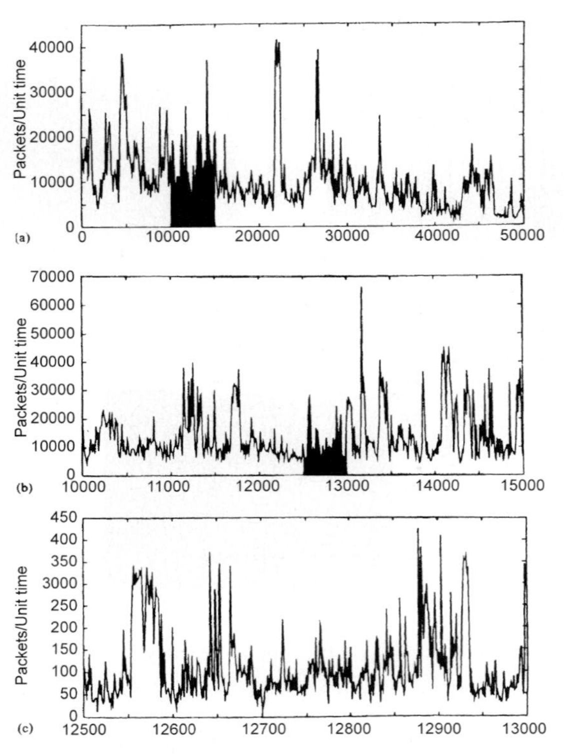

Fig. 8.16. Self-similarity of an information packet's fluctuation on different time scales ("strange attractor") [8.33]

preferences are weighted and propagated back to the term and input layer ("back propagation") and are improved during several iterations.

Neuronal maps in brains (e.g., visual cortex) generate representations of the world by neuronal clusters. Kohonen's self-organizing maps work in the same manner (Fig. 4.14). They can be applied to detect semantic similarities between documents, self-organizing in clusters. Each neuron of the neuronal net (output space) is connected to a weight of neuronal activity (input space). As an example, 245 documents of countries were characterized by 952 weighted key terms. In a 10×10 map, neurons were connected with names of countries and regions according to the input pattern of documents [8.38]. Countries and regions with similar properties self-organize in neighbouring clusters (e.g., islands, Western Europe, Eastern Europe, South America, and Africa).

In sociobiology, we can learn from populations of ants and termites how to organize traffic and information processing through swarm intelligence. From a technical point of view, we need intelligent programs distributed throughout the nets. There are already more or less intelligent virtual organisms (agents), learning, self-organizing, and adapting to our individual preferences of information, to select our e-mails, to prepare economic transactions, or to defend against the attacks of hos-

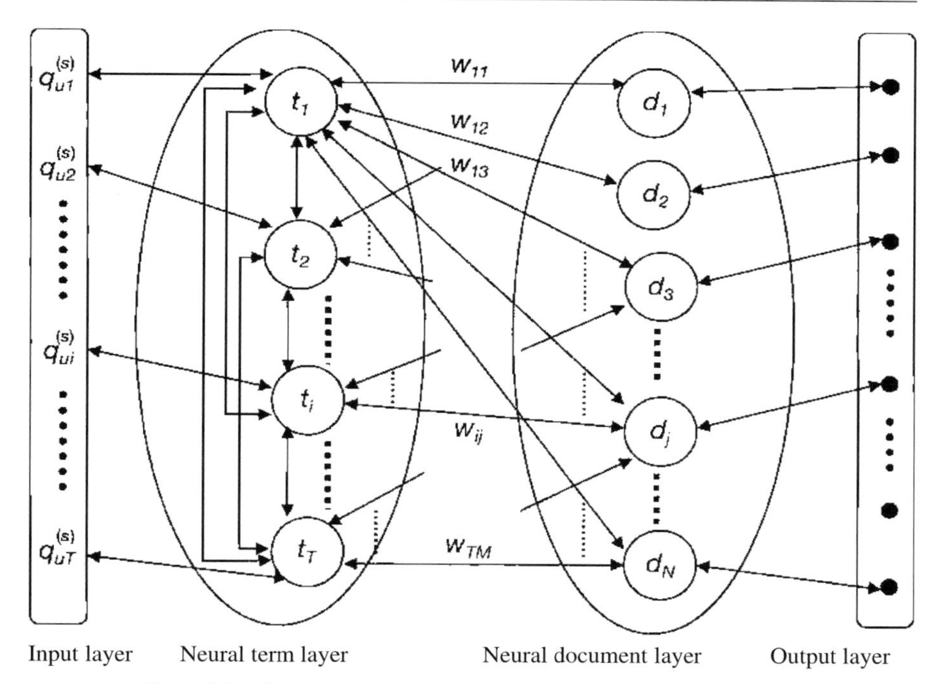

Fig. 8.17. Information retrieval with multi-layered neural nets [8.37]

tile computer viruses, like the human immune system. Virtual agents are designed with different degrees of autonomy, mobility, reactivity, and learning capabilities for communicating. They communicate and cooperate with their virtual environment as local spheres of influence.

Stationary agents localized in special servers or mobile agents can be sent as byte codes into the World Wide Web, performing their services without an online connection between client and server. E-commerce is a challenge for complexity research, as it can only be managed with the help of virtual agents, which support economic transactions. In the future, genetic algorithms will enable us to breed populations of agents in a complex evolution of virtual life. Populations of agents can reproduce themselves by genetic algorithms in order to optimize their information retrieval according to the queries of a user. Agents start with a user's profile and weight the relevance of a document, e.g., by determining the distance (number of links) between keywords and the keywords of a query. The "energy of life" of an agent increases or decreases according to the success or failure of the query. Successful agents are selected, mutate their genotype, and reproduce themselves [8.39].

Initialize agents;
Obtain queries from user;
while (there is an alive agent){
    Get document D_a pointed by current agent;
    Pick an agent a randomly;

```
        Select a link and fetch select document D'_a;
        Compute the relevancy of document D'_a;
        Update energy (E_a) according to the document relevancy;
        if (E_a > ε)
                Set parent and offspring's genotype appropriately;
                Mutate offspring's genotype;
        else if (E_a < 0)
                Kill agent a;
}
Update user profile;
```

The analogy between computational ecosystems and biological ecosystems [8.40] does not imply any reductionism or biologism. In the framework of complex systems, computational ecosystems and biological ecosystems are only models of mathematical evolution equations characterizing the nonlinear dynamics of complex systems. According to Minsky, the computer-assisted networks all over the world can be interpreted as a "marketplace" or "society of mind" [8.41].

Philosophically, this new kind of worldwide "knowledge medium" or computer-assisted "intelligence" may remind us of Hegelian ideas concerning the emergence of an "objective mind" embodied in human societies and their legal systems, economies, and bureaucracies, that overcomes the "subjective mind" of individual humans [8.42]. But these computational ecosystems have neither the consciousness nor the intentionality possessed by the cerebral neural networks of human individuals (compare Chap. 4).

Nevertheless, the question arises of whether computational ecosystems may be called "intelligent" to some degree. Individuals taking intelligence tests are judged by their ability to achieve goals set by a test-giver using time provided for the purpose. In this context, intelligence is no metaphysical universal concept; instead, there are several more or less well defined and testable standards of behavior. Some authors have suggested judging the "intelligence" of a society likewise by its ability to achieve goals set by certain legitimated subgroups (for instance, parliaments), using resources provided for the purpose. The degree of "intelligence" would depend on the range of goals that can be achieved, the speed with which they can be achieved, and the efficiency of the means used. The details of these definitions may differ, but "intelligence" based on these concepts would actually refer to a macrofeature of the whole system. The practical purpose of these definitions would be to compare computational ecosystems and their degrees of successful problem solving. In contrast, terms like the collective "intelligence" of a nation are ideologically dangerous. Furthermore, we should be aware that technical standards of intelligence are to be distinguished from concepts like consciousness which have actually been realized by the self-referentiality of certain brain networks in biological evolution.

The growth of informational and computational ecosystems is connected with a fundamental change of society characterized by a switch from traditional industries handling goods to knowledge industries for information and an economy of information services. The production, distribution, and administration of information have become the main acitivites of modern knowledge based societies. Thus, the interface between humans and information systems must be continously improved in order to realize the ideal of worldwide human communication. Human means of expression like speech, gestures, or handwriting should be understood immediately by computation and information systems. The "whole-person paradigm" and the "human-machine network" are highlights in the future world of communication.

Human communication involves not only bit strings of information, but also intuition, feeling, and emotion. The future world of communication is sometimes called a "global village" to emphasize the degree of familiarization created by the high-tech environment. But its acceptance decisively depends on the realization of human-friendly interfaces. We have to consider a new kind of complexity, referring to human intuition and emotion. The old ideals of rationality abstracting from these essentials of human life are absolutely ignorant of the human world. Even the process of scientific research is inspired by human intuition and driven by human emotions that must be considered in the future world of communication.

Some people are afraid that the final attractor of sociocultural evolution will not be a global village or worldwide *polis*, but a gigantic "Leviathan" dominating mankind with the effectiveness of modern high-tech procedures. In the computerized world of virtual realities, all human means of expressions would be digitized without any niche for personal intimacy. But the complexity of sociocultural evolution could allow several attractors. They cannot be forecast or determined by human decisions, but they may be influenced by conditions and constraints humans are able to achieve. What is the chance of human freedom in a world of high complexity? What is the degree of individual responsibility in a complex world of collective effects with high nonlinearity? These questions lead to an epilogue on ethics in a complex and nonlinear world.

The bounded rationality of users of the World Wide Web is a challenge to informatics, especially when the complex information associated with web-based applications and services that offer access to a vast variety of information sources is considered. Obtaining a precise account of what the individual user wants or means is a mission-critical task, particularly in areas of computer science that rely on users expressing their individual needs, such as when searching databases and information systems according to user queries, retrieving media in document collections, or encountering selection problems in Web services or e-commerce work-flows.

In practical system implementations, such information is usually deduced from a user's profile or some explicitly stated preferences. User modeling is concerned with trying to fully describe which of the users' interests should influence a computer application. However, psychological research shows that users are usually not fully conscious of their exact wishes, even in purposeful tasks [8.43]. A tedious process like the manual selection of services or areas of interest is often required to personalize publish/subscribe systems. Moreover, given that some knowledge is incorporated, the elicited information will naturally be incomplete, and so simply logging and storing and using user-stated keywords/behavior will sometimes lead to counter-intuitive results. Therefore, in order to improve the relevance of a personalized system's performance, a system must not only focus on explicit user specifications, but it should also take information specified by the user's implicit notions, situation, or assumed common knowledge into account. This information can be garnered from four main sources [8.44]:

Long-term preferences: The notion of relevance from previous interactions or generally applicable knowledge about a user is used.

Intention: The specific user's reason for interacting is included when personalizing the system.

Situation: The present state and environment of the user is used to decide whether specific preferences or rules are applicable.

Domain: Knowledge of the specific domain (often referred to as expert knowledge) is used within an interaction.

Typical personalized information systems are considered in the following examples. Long-term preferences include typical recurring individual preferences that represent individual tastes, like colors, general areas of interest, or preferred layout settings. This kind of preference can generally be used to personalize the system for individual users and so it is the kind of data often stored in user profiles. Systems, however, cannot always rely on these preferences, since different preferences may apply to different categories or the preferences simply may not be applicable in a certain context. Consider, for instance, a set of color preferences in an e-commerce setting. Though a user can be assumed to have a certain favourite color that will apply to shopping decisions, the preferences might be different for clothing and cars: the user may like driving red cars, for example, but the user may not like wearing red clothes. Moreover, when shopping for books the color preferences becomes entirely inapplicable, since it is somewhat rare to buy a book based on its color [8.45].

Less general, but more interesting, types of personalization tasks are covered by the last three categories. Consider the intention in a real world application like book shopping. Personalized book stores (including Internet portals such as www.amazon.com) will usually maintain a list of recommended books that is compiled based on the topics a user expressed interest in during previous interactions (i.e., a long-term profile of topical categories). Now assume that the same user accesses the book store with the intention of buying a present for an acquaintance. In most cases this present will be purchased based on the preferences of the acquaintance; hence the user's profile is not applicable in this case, and the interaction and the topics accessed should not be used to update the user's personal profile. In this example, the intention of the user – to buy a book for him/herself or for a different person – makes all the difference. Thus, the (assumed) intention of a user will help to specify the choices that the user should be offered in personalization and which characteristics the user will require at a certain point in time during his/her interaction with the system. Typical examples of the incorporation of intent into Web-based systems are adaptive hypertext applications, where the user's previous interactions and current navigation patterns are used to personalize the environment for the user [8.46].

The current situation also impacts on how the system should be personalized. Context-aware systems use clues from the user's direct environment (like the time or location), personal characteristics like emotional states, technical characteristics like client device capabilities, or certain high-level situational information like "user in a business meeting" or "user at home" (both of which are examples of social situations) when performing personalization. Examples of systems that use this approach are location-based services, situation-based communication routing, or the context-aware synthesis of multimedia content [8.47].

The most renowned realizations of the last kind of personalization preferences, those based on domain knowledge, are the so-called "expert systems" that contain domain knowledge elicited from domain experts, often along with rules for deducing how this knowledge should be applied to a particular situation. However [8.48], most expert knowledge is not represented by rules, but embodied in the experts themselves. Thus, we cannot simply consider domain preferences in the sense of expert systems; we must instead rely on domain-specific heuristics, such as which general preferences may be applicable (and in which combination), or on what users generally care about in a certain domain. The idea behind this is often referred to as common knowledge or world knowledge – knowledge about the environment that all humans interacting in a certain domain are assumed to know or implicitly share. Ontologies provide a good way of presenting some of this knowledge for use by those who are not experts in the domain. In today's systems, the latter three kinds of preferences – if included at all – are mostly built directly into the application logic and represent the embodied mind, as opposed to the collected individual long-term preferences, which form the user's individual profile.

Let us consider two specific application studies where we can see parts of the embodied mind, as represented by adequate preferences, being used for improved system personalization, in order to handle the complexity of the information flood. To achieve effective personalization, knowledge from all four sources discussed above must be blended with the specific user-provided details/keywords for an individual interaction. Though not all embodied knowledge can be captured in this way, this method nevertheless provides a useful way of personalizing systems under the notion of bounded rationality.

First consider personalized information searches of and retrievals from databases and information systems. The classical relational model that is still predominant in today's practical database applications uses relational algebra to specify rigid selection predicates that allow objects to be selected with certain characteristics from (usually) large data sets. Though this model is applicable to a variety of simple cases, for example when customer information for a bank account with a certain account number must be retrieved, modern information-driven environments demand somewhat fuzzier capabilities when specifying the information that the user needs to accomplish his/her task. In most practical applications, such as Web search engines, information searches will lead to either empty or too many results. If a user asks a very specific query during a necessarily fuzzy task like an information search, the query may be overspecified (for instance, overly specific keywords may be chosen), leading to an empty result that is not very helpful to the user. On the other hand, asking rather unspecific (i.e., underspecified) queries is bound to lead to a flood of information. In this case, lots of items more or less closely match the user's needs. However, since the user does not know the contents of the underlying database or information collection, he or she simply cannot be expected to know how specific their query needs to be to retrieve a helpful yet manageable set of results.

Preferences that show the structure of interests for an individual user or ontologies that model the common level of understanding of topics in a certain domain can help to tackle this problem. The expansion of queries using user-specific preferences is rooted in the notion of cooperative answering [8.49]. The basic idea behind cooperative retrieval systems is to relax the user-specified terms until a match can be found in a collection of data [8.50]. In this case, even an overspecified query will lead to some useful results and avoid empty result sets as well as bypassing the need for any tedious manual refinement of queries. This way of dealing with query predicates, as soft constraints, is also necessary when performing personalization tasks using individual preferences that have not been explicitly stated for a specific interaction. Since they are implicitly assumed by the system to represent common or embodied knowledge ascertained from either long-term profiles, intentions, situations, or domains, they must be considered to be less important than any explicitly provided terms (i.e., they are soft constraints that can be used to refine result sets that are too large, or they can be relaxed when empty result sets are retrieved). A system of preferences can be introduced in the form of strict partial orders with simple "I like A better than B" semantics into database queries [8.51]. For example, when searching for a rental car, the user could state that cars with automatic transmission are preferred over those with manual transmissions. If two offers are retrieved that meet all of the basic requirements, the result set can be ordered by or even limited to those objects that also fulfil the transmission preference. Single preferences such as type of transmission in a car can be modeled and combined into more complex queries using operators for deriving Pareto sets (i.e., all preferences are considered as equally important, following the Pareto principle from economics), prioritized sets (i.e., order is imposed on the preferences, such as lexicographic order), and ranked result sets (i.e., preferences associated with numerical domains that are aggregated using suitable utility functions).

As a second example, let us consider the discovery of useful Web services or the selection of suitable services for constructing complex workflows in a personalized manner [8.52]. Service-oriented application infrastructures are becoming more and more common. As the Internet becomes ubiquitous, the Web service paradigm is expected to substantially alter the modern business world. The essential components of this emerging service paradigm are Internet-based, modular applications that provide standard interfaces and communication protocols for efficient and effective service provisioning between different business units or businesses and customers. The reusability of basic building blocks (or implementations) that are common in some workflows and the easy customization possible within complex workflows are particularly appealing features of these services. However, just as we noted for information searches above, a retrieval model based on exact matches alone is unlikely to succeed. Users are generally much more interested in accomplishing high-level tasks than the exact intermediate steps required to do so. Thus, exactly specifying all characteristics of the services needed instead of using a more fuzzy understanding of the workflow in question is usually a counterproductive approach. As the complexity and diversity of Web services are expected to grow, enhanced techniques for service discovery and selection will be needed.

When they start to design a service, such as a restaurant reservation or flight booking service, providers already have quite specific ideas about what the service should provide and about what types of interactions to expect. Thus providers are domain experts that can provide a set of useful domain preferences and even ontologies for categorization that foster the successful execution/composition of services, even for non-expert users. Moreover, providers may also anticipate different ways in which the service could be used (possibly in different situations). Generally, only a certain number of typical requests/business processes will exist in a well-defined service. These typical interactions for different users/groups are also preference patterns or usage patterns that allow for a personalization approach. A usage pattern may for instance depend on the basic intentions of a significant group of users. Different intentions will require different patterns that depend on both the user's profile, which states his or her notion of a service's usefulness or desired characteristics (like execution costs, quality guarantees, etc.), as well as the service profiles that are employed to carry out the actual business task. The basic approach where demands are relaxed when empty result sets are obtained is also necessary in this case, in order to support users in a cooperative fashion.

In order to illustrate some examples of the different types of preferences and how to derive them, consider a restaurant booking service. One of the most important parameters to consider when choosing a restaurant is the type of cuisine. Any restaurant can be characterized adequately using a cuisine parameter, but querying for it becomes a tedious process if it is not cooperatively supported. For example, if a user is interested in eating at a restaurant offering Sichuan cuisine located in a particular district, such a query might be too specific and deliver an empty result

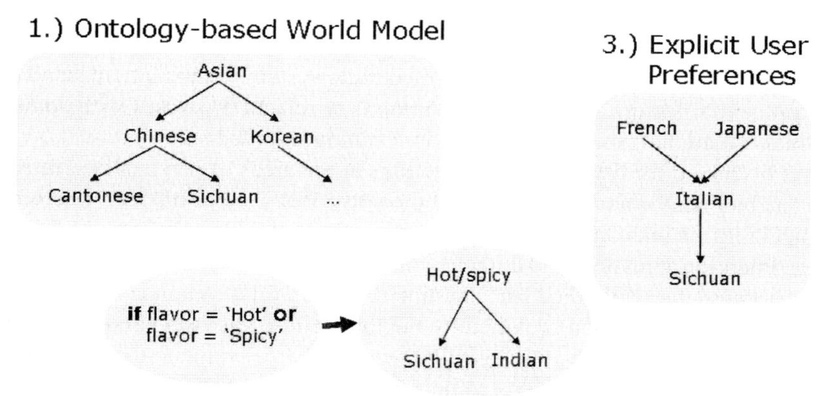

Fig. 8.18. Preferences derived from different sources in complex information worlds by personalized information systems [8.53]

set. On the other hand, just asking for a restaurant that offers Asian cuisine might result in a large set of rather unspecific choices offering Chinese, Japanese, Indian, or Thai cuisine. One thing a cooperative system could use is an explicit preference (either explicitly specified with the current service request or derived from previous interactions of the same user). Such a preference could, for example, specify that a user prefers Sichuan cuisine over Italian and over French and Japanese (see Fig. 8.18).

When such a preference is not explicitly given, the service provider can assume some information. For example, the general notion of a similarity between cuisines from geographically close regions could be assumed, based on the reasoning that cuisines in similar geographic regions involve the same kind of ingredients, herbs, and spices. One possible assumption would thus be that, since Sichuan is a specific Chinese cuisine, other Chinese cuisines like Cantonese could be acceptable alternatives to a user asking for Sichuan cuisine. To express this knowledge, a domain-specific ontology like that shown in Fig. 8.18 could be provided, and in the case of empty results and no explicitly stated preferences it could be used to relax a user's request. As an example of the application of user intentions, consider the usage pattern (or conceptual view) shown in Fig. 8.18. If we assume that a user is interested in hot and spicy food (i.e., a user has a rather taste-based notion of the similarities between types of restaurant, instead of the geographical one presented above), some cuisines like Indian or Mexican are closer to Sichuan cuisine (known for its spiciness) than those that are geographically closer. Thus, if a user is known to subscribe to a specific taste, a more specific pattern must be used for cooperative request relaxation.

In this section we focused on the representation of knowledge used for personalization tasks in complex information worlds. Starting from the notion that not all of the most relevant information used for personalization tasks can be elicited as expert knowledge, since it is embodied in the individual user (which is also consistent with current brain research), we proposed the use of flexible preference-based frameworks when personalizing computer systems under the paradigm of bounded rationality. This means that although an electronic system cannot anticipate all possible influential factors, it can at least enrich user-related processes with some intentional, situational, and domain-specific common knowledge.

As discussed for typical user interactions in the areas of personalized retrieval from databases/information systems and proactive Web service discovery/selection, personalizing the interaction with preferences from each individual user's long-term profile, intention, situation, and domain will result in improved system effectiveness. This is because the "embodied" information necessary for a certain task can be combined with the information provided by the user. Since this embodied information is not provided consciously, this embellishment adds real value to the personalized task. However, since all of the preferences that are combined with the user information have a lower level of importance than the explicitly provided information (and so can be ignored – relaxed – if necessary), this enhancement of the user information respects the individual user's needs and does not violate explicit constraints. Obviously, personalized computer systems do not aim to be a complete computational model of the embodied human mind, which was an impractical illusion of traditional AI and expert systems. In a complex information world, we want to construct effective and appropriate tools and service systems under the conditions of bounded knowledge, which requires interdisciplinary cooperation, especially with cognitive science and philosophy of mind.

8.4 Complex Mobile Networks, Ubiquitous Computing, and Adaptive E-Learning

Complexity of global networking is not restricted to increasing numbers of PC's, workstations, servers, and supercomputers interacting via data traffic in the Internet. With less complexity than a PC, cheap and smart devices of low-power are distributed in intelligent environments throughout our everyday world [8.54]. Examples are tabs, pads, and boards: inch-scale machines that approximate active Post-It notes, foot-scale ones that behave something like a sheet of paper, a book or a magazine, and yard-scale displays that are the equivalent of a blackboard or bulletin board. Tabs, pads, and boards are just the beginning of ubiquitous computing. Smart devices are intelligent microprocessors embedded in an alarm clock, the microwave oven, the TV remote controls, the stereo and TV systems, the kids' toys, etc. Ubiquitous computing makes 'things that think' [8.55], not only highly intelligent supercomputers, but an intelligent superorganism with 'swarm intelligence'. The third generation (3G) services of wireless communication include packet networks and interconnectivity of computerized appliances, such as phones, faxes, printers, software radio, etc. The enabling technologies demand faster data converters, more powerful processors, Java and other forms of downloadable software. The technical development of 3G-Communicators is an interdisciplinary system engineering task.

Like a GPS (Global Position System) in car traffic, things of everyday life could interact telematically by more or less intelligent sensors. Global position systems are nice examples of complex information systems realizing the local activity concept. A car driver using GPS is telematically guided by a network of neighbor GPS stations. In future, the processors, chips, and displays of these smart devices will not need a user interface such as a mouse, or keyboards, only a pleasant and effective place to get things done. Wireless computing devices of all scales become more and more invisible to the user [8.56]. Ubiquitous computing enables people to live, work, use, and enjoy things without being aware of their computing devices.

From a technical point of view, ubiquitous computing is a challenge for global networking using wireless media access, wide-bandwidth range, real-time capabilities for multimedia over standard networks, and data packet routing. Not only millions of PC'c, but billions of smart devices are interacting via the Internet. They are real, physical things of varying scale, but with virtual data shadows in the Internet, the control of which requires powerful, complexity aware, data tracking management. The overwhelming flow of data and information enforces us to operate at the edge of chaos.

Global networking is becoming one of the exciting challenges of complexity research. Understanding complex systems in nature and society supports the effective management of communication networks. In the 21st century, information, communication, and biotechnology are growing together. Therefore, information processing requires learning from nature. Information can be generated, transmitted, stored, processed, and represented in nature by sense organs, the nervous system and the brain. Cognitive processes like learning and thinking, language, motorics, perception, and communication, are simulated using technology by physical, chemical, and biological sensors, light-wave conductors, electronic, optical stores, microprocessors, neural nets, robotics, virtual reality, ubiquitous computing, artificial life and intelligence. Together they aim at producing learning, adapting, and self-organizing evolutionary complex systems.

An exciting example of current research is the evolutionary architecture of future automobiles, integrating all aspects of complexity and self-organization. The automobile industry is still one of the driving and dominating engines of the global economy. Thus, complexity research finds a realistic application in the production of future cars as learning, adapting, and self-organizing evolutionary complex systems. A challenge of the automobile industry is the increasing complexity of electronic systems. If we consider the electronic cable systems of automobiles from the beginning through to today, there will be a surprising similarity to neural networks of organisms which increase in complexity during evolution. Contrary to biological evolution, electronic systems of today are rigid, compact, and unflexible. So tiny failures can lead to a collapse of the whole system. In an evolutionary architecture (EvoArch) the nervous system of an automobile is devided into autonomous units (carlets) which can configurate themselves in cooperative functions, in order to solve intelligent tasks. Examples are the complex functions of motor, brake, and light, wireless guide systems like GPS, smart devices for information processing, and the electronic infrastructure of entertainment.

In the case of injuries or accidents, living organisms have the remarkable capability of self-healing and flexible change of units, in order to support vital functions via different organic cooperations. If, for example, people loose their ability to speak by a stroke, the damaged parts of the brain will sometimes be replaced by new configurations and connections of neural areas. If in an automobile, for example, the autonomous unit “lamp” will fail, then other autonomous units connected with the damaged one will search for substitutes by self-organization. They will arrange themselves with other units, in order to guarantee the vital functions of a car. In an evolutionary electronic architecture (EvoArch) [8.57], there are several “self-x-features” with great similarity to self-organizing organic systems in biological evolution (Fig. 8.19a): Self-healing demands self-configuration and self-diagnosis. Self-diagnosis means error recognition and self-reflexion. In a car, self-configuration consists of self-reflexion, an update mechanism, and extension. Self-advisory means self-reflexion, situation recognition, and a knowledge book. Self-adaption is based on situation recognition and self-configuration. The relations of self-x-features are illustrated in a taxonomy:

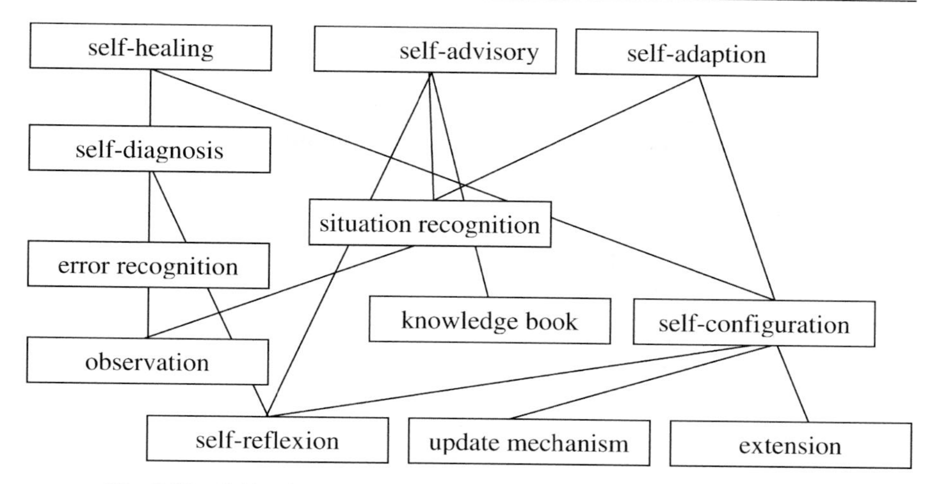

Fig. 8.19a. Self-x-features in an evolutionary architecture (EvoArch) [8.57]

According to the complex systems approach, the functions of a car are considered as macro-features which emerge from self-organizing interactions and the cooperation of autonomous units on the micro-level. Examples of autonomous units of a car (carlets) are, e.g., switches, lamps, tuners, controlers, regulators, horn. A car function like "air conditioning", "turn indication" or "hazard warning" needs one or more switches which must be selected among more than hundred candidates. Actually, a car function like "turn indication" needs carlets for a turn indicator switch, a terminal switch, turn indicator flashing, and several turn indicator lamps. In an evolutionary architecture, cooperations are realized in the EvoArch-arena (Fig. 8.19b), where active autonomous units ask for cooperation with passive autonomous units which have the appropriate features to execute a car function. Each unit (carlet) has an ID-number for self-identification. It can declare its property (e.g., turn indication) and its intention (e.g., search for a switch). The interaction of units (carlets) is made possible by a communication system (carCom) with information retrieval procedures, protocols, and contracts of cooperation which are well-known in the internet like RMI (Remote Method Invocation) and RPC (Remote Procedure Call). As in the internet, the network-management is based on the middleware of routing-procedures with routing-protocols and routing-tables.

After a call for cooperation, an active autonomous unit gets a list of possible passive candidates to replace, e.g., a damaged carlet. In the next step, exactly one candidate must be selected in order to realize a car function through cooperation. In the case of damaged units and failures of functions, the flexibility of a car and potentiality of its self-x-features increases with the number of possible candidates. But, on the other hand, the amount of computational time which is needed for selection increases nonlinearly with the size of the list of candidates. In a running car of today, approx. 30 functions need one or more switches which must be selected among more than 100 candidates. In general, if n is the number of active autonomous units,

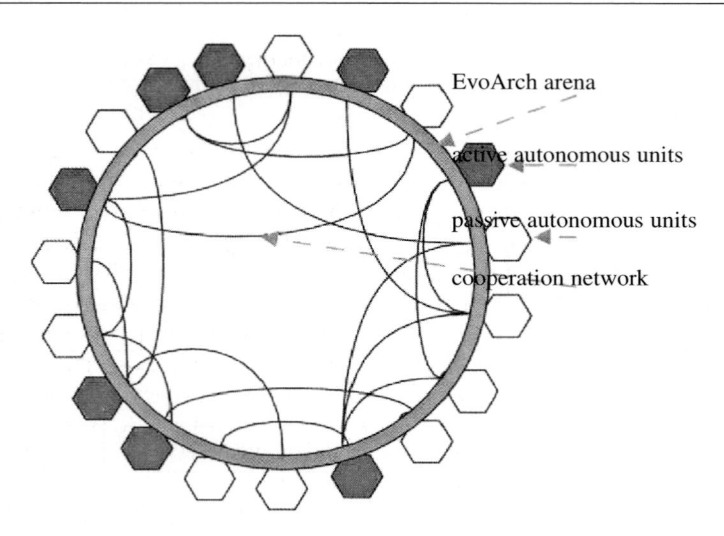

Fig. 8.19b. Self-organization of car functions in an evolutionary architecture (EvoArch) [8.57]

m the number of passive units required, and M the number of possible passive units with $M - m > mn$, then the number $\mathrm{Con}(m, n)$ of possible configurations of partners is given by the formula

$$\mathrm{Con}(m, n) = n! \sum_i \binom{M - m(n-i)}{m}$$

The reason is that $n!$ combinations are necessary to determine the succession of active autonomous units searching for passive ones. Without taking succession into account, $\binom{M}{m}$ combinations are possible. The selection among M possible units is diminished after each step by m, i.e. in the second step there are only $\binom{M-m}{m}$ combinations possible, in the third step only $\binom{M-2m}{m}$ combinations, etc. For $n = 30$ active autonomous units, $m = 2$ necessary passive units for a car function, and $M = 100$ possible passive units, we get 2.09×10^{17} combinations. If a combination corresponds to one computational instruction, a processor with 10^2 MIPS (million instructions per second) will need 2.42×10^{24} days for checking all possible combinations. Obviously, information retrieval is needed to reduce complexity.

As in the internet, the quality of information retrieval depends on the precision of calls for necessary partners. An unprecise call like "search for a switch" leads to a too long list of possible candidates. A more precise call "search for a switch for the left window in the left door" confines the possiblities. We need procedures for weighting and giving preferences which are well known in soft computing and artificial intelligence. In soft computing features are neither relevant nor not relevant as in Boolean logic. They are more or less relevant according to the degrees of their similarity. In a taxonomy, features and functions of a car can be classified by categories of similarity: On the bottom, the hierarchy consists of carlets as, e.g., lamps and switches. On the next level, a carlet like a lamp may belong to the category "lamps left behind" or "yellow lamps" which may belong to the more abstract categories "functions outside the car" or "functions in front of the car" on the following level up to abstract categories such as "position", "indication", "input", "output" etc. on the top level. In the sense of object-oriented programming languages (e.g., Java or C^{++}), these categories may be considered as classes with attributes and methods which are represented in a diagrame of nodes and edges. The arrows of edges indicate relations of inheritable features which can be implemented in a computational model of evolutionary architecture. If a carlet searches for a possible partner to realize a car function, its features will be confined according to the attributes of its call for cooperation localizing the position of a carlet as partner in the diagrame of taxonomy. T(taxonomy)-selection reduces the number of possibilities and enables the emergence of self-x-features such as self-healing or self-adaption in proper time.

Self-healing and self-adaption does not only mean replacing damaged hardware units or extending the architecture by new hardware units. In biological evolution, new physiological arrangements can be developed by an organism to guarantee vital functions or to introduce new abilities. In evolutionary architecture, the computational system of a car makes it possible to design virtual units to replace demaged hardware carlets or to introduce new functions requested by a driver. This enormous progress is based on mobile networks and ubiquitous computing. In Bluetooth technology, complex networks of cables between carlets are replaced by wireless communication. Each unit with an appropriate interface can take part in the wireless interaction. New electronic connections for, e.g., mobile telefones, GPS, or sets of electronic entertainment can be embedded by specific communication protocols (profiles). But it is also possible to control car functions from outside by mobile smart devices such as a PDA (Portable Digital Assistant). Drivers and engineers can influence all kinds of autonomous units in the same way physicians influence the organs of a living body. Thus, a car-gateway is introduced as an autonomous unit to realize the interface between the IT-world inside a car and the IT-world outside with mobile and smart devices of ubiquitous computing.

Bluetooth technology opens completely new avenues of utilities. In an evolutionary architecture (EvoArch), new functions can be adapted or even created according to the desires of a user. If, e.g., the joystick of a radio panel fails, the Bluetooth-car-gateway can design a virtual model of the joystick with the same logical procedures as the hardware version. The virtual joystick is represented on the touch screen of a PDA which can be used by a driver to operate the radio. The function "radio behavior" with carlets like "display" or "loudspeaker" is enlarged by the new autonomous unit "virtual joystick" with a wireless connection to a PDA. In EvoArch, an automobile is transformed into a self-organizing information processing system which is embedded into the world of wireless mobile networks and ubiquitous computing.

In the traditional automobile industry, the car architecure remains unchanged during the "lifetime" of a car after the end of its production. Future architecures must be open and flexible like evolutionary organisms. There are hard facts of global markets also demanding flexibility of car architecture. An automobile company cannot be restricted to one singular series of car model for several years. The innovation cycles and competition with other companies are accelerating more and more. If companies decide to cooperate with one another on global markets, then autonomous units of car series must be interchangeable. Later on, companies must take care of the political, economic, and ecological conditions of markets. Demands of consumers, fashions, and trends are changing quickly. Thus, autonomous units must be replaced or extended during the lifetime of a car like reproducing cells in living organisms. Self-adaption, self-configuration, and self-explanation are self-x-features of car production related to the whole life time of a series like the biological evolution of a population. Like living organisms, evolutionary architectures of automobiles are examples of information dynamics of complex systems self-organizing via information processing and communication of cellular units.

The information dynamics in nature and society have been analyzed in the framework of the theories of information, computability, and nonlinear dynamics. We have considered the dynamics of natural systems (e.g., atomic, molecular, genetic, neuronal systems), computational systems (e.g., quantum, molecular, DNA-, bio-, neurocomputing systems), global networking (e.g., internet, routing, information retrieval, multiagent systems), and ubiquitous computing (e.g., mobile phones, GPS, PDA, smart devices, intelligent environments).

Global networking is no longer only a challenge of technical development. Ubiquitous computing could improve the human interface with information systems, but it must not perplex people with a diversity of technical equipment. Global networking must be developed as a calm and invisible technology. Calm and invisible computing tries to integrate global networking and information processing in human environments and daily life without enslaving people with technical scenarios. Global networking must be developed as a technical service to mankind, no more no less. Thus, information processing in global networks cannot only be pushed by the technical sciences. It must be an interdisciplinary task of microelectronics, computer science, information science, and also a challenge of cognitive science, sociology, and the humanities.

In this age of globalization, people are nomadic and must adapt to changing information environments. In nomadic and pervasive computing, personalized digital profiles are automatically matched to local information services. Examples of this include personalized information services that are used when traveling and in traffic. Travellers can communicate by their PDAs (personal digital assistants) with the information system of, say, a railway station or airport. Tickets, reservations, and payments are handled electronically and wirelessly without a PC or a booking office. Information is automatically actualized. The traveler is guided by electronic advice to the seat that they booked.

In a nomadic world, a mobile user must be able to access their information at their present location, wherever that may be in the world [8.58]. The user's portable PDA is equipped with a personalized user profile, personal data, and preferences. Local servers are contacted automatically, personalized user profiles are matched and adapted to changing locales, allowing the preferences and services desired (e.g., information on hotels, restaurants) to be realized. Most of the information retrieved by humans is obtained subconsciously (e.g., via body language and gestures). Thus, as well as a preference manager for explicit query preferences, the derivation of preferences requires interdisciplinary cognitive research to build up an adequate rule base for the expected usage patterns, domains, and situations. Cognitive expansion of queries aims at the implementation of a knowledge-based query builder that allows complex query building in cooperation with the user.

Globalization requires mobility and steady adaptation to changing environments in everyday life. Several questions must be addressed in this regard. How do we coordinate the technical infrastructure and the heterogeneity of mobile equipment? Does the nomadization of information technology correspond to a nomadization of society, with the erosion of social structures and the loss of local rootedness? Personalized information systems should be adapted to the conditions and needs of human beings. In nomadic and ubiquitous computing, personalized information devices are wireless and pervasive. However, the technical equipment required to enable this computing should be hidden in the background (invisible computing) and used to augment our lives (augmented reality).

Further examples are personalized information services in a supermarket [8.59]. In nomadic and ubiquitous computing, even cheap supermarket products could be represented electronically in databases with high computational power. Therefore, supermarkets could introduce personalized and dynamic prices depending on personal customer data (e.g., frequency and amount of shopping) and products (e.g., the "best before" date). What social consequences would follow from personalized prices? Social discrimination or economic competition? How can the abuse of personal customer profiles be prevented? Personalized information services may also improve the complex communications that occur in a hospital [8.60]. In ubiquitous computing, physical reality is mapped onto a world of digital shadows. In a hospital, physicians and patients could be electronically monitored and controlled by pervasively distributed sensors and microprocessors. The schedules and activities of patients and physicians could be coordinated automatically, improving time management. But how can privacy and security be guaranteed? Who is responsible for failures in a complex information and computational system with automatic services?

Global information worlds can only be handled by knowledgeable and well-educated people. However, people have different intellectual capacities and live and work under different conditions [8.61]. Thus, knowledge must be tailored to their different and changing intellectual capacities and living conditions. How can we apply personalized information services to education? In today's knowledge based economies, the knowledge and skills of citizens are becoming increasingly important. Based on this notion, a new paradigm of personalized on-demand learning emerges, where the anyone, anytime, anywhere delivery of education and training is adapted to the requirements and preferences of each individual citizen within different e-learning and e-working settings. People learn in different ways. Thus, learning

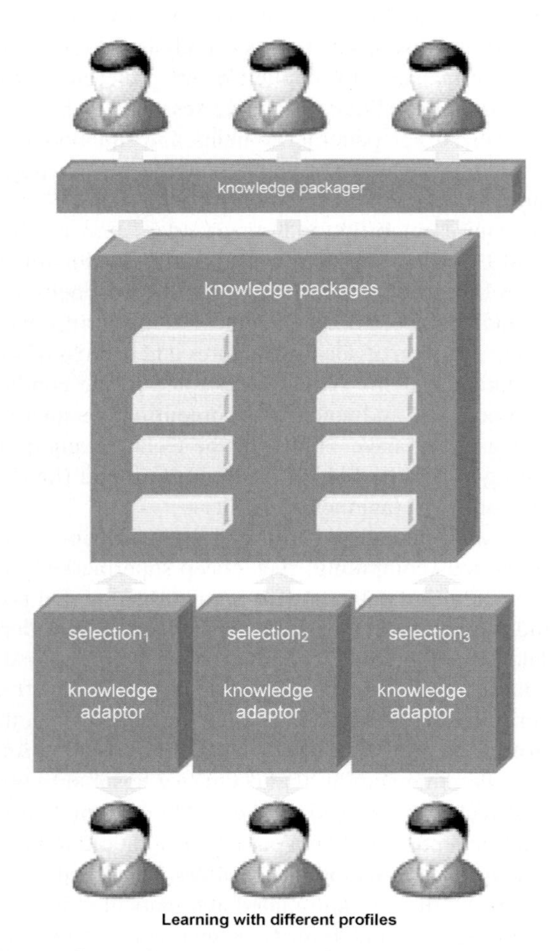

Fig. 8.20. Personalized information systems in global knowledge society [8.62]

materials must be packaged by personalized information systems according to the requirements of those who wish to learn (Fig. 8.20).

In the global knowledge society, different e-learning and e-working settings must be developed for different users, such as service providers, authors and publishers of learning materials, platform providers, and citizens, in a personalized way. In this age of globalization, nomadic computing is leading to an increased need to learn on the move. A mobile learning environment requires that appropriate learning resources are available to the learner anytime and anywhere. This can be achieved by adapting the content to the preferences of the learner, the characteristics of the device used to interact with the learning material (e.g., a PDA, mobile phone, or laptop), and an appropriately selected narrative template. In the field of artificial intelligence, adaptive and personalized agents and robots will be developed to support adaptive and personalized learning environments. They must not only understand natural human language, but also human emotions, motivations, preferences, and the different forms of human creativity. With the increasing complexity and specialization of human knowledge, they will be an essential part of adaptive and personalized learning.

In this age of globalization, firms and organizations must be learning, adaptive, and self-organizing information systems if they are to remain competitive in the marketplace. Learning organizations are interested in providing their employees with personalized and tailored internal access to organizational e-knowledge materials that promote life-long learning and training scenarios. Only personalized access guarantees that all of the human resources of the organization, with their individual competencies for creative innovations, will be utilized [8.63].

Humans in organizations (e.g., firms, administrations, universities) are not just reacting agents but also embodied minds with perception, cognition, and emotions. Their intentions and actions are determined by conscious and subconscious images of an actual situation. Management must recognize and use the creativity and tacit knowledge of people for innovation coaching. We distinguish between tacit and explicit knowledge based on brain dynamics. Tacit knowledge means knowledge of experiences and the body, simultaneous knowledge (here and now), analog knowledge, and practice. Explicit knowledge is mainly sequential symbolic knowledge or digitally declared knowledge. In complex human organizations, there is a knowledge-conversion process with several transitional steps [8.64]. The first step, socialization, means the transition from tacit knowledge to tacit knowledge – the accumulation of tacit knowledge, the extra-organizational collection of social information, the intra-organizational collection of social information, and the transfer of tacit knowledge. The next step refers to the transition from tacit knowledge to explicit knowledge. To achieve this, managers should facilitate creative and essential dialogues, hypothetical thinking, the use of metaphors, and include organizational designers. In the third step, combination, explicit knowledge is connected with explicit knowledge via acquisition and integration (plans, computer simulations), synthesis and processing (manuals, documents, databases), and dissemination (presentation). The last step, internalization, changes explicit to tacit knowledge through personal experience, acquisition of real-world knowledge, simulation and experimentation, and the acquisition of virtual-world knowledge.

One can then distinguish four kinds of knowledge assets in organizations, corresponding to these four phases of the knowledge-conversion process. Experimental knowledge assets mean tacit knowledge shared through common experiences, such as skills and know-how of individuals, care, trust, and security, or energy, passion, and tension. Conceptual knowledge assets refer to explicit knowledge articulated through images, symbols, and language. Examples are product concepts, and design or brand equity in firms. Systemic knowledge assets contain systemized and packed explicit knowledge, like documents, specifications, manuals, databases, patents, and licenses. Routine knowledge assets are incorporated in the tacit knowledge that is routinized and embedded in the actions and practices of coworkers.

When organizations (e.g., firms, universities, research laboratories, suppliers, and customers) become allies in a partnership, new platforms for learning are added. These are called interorganizational [8.65]. When organizations begin to adapt to larger networks, platforms for learning multiply. Virtual organizations are systems of networking partners who coordinate their activities through shared missions, visions, values, projects, and products. Organizational learning is derived from transformations of individual knowledge into artificial memories, rules, and routines. Explicit rules (e.g., written regulations, memoranda) have an objective existence, whereas tacit knowledge exists only in the practices of interesting people. Explicit and tacit knowledge are not only the abilities of individuals, but the potentials of virtual networks. In this age of globalization, learning organizations compete in global networks or become allies. Therefore, we need global self-organizing and intelligent knowledge systems in order to find tailored solutions for staff, partners, and clients, and to manage complex competitive situations.

During evolution, new forms of information storage have developed from genetic information up to neuronal information and finally extrasomatic information. In humans, the storage capacity of approximatly 10^{10} bit genetic information is overcome by approximatly 10^{14} bit neuronal information in brains, corresponding roughly to the possible number of synaptic connections. For the past 10^3 years, mankind has developed extrasomatic information storage (e.g., libraries, data bases, the Internet). Their information capacity has overcome the capacity of single brains. At the beginning of the 21st century, the exponential dynamics of key technologies is dominated by the miniaturization of computers, gene technology, and the growth of the Internet. According to J. Schumpeter (1883–1950), economic growth is driven by basic innovations of science and technology initiating long-term Kondratieff-cycles of business and enterpreneurship. After four cycles of the industrial society, depending mainly on natural resources, the 5th Kondratieff has generated the information society with mainly non-material products of information services.

In industrial societies economies have been characterized by the steps of production, logistics, distribution, marketing and sale of material goods. In information societies, there is an offer and demand of virtual information products and services with steps of information collection, information systematization, information retrieval, production and trade of information-based systems. Therefore, economists distinguish material chains of value in industral societies and virtual chains of value in information societies. According to Shannon, the content of information goods is measured by the degree of news for a receiver. But it is not sufficient to be well informed in order to settle our life. In a next step, information of high value must be evaluated and applied to solve problems. Information must be transformed to knowledge in the sense of know-how for problem solving.

Besides matter and life, the chief ingredients of the 21st century are information and knowledge. In a knowledge society, science is a productive power of economic and social growth which needs new strategies of cooperation with economy and politics [8.66]. The "wealth of nations" (A. Smith) is the knowledge of their people.

Therefore in the process of globalization with competing nations and societies, education must secure the sustainable future of the knowledge society. The technical evolution of information and knowledge is the fundamental challenge of mankind in the 21st century. Humans will no longer be only products of a blind evolution, but will try to influence their development by use of information and knowledge.

But, knowledge is not sufficient in order to decide about the goals of our problem solving. The experience with science and technology in the last century undoubtedly underlines the demand for ethical evaluations of our technical developments. Thus, the future of the information and knowledge society cannot be separated from ethics and politics. In the tradition of philosophy, the connection of knowledge with ethics has been called wisdom. The transformation from data to information, knowledge, and wisdom is the interdisciplinary challenge of a knowledge society [8.67].

9 Epilogue on Future, Science, and Ethics

The principles of complex systems suggest that the physical, social, and mental world is nonlinear, complex, random. This essential result of epistemology has important consequences for our present and future behavior. Science and technology will have a crucial impact on future developments. Thus this book finishes with an outlook on future, science, and ethics in a nonlinear, complex, and random world. What can we know about its future? What should we do?

9.1 Complexity, Forecasts, and the Future

In ancient times the ability to predict the future seemed to be a mysterious power of prophets, priests and astrologists. In the oracle of Delphi, for example, the seer Pythia (6th century B.C.) revealed the destiny of kings and heroes in a state of trance (Fig. 9.1) In modern times people came to believe in the unbounded capabilities of Laplace's demon: Forecasting in a linear and conservative world without friction and irreversibility would be perfect. We only need to know the exact initial conditions and equations of motion of a process in order to predict the future events by solving the equations for future times. Philosophers of science have tried to analyze the logical conditions of forecasting in the natural and social sciences [9.1]. Belief in man's forecasting power has been shaken over the course of this century by several scientific developments. Quantum theory teaches us that, in general, we can only make predictions in terms of probabilities (cf. Sect. 2.3). A wide class of phenomena is governed by deterministic chaos: Although their motions obey the laws of Newtonian physics, their trajectories depend sensitively on their initial conditions and thereby exclude predictions in the long run. In dissipative systems, such as the fluid layer of a Bénard experiment (Fig. 2.20), the emergence of order depends on microscopically small initial fluctuations. A tiny event, such as the stroke of a butterfly's wing, can, in principle, influence the global dynamics of weather. In chaotic systems, the prediction of future events is restricted, because the information flow from past to future decreases: The Kolmogorov–Sinai entropy has a finite value. But, in the case of random and noise, every correlation of past and future decays and the Kolmogorov-Sinai entropy is running to infinity: No prediction is possible. Obviously, the randomness of human fate was the challenge of ancient prophets, priests, and astrologists. In Chap. 7 we have learnt that patterns and relationships in economics, business, and society sometimes change dramatically. Going beyond

Fig. 9.1. Aigeas, king of Athens, asking the Oracle at Delphi about his future (Greek bowl: 440–430 B.C.)

the natural sciences, people's actions, which are observed in the social sciences, can and do influence future events. A forecast can, therefore, become a self-fulfilling or self-defeating prophecy that itself changes established patterns or relationships of the past. Is forecasting nothing more than staring into a crystal ball?

But nearly all our decisions are related to future events and require forecasts of circumstances surrounding that future environment. This is true for personal decisions, such as when and whom to marry or when and how to invest savings, and for complex decisions affecting an entire organization, firm, society, or the global state of the earth. In recent years increased emphasis has been placed on improving forecasting and decision making in economy and ecology, management and politics. Economic shocks, ecological catastrophes, political disasters, but also chances such as new markets, new technological trends, and new social structures, should no longer be random and fateful events sent by the gods. People want to be prepared and have thus developed a variety of quantitative forecasting methods for different situations, e.g., in business and management. From a methodological point of view, every quantitative forecasting instrument can be characterized by a particular predictability horizon which limits its reliable application. Let us have a look at the strengths and weakness of some forecasting instruments.

The most common quantitative methods of forecasting are the time-series procedures [9.2]. They assume that some pattern in a data series is recurring over time and can be extrapolated to future periods. Thus, a time-series procedure may be appropriate for forecasting environmental factors such as the level of employment or the pattern of weekly supermarket sales where individual decisions have little impact. But time-series methods cannot explain the causes behind the data patterns. In historical times, the method was used by the Babylonian astronomers who extrapolated the data pattern of moonrise into the future without any explanation based on models of planetary motion. In the 18th century physicists knew little about the causes of sunspots. But in the observations of sunspots a pattern of frequency and magnitude was found and predictions were possible by its continuation through time-series analysis. In business and economics, there are various underlying patterns in data series. A horizontal pattern exists where there is no trend in the data (e.g., products with stable sales). A seasonal pattern exists when a series fluctuates according to some seasonal factor such as products whose sale depends on the weather. A cyclic pattern may not repeat itself at constant intervals of time, e.g., the price of metals or the gross national product. A trend pattern exists when there is a general increase or decrease in the value of the variable over time. When an underlying pattern exists in a data series, that pattern must be distinguished from randomness by averaging and weighting ("smoothing") the past data values. Mathematically, a linear smoothing method can be used effectively with data that exhibit a trend pattern. But smoothing methods make no attempt to identify individual components of the basic underlying patterns. There may be subpatterns of trend, cycle, and seasonal factors, which must be separated and decomposed in analyzing the overall pattern of the data series.

While in time-series procedures some data pattern from the past is merely extrapolated to the future, an explanatory model assumes a relationship between the ("dependent") variable y that we want to forecast and another ("independent") variable x. For example, the dependent variable y is the cost of production per unit, and the independent variable x determining the cost of production is the number of units produced. In this case, we can model the relationship in a two-dimensional coordinate system of y and x and draw a straight line that in some sense will give the best linear approximation of the relationship. Regression analysis uses the method of least squares in order to minimize the distance between the actual observations y and the corresponding points $\hat{y}$ on the straight line of linear approximation. Obviously, there are many situations in which this is not a valid approach. An example is the forecast of monthly sales varying nonlinearily according to the seasons of the year. Furthermore, every manager knows that sales are not influenced by time alone, but by a variety of other factors such as the gross national product, prices, competitors, production costs, taxes, etc. The linear interaction of two factors only is a simplification in economy similar to the two-body problems in the linear and conservative world of classical physics.

But, of course, a complex model that is more accurate requires a larger amount of effort, greater expertise and more computational time. In many decision-making situations more than one variable can be used to explain or forecast a certain dependent variable. An ordinary example is a marketing manager who wants to forecast corporate sales for the coming year and to better understand the factors that influence them. Since he has more than one independent variable, his analysis is known as multiple regression analysis. Nevertheless, the dependent variable he wishes to forecast is expressed as a linear function of the independent variables. The computation of the coefficients in the regression equation is based on the use of a sample of past observations. Consequently the reliability of forecasts based on that regression equation depends largely on the specific sample of observations that were used. Therefore degrees of reliability must be measured by tests of statistical significance. While multiple regression involves a single equation, econometric models can include any number of simultaneous multiple regression equations [9.3]. In the case of linear equations, the mathematical methods of solution are based on linear algebra and linear optimization methods (e.g., simplex method). In spite of their linearity, the econometric models may be highly complicated with many variables which can only be mastered by computer programs and machines. The solution strategy of nonlinear programming in economics often decomposes complex problems into subproblems which can be approximately treated as linear.

An implicit assumption in using these methods is that the model best fitting the available historical data will also be the best model to predict the future beyond these data. But this assumption does not hold true for the great majority of real-world situations. Furthermore, most data series used in economics and business are short, measurement errors abound, and controlled experimentation is not possible. It is therefore necessary to understand how various forecasting methods succeed when changes in the established patterns of the past take place. The predictions are different at the various forecasting horizons characterizing each method. Obviously, there is no unique method that can forecast best for all series and forecasting horizons. Sometimes there is nothing in the past data to indicate that a change will be forthcoming. Thus, it may be impossible to anticipate a pattern change without inside knowledge. Pattern shifts or the "change of paradigms" is an everyday experience of business people and managers and by no means an extraordinary insight of some philosophers of science in the tradition of Kuhn et al.

Are there quantitative procedures for determining when a pattern or relationship in a data series has changed? Such methods indeed exist and use a tracking signal to identify when changes in the forecasting errors indicate that a nonrandom shift has occurred. In a quality control chart of, e.g., a production series of cars, the output of the equipment is sampled periodically. As long as that sample mean is within the control limits, the equipment is operating correctly. When this is not the case, the production is stopped and an appropriate action is taken to return it to correct operation. In general, automatic monitoring of quantitative forecasting methods follows the concept of a quality control chart. Every time a forecast is made, its error (i.e., actual minus predicted value) is checked against the upper and lower control limits. If it is within an acceptable range, the extrapolated pattern has not changed. If the forecasting error is outside the control limits, there has probably been some systematic change in the established pattern. Automatic monitoring through tracking signals may be appropriate when large numbers of forecasts are involved. But in the case of one or only a few series, one must still play a waiting game to discover whether changes in the trends of business data are occurring.

Forecasting the future of technological trends and markets, the profitability of new products or services, and the associated trends in employment and unemployment is one of the most difficult, but also most necessary tasks of managers and politicians. Their decisions depend on a large number of technological, economic, competitive, social, and political factors. Since the emergence of commercial computers in the 1950s there has been hope that one might master these complex problems by increasing computational speed and data memory. Indeed, any quantitative forecasting method can be programmed to run on a computer. As no single forecasting method is appropriate for all situations, computer-based multiple forecasting systems have been developed in order to provide a menu of alternative methods for a manager. An example is the forecasting system SIBYL which is named after the ancient seer Sibyl. The story goes that Sibyl of Cumae sold the famous Sibylian books to the Roman king Tarquinius Superbus.

Indeed, SIBYL is a knowledge-based system (cf. Sect. 6.1) for a computerized package of forecasting methods [9.4]. It provides programs for data preparation and data handling, screening of available forecasting methods, application of selected methods, and comparing, selecting, and combining of forecasts. In screening alternative forecasting techniques, the inference component of the knowledge based system suggests those methods that most closely match the specific situation and its characteristics based on a broad sample of forecasting applications and decision rules. The final function of SIBYL is that of testing and comparing which method provides the best results. The interface of user and system is as friendly and efficient as possible, in order to suit a forecasting expert as well as a novice. Nevertheless, we must not forget that SIBYL can only optimize the application of stored forecasting methods. In principle, the predictability horizon of forecasting methods cannot be enlarged by the application of computers. Contrary to the learning ability of a human expert, forecasting systems such as SIBYL are still program-controlled with the typical limitations of knowledge-based systems.

In general, the computer-based automation of forecasting followed along the lines of linear thinking. On the other hand, the increasing capability of modern computers encouraged researches to analyze nonlinear problems. In the mid-1950s meteorologists preferred statistical methods of forecasting based on the concept of linear regression. This development was supported by Norbert Wiener's successful predicting of stationary random processes. Edward Lorenz was sceptical about the idea of statistical forecasting and decided to test its validity experimentally against a nonlinear dynamical model (cf. Sect. 2.4). Weather and climate is an example of an open system with energy dissipation. The state of such a system is modeled by a point in a phase space, the behavior of the system by a phase trajectory. After some transient process a trajectory reaches an attracting set ("attractor") which may be a stable singular point of the system (Fig. 2.14a or 3.11c), a periodic oscillation called a limit cycle (Fig. 3.11d) or a strange attractor (Fig. 2.21). If one wants to predict the behavior of a system containing a stable singular point or a limit cycle, one may observe that the divergence of nearby trajectories appears not to be growing and may even diminish (Fig. 9.2). In this case, a whole class of initial conditions will be able to reach the steady state and the corresponding systems are predictable. An example is an ecological system with periodic trajectorties of prey and predator populations modeled by nonlinear Lotka–Volterra equations. The divergence or

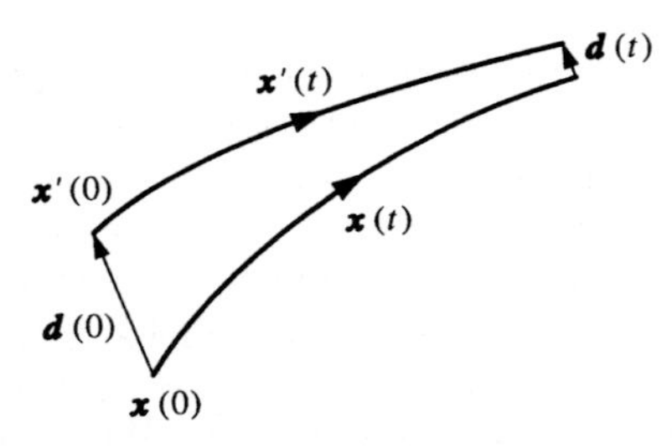

Fig. 9.2. Predictable system with stable point attractor or limit cycle and convergence of nearby trajectories [9.5]

convergence of nearby trajectories can be measured numerically by the so-called Lyapunov exponent:

Let us consider two nearby trajectories $\boldsymbol{x}(t)$ and $\boldsymbol{x}'(t)$ with the initial states $\boldsymbol{x}(0)$ and $\boldsymbol{x}'(0)$ at time $t = 0$ and the length $d(t) = |\boldsymbol{x}'(t) - \boldsymbol{x}(t)|$ of the vector $\boldsymbol{d}(t)$. If the trajectories converge, then $d(t) \approx e^{\Lambda t}$ and $\Lambda < 0$. The quantity Λ is called Lyapunov exponent and defined as $\Lambda(\boldsymbol{x}(0), \boldsymbol{d}(0)) = \lim_{t\to\infty} \lim_{d(0)\to 0} [(1/t) \ln(d(t)/d(0))]$. If it is positive, the Lyapunov exponent gives the rate of divergence. In Fig. 9.2, the model process $\boldsymbol{x}'(t)$ delivers reliable predictions of the real process $\boldsymbol{x}(t)$, because the system is assumed to have converging trajectories independent of their initial conditions.

A phase portrait of a nonlinear system may have a number of attractors with different regions ("separatrices") of approaching trajectories (cf. Fig. 2.10). For forecasting the future of the evolving system it is not sufficient to know all possible attractors and the initial state $\boldsymbol{x}(0)$. What we need to know in addition are the separatrices for attraction basins of the different attractors. If the initial state of a system happens to be far away from the basin of a certain attractor, the final state of the corresponding attractor cannot be predicted.

In Fig. 2.22a-c, the nonlinear logistic map describes a transition from order to chaos depending on an increasing control parameter. Figure 2.23a,b illustrates the corresponding sequence of bifurcations with the chaotic regime occurring beyond a critical threshold. If the corresponding Lyapunov exponent is positive, the behavior of the system is chaotic. If it is zero, the system has a tendency to bifurcate. If it is negative, the system is in a stable state or branch of the bifurcation tree. In this case the system is predictable. In the other cases the sensitivity to initial conditions comes into play. It is remarkable that a nonlinear system in the chaotic regime is nonetheless not completely unpredictable. The white stripes or "windows" in the grey veil of a chaotic future (Fig. 2.23b) indicate local states of order with negative Lyapunov exponents. Thus, in a sea of chaos we may find predictable islands of order. In this case the system is at least predictable for characteristic short intervals of time.

In general, the degree of predictability is measured by a statistical correlation between the observed process and the model at the particular time since the start of the observation. Values close to unity correspond to a satisfactory forecast, while small values indicate a discrepancy between observation and prediction. Every forecasting model has a certain time of predictable behavior after which the degree of predictability decreases more or less rapidly to zero. With improvement of the model the time of predictable behavior may be enlarged to some extent. But the predictability range depends upon fluctuational parameters. Weak microscopic perturbations of locally unstable chaotic systems can reach a macroscopic scale in a short time. Thus, local instabilities reduce the improvement of predictable behavior drastically. The predictability horizon of a forecasting system means a finite timespan of predictable behavior that cannot be surpassed by either improved measuring instruments or a refined prediction model. When we remember that the atmosphere is modeled, following Lorenz, by nonlinear systems with local and global instabilities, we realize the difficulties encountered by meteorologists in obtaining efficient long- or even medium-term forecasting. The belief in a linear progress of weather forecasting by increasing computational capacities was an illusion of the 1950s.

As nonlinear models are applied in different fields of research, we gain general insights into the predictable horizons of oscillatory chemical reactions, fluctuations of species, populations, fluid turbulence, and economic processes. The emergence of sunspots, for instance, which was formerly analyzed by statistical methods of time-series is by no means a random activity. It can be modeled by a nonlinear chaotic system with several characteristic periods and a strange attractor only allowing bounded forecasts of the variations. In nonlinear models of public opinion formation, for instance, we may distinguish a predictable stable state before the public voting ("bifurcation") when neither of two possible opinions is preferred, the short interval of bifurcation when tiny unpredictable fluctuations may induce abrupt changes, and the transition to a stable majority. The situation reminds us of growing air bubbles in turbulently boiling water: When a bubble has become big enough, its steady growth on its way upward is predictable. But its origin and early growth is a question of random fluctuation. Obviously, nonlinear modeling explains the difficulties of the modern Pythias and Sibyls of demoscopy.

Today, nonlinear forecasting models do not always deliver better and more efficient predictions than the standard linear procedures. Their main advantage is the explanation of the actual nonlinear dynamics in real processes, the identification and improvement of local horizons with short-term predictions. But first of all an appropriate dynamical equation governing an observation at time t must be reconstructed, in order to predict future behavior by solving that equation. Even in the natural sciences, it is still unclear whether appropriate equations for complex fields such as earthquakes can be derived. We may hope to set up a list in a computer memory with typical nonlinear equations whose coefficients can be automatically adjusted for the observed process. Instead, to make an exhaustive search for all possible relevant parameters, a learning strategy may start with a crude model operating over relatively short times and then specify a smaller number of parameters in a relatively narrow range of values. An improvement of short-term forecasting has been realized by the learning strategies of neural networks. On the basis of learned data, neural nets can weight the input data and minimize the forecasting errors of short-term stock quotations by self-organizing procedures (Fig. 6.14a,b). So long as only some stock market advisors use this technical support, they may do well. But if all agents in a market use the same learning strategy, the forecasting will become a self-defeating prophecy.

The reason is that human societies are not complex systems of molecules or ants, but the result of highly intentional acting beings with a greater or lesser amount of free will [9.6]. A particular kind of self-fulfilling prophecy is the Oedipus effect in which people like the legendary Greek king try, in vain, to change their future as forecasted to them. From a macroscopic viewpoint we may, of course, observe single individuals contributing with their activities to the collective macrostate of society representing cultural, political, and economic order ("order parameters"). Yet, macrostates of a society, of course, do not simply average over its parts. Its order parameters strongly influence the individuals of the society by orientating ("enslaving") their activities and by activating or deactivating their attitudes and capabilities. This kind of feedback is typical for complex dynamical systems. If the control parameters of the environmental conditions attain certain critical values due to internal or external interactions, the macrovariables may move into an unstable domain out of which highly divergent alternative paths are possible. Tiny unpredictable microfluctuations (e.g., actions of very few influential people, scientific dicoveries, new technologies) may decide which of the diverging paths in an unstable state of bifurcation society will follow.

One of the deepest insights into complex systems is the fact that even complete knowledge of microscopic interactions does not guarantee predictions of the future. In this book, we have learnt that simple rules of physical, genetic, neural, or social dynamics can generate very complex and even random patterns of material formation, organic growth, mental recognition, and social behavior. Randomness, in a practical sense, only means that future formation or behavior cannot be detected by familiar and well-known patterns or programs. In this case, the computability of the future is not reducible relative to certain patterns and programs. Randomness, in principle, implies computational irreducibility: Then, there is no finite method of predicting how the system will behave except by going through nearly all the steps of actual development. In the case of randomness, there is no shortcut to evolution. Mathematical systems like cellular automata (CA) or technical systems like cellular neural/nonlinear networks (CNN) can achieve exactly the same level of complexity and randomness of nature and society. Thus, the traditional view of science – that precise knowledge of laws allows precise forecasting – fails in the case of nonlinear and random dynamics.

9.2 Complexity, Science, and Technology

Despite the difficulties referred to above, we need reliable support for short-, medium-, and long-term forecasts of our local and global future. A recent demand from politics is the modeling of future developments in science and technology which have become a crucial factor of modern civilization. Actually, this kind of development seems to be governed by the complex dynamics of scientific ideas and research groups which are embedded in the complex network of human society. Common topics of research groups attract the interest and capacity of researchers for longer or shorter periods of time. These research "attractors" seem to dominate the activities of scientists like the attractors and vortices in fluid dynamics. When states of research become unstable, research groups may split up into subgroups following particular paths of research which may end with solutions or may bifurcate again, and so forth. The dynamics of science seems to be realized by phase transitions in a bifurcation tree with increasing complexity. Sometimes scientific problems are well-defined and lead to clear solutions. But there are also "strange" and "diffuse" states like the strange attractors of chaos theory.

Historically, quantitative inquiries into scientific growth started with statistical approaches such as Rainoff's work on "Wave-like fluctuations of creativity in the development of West-European physics in the 18th and 19th century" (1929). From a sociological point of view Robert Merton discussed "Changing foci of interest in the sciences and technology", while Pitirim Sorokin analyzed the exponential increase of scientific discoveries and technological inventions since the 15th century. He argued that the importance of an invention or discovery does not depend on subjective weighting, but on the amount of subsequent scientific work inspired by the basic innovation. As early as 1912 Alfred Lotka had the idea of describing true epidemic processes like the spread of malaria and chemical oscillations with the help of differential equations. Later on, the information scientist William Goffman applied the epidemic model to the spread of scientific ideas. There is an initial focus of "infectious ideas" infecting more and more people in quasi-epidemic waves. Thus, from the viewpoint of epidemiology, the cumulation and concentration in a scientific field is modeled by so-called Lotka- and Bradford-distributions, starting with a few articles of some individual authors which are the nuclei of publication clusters [9.7]. The epidemic model was also applied to the spread of technical innovations. In all these examples we find the well-known S-curve of a logistic map (Fig. 2.22a) with a slow start followed by an exponential increase and then a final slow growth towards saturation. Obviously a learning process is also described in the three phases of an S-curve with slow learning success of an individual in the beginning, then a rapid exponential increase and finally a slow final phase approaching saturation.

The transition from statistical analysis to dynamical models has the great methodological advantage that incomprehensible phenomena such as strange fluctuations or statistical correlations of scientific activities can be illustrated in computer-assisted simulation experiments with varying dynamical scenarios. The epidemic model and Lotka–Volterra equation were only a first attempt to simulate coupled growth processes of scientific communities. However, essential properties of evolutionary processes like creation of new structural elements (mutation, innovation, etc.) cannot be reflected. Evolutionary processes in social systems have to be pictured through unstable transitions by which new ideas, research fields, and technologies (like new products in economic models) replace already existing ones and thereby change the structure of the scientific system. In a generalization of Eigen's equation of prebiotic evolution (cf. Sect. 3.3), the scientific system is described by an enumerable set of fields (i.e., subdisciplines of a scientific research field), each of which is characterized by a number of occupying elements (i.e., scientists working in the particular subdiscipline). Elementary processes of self-reproduction, decline, exchange, and input from external sources or spontaneous generation have to be modeled. Each self-replication or death process changes only the occupation of a single field. For simple linear self-reproduction processes without exchange, the selection value of a field is given by the difference between the "birth" and "death" rates of the field. When a new field is first populated, it is its selection value that decides whether the system is stable or unstable with respect to the innovation. If its selection value is larger than any other selection value of existing fields, the new field will outgrow the others, and the system may become unstable. The evolution of new fields with higher selection values characterizes a simple selection process according to Darwinian "survival of the fittest".

But we must not forget that such mathematical models do not imply the reduction of scientific activities to biological mechanisms. The variables and constants of the evolution equation do not refer to biochemical quantities and measurements, but to the statistical tables of scientometrics. Self-reproduction corresponds to young scientists joining the field of research they want to start working in. Their choice is influenced by education processes, social needs, individual interest, scientific schools, etc. Decline means that scientists are active in science for a limited number of years. The scientists may leave the scientific system for different reasons (e.g., age). Field mobility means the process of exchange of scientists between research fields according to the model of migration. Scientists might prefer the direction of higher attractiveness of a field expressed by a higher self-reproduction rate. When processes include exchange between fields with nonlinear growth functions of self-reproduction and decline, then the calculation of selection values of an innovation is a rather complicated mathematical task. In general, a new field with higher selection value is indicated by the instability of the system with respect to a corresponding perturbation.

Actually, scientific growth is a stochastic process. When, for example, only a few pioneers are working in the initial phase of a new field, stochastic fluctuations are typical. The stochastic dynamics of the probable occupation density in the scientific subfields is modeled by a master equation with a transition operator which is defined by transition probabilities of self-reproduction, decline, and field mobility. The stochastic model provides the basis for several computer-assisted simulations of scientific growth processes. The corresponding deterministic curves, as average over a large number of identical stochastic systems, are considered for trend analysis, too. As a result, the general S-shaped growth law for scientific communities in subdisciplines with a delayed initial phase, a rapid growth phase, and a saturation phase has been established in several simulations. In a series of simulations (Fig. 9.3), a research field was assumed to comprise about 120–160 members. For five fields, 100 scientists were chosen as initial condition with the saturation domain near the initial conditions. A sixth field is not yet set up (with the initial condition of zero members). In a first example, the influence of the self-reproduction process on the growth curve of the new field was simulated for several cases. With increas-

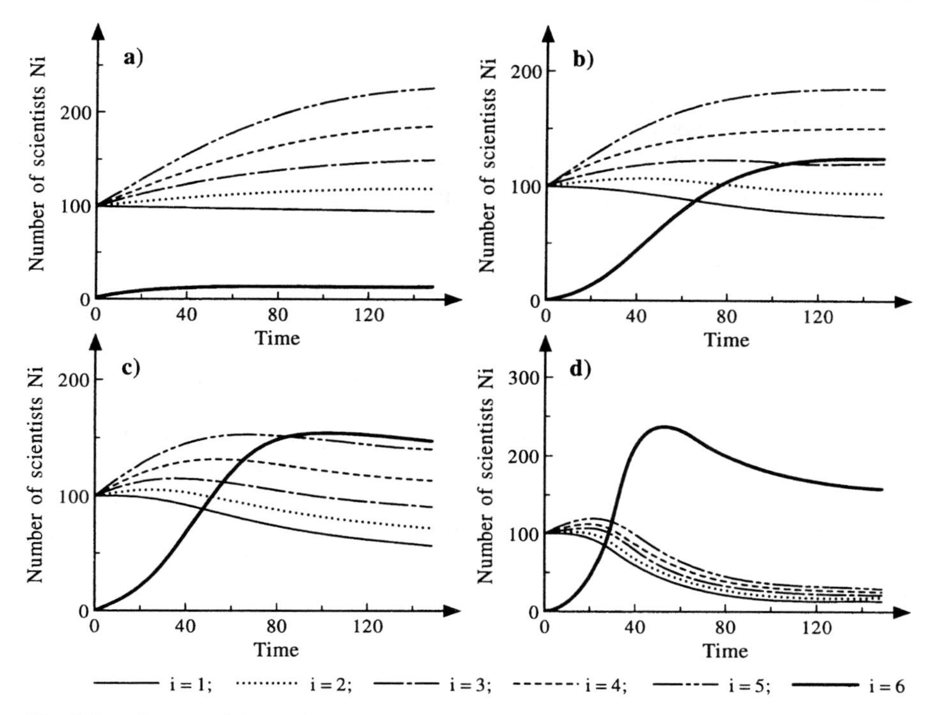

Fig. 9.3. Influence of the self-reproduction rate of a new scientific field on the growth curves of neighboring fields [9.8]

ing self-reproduction rates the new field grows ever more rapidly at the expense of neighboring fields.

The emergence of a new field may have a tendency to more coexistence or selection. The growth of the initial phase may be more or less rapid or can also be delayed. A famous example of delayed growth in the history of science is chaos theory itself, which was treated by only very few scientists (e.g., Poincaré) in its initial phase. Although the mathematical principles of the new field were quite clear, its exponential growth began only some years ago when computational technology could handle nonlinear equations. Sometimes an emerging field cannot expand to a real domain of science, because it has only a weak selection advantage in comparison with mighty surrounding fields. It is a pity that some technological fields such as alternative energies (e.g., wind, solar) are still in such a poor state, surrounded by the powerful industries of traditional or nuclear energy. If a new attractive field emerges, a strong influx of scientists from the surrounding fields can be observed. These people are adapting to the style and problem solving pattern of the new field. This kind of directed field mobility sometimes leads to the phenomena of fashion in science.

It is well-known that the S-shaped nonlinear logistic map gives rise to a variety of complex dynamical behaviors such as fixed points, oscillations, and deterministic chaos, if the appropriate control parameters increase beyond certain critical values

(Fig. 2.22). Obviously, both the stochastic and the deterministic models reflect some typical properties of scientific growth. Such effects are structural differentiation, deletion, creation, extension of new fields with delay, disappearance, rapid growth, overshooting fashions, and regression. The computer-assisted graphic simulations of these dynamical effects allow characterization by appropriate order parameters which are testable on the basis of scientometric data. Possible scenarios under varying conditions can be simulated, in order to predict the landmarks and the scope of future developments.

But so far, the evolution of scientific research fields has been considered in the model only in terms of changes of the scientific manpower in the selected fields. A more adequate representation of scientific growth must take account of the problem-solving processes of scientific endeavors. But it is a difficult methodological problem to find an adequate state space representing the development of problem solving in a scientific field. In the mathematical theory of biological evolution, the species can be represented by points in a high-dimensional space of biological characters (Fig. 3.4). The evolution of a species corresponds to the movement of a point through the phenotypic character space. Analogously, in the science system, a high-dimensional character space of scientific problems has to be established. Configurations of scientific articles which are analyzed by the technique of multidimensional scaling in co-citation clusters can be represented by points in a space of two or three dimensions. Sometimes research problems are indicated by sequences of keywords ("macro-terms") which are registered according to the frequency of their occurrence or co-occurrence in a scientific text.

In a continuous evolution model each point of the problem space is described by a vector corresponding to a research problem (Fig. 9.4a). The problem space consists of all scientific problems of a scientific field, of which some are perhaps still unknown and not under investigation. This space is metric, because the distance between two points corresponds to the degree of thematic connection between the problems represented. The scientists working on problem q at time t distribute themselves over the problem space with the density $x(q, t)$. In the continuous model $x(q, t)\,\mathrm{d}q$ means the number of scientists working at time t in the "problem element" $\mathrm{d}q$ (Fig. 9.4b).

Thus the research fields may correspond to more or less closely connected point clouds in the problem space. Single points between these areas of greater density correspond to scientists working on isolated research problems which may represent possible nuclei of new research fields. History of science shows that it may take decades before a cluster of research problems grows up into a research field. In the continuous model, field mobility processes are reflected by density change: If a scientist changes from problem q to problem q', then the density $x(q, t)$ will get smaller and $x(q', t)$ will increase. The movement of scientists in the problem space is modeled by a certain reproduction–transport equation. A function $a(q)$ expresses the rate at which the number of scientists in field q is growing through self-reproduction and decline. Thus, it is a function with many maxima and minima over the problem space, expressing the increasing or decreasing attractiveness of the problems in a scientific field. In analogy to physical potentials (e.g., Fig. 4.10), one

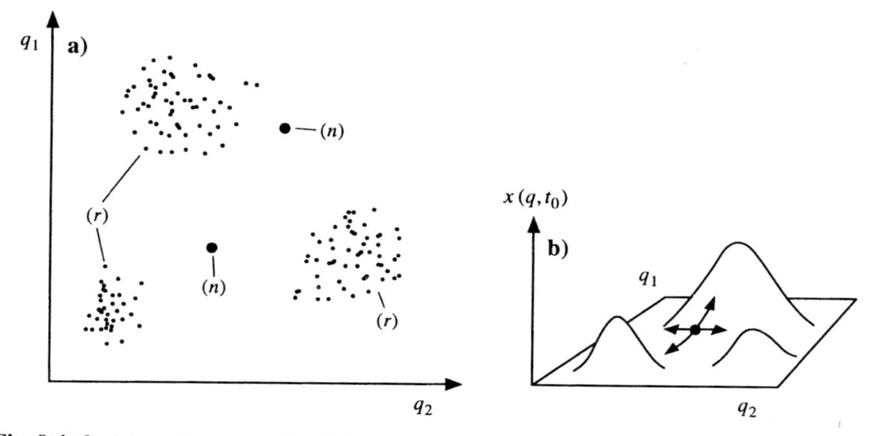

Fig. 9.4a,b. A two-dimensional problem space (**a**) with research fields (r) as clouds of related problems, possible nuclei (n) of new research fields, and a potential landscape (**b**) of research activities $x(q, t)$ in problem $q = (q_1, q_2)$ of the problem space at time t [9.9]

may interpret $a(q)$ as a potential landscape of attractiveness with hills and valleys, representing the attractors and deadlocked areas of a research field (Fig. 9.4b).

Dynamical models of the growth of knowledge become testable by scientometrics. Thus, they may open a bridge between philosophy of science with its conceptual ideas of scientific growth and history of science with its evaluation of scientific documents. In cognitive scientometrics an attempt has recently been made to quantify the concept of research problems and to represent them in appropriate problem spaces by bibliometric, cognitive, and social characteristics. The simplified schemes of the history of science which have been suggested by Popper, Kuhn, and others, could perhaps be replaced by testable hypotheses. Kuhn's discontinuous sequence with phases of "normal" and "revolutionary" science is obviously not able to tackle the growth of knowledge. On the other hand, the naive belief of some historians that the growth of science is a continuous cumulation of eternal truths is not appropriate to the complex dynamics of research in any way. Even Popper's sophisticated late philosophy that science does not grow through a monotonic increase of the number of indubitably established laws but through learning strategies of hypotheses and criticism needs more precision and clarification with reference to the changing historical standards of methodology, institutionalization, and organization. The increasing computational capacities of modern computers enable a new quantitative approach with simulation experiments in social sciences. The great advantage of dynamical models is their computer-assisted graphic illustration of several scenarios with varying parameters. These scenarios may confirm, restrict, or refute the chosen model. Last but not least, we need reliable support for decisions in science policy. Different scenarios of future developments may help us to decide where to invest our limited resources of research budget and how to realize desirable future states of society.

Thus nonlinear modeling and computer-assisted simulation may enable us to derive multiple futures, but provides no algorithm to decide between them. Normative goals must be involved in order to realize desirable future states of society. Since the 1960s the reports of the Club of Rome have tried to initiate international debates about the goals and alternative futures of mankind, supported by quantitative long-term forecasts. In Sect. 7.2, we saw the limitations of quantitative long-term forecasting in a nonlinear world. Consequently, scientific ideas and technological innovations cannot be forced into being by political decisions. But they must no longer be fateful random events which may or may not happen. We need instruments to evaluate desirable goals and their chance of realization.

A nonquantitative approach is the so-called Delphi method, used to prepare decisions and forecasts of scientific and technological trends by a panel of experts. The name "Delphi" is a reference to the legendary Pythia (Fig. 9.1) who was said to prepare her prophecies by gathering information about her clients. The Delphi method of today uses the estimates of scientific experts. The individual experts are kept apart so that their judgement is not influenced by social pressure or group behavior. The experts were asked in a letter to name and to weight inventions and scientific breakthroughs that are possible and/or desirable in a certain period of time. Sometimes they are not only asked for the probability of each development: Additionally, they are asked to estimate the probability that the occurrence of any one of the potential developments will influence the likelihood of occurrence of each of the others. Thus one gets a correlated network of future developments which can be represented by a matrix of subjective conditional probabilities. In the next phase the experts are informed about the items with general consensus. When they are asked to state the reasons for their disagreement with the majority, several of the experts re-evaluate their time estimates, and a narrower range for each breakthrough may be arranged.

The Delphi method, of course, cannot deliver a single answer. But the spread of expert opinions gathers considerable information about potential major breakthroughs. The average deviations from the majority should be narrowed down without pressuring the experts with extreme responses. But the Delphi method therefore cannot predict the unexpected. Sometimes the Delphi method is supported by the relevance-tree method, in order to select the best actions from alternatives by constructing decision trees. The relevance-tree method uses the ideas of decision theory to assess the desirability of a certain future and to select those areas of science and technology whose development is necessary for the achievment of those goals.

Obviously there is no single method of forecasting and deciding in a complex nonlinear world. We need an integrative ("hybrid") network of quantitative and qualitative methods. Finally, we need ethical landmarks to guide us in applying these instruments and in mastering the future.

9.3 Complexity, Responsibility, and Freedom

In recent years, ethics has become a major topic attracting increasing interest from a wide variety of professionals including engineers, physicians, scientists, managers, and politicians. The reasons for this interest are the growing problems with the environment, the economy, and modern technologies, questions of responsibility, increasing alarm, and decreasing acceptance of critical consequences in a highly industrialized world. But we must be aware that our standards of ethical behavior have not fallen to Earth from Heaven and have not been revealed by some mysterious higher authority. They have changed and they will continue to change, because they are involved in the evolution of our sociocultural world.

In modeling human society we must not forget the highly nonlinear selfreferentiality of a complex system with intentionally acting beings. There is a particular measurement problem in social sciences arising from the fact that scientists observing and recording behavior of society are themselves members of the social system they observe. Well-known examples are the effects of demoscopic opinion-polling during political elections. Furthermore, theoretical models of society may have a normative function influencing the future behavior of its agents. A well-known example was the social Darwinism of the 19th century which tried to explain the social development of mankind as a linear continuation of biological evolution. Actually, that social theory initiated a brutal ideology legitimating the ruthless selection of the social, economic, and racial victors of history. Today, it is sometimes fashionable to legitimate political ideas of basic democracy and ecological economy by biological models of self-organization [9.10]. But nature is neither good nor bad, neither peaceful nor militant. These are human evaluations. Biological strategies over millions of years have operated at the expense of myriads of populations and species with gene defects, cancer, etc., and have, from a human point of view, perpetrated many other cruelties. They cannot deliver the ethical standards for our political, economic, and social developments.

In this book we have seen that the historical models of life, mind, and society often depend on historical concepts of nature and historical standards of technology. Especially the linear and mechanistic view of causality was a dominant paradigm in the history of natural, social, and technical sciences. It also influenced ethical norms and values which cannot be understood without the epistemic concepts of the historical epochs in which they arose. The historical interdependence of epistemology and ethics does not mean any kind of relativism or naturalism. As in the case of scientific theories and hypotheses we have to distinguish their context of historical and psychological invention and discovery from their context of justification and validity. Even human rights have a historical development with changing meanings [9.11]. Hegel once mentioned that the history of mankind can be understood as a "development to freedom". Thus, before we discuss possible ethical consequences in a complex, nonlinear, and random world, we should take a short glance at the historical development of ethical standards.

Ethics is a discipline of philosophy like logic, epistemology, philosophy of science, language, law, religion, and so on [9.12]. Historically, the word "ethics" stems back to the Greek word ἦθος, which means custom and practice. Originally, ethics was understood as a doctrine of moral customs and institutions teaching people how to live. The central problem of ethics has become that of finding a good moral code with advice on how to live well, how to act justly, and how to decide rationally. Some essential concepts of ethics were already discussed in Greek philosophy following Socrates. His student Plato generalized the Socratic quest for a good life to the universal idea of the greatest good, which is eternal and independent of the historical life behind the transistory and continously changing world of matter [9.13].

Aristotle criticized his teacher's doctrine of eternal values as ignorant of real human life. For Aristotle the validity of the good, the just, and the rational is referred to the political society (*polis*), the family, and the interaction of single persons [9.14]. Justice in the polis is realized by the proportionality or the equilibrium of natural interest of free men. The greatest good of man is happiness, which is realized by a successful life according to the natural customs and practice in the polis and the family. Obviously, Aristotle's concept of ethics corresponds to his organic view of a nature filled with growing and maturing organisms like plants, animals, and humans.

After the dissolution of the Greek polis, ethics needed a new framework of standards. In Epicurean ethics, the internal equality of individual life, action, and feeling was emphasized, while the ethics of the Stoics underlined the external equality of all people achieved by nature. In the Christian Middle Ages a hierarchy of eternal values was guaranteed by the divine order of the world. At the beginning of modern times the theological framework as a universally accepted foundation of ethics was ripe for dissolution.

Descartes not only suggested a mechanistic model of nature but also demanded a moral system founded upon scientific reason. Baruch Spinoza derived an axiomatized system of rationalist morality corresponding to the deterministic and mechanistic model of nature. As the laws of nature are believed to be identical with the laws of rationality, human freedom only could mean acting according to deterministic laws which were recognized as rational. The greatest good meant the dominance of rationality over the affects of the material human body. Hobbes defended a mechanistic view of nature and society, but he doubted human rationality. Political laws and customs can only be guaranteed by the centralized power of "Leviathan". The greatest good is peace as a fixed and final equilibrium in an absolutist state.

The liberal society of Locke, Hume, and Smith was understood by analogy with Newton's model of separable forces and interacting celestial bodies. In the American and French revolutions, individual freedom was proclaimed as a natural right [9.15]. But how to justify individual freedom in a mechanistic world with deterministic causality? Every natural event is the effect of a linear chain of causes which in principle can be derived by mechanistic equations of motion. Only humans are assumed to be capable of spontaneous and free decisions initiating causal chains of actions without being influenced by outer circumstances. Kant called this human feature the "causality of freedom".

As nobody's opinions and desires are privileged, only advice which is acceptable for everybody is justified as reasonable. In the words of Kant, only those generally accepted "maxims" can be justified as universal moral laws. This formal principle of moral universality is Kant's famous categorical imperative of reason: we should act according to imperatives which can be justified as general laws of morality. The freedom of a person is limited by the freedom of his or her neighbor. In another famous formulation, Kant said that humans as free beings should not be misused as instruments for the interests of other humans. Thus, besides the mechanistic world of nature, which is ruled by deterministic laws, there is the internal world of reason with the laws of freedom and morality. Kant's ethic of freedom has been incorporated in the formal principles of every modern constitutional state [9.16].

But how can the laws of freedom be realized in the real world of politics and economics? During the initial process of industrialization the ethics of Anglo-American utilitarianism (due to Bentham and Mill) required an estimation of personal happiness. The happiness of the majority of people was declared to be the greatest good of ethics. While Kant suggested a formal principle of individual freedom, the utilitarian principle of happiness can be interpreted as its material completion. It was explicitly demanded as a natural human right in the American constitution. The utilitarian philosophers and economists defined the demand for happiness as an utility function that must be optimized with as few costs as possible in order to realize the greatest welfare for the majority of people. The principles of utilitarianism have become the ethical framework of welfare economics [9.17].

Modern philosophers like John Rawls have argued that the utilitarian principle in combination with Kant's demand for ethical universality can help to realize the demand for a just distribution of goods in modern welfare politics [9.18]. From a methodological point of view, the ethical, political, and economic model of utilitarianism corresponds to a self-organizing complex system with a single fixed point of equilibrium which is realized by an optimization of the utility function of a society and which is connected with a just distribution of goods to the majority of people.

Obviously, Kant's ethics as well as Anglo-American utilitarianism are normative demands to judge our actions. They may be accepted or not by individuals. Hegel argued that the subjective ethical standards of individuals were products of objective historical processes in history which were realized by the institutions of a society. Thus, he distinguished between the subjective morality and subjective reason of individuals and the objective morality and objective reason of institutions in a society. Historically, Hegel's foundation of ethics in the actual customs and morality of a civil society reminds the reader of Aristotle's realistic ethics of the Greek polis. But Aristotle's order of society was static, while Hegel assumed a historical evolution of states and their institutions.

From a methodological point of view, it is remarkable that Hegel already distinguished between the micro-level of individuals and a macro-level of societies and their institutions which is not only the sum of their citizens. Furthermore, he described an evolution of society which is not determined by the intentions and the subjective reason of single individuals, but by the self-organizing process of a collective reason. Nevertheless, Hegel believed in a rather simplified model of evolution with sequential states of equilibrium leading to a final fixed point which is realized by the attractor of a just civil society. Actual history after Hegel showed that his belief in the rational forces of history driving human society to a final state of justice by self-organization was a dangerous illusion. It is well known that his model was modified and misused by totalitarian politicians of the political right and left.

Friedrich Nietzsche attacted the belief in objective reason as well as in eternal ethical values as idealistic ideologies which were contrary to the real forces of life. Nietzsche's philosophy of life was influenced by Darwin's biology of evolution, which had become a popular philosophy in the late 19th century. Although Nietzsche had criticized nationalism and racism in his writings, his glorification of life and the victors in the struggle of life was terribly misused in the politics of the 20th century. Nevertheless, he is another example to show that concepts from the natural sciences have influenced political and ethical ideas [9.19].

Nietzsche's nihilism and his critique of modern civilization were continued by Martin Heidegger in our century. In Heidegger's view, the technical evolution of mankind is an automatism without orientation which has forgotten the essential foundation of man and humanity. A philosopher like Heidegger cannot and will not change or influence this evolution. He only has the freedom to bear this fate with composure. But in what way is Heidegger's attitude against technology and civilization more than resignation, fatalism, and an escape into an idyllic utopia without technology which has never existed in history? It seems to be the extreme counterposition to the Laplacean belief in an omnipotent planning and controlling capacity in nature and society [9.20].

What are the ethical consequences of the complex system approach which has been discussed in this book? First, we must be aware that the theory of complex systems is not a metaphysical process ontology. It is not an epistemic doctrine in the traditional sense of philosophy. The principles of this methodology deliver a heuristic scheme for constructing models of nonlinear complex systems in the natural and social sciences. If these models can be mathematized and their properties quantified, then we get empirical models which may or may not fit the data. Moreover, it tries to use a minimum of hypotheses in the sense of Ockham's razor. Thus it is a *mathematical, empirical, testable*, and *heuristically economical methodology*. Furthermore, it is an *interdisplinary research program* in which several natural and social sciences are engaged. However, it is not an ethical doctrine in the traditional sense of philosophy.

Nevertheless, our models of complex, nonlinear, and random processes in nature and society have important consequences for our behavior. In general, linear thinking may be dangerous in a nonlinear complex reality. We have seen that traditional concepts of freedom were based on linear models of behavior. In this framework every event is the effect of a well defined initial cause. Thus, if we assume a linear model of behavior, the responsibility for an event or effect seems to be uniquely decidable. But what about the global ecological damage which is caused by the local nonlinear interactions of billions of self-interested people? Recall, as one example, the demand for a well balanced complex system of ecology and economics. As ecological chaos can be global and uncontrollable, some philosophers like Hans Jonas have proposed that we stop any activity which could perhaps have some unknown consequences [9.21]. But we can never forecast all developments of a complex system in the long run. Must we therefore retire into a Heidegger-like attitude of resignation? The problem is that doing nothing does not necessarily stabilize the equilibrium of a complex system and can drive it to another metastable state. In chaotic situations, short-term forecasting is possible in complex systems, and attempts are being made to improve it in economics, for instance. But in the case of randomness and information noise, any kind of forecasting fails, eventhough we may be completely informed about the local rules of interaction in a complex system.

In a linear model, the extent of an effect is believed to be similar to the extent of its cause. Thus, a legal punishment of a punishable action can be proportional to the degree of damage effected. But what about the butterfly effect of tiny fluctuations which are initiated by some persons, groups, or firms, and which may result in a global crisis in politics and economics? For instance, consider the responsibility of managers and politicians whose failure can cause the misery of thousands or millions of people [9.22]. But what about responsibility in the case of random events? Information noise in the Internet, for example, must be prevented beforehand. If random happens, it is too late.

As the ecological, economic, and political problems of mankind have become global, complex, nonlinear, and random, the traditional concept of individual responsibility is questionable. We need new models of collective behavior depending on the different degrees of our individual faculties and insights. Individual freedom of decision is not abolished, but restricted by collective effects of complex systems in nature and society which cannot be forecast or controlled in the long run. Thus, it is not enough to have good individual intentions. We have to consider their nonlinear effects. Global dynamical phase portraits deliver possible scenarios under certain circumstances. They may help to achieve the appropriate conditions for fostering desired developments and preventing evil ones.

The dynamics of globalization is surely the most important political challenge of complexity for the future of mankind. After the fall of the Berlin wall, politicians believed in the linear assumption that coupling the dynamics of free markets and democracy would automatically lead to a community of modernized, peace-loving nations with civic-minded citizens and consumers. This was a terrible error in a complex world! From our point of view, complexity is driven by multi-component dynamics. Politicians and economists forgot that there are also ethnic and religious, psychological and social forces which can dominate the whole dynamics of a nation at a critical point of instability. As we all know from complex dynamical systems, we must not forget the initial and secondary conditions of dynamics. Instability emerges if free markets and elections are implemented under conditions of under-development.

Recent studies [9.23] demonstrate that in many countries of Southeast Asia, South America, Africa, Southeast Europe, and the Middle East the coupling of laissez-faire economics and electoral freedom did not automatically lead to more justice, welfare, and peace, but tipped the balance in these regions toward disintegration and strife. One reason is that these countries mainly have no broad majority of well educated people. Thus, minorities of clever ethnic groups, tribes, and clans come to power and dominate the dynamics of markets and politics. In the terminology of complex dynamics, they are the order parameters dominating ("enslaving") the whole dynamics of a nation. Again, the good intentions of democracy and free markets are not sufficient. We must consider the local conditions of countries and regions.

In classical philosophy, the transition from an intended development to a development contrary to the spirit of the philosophy has become famous as a contradiction of dialectics (e.g., Hegel). Good intentions may lead to bad effects. But sometimes human agents are driven by history to good effects without their subjective intentions. Hegel called it a "stratagem of reason" (List der Vernunft). Actually, it is a well-known effect of nonlinear dynamics. Therefore, market-dominant minorities are not a priori evil. Minorities are also the driving forces of activity. If they are open-minded and flexible, they prevent narrow-minded "enslaving" which may be successful only for a short time. In their own interest, they must try to stabilize the whole system in the long run. Therefore they should help dampening the social effects of free markets, bridging social cleavages, and transcending class division during a phase transition to democracy and welfare for the majority of the people. But these phase transitions may be different from region to region in the world. Responsible decisions require sensativity to local conditions in light of the butterfly effect.

There are not only local minorities in regions and countries. During the process of globalization, a minority of nations, institutions, and companies can come to power and dominate the whole dynamics of global economics and politics. Recent discussions on globalization show that a lot of people are not happy with the results of globalization. But it is necessary to understand that globalization means nothing more than the gobal dynamics of political and economic systems in the world. So, in a first run, it is neither good nor evil like the dynamics of weather. But contrary to weather, the dynamics of globalization is generated by the interactions of humans and their institutions. Thus, there will be a chance to influence globalization if we take into account the dynamical laws of complexity and nonlinearity.

It is a hard fact that the order parameters of globalization have been defined by a minority of nations. They are the world's preeminent political, economic, military, and technological powers whether we like it or not. Philosophers, mathematicians, and systems scientists have no power. But, again, we should use Hegel's "stratagem of reason": Minorities are also the centers of driving power which enables chances for change. Concepts and ideas without political power have no chance. If the dominating minorities of globilization are open-minded and flexible, they will prevent narrow-minded "enslaving" which may be successful only for a short time. In their own interest, they must try to stabilize the whole system in the long run. Therefore they should help dampening the social effects of global free markets, bridging social cleavages, and transcending class division during a phase transition to global democracy and welfare for the majority of the people.

Globalization means the critical phase transion to global governance in the world. We need new global structures to manage the political, economic, military, and technological power in the world according to the interests of the majority of people on earth. Global structures emerge from the nonlinear interactions of peoples, nations, and systems. At the end of the 18th century, Kant already demanded a law of nations leading "To Eternal Peace" (1795) [9.24]. After the 1st World War, president Wilson of the United States strongly influenced the foundation of the League of Nations. After the 2nd World War, the United Nations (UN) presented a new chance to handle international conflicts, but they often fail because of their lack of power. The dilemma of international law is that law needs power to enforce rights and ethical norms. Therefore nations have to give up parts of their sovereignty, in order to be dominated by commonly accepted "order parameters". After September 11 2001, a global network of terrorism threatens the preeminent political and economic nations of the world. This is the reason why especially the United States, which historically helped found the League of Nations as well as the Unitied Nations, now hesitates to restrict its national sovereignty and prefers to organize its own national security through global military defense.

Clearly it is a long way to global governance among autonomous nations. On the other hand, we must not forget the practical progress made by new social and humanitarian institutions of the UN. New economic, technological, and cultural networks of cooperation emerge and let people grow slowly together in spite of reactions and frictions in political reality. On the way to "eternal peace", Kant described a federal (multi-component) community of autonomous nations self-organizing their political, economic, and cultural affairs without military conflicts. But an eminent working condition of his model is the demand that states organize their internal affairs according to the civil laws of freedom. It is a hard fact of historical experience that civic-mindedness and humanization have sometimes not only be defended, but also enforced by military power. As long as the demand for civil laws of freedom is not internationally fulfilled, the organization of military power is an urgent challenge to globalization.

Globalization and international cooperation is accelerated by the growth of global information and computational networks like the internet and wireless mobile communication systems. On the other hand, the electronic vision of a global village implies a severe threat to personal freedom. If information about citizens can easily be gained and evaluated in large communication networks, then the danger of misuse by interested institutions must to be taken in earnest. As in the traditional economy of goods, there may arise information monopolies acting as dominating minorities prejudicing other people, classes, and countries. For instance, consider the former "Third World" or the "South" with its less developed systems for information services which would have no fair chance against the "North" in a global communication village.

Our physicians and psychologists must learn to consider humans as complex nonlinear entities of mind and body. Linear thinking may fail to yield a successful diagnosis. Local, isolated, and "linear" therapies of medical treatment may cause negative synergetic effects. Thus, it is noteworthy that mathematical modeling of complex medical and psychological situations advises high sensitivity and cautiousness, in order to heal and help ill people. The complex system approach cannot explain to us *what* life is. But it can show us *how* complex and sensitive life is. Thus, it can help us to become aware of the value of our life.

But what about the value of our life if it is computable? One of the most essential insights of this book is that the dynamics of nature and society are not only characterized by nonlinearity and chaos, but by randomness, too. Only in randomness can human free will have a real chance [9.25]. In the completely deterministic and computational world of a mechanical nature, Kant had to postulate a transcendental world in order to make free will, ethical duties, and responsibility possible. In random states of nature and society, the behavior of a system is not determined in any way. Random dynamics can be generated even if all rules of interaction of the elements in a dynamical system are known. In this case, the dynamics of a system correspond to irreducible computation, which means that there is no chance of forecasting. The only way to learn anything about the future of the system is to perform the dynamics. The macrobehavior of a brain, for example, could correspond to an irreducible computation, although we know all the rules of synaptic interactions. In this case, there is no shortcut or finite program for our life. We have to live our life in order to experience it. It is amazing that human free will seems to be supported just by the mathematical theory of computability.

Obviously, the theory of complex systems has consequences for the ethics of politics, economics, ecology, medicine, and biological, computational, and information sciences. These ethical consequences strongly depend on our knowledge about complex nonlinear dynamics in nature and society, but they are not derived from the principles of complex systems. Thus, we do not defend any kind of ethical naturalism or reductionism. Dynamical models of urban developments, global ecologies, human organs, or information networks only deliver possible scenarios with different attractors. It is a question for us to evaluate which attractor we should prefer ethically and help to realize by achievement of the appropriate conditions. Immanuel Kant summarized the problems of philosophy in the three famous questions [9.26]:

What can I know?
What must I do?
What may I hope?

The first question concerns epistemology with the possibilities and limitations of our recognition. The theory of complex systems explains what we can know and what we cannot know about nonlinear dynamics in nature and society. In general, the question invites a demand for scientific research, in order to improve our knowledge about complexity and evolution.

The second question concerns ethics and the evaluation of our actions. In general, it invites a demand for sensitivity in dealing with highly sensitive complex systems in nature and society. We should neither overact nor retire, because overaction as well as retirement can push the system from one chaotic state to another.

We should be both cautious and courageous, according to the conditions of nonlinearity and complexity in evolution. In politics we should be aware that any kind of mono-causality may lead to dogmatism, intolerance, and fanaticism.

Kant's last question "What may we hope?" concerns the Greatest Good, which has traditionally been discussed as *summum bonum* in the philosophy of religion. At first glance, it seems to be beyond the theory of complex systems, which only allows us to derive global scenarios in the long run and short-term forecasts under particular conditions. But when we consider the long sociocultural evolution of mankind, the greatest good that people have struggled for has been the dignity of their personal life. This does not depend on individual abilities, the degree of intelligence, or social advantages acquired by the contingencies of birth. It has been a free act of human self-determination in a stream of nonlinearity and randomness in history. We have to project the Greatest Good on an ongoing evolution of increasing complexity.

References

Chapter 1

1.1 Mainzer, K./Schirmacher, W. (eds.): Quanten, Chaos und Dämonen. Erkenntnistheoretische Aspekte der modernen Physik. B.I. Wissenschaftsverlag: Mannheim (1994)

1.2 Stein, D.L. (ed.): Lectures in the Sciences of Complexity, Santa Fe Institute Studies in the Sciences of Complexity vol. 1. Addison-Wesley: Redwood City, CA, (1989); Jen, E. (ed.): Lectures in Complex Systems, Santa Fe Institute Studies in the Sciences of Complexity vol. 2 (1990); Stein, D.L. (ed.): Lectures in Complex Systems, Santa Fe Institute Studies in the Sciences of Complexity vol. 3 (1991); Kurdyumov, S.P.: Evolution and self-organization laws in complex systems, Intern. J. Modern Physics C vol. 1, no. **4** (1990) 299–327

1.3 Nicolis, G./Prigogine, I.: Exploring Complexity. An Introduction. W.H. Freeman: New York (1989)

1.4 Haken, H.: Synergetics. An Introduction, 3rd Edn. Springer: Berlin (1983)

1.5 Mainzer, K.: Symmetries in Nature. De Gruyter: New York (1995) (German original: Symmetrien der Natur 1988); Mainzer, K.: Symmetry and Complexity: The Spirit and Beauty of Nonlinear Science. World Scientific: Singapore (2005)

1.6 Chua, L.O.: CNN: A Paradigm for Complexity. World Scientific: Singapore (1998)

Chapter 2

2.1 For historical sources of Sect. 2.1 compare Mainzer, K.: Symmetries in Nature. De Gruyter: New York (1994) (German original 1988) Chapter 1

2.2 Diels, H.: Die Fragmente der Vorsokratiker, 6th ed., revised by W. Kranz, 3 vol. Berlin (1960/1961) (abbrev.: Diels-Kranz), 12 A 10 (Pseudo-Plutarch)

2.3 Diels-Kranz 13 A 5, B 1

2.4 Diels-Kranz 22 B 64, B 30

2.5 Heisenberg, W.: Physik und Philosophie. Ullstein: Frankfurt (1970) 44

2.6 Diels-Kranz 22 B8

2.7 Diels-Kranz 31 B8

2.8 Heisenberg, W.: Die Plancksche Entdeckung und die philosophischen Grundlagen der Atomlehre, in: Heisenberg, W.: Wandlungen in den Grundlagen der Naturwissenschaften. S. Hirzel: Stuttgart (1959) 163

2.9 Cf. also Hanson, N.R.: Constellations and Conjectures. Boston (1973) 101

2.10 Hanson, N.R. (see Note 9, 113) carried out corresponding calculations.

2.11 Bohr, H.: Fastperiodische Funktionen. Berlin (1932)

2.12 Forke, A.: Geschichte der alten chinesischen Philosophie. Hamburg (1927) 486; Fêng Yu-Lan: A History of Chinese Philosophy vol. 2: The Period of Classical Learning. Princeton NJ (1953) 120
2.13 Mainzer, K.: Geschichte der Geometrie. B. I. Wissenschaftsverlag: Mannheim/ Wien/ Zürich (1980) 83; Edwards, C.H.: The Historical Development of the Calculus. Springer: Berlin (1979) 89
2.14 Mainzer, K.: Geschichte der Geometrie (see Note 13) 100; Abraham, R.H./Shaw, C.D.: Dynamics – The Geometry of Behavior Part 1. Aerial Press: Santa Cruz (1984) 20
2.15 Audretsch, J./Mainzer, K. (eds.): Philosophie und Physik der Raum-Zeit. B.I. Wissenschaftsverlag: Mannheim (1988) 30
2.16 Audretsch, J./Mainzer, K. (eds.): Philosophie und Physik der Raum-Zeit (see Note 15) 40; Weyl, H.: Raum, Zeit, Materie. Vorlesung über Allgemeine Relativitätstheorie. Wissenschaftliche Buchgesellschaft: Darmstadt (1961) (Reprint of the 5th Edition (1923)) 141
2.17 Mach, E.: Die Mechanik. Historisch-kritisch dargestellt. Wissenschaftliche Buchgesellschaft: Darmstadt (1976) (Reprint of the 9th Edition (1933)) 149; Abraham, R.H./Shaw, C.D.: Dynamics – The Geometry of Behavior (see Note 14) 57
2.18 Ruelle, D.: Small random pertubations of dynamical systems and the definition of attractors. Commun. Math. Phys. **82** (1981) 137–151; Abraham, R.H./ Shaw, C.D.: Dynamics – The Geometry of Behavior (see Note 14) 45
2.19 For an analytical elaboration cf. Stauffer, D./ Stanley, H.E.: From Newton to Mandelbrot. A Primer in Theoretical Physics. Springer: Berlin (1990) 26
2.20 Nicolis, G./Prigogine, I.: Die Erforschung des Komplexen (see Chapter 1, Note 3) 132; Abraham, R.H./Shaw, C.D.: Dynamics – The Geometry of Behavior (see Note 14) 168, 174
2.21 For an analytical elaboration cf. Mainzer, K.: Symmetries in Nature (see Note 1) Chapter 3.31; Stauffer, D./Stanley, H.E.: From Newton to Mandelbrot (see Note 19) 24
2.22 Arnold, V.I.: Mathematical Methods of Classical Mechanics. Springer: Berlin (1978); Davies, P.C.W.: The Physics of Time Asymmetry. Surrey University Press: London (1974); Penrose, R.: The Emperor's New Mind. Oxford University Press: Oxford (1989) 181
2.23 Lichtenberg, A.J./Liebermann, M.A.: Regular and Stochastic Motion. Springer: Berlin (1982); Schuster, H.G.: Deterministic Chaos. An Introduction. Physik-Verlag: Weinheim (1984) 137
2.24 Poincaré, H.: Les Méthodes Nouvelles de la Méchanique Céleste. Gauthier-Villars: Paris (1892)
2.25 Arnold, V.I.: Small Denominators II, Proof of a theorem of A.N. Kolmogorov on the preservation of conditionally-periodic motions under a small perturbation of the Hamiltonian, Russ. Math. Surveys **18** (1963) 5; Kolmogorov, A.N.: On Conservation of Conditionally-Periodic Motions for a Small Change in Hamilton's Function, Dokl. Akad. Nauk. USSR **98** (1954) 525; Moser, J.: Convergent series expansions of quasi-periodic motions, Math. Anm. **169** (1967) 163
2.26 Cf. Arnold, V.I.: Mathematical Methods of Classical Mechanics (see Note 22); Schuster, H.G.: Deterministic Chaos (see Note 23), 141
2.27 Hénon, M./Heiles, C.: The applicability of the third integral of the motion: Some numerical experiments, Astron. J. **69** (1964) pp. 73; Schuster, H.G.: Deterministic

Chaos (see Note 23), 150; Figures 2.16a–d from M.V. Berry in S. Jorna (ed.), Topics in nonlinear dynamics, Am. Inst. Phys. Conf. Proc. vol. **46** (1978)

2.28 For mathematical details compare, e.g. Staufer, D./Stanley, H.E.: From Newton to Mandelbrot (see Note 19), 83

2.29 Mainzer, K.: Symmetrien der Natur (see Note 1), 423; Primas, H./Müller-Herold, U.: Elementare Quantenchemie. Teubner: Stuttgart (1984) with an elementary introduction to the Galileo-invariant quantum mechanics (Chapter 3)

2.30 Audretsch, J./Mainzer, K. (eds.): Wieviele Leben hat Schrödingers Katze? B.I. Wissenschaftsverlag: Mannheim (1990)

2.31 Gutzwiller, M.C.: Chaos in Classical and Quantum Mechanics. Springer: Berlin (1990)

2.32 Friedrich, H.: Chaos in Atomen, in: Mainzer, K./Schirmacher, W. (eds.): Quanten, Chaos und Dämonen (see Note 1 of Chapter 1); Friedrich, H./Wintgen, D.: The hydrogen atom in a uniform magnetic field, Physics Reports **183** (1989) 37–79

2.33 Birkhoff, G.D.: Nouvelles recherches sur les systèmes dynamiques, Mem. Pont. Acad. Sci. Novi Lyncaei **1** (1935) 85

2.34 Enz, C.P.: Beschreibung nicht-konservativer nicht-linearer Systeme I–II, Physik in unserer Zeit **4** (1979) 119–126, **5** (1979) 141–144 (II)

2.35 Lorenz, E.N.: Deterministic nonperiodic flow, J. Atoms. Sci. **20** (1963) 130; Schuster, H.G.: Deterministic Chaos (see Note 23) 9

2.36 Eckmann, J.P.: Roads to turbulence in dissipative dynamical systems, Rev. Mod. Phys. **53** (1981) 643; Computer simulation of Fig. 2.21 from Lanford, O.E., Turbulence Seminar, in: Bernard, P./Rativ, T. (eds.): Lecture Notes in Mathematics 615, Springer: Berlin (1977) 114

2.37 Mandelbrot, B.B.: The Fractal Geometry of Nature, Freeman: San Fransisco (1982); Grassberger, P.: On the Hausdorff dimension of fractal attractors, J. Stat. Phys. **19** (1981) 25; Lichtenberg, A.J./Liebermann, M.A.: Regular and Stochastic Motions (see Note 23)

2.38 Collet, P./Eckmann, J.P.: Iterated Maps of the Interval as Dynamical Systems, Birkhäuser: Boston (1980) (see Figures 2.22–24)

2.39 Großmann, S./Thomae, E.: Invariant distributions and stationary correlation functions of one-dimensional discrete processes, Z. Naturforsch. **32 A** (1977) 353; Feigenbaum, M.J.: Quantitative universality for a class of nonlinear transformations, J. Stat. Phys. **19** (1978) 25

2.40 Mainzer, K.: Symmetrien der Natur (see Note 1)

2.41 Cf. Nicolis, G./Prigogine, I.: Die Erforschung des Komplexen (see Note 3, Chapter 1) 205

2.42 Cf. Prigogine, I.: From Being to Becoming – Time and Complexity in Physical Sciences, Freemann: San Fransisco (1980); Introduction to Thermodynamics of Irreversible Processes, Wiley: New York (1961)

2.43 Fig. 2.26 from Feynman, R.P./Leighton, R.B./Sands, M.: The Feynman Lectures of Physics vol. II., Addison-Wesley (1965)

2.44 Haken, H.: Synergetics (see Note 4, Chapter 1) 5

2.45 Haken, H.: Synergetics (see Note 4, Chapter 1) 202; Haken, H.: Advanced Synergetics. Instability Hierarchies of Self-Organizing Systems and Devices. Springer: Berlin (1983) 187; Weinberg, S.: Gravitation and Cosmology. Principles and Applications of the General Theory of Relativity. Wiley: New York (1972)

2.46 Cf. Mainzer, K.: Symmetrien der Natur (see Note 1) Chapter 4

2.47 Curie, P.: Sur la Symétrie dans les Phénomènes Physiques, Journal de Physique **3** (1894) 3

2.48 Audretsch, J./Mainzer, K. (eds.): Vom Anfang der Welt. C.H. Beck: München (^{2}1990); Mainzer, K.: Symmetrien der Natur (see Note 1) 515; Fritzsch, H.: Vom Urknall zum Zerfall. Die Welt zwischen Anfang und Ende. Piper: München (1983) 278

2.49 Hawking, S.: A Brief History of Time. From the Big Bang to Black Holes. Bantam Press: London (1988); Hoyle, F./Burbridge, G./Narlikar, J.V.: A quasi-steady state cosmological model with creation of matter. Astrophys. Journal **410** (1993) 437–457

2.50 Hartle, J.B./Hawking, S.W.: Wave function in the universe. Physical Review D 28 (1983) 2960–2975; Mainzer, K.: Hawking. Herder: Freiburg (2000)

2.51 Audretsch, J./Mainzer, K. (eds.): Vom Anfang der Welt (see Note 48) 165

2.52 Greene, B.: The Elegant Universe: Superstrings, Hidden Dimensions, and the Quest for the Ultimate Theory. W.W. Norton & Co: New York (1999); Hawking, S.W.: The Universe in a Nutshell. Bantam Books: New York (2001); Mainzer, K.: The Little Book of Time. Copernicus Books: New York (2002)

2.53 Whitesides, G.M./Mathias, J.P./Seto, C.T.: Molecular self-assembly and nanochemistry: A chemical strategy for the synthesis of nanostructures. Science **254** (1991) 1312–1319

2.54 Feynman, R.: There's plenty of room at the bottom. Miniaturization 282 (1961) 295–296

2.55 Drexler, K.E.: Nanotechnology summary. Encyclopedia Britannica Science and the Future Yearbook 162 (1990); compare also Drexler, K.E.: Nanosystems: Molecular Machinery, Manufacturing, and Computation. John Wiley & Sons: New York (1992)

2.56 Whitesides, G.M.: The once and future nanomachine. Scientific American 9 (2001) 78–83

2.57 Newkome, G.R. (ed.): Advances in Dendritic Macromolecules. JAI Press: Greenwich, Conn. (1994)

2.58 Curl, R.F./Smalley, R.E.: Probing C_{60}. Science **242** (1988) 1017–1022; Smalley, R.W.: Great balls of carbon: The story of Buckminsterfullerene. The Sciences **31** (1991) 22–28

2.59 Müller, A.: Supramolecular inorganic species: An expedition into a fascinating rather unknown land mesoscopia with interdisciplinary expectations and discoveries, J. Molecular Structure **325** (1994) 24; Angewandte Chemie (International Edition in English) **34** (1995) 2122–2124; Müller, A./Mainzer, K.: From molecular systems to more complex ones. In: Müller, A./Dress, A./Vögtle, F. (Eds.): From Simplicity to Complexity in Chemistry – and Beyond. Vieweg: Wiesbaden (1995) 1–11

2.60 Fig. 2.32 with drawings of Bryan Christie: Spektrum der Wissenschaft Spezial 2 (2001) 22

2.61 Dry, C.M.: Passive smart materials for sensing and actuation. Journal of Intelligent Materials Systems and Structures **4** (1993) 415

2.62 Amato, I.: Animating the material world. Science **255** (1992) 284–286

2.63 Joy, B.: Why the future doesn't need us. Wired 4 (2000)

2.64 Smalley, R.E.: Of chemistry, love and nanobots. Scientific American 9 (2001) 76–77

2.65 Abarbanel, H.D.I.: Analysis of Observed Data. Springer: New York (1996); Kanz, H./Schreiber, T.: Nonlinear Time Series Analysis. Cambridge University Press: Cambridge (1997)

2.66 Takens, F.: Detecting strange attractors in turbulence. In: Rand, D.A./Young, L.S. (eds.): Dynamical Systems and Turbulence. Springer: Berlin (1981) 366–381

2.67 Kaplan, D./Glass, L.: Understanding Nonlinear Dynamics. Springer: New York (1995) 310 (Fig. 6.20)

2.68 Kaplan, D./Glass, L.: Understanding Nonlinear Dynamics (see Note 67) 310 (Fig. 6.21), 311 (Fig. 6.22)

2.69 Kaplan, D./Glass, L.: Understanding Nonlinear Dynamics (see Note 67) 316 (Fig. 6.26), 317 (Fig. 6.28)

2.70 Grassberger, P./Procaccia I.: Characterization of strange attractors. Physical Review Letters 50 (1983) 346–349

2.71 Deco, G./Schürmann, B.: Information Dynamics: Foundations and Applications. Springer: New York (2001) 17 (Fig. 2.6)

2.72 Chen, G./Moiola, J.L.: An overview of bifurcation, chaos and nonlinear dynamics in control systems. In: Chua, L.O. (ed.): Journal of the Franklin Institute Engineering and Applied Mathematics: Philadelphia (1995) 838

2.73 Peitgen, H.-O., Richter, P.H.: The Beauty of Fractals: Images of Complex Dynamical Systems. Springer: Berlin (1986) 32

2.74 Mandelbrot, B.B.: Multifractals and $1/f$ Noise. Springer: Berlin (1997)

Chapter 3

3.1 For historical sources of Sect. 3.1 compare Mainzer, K.: Die Philosophen und das Leben. In: Fischer, E.P./Mainzer, K. (eds.): Die Frage nach dem Leben. Piper: München (1990) 11–44

3.2 Diels-Kranz (see Note 2, Chapter 2) 12 A 30

3.3 Aristotle: Historia animalium 588 b 4

3.4 Aristotle: De generatione animalium II 736 b 12-15. a 35-b2

3.5 Descartes, R.: Discours de la méthode. Leipzig (1919/20) 39

3.6 Borelli, G.A.: De motu animalium. Leipzig (1927) 1

3.7 Leibniz, G.W.: Monadology §64

3.8 Bonnet, C.: Contemplation de la nature (1764). Oeuvres VII, 45

3.9 Kant, I.: Kritik der Urteilskraft. Ed. G. Lehmann, Reclam: Stuttgart (1971) 340

3.10 Goethe, J.W.: Dichtung und Wahrheit. In: Werke (Hamburger Ausgabe) Bd. IX 490

3.11 Schelling, F.W.J.: Sämtliche Werke Bd.II (ed. Schröter, M.), München (1927) 206

3.12 Darwin, C.: On the Origin of Species by Means of Natural Selection, or the Preservation of Favoured Races in the Struggle for Life. London (1859); The Descent of Man, and Selection in Relation to Sex. London (1871); Spencer, H.: Structure, Function and Evolution (ed. Andrenski, S.), London (1971); For a modern evaluation of Darwin's position compare Richards, R.: The Meaning of Evolution. University of Chicago Press: Chicago (1992)

3.13 Boltzmann, L.: Der zweite Hauptsatz der mechanischen Wärmetheorie. In: Boltzmann, L. (ed.): Populäre Schriften. Leipzig (1905) 24–46

3.14 Cf. Schneider, I.: Rudolph Clausius' Beitrag zur Einführung wahrscheinlichkeitstheoretischer Methoden in die Physik der Gase nach 1856. Archive for the History of Exact Sciences **14** (1974/75) 237–261

3.15 Prigogine, I.: Introduction to Thermodynamics of Irreversible Processes (see Note 42, Chapter 2)

3.16 Cf. Boltzmann, L.: Über die mechanische Bedeutung des zweiten Hauptsatzes der Wärmetheorie (1866). In: Boltzmann, L.: Wissenschaftliche Abhandlungen (ed. Hasenöhrl, F.) vol. **1** Leipzig (1909), repr. New York (1968) 9–33; Analytischer Beweis

des zweiten Hauptsatzes der mechanischen Wärmetheorie aus den Sätzen über das Gleichgewicht der lebendigen Kraft (1871) 288–308
3.17 Cf., e.g., Einstein's famous article 'Über die von der molekularkinetischen Theorie der Wärme geforderte Bewegung von in ruhenden Flüssigkeiten suspendierten Teilchen'. Annalen der Physik **17** (1905) 549–560
3.18 Poincaré, H.: Sur les tentatives d'explication méchanique des principes de la thermodynamique. Comptes rendus de l'Académie des Sciences **108** (1889) 550–553; Zermelo, E.: Über einen Satz der Dynamik und die mechanische Wärmetheorie. Annalen der Physik **57** (1896) 485
3.19 Cf. Popper, K.R.: Irreversible processes in physical theory. Nature **181** (1958) 402–403; Reichenbach, H.: The Direction of Time. Berkeley (1956); Grünbaum, A.: Philosophical Problems of Space and Time. Dordrecht (1973); Hintikka, J./Gruender, D./Agazzi, E. (eds.): Probabilistic Thinking, Thermodynamics and the Interaction of the History and Philosophy of Science II. Dordrecht/Boston/ London (1978)
3.20 Boltzmann, L.: Der zweite Hauptsatz der mechanischen Wärmetheorie. In: Boltzmann, L.: Populäre Schriften (see Note 13) 26–46
3.21 Boltzmann, L.: Über die Frage nach der objektiven Existenz der Vorgänge in der unbelebten Natur. In: Boltzmann, L.: Populäre Schriften (see Note 13) 94–119
3.22 Monod, J.: Le Hasard et la Nécessité. Editions du Seuil: Paris (1970)
3.23 Primas, H.: Kann Chemie auf Physik reduziert werden? Chemie in unserer Zeit **19** (1985) 109–119, 160–166
3.24 Bergson, H.L.: L'évolution créative. Paris (1907); Heitler, W.H.: Über die Komplementarität von lebloser und lebender Materie. Abhandlungen der Math.-Naturw. Klasse d. Ak. d. Wiss. u. Lit. Mainz Nr. **1** (1976) 3–21; Driesch, A.: Philosophie des Organischen. Leipzig (1909); Whitehead, A.N.: Process and Reality. An Essay in Cosmology. New York (1978)
3.25 Schrödinger, E.: Was ist Leben? Piper: München (1987) 133
3.26 Schrödinger, E.: Was ist Leben? (see Note 25) 147
3.27 Thompson, W.: The Sorting Demon of Maxwell (1879). In: Thompson, W.: Physical Papers I–VI, Cambridge (1882–1911), V, 21–23
3.28 Prigogine, I.: Time, irreversibility and structure. In: Mehra, J. (ed.): The Physicist's Conception of Nature. D. Reidel: Dordrecht/Boston (1973) 589
3.29 Eigen, M.: The origin of biological information. In: Mehra, J. (ed.): The Physicist's Conception of Nature (see Note 28) 607; Fig. 3.2 shows a so-called coat gene obtained by nuclease digestion of phage MS2-RNA (Min Jou, W./Haegemann, G./Ysebaert, M./Fiers, W.: Nature **237** (1972) 82). This gene codes for a sequence of 129 amino acids. The structure is further spatially folded. Also compare Perelson, A.S./Kauffman, S.A. (eds.): Molecular Evolution on Ragged Landscapes: Proteins, RNA, and the Immune System. Santa Fé Institute Studies in the Sciences of Complexity, Proceedings vol. **9**. Addison-Wesley: Redwood City (1990)
3.30 For a survey cf. Depew, D.J./Weber, B.H.: Evolution at a Crossroads. The New Biology and the New Philosophy of Science. MIT Press: Cambridge, MA (1985); Ebeling, W./Feistel, R.: Physik der Selbstorganisation und Evolution. Akademie-Verlag: Berlin (1982); Haken, H./Haken-Krell, M.: Entstehung von biologischer Information und Ordnung. Wissenschaftliche Buchgesellschaft: Darmstadt (1989); Hofbauer, L.: Evolutionstheorie und dynamische Systeme. Mathematische Aspekte der Selektion. Springer: Berlin (1984)
3.31 Eigen, M.: Homunculus im Zeitalter der Biotechnologie – Physikochemische Grundlagen der Lebensvorgänge. In: Gross, R. (ed.): Geistige Grundlagen der Medizin.

Springer: Berlin (1985) 26, 36 for Fig. 3.4a–d; Maynard Smith, J.: Optimization theory in evolution. Annual Review of Ecological Systems **9** (1978) 31–56; Mainzer, K.: Metaphysics of nature and mathematics in the philosophy of Leibniz. In: Rescher, N. (ed.): Leibnizian Inquiries. University Press of America: Lanham/New York/London (1989) 105–130

3.32 Dyson, F.: Origins of Life. Cambridge University Press: Cambridge (1985)

3.33 Kauffman, S.: Autocatalytic sets of proteins. Journal of Theoretical Biology **119** (1986) 1–24

3.34 For a survey cf. Kauffmann, A.S.: Origins of Order: Self-Organization and Selection in Evolution. Oxford University Press: Oxford (1992)

3.35 Haken, H.: Synergetics (see Note 4, Chapter 1) 310

3.36 Hess, B./ Mikhailov, A.: Self-organization in living cells. In: Science **264** (1994) 223–224; Ber. Bunsenges. Phys. Chem. **98** (1994) 1198–1201 (extended version)

3.37 Susman, M.: Growth and Development. Prentice-Hall: Englewood Cliffs, NJ (1964); Prigogine, I.: Order through fluctuation. In: Jantsch, E./Waddington, C.H. (eds.): Evolution and Consciousness. Human Systems in Transition. Addison-Wesley: London (1976) 108

3.38 Gerisch, G./Hess, B.: Cyclic-AMP-controlled oscillations in suspended dictyostelium cells: Their relation to morphogenetic cell interactions. Proc. Natl. Acad. Sci. **71** (1974) 2118

3.39 Rosen, R.: Dynamical System Theory in Biology. Wiley-Interscience: New York (1970); Abraham, R.H./Shaw, C.D.: Dynamics – The Geometry of Bahavior (see Note 14, Chapter 2) 110 for Figs. 3.6

3.40 Meinhardt, H./Gierer, A.: Applications of a theory of biological pattern formation based on lateral inhibition. J. Cell. Sci. **15** (1974) 321 (Figs. 3.7–8); Meinhardt, M.: Models of Biological Pattern Formation. Academic Press: London (1982)

3.41 For a survey compare Gerok, W. (ed.): Ordnung und Chaos in der belebten und unbelebten Natur. Verhandlungen der Gesellschaft Deutscher Naturforscher und Ärzte. 115. Versammlung (1988), Stuttgart (1989); Mainzer, K.: Chaos und Selbstorganisation als medizinische Paradigmen. In: Deppert, W./Kliemt, H./Lohff, B./Schaefer, J. (eds.): Wissenschaftstheorien in der Medizin. De Gruyter: Berlin/New York (1992) 225–258

3.42 Bassingthwaighte, J.B./van Beek, J.H.G.M: Lightning and the heart: Fractal behavior in cardiac function. Proceedings of the IEEE **76** (1988) 696

3.43 Goldberger, A.L./Bhargava, V./West, B.J.: Nonlinear dynamics of the heartbeat. Physica **17D** (1985) 207–214; Nonlinear dynamics in heart failure: Implications of long-wavelength cardiopulmonary oscillations. American Heart Journal **107** (1984) 612–615; Ree Chay, T./Rinzel, J.: Bursting, beating, and chaos in an excitable membrance model. Biophysical Journal **47** (1985) 357–366; Winfree, A.T.: When Time Breaks Down: The Three-Dimensional Dynamics of Electrochemical Waves and Cardiac Arrhythmias. Princeton (1987); Guevara, M.R./Glass, L./Schrier, A.: Phase locking, period-doubling bifurcations, and irregular dynamics in periodically stimulated cardiac cells. Science **214** (1981) 1350

3.44 Cf. Johnson, L.: The thermodynamic origin of ecosystems: a tale of broken symmetry. In: Weber, B.H./Depew, D.J./Smith, J.D. (eds.): Entropy, Information, and Evolution. New Perspectives on Physical and Biological Evolution. MIT Press. Cambridge, MA (1988) 75–105; Schneider, E.D.: Thermodynamics, ecological succession, and natural selection: a common thread. In: Weber, B.H./Depew, D.J./Smith, J.D. (eds.): Entropy, Information, and Evolution (see Note 44) 107–138

3.45 Odum, E.P.: The strategy of ecosystem development. Science **164** (1969) 262–270; Margalef, R.: Perspectives in Ecological Theory. University of Chicago Press: Chicago (1968)
3.46 Lovelock, J.E.: The Ages of Gaia. Bantam (1990); Schneider, S.H./Boston, P.J. (eds.): Scientists on Gaia. MIT Press: Cambridge, MA (1991); Pimm, S.: The Balance of Nature, University of Chicago Press: Chicago (1991)
3.47 Cf. Rosen, R.: Dynamical System Theory in Biology (see Note 37); Freedmann, H.I.: Deterministic Mathematical Models in Population Ecology. Decker: New York (1980); Abraham, R.H./Shaw, C.D.: Dynamics – The Geometry of Behavior (see Note 14, Chapter 2) 85
3.48 Lotka, A.J.: Elements of Mathematical Biology. Dover: New York (1925); Volterra, V.: Leçons sur la théorie mathématique de la lutte pour la vie. Paris (1931); Haken, H.: Synergetics (see Note 4, Chapter 1) 130, 308
3.49 Rettenmeyer, C.W.: Behavioral studies of army ants. Kansas Univ. Bull. **44** (1963) 281; Prigogine, I.: Order through Fluctuation: Self-Organization and Social System. In: Jantsch, E./Waddington, C.H. (eds.): Evolution and Consciousness (see Note 37) 111
3.50 Prigogine, I./Allen, P.M.: The challenge of complexity. In: Schieve, W.C./Allen, P.M.: Self-Organization and Dissipative Structures. Applications in the Physical and Social Sciences. University of Texas Press: Austin (1982) 28; Wicken, J.S.: Thermodynamics, evolution, and emergence: Ingredients of a new synthesis. In: Weber, B.H./Depew, D.J./Smith, J.D. (eds.): Entropy, Information, and Evolution (see Note 44) 139–169
3.51 Bassingthwaighte, J.B., Liebovitch, L.S., West, B.J.: Fractal Physiology. Oxford University Press: Oxford (1994) 16
3.52 West, B.J., Brown, J.H., Enquist, B.J.: A general model for the origin of allometric scaling laws in biology. In: Science **276** (1997) 122
3.53 West, B.J., Brown, J.H., Enquist, B.J.: The fourth dimension of life: Fractal geometry and allometric scaling of organisms. In: Science **284** (1999) 1677
3.54 West, B.J., Deering, B.: The Lure of Modern Science: Fractal Thinking. World Scientific: Singapore (1995) 219

Chapter 4

4.1 Diels-Kranz: B 36
4.2 Cf. Guthrie, W.K.C.: A History of Greek Philosophy vol. I: The Earlier Presocratics and the Pythagoreans. Cambridge University Press: Cambridge (1962) 349; Popper, K.R./Eccles, J.C.: The Self and its Brain. Springer: Berlin (1977) 161
4.3 Aristotle: De anima 403 b 31
4.4 Plato: Menon
4.5 Cf. Galen: Galen on Anatomical Procedures. Translation of the Surviving Books with Introduction and Notes. Oxford University Press: London (1956)
4.6 Wickens, G.M: Avicenna. Scientist and Philosopher. A Millenary Symposium: London (1952)
4.7 Descartes, R.: Meditations (1641). Eds. E. Haldane, G. Ross. Cambridge University Press: Cambridge (1968) 153
4.8 Descartes, R.: Treatise on Man (1664). Harvard University Press: Cambridge, Mass. (1972)

4.9 Spinoza, B.: Ethics
4.10 Leibniz, G.W.: Monadology; Rescher, N.: Leibniz: An Introduction to his Philosophy. Basil Blackwell: Oxford (1979)
4.11 Hume, D.: A Treatise of Human Nature (1739). Penguin: Harmondsworth (1969) 82
4.12 Mainzer, K.: Kants Begründung der Mathematik und die Entwicklung von Gauß bis Hilbert. In: Akten des V. Intern. Kant-Kongresses in Mainz 1981 (ed. Funke, G.). Bouvier: Bonn (1981) 120–129
4.13 Brazier, M.A.B.: A History of Neurophysiology in the 17th and 18th Centuries. Raven: New York (1984); Cf. Clarke, E./O'Malley, C.D.: The Human Brain and Spinal Cord: A Historical Study illustrated by Writings from Antiquity to the Twentieth Century. University of California Press: Berkeley (1968)
4.14 Helmholtz, H.v.: Schriften zur Erkenntnistheorie (eds. Hertz, P./Schlick, M.). Berlin (1921); Mainzer, K.: Geschichte der Geometrie (see Note 13 Chapter 2) 172
4.15 Müller, J.: Handbuch der Physiologie des Menschen. Koblenz (1835)
4.16 Helmholtz, H.v.: Vorläufiger Bericht über die Fortpflanzungsgeschwindigkeit der Nervenreizung. Archiv für Anatomie, Physiologie und wissenschaftliche Medizin (1850) 71–73
4.17 James, W.: Psychology (Briefer Course). Holt: New York (1890) 3
4.18 James, W.: Psychology (see Note 17) 254
4.19 James, W.: Psychology (see Note 17) Fig. 57
4.20 Cf. Baron, R.J.: The Cerebral Computer. An Introduction to the Computational Structure of the Human Brain. Lawrence Erlbaum: Hillsdale N.J. (1987); Braitenberg, V.: Gehirngespinste. Neuroanatomie für kybernetisch Interessierte. Springer: Berlin (1973)
4.21 Churchland, P.S./Sejnowski, T.J.: Perspectives in cognitive neuroscience. Science **242** (1988) 741–745. The subset of visual cortex is adapted from van Essen, D./Maunsell, J.H.R.: Two-dimensional maps of the cerebral cortex. Journal of Comparative Neurology **191** (1980) 255–281. The network model of ganglion cells is given in Hubel, D.H./Wiesel, T.N.: Receptive fields, binocular interaction and functional architecture in the cat's visual cortex. Journal of Physiology **160** (1962) 106–154. An example of chemical synapses is shown in Kand, E.R./Schwartz J.: Principles of Neural Science. Elsevier: New York (1985)
4.22 Cf. Churchland, P.M.: A Neurocomputational Perspective: The Nature of Mind and the Structure of Science. MIT Press: Cambridge, Mass./London (1989) 99
4.23 Pellionisz, A.J.: Vistas from tensor network theory: A horizon from reductionalist neurophilosophy to the geometry of multi-unit recordings. In: Cotterill, R.M.J. (ed.): Computer Simulation in Brain Science. Cambridge University Press: Cambridge/New York/Sydney (1988) 44–73; Churchland, P.M.: A Neurocomputational Perspective (see Note 22) 83, 89
4.24 Cf. Schwartz, E.L. (ed.): Computational Neuroscience. MIT Press: Cambridge, Mass. (1990)
4.25 Cf. Churchland, P.S./Sejnowski, T.J.: The Computational Brain. MIT Press: Cambridge, Mass. (1992) 169
4.26 Hebb, D.O.: The Organization of Behavior. Wiley: New York (1949) 50
4.27 Kohonen, T.: Self-Organization and Associative Memory. Springer: Berlin (1989) 105; Churchland, P.S./Sejnowski, T.J.: The Computational Brain (see Note 25) 54; Ritter, H./Martinetz, T./Schulten, K.: Neuronale Netze. Eine Einführung in die Neuroinformatik selbstorganisierender Netzwerke. Addison-Wesley: Reading, Mass. (1991) 35

4.28 Hopfield, J.J.: Neural Network and physical systems with emergent collective computational abilities. Proceedings of the National Academy of Sciences **79** (1982) 2554–2558

4.29 Hertz, J./Krogh, A./Palmer, R.G.: Introduction to the Theory of Neural Computation. Addison-Wesley: Redwood City (1991)

4.30 Serra, R./Zanarini, G.: Complex Systems and Cognitive Processes. Springer: Berlin (1990) 78

4.31 Hertz, J./Krogh, A./Palmer, R.G.: Introduction to the Theory of Neural Computation (see Note 29); Hopfield, J.J./Tank, D.W.: Computing with neural circuits: A model. Science **233** (1986) 625–633

4.32 Ackley, D.H./Hinton, G.E./Sejnowski, T.J.: A learning algorithm for Boltzmann machines. Cognitive Science **9** (1985) 147–169

4.33 A mathematical elaboration of the learning algorithm for a Boltzmann machine is given in Serra, R./Zanarini, G.: Complex Systems and Cognitive Processes (see Note 30) 137. An illustration is shown in Churchland, P.S./Sejnowski, T.J.: The Computational Brain (see Note 25) 101

4.34 Rumelhart, D.E./Zipser, D.: Feature discovery by competitive learning. In: McClelland, J.L./Rumelhart, D.E. (eds.): Parallel Distributed Processing. MIT Press: Cambridge, Mass. (1986)

4.35 Kohonen, T.: Self-Organization and Associative Memory (see Note 27) 123

4.36 Kohonen, T.: Self-Organization and Associative Memory (see Note 27) 125

4.37 Ritter, H./Martinetz, T./Schulten, K.: Neuronale Netze (see Note 27) 75

4.38 Suga, N./O'Neill, W.E.: Neural axis representing target range in the auditory cortex of the mustache Bat. Science **206** (1979) 351–353; Ritter, H./Martinetz, T./Schulten, K.: Neuronale Netze (see Note 27) 88

4.39 Widrow, B./Hoff, M.E.: Adaptive switching circuits. 1960 IRE WESCON Convention Record. IRE: New York (1960) 36–104

4.40 Cf. Churchland, P.S./Sejnowski, T.J.: The Computational Brain (see Note 25) 106

4.41 Rumelhart, D.E./Hinton, G.E./Williams, R.J.: Learning representations by back-propagating errors. Nature **323** (1986) 533–536; Arbib, M.A.: Brains, Machines, and Mathematics. Springer: New York (1987) 117

4.42 Köhler, W.: Die physischen Gestalten in Ruhe und im stationären Zustand. Vieweg: Braunschweig (1920); Jahresberichte für die ges. Physiol. und exp. Pharmakol. **3** (1925) 512–539; Stadler, M./Kruse, P.: The self-organization perspective in cognitive research: Historical remarks and new experimental approaches. In: Haken, H./Stadler, M. (eds.): Synergetics of Cognition. Springer: Berlin (1990) 33

4.43 Cf. Churchland, P.M.: A Neurocomputational Perspective (see Note 22) 209

4.44 Cf. Churchland, P.M.: A Neurocomputational Perspective (see Note 22) 211

4.45 Cf. Feigl, H./Scriven, M./Maxwell, G. (eds.): Concepts, Theories and the Mind-Body Problem. University of Minnesota Press: Minneapolis (1958); Marcel, A.J./Bisiach, E. (eds.): Consciousness in Contemporary Science. Clarendon Press: Oxford (1988); Bieri, P.: Pain: A case study for the mind-body problem. Acta Neurochirurgica **38** (1987) 157–164; Lycan, W.G.: Consciousness. MIT Press: Cambridge, Mass. (1987)

4.46 Flohr, H.: Brain processes and phenomenal consciousness. A new and specific hypothesis. Theory & Psychology **1**(2) (1991) 248

4.47 von der Malsburg, C.: Self-organization of orientation sensitive cells in the striate cortex. Kybernetik **14** (1973) 85–100; Wilshaw, D.J./von der Malsburg, C.: How patterned neural connections can be set up by self-organization. Proceedings of the Royal Society Series B **194** (1976) 431–445

4.48 Barlow, H.B.: Single units and sensatioperceptual psychology. Perception 1 (1972) 371

4.49 Singer, W.: The role of synchrony in neocortical processing and synaptic plasticity. In: Domany, E./Van Hemmen, L./Schulten, K. (eds.): Model of Neural Networks II. Springer: Berlin (1994)

4.50 Deco, G./Schürmann, B.: Information Dynamics: Foundations and Applications (see Note 2.71) 229 (Fig. 10.1)

4.51 Cf. Pöppel, E. (ed.): Gehirn und Bewußtsein. VCH Verlagsgesellschaft: Weinheim (1989); Singer, W. (ed.): Gehirn und Kognition. Spektrum der Wissenschaft: Heidelberg (1990)

4.52 Haken, H./Stadler, M. (eds.): Synergetics of Cognition (see Note 42) 206

4.53 Haken, H./Stadler, M. (eds.): Synergetics of Cognition (see Note 42) 204

4.54 Pöppel, E.: Die neurophysiologische Definition des Zustands "bewußt". In: Pöppel, E. (ed.): Gehirn und Bewußtsein (see Note 51) 18

4.55 Searle, J.R.: Intentionality. An Essay in the Philosophy of Mind. Cambridge University Press: Cambridge (1983); Dennett, D.: The Intentional Stance, MIT Press: Cambridge, Mass. (1987)

4.56 Shaw, R.E./Kinsella-Shaw, J.M.: Ecological mechanics: A physical geometry for intentional constraints. Hum. Mov. Sci. **7** (1988) 155

4.57 For Figs. 4.23a–d, Kugler, P.N./Shaw, R.E.: Symmetry and symmetry breaking in thermodynamic and epistemic engines: A coupling of first and second laws. In: Haken, H./Stadler, M. (eds.): Synergetics of Cognition (see Note 42) 317, 318, 319, 328

4.58 Kelso, J.A.S./Mandell, A.J./Shlesinger, M.F. (eds.): Dynamic Patterns in Complex Systems. World Scientific: Singapore (1988); For Figs. 4.24a–b, 4.25 compare Haken, H./Haken-Krell, M.: Erfolgsgeheimnisse der Wahrnehmung. Deutsche Verlags-Anstalt: Stuttgart (1992) 36, 38

4.59 Kelso, J.A.S.: Phase transitions: Foundations of behavior. In: Haken, H./Stadler, M. (eds.): Synergetics of Cognition (see Note 42) 260

4.60 Searle, J.R.: Mind, brains and programs. Behavioral and Brain Science **3** (1980) 417–424; Intrinsic intentionality. Behavioral and Brain Science **3** (1980) 450–456; Analytic philosophy and mental phenomena. Midwest Studies in Philosophy **5** (1980) 405–423. For a critique of Searle's position compare Putnam, H.: Representation and Reality. MIT Press: Cambridge, Mass. (1988) 26

4.61 Eccles, J.C.: The Neurophysiological Basis of Mind. Clarendon Press: Oxford (1953); Facing Reality. Springer: New York (1970); Eccles, J.C. (ed.): Mind and Brain, Paragon: Washington, D.C. (1982)

4.62 Palm, G.; Assoziatives Gedächtnis und Gehirn. In: Singer, W. (ed.): Gehirn und Kognition (see Note 51) 172; Palm, G. (ed.): Neural Assemblies: An Alternative Approach to Artificial Intelligence. Springer: Berlin (1984)

4.63 Merleau-Ponty, M.: Phenomenology of Perception. Routledge & Kegan Paul: London (1962)

4.64 Freeman, W.J.: How and why brains create sensory information. In: International Journal of Bifurcation and Chaos **14** (2004) 515–530

4.65 Förstl, H. (Ed.): Theory of Mind. Neurobiologie und Psychologie sozialen Verhaltens. Springer: Berlin (2007)

4.66 Dreyfus, H.L.: Husserl, Intentionality, and Cognitive Science. MIT Press: Cambridge MA (1982); Searle, J.R.: Intentionality. An Essay in the Philosophy of Mind. Cambridge University Press: Cambridge (1983)

Chapter 5

5.1 For this chapter compare Mainzer, K.: Computer – Neue Flügel des Geistes? De Gruyter: Berlin/New York (1993); Mainzer, K.: Die Evolution intelligenter Systeme. Zeitschrift für Semiotik **12** (1990) 81–104

5.2 Feigenbaum, E.A./McCorduck, P.: The Fifth Generation. Artificial Intelligence and Japan's Computer Challenge to the World. Michael Joseph: London (1984)

5.3 Cf. Williams, M.R.: A History of Computing Technology. Prentice-Hall: Englewood Cliffs (1985)

5.4 Cohors-Fresenborg, E.: Mathematik mit Kalkülen und Maschinen. Vieweg: Braunschweig (1977) 7

5.5 Herrn von Leibniz' Rechnung mit Null und Eins. Siemens AG: Berlin (1966); Mackensen, L. von: Leibniz als Ahnherr der Kybernetik – ein bisher unbekannter Leibnizscher Vorschlag einer "Machina arithmetica dyadicae". In: Akten des II. Internationalen Leibniz-Kongresses 1972 Bd. **2**. Steiner: Wiesbaden (1974) 255–268

5.6 Scholz, H.: Mathesis Universalis. Schwabe: Basel (1961)

5.7 Babbage, C.: Passages from the Life of a Philosopher. Longman and Co.: London (1864); Bromley, A.G.: Charles Babbage's Analytical Engine 1838. Annals of the History of Computing **4** (1982) 196–219

5.8 Cf. Minsky, M.: Recursive unsolvability of Post's problem of "tag" and other topics in the theory of Turing machines. Annals of Math. **74** 437–454; Sheperdson, J.C./Sturgis, H.E.: Computability of recursive functions. J. Assoc. Comp. Mach. **10** (1963) 217–255. For the following description of register machines compare Rödding, D.: Klassen rekursiver Funktionen. In: Löb, M.H. (ed.): Proceedings of the Summer School in Logic. Springer: Berlin (1968) 159–222; Cohors-Fresenborg, E.: Mathematik mit Kalkülen und Maschinen (see Note 4)

5.9 Turing, A.M.: On computable numbers, with an application to the "Entscheidungsproblem". Proc. London Math. Soc., Ser. **2** **42** (1936) 230–265; Post, E.L.: Finite combinatory processes – Formulation I. Symbolic Logic I (1936) 103–105; Davis, M.: Computability & Unsolvability. McGraw-Hill: New York (1958) 3

5.10 Arbib, M.A.: Brains, Machines, and Mathematics. Springer: New York (1987) 131

5.11 Mainzer, K.: Der Konstruktionsbegriff in der Mathematik. Philosophia Naturalis **12** (1970) 367–412

5.12 Cf. Knuth, D.M.: The Art of Computer Programming. Vol. 2. Addison-Wesley: Reading, MA (1981); Börger, E.: Berechenbarkeit, Komplexität, Logik. Vieweg: Braunschweig (1985)

5.13 Grötschel, M./Lovásc, L./Schryver, A.: Geometric Algorithms and Combinatorial Optimization. Springer: Berlin (1988)

5.14 Gardner, A.: Penrose Tiles to Trapdoor Ciphers. W.H. Freeman: New York (1989)

5.15 Cf. Arbib, M.A.: Speed-up theorems and incompleteness theorem. In: Caianiello, E.R. (ed.): Automata Theory. Academic Press (1966) 6–24; Mostowski, A.: Sentences Undecidable in Formalized Arithmetic. North-Holland: Amsterdam (1957); Beth, E.W.: The Foundations of Mathematics, North-Holland: Amsterdam (1959); Kleene, S.C.: Introduction to Metamathematics. North-Holland: Amsterdam (1967)

5.16 Gödel, K.: On formally undecidable propositions of Principia Mathematica and related systems I. Monatshefte für Mathematik und Physik 38 (1931) 173–198

5.17 Turing, A.: On computable numbers, with an application to the entscheidungsproblem. London Mathematical Society Series 2 42 (1936–1937) 230–265; A correction. Ibid. 43 (1937) 544–546; cf. Chaitin, G.J.: The Limits of Mathematics: A Course on Information Theory and the Limits of Formal Reasoning. Springer: New York (1998)

5.18 Chaitin, G.J.: On the length of programs for computing finite binary sequences. Journal of the ACM 13 (1966) 547–569; On the length of programs for computing finite binary sequences: statistical considerations. Journal of the ACM 16 (1969) 145–159

5.19 Chaitin, G.J.: The Limits of Mathematics: A Course on Information Theory and the Limits of Formal Reasoning (see Note 17) 60

5.20 Shannon, C.E./Weaver, W.: The Mathematical Theory of Communication. University of Illinois Press: Chicago (1949); cf. also Cover, T./Thomas, J.: Elements of Information Theory. John Wiley & Sons: New York (1991)

5.21 Bennett, C.H.: Quantum Information and Computation. Physics Today 10 (1995)

5.22 Deco, G./Schlittenkopf, C./Schürmann, B.: Determining the information flow of dynamical systems from continuous probability distributions. Physical Review Letters 78 (1997) 2345

5.23 Deco, G./Schürmann, B.: Information Dynamics: Foundations and Applications. (see Note 2.71) 111 (Fig. 5.2)

5.24 Beltrami, E.: What is Random? Chance and Order in Mathematics and Life. Springer: New York (Copernicus)(1999) 133

5.25 Beltrami, E.: What is Random? (see Note 24) 133

5.26 Press, W.H.: Flicker noise in astronomy and elsewhere. In: Comments Astrophys. **7** (1978) 103–119

5.27 Mandelbrot, B.B.: Multifractals and $1/f$ Noise. Springer: Berlin (1997)

5.28 Gnedenko, B.V., Kolmogorov, A.N.: Limit Distributions for Sums of Independent Random Variables. Addison-Wesley: Cambridge MA (1954)

5.29 Mantegna, R.N., Stanley, H.E.: An Introduction to Econophysics. Correlations and Complexity in Finance. Cambridge University Press: Cambridge (2000) 16 (Fig. 3.1)

5.30 Mantegna, R.N., Stanley, H.E.: An Introduction to Econophysics (see Note 29) 16 (Fig. 3.2)

5.31 Feller, W.: An Introduction to Probability Theory and Its Applications, Vol. 1. J. Wiley & Sons: New York (1968)

5.32 Mantegna, R.N., Stanley, H.E.: An Introduction to Econophysics (see Note 29) 19 (Fig. 3.4)

5.33 Lévy, P.: Calcul des probabilités, Gauthier-Villars: Paris (1925)

5.34 Mantegna, R.N., Stanley, H.E.: An Introduction to Econophysics (see Note 29) 28 (Fig. 4.1)

5.35 Lévy, M., Solomon, S.: Power laws are logarithmic Boltzmann laws. In: Intl. J. Mod. Phys. C **7** (1996) 595–601

5.36 Audretsch, J./Mainzer, K. (eds.): Wieviele Leben hat Schrödingers Katze? B.I. Wissenschaftsverlag: Mannheim (1990) 273 (Fig. 5.14a), 274 (Fig. 5.14b), 276 (Fig. 5.14c)

5.37 Penrose, R.: Newton, quantum theory and reality. In: Hawking, S.W./Israel, W.: 300 Years of Gravity. Cambridge University Press: Cambridge (1987)

5.38 Deutsch, D.: Quantum theory, the Church-Turing principle and the universal quantum computer. Proc. Roy. Soc. **A400** (1985) 97–117

5.39 Nielsen, M.A., Chuang, I.L.: Quantum Computation and Quantum Information. Cambridge University Press: Cambridge (2001)

5.40 Keyl, M.: Fundamentals of quantum information theory. In: Physics Report **369** (2002) 435

5.41 Keyl, M.: Fundamentals of quantum information theory (see Note 40) 436

5.42 Bouwmeester, D., Ekert, A., Zeilinger, A. (eds.): The Physics of Quantum Information. Quantum Cryptography, Quantum Teleportation, Quantum Computation. Springer: Berlin (2000) 12
5.43 Bouwmeester, D., Ekert, A., Zeilinger, A. (eds.): The Physics of Quantum Information (see Note 5.42) 50
5.44 Mainzer, K.: Symmetry and Complexity. The Spirit and Beauty of Nonlinear Science. World Scientific: Singapore (2005) 110
5.45 Horowitz, G.T., Maldacena, J.: The black hole final state. In: Journal of High Energy Physics **2** (2004) 8
5.46 For this chapter compare Mainzer, K.: Computer – Neue Flügel des Geistes? De Gruyter: Berlin (1993); Mainzer, K.: Philosophical concepts of computational neuroscience. In: Eckmiller, R./Hartmann, G./Hauske, G. (eds.): Parallel Processing in Neural Systems and Computers. North-Holland: Amsterdam (1990) 9–12
5.47 McCullock, W.S./Pitts, W.: A logical calculus of the ideas immanent in nervous activity. In: Bulletin of Mathematical Biophysics **5** (1943) 115–133; Arbib, M.A.: Brains, Machines and Mathematics (see Note 10) 18
5.48 Neumann, J. von: The Computer and the Brain. Yale University Press: New Haven (1958)
5.49 Cf. Burks, A.W. (ed.): Essays on Cellular Automata. University of Illinois Press (1970); Myhill, J.: The abstract theory of self-reproduction. In: Mesarovic, M.D. (ed.): Views on General Systems Theory. John Wiley (1964) 106–118
5.50 Wolfram, S.: Universality and complexity in cellular automata. In: Farmer, D./Toffoli, T./Wolfram, S. (eds.): Cellular Automata. Proceedings of an Interdisciplinary Workshop. North-Holland: Amsterdam (1984) 1–35; Wolfram, S.: Theory and Applications of Cellular Automata. World Scientific: Singapore (1986); Demongeot, J./Golés, E./Tchuente, M. (eds.): Dynamical Systems and Cellular Automata. Academic Press: London (1985); Poundstone, W.: The Recursive Universe. Cosmic Complexity and the Limits of Scientific Knowledge. Oxford University Press: Oxford (1985)
5.51 Mainzer, K.: Komplexität in der Natur. In: Nova Acta Leopoldina NF 76 Nr. 303 (1997) 165–189
5.52 Wolfram, S.: A New Kind of Science. Wolfram Media, Inc.: Champaign, Ill. (2002) 443
5.53 Wolfram, S.: A New Kind of Science (see Note 52) 737

Chapter 6

6.1 Pfeifer, R., Scheier, C.: Understanding Intelligence. M.I.T. Press: Cambridge MA (2001) 45 (Fig. 2.5)
6.2 Mainzer, K.: Rationale Heuristik und Problem Solving. In: Burrichter, C. Inhetveen, R. Kötter, R. (eds.): Technische Rationalität und rationale Heuristik. Schöningh: Paderborn (1986) 83–97
6.3 Mainzer, K.: Knowledge-based systems. Remarks on the philosophy of technology and Artificial Intelligence. Journal for General Philosophy of Science **21** (1990) 47–74
6.4 Bibel, W. Siekmann, J.: Künstliche Intelligenz. Springer: Berlin (1982); Nilson, N.J.: Principles of Artificial Intelligence. Springer: Berlin (1982); Kredel, L.: Künstliche Intelligenz und Expertensysteme. Droemer Knaur: München (1988)
6.5 Puppe, F.: Expertensysteme. Informatik-Spektrum **9** (1986) 1–13

6.6 For instance, one of the constructors of DENDRAL and MYCIN received a Ph.D. in philosophy for his dissertation "Scientific Discoveries", supervised by Gerald Massey (today Center for Philosophy of Science, Pittsburgh). Also refer to Glymour, C.: AI is philosophy. In: Fetzer, J.H. (ed.): Aspects in Artificial Intelligence. Kluwer Academic Publisher: Boston (1988) 195–207

6.7 Buchanan, B.G. Sutherland, G.L. Feigenbaum, E.A.: Heuristic DENDRAL: A program for generating processes in organic chemistry. In: Meltzer, B. Michie, D. (eds.): Machine Intelligence **4**. Edinburgh (1969); Buchanan, B.G. Feigenbaum, E.A.: DENDRAL and META-DENDRAL: Their applications dimension. In: Artificial Intelligence **11** (1978) 5–24

6.8 McCarthy, J.: LISP Programmer's Manual (Rep. MIT Press). Cambridge, MA (1962); Stoyan, H. Goerz, G.: LISP – Eine Einführung in die Programmierung. Springer: Berlin (1984)

6.9 Randall, D. Buchanan, B.G. Shortliffe, E.H.: Production rules as a representation for a knowledge-based consultation program. Artificial Intelligence **8** (1977); Shortliffe, E.H.: MYCIN: A rule-based computer program for advising physicians regarding antimicrobial therapy selection. AI Laboratory, Memo 251, STAN-CS-74-465, Stanford University; Winston, P.H.: Artificial Intelligence. Addison-Wesley: Reading, MA (1977) Chap. 9

6.10 Doyle, J.: A truth maintenance system. Artificial Intelligence **12** (1979) 231–272

6.11 Lenat, D.B. Harris, G.: Designing a rule system that searches for scientific discoveries. In: Waterman, D.A. Hayes-Roth, F. (eds.): Pattern Directed Inference Systems. New York (1978) 26; Lenat, D.B.: AM: Discovery in mathematics as heuristic search. In: Davis, R. Lenat, D.B. (eds.): Knowledge-Based Systems in Artificial Intelligence. New York (1982) 1–225

6.12 Langley, P.: Data-driven discovery on physical laws. Cognitive Science **5** (1981) 31–54

6.13 Langley, P. Simon, H.A. Bradshaw, G.L. Zytkow, J.M.: Scientific Discovery: Computational Explorations of the Creative Processes. Cambridge, MA (1987)

6.14 Kulkarni, D. Simon, H.A.: The process of scientific discovery: The strategy of experimentation. Cognitive Science **12** (1988) 139–175

6.15 Cf. Winston, P.H.: Artificial Intelligence (see Note 9)

6.16 Rosenblatt, F.: The perceptron: A probabilistic model for information storage and organization in the brain. Psychological Review **65** (1958) 386–408

6.17 Minsky, M. Papert, S.A.: Perceptrons. MIT Press: Cambridge MA (1969) (expanded edition 1988)

6.18 Gorman, R.P. Sejnowski, T.J.: Analysis of hidden units in a layered network trained to classify sonar targets. Neural Networks **1** (1988) 75–89; Churchland, P.M.: A Neurocomputational Perspective. The Nature of Mind and the Structure of Science. MIT Press: Cambridge MA (1989)

6.19 Sejnowski, T.J. Rosenberg, C.R.: NETtalk: A parallel network that learns to read aloud. The Johns Hopkins University Electrical Engineering and Computer Science Technical Report IHU/EECS-86/01 (1986) 32; Cf. Kinzel, W. Deker, U.: Der ganz andere Computer: Denken nach Menschenart. Bild der Wissenschaft **1** (1988) 43

6.20 Schöneburg, E. Hansen, N. Gawelczyk: Neuronale Netzwerke. Markt & Technik: München (1990) 176, 177

6.21 Rumelhart, D.E. Smolensky, P. McClelland, J.L. Hinton, G.E.: Schemata and sequential thought processes. In: McClelland, J.L. Rumelhart, D.E. (eds.): Parallel Distributed Processing: Explorations in the Microstructure of Cognition vol. 2: Applications. MIT Press: Cambridge MA (1986); Churchland, P.S.: Neurophilosophy. Toward a Unified Science of the Mind-Brain. MIT Press: Cambridge MA (1988) 465

6.22 Haken, H. (ed.): Neural and Synergetic Computers. Springer: Berlin (1988) 5–7

6.23 Haken, H.: Synergetics as a tool for the conceptualization and mathematization of cognition and behaviour – How far can we go? In: Haken, H. Stadler, M. (eds.): Synergetics of Cognition. Springer: Berlin (1990) 23–24

6.24 Chua, L.O.: CNN: A Paradigm for Complexity. World Scientific: Singapore (1998); Huertas, J.L. Chen, W.K. Madan, R.N. (eds.): Visions of Nonlinear Science in the 21st Century. Festschrift Dedicated to Leon O. Chua on the Occasion of his 60th Birthday. World Scientific: Singapore (1999); Mainzer, K.: CNN and the evolution of complex information systems in nature and society. In: Tetzlaff, R. (ed.): Cellular Neural Networks and their Applications. Proc. 7th IEEE International CNN Workshop. World Scientific: Singapore (2002) 483–497

6.25 Chua, L.O. Yang, L.: Cellular neural networks: Theory. IEEE Transactions on Circuits and Systems 35 (1988) 1257–1272; Cellular neural networks: applications. IEEE Transactions on Circuits and Systems 35 (1988) 1273–1290

6.26 Chua, L.O. Gulak, G. Pierzchala, E. Rodriguez-Vázquez (eds.): Cellular neural networks and analog VLSI. Analog Integrated Circuits and Signal Processing: An International Journal 15 No. 3 (1998)

6.27 Chua, L.O. Roska, T.: Cellular Neural Networks and Visual Computing: Foundations and Applications. Cambridge University Press: Cambridge (2002) 7 (Fig. 2.1), 8 (Fig. 2.2); cf. also Mainzer, K.: Cellular neural networks and visual computing: foundations and applications according to the book of Leon O. Chua and Tamás Roska. International Journal of Bifurcation and Chaos in Applied Sciences and Engineering 13 (2003)

6.28 Chua, L.O. Roska, T.: Cellular Neural Networks and Visual Computing (see Note 27) 36 (Fig. 3.1)

6.29 Chua, L.O.: CNN: A Paradigm for Complexity (see Note 24) 48 (Fig. 2.6.1)

6.30 Chua, L.O. Roska, T.: Cellular Neural Networks and Visual Computing (see Note 27) 119–120 (Example 5.1)

6.31 Chua, L.O. Roska, T.: Cellular Neural Networks and Visual Computing (see Note 27) 121 (Example 5.2)

6.32 Chua, L.O. Roska, T.: Cellular Neural Networks and Visual Computing (see Note 27) 206 (Fig. 8.1)

6.33 Chua, L.O. Roska, T.: Cellular Neural Networks and Visual Computing (see Note 27) 207 (Example 8.1)

6.34 Chua, L.O. Roska, T.: Cellular Neural Networks and Visual Computing (see Note 27) 211 (Example 8.2), 213 (Example 8.4)

6.35 Chua, L.O. Roska, T.: Cellular Neural Networks and Visual Computing (see Note 27) 289 (Fig. 12.9)

6.36 Chua, L.O. Roska, T.: Cellular Neural Networks and Visual Computing (see Note 27) 243 (Fig. 9.2)

6.37 Chua, L.O. Roska, T.: Cellular Neural Networks and Visual Computing (see Note 27) 254 (Fig. 9.13)

6.38 Chua, L.O.: CNN: A Paradigm for Complexity (see Note 24) 159 (Fig. 52)

6.39 Chua, L.O. Yoon, S. Dogaru, R.: A nonlinear dynamics perspective of Wolfram's new kind of science. Part I: Threshold of complexity. International Journal of Bifurcation and Chaos in Applied Sciences and Engineering 12 (2002) Fig. 2

6.40 Chua, L.O. Yoon, S. Dogaru, R.: A nonlinear dynamics perspective of Wolfram's new kind of science (see Note 39) Fig. 9, Fig. 12, Fig. 14
6.41 Chua, L.O. Yoon, S. Dogaru, R.: A nonlinear dynamics perspective of Wolfram's new kind of science (see Note 39) Table 2
6.42 Chua, L.O. Yoon, S. Dogaru, R.: A nonlinear dynamics perspective of Wolfram's new kind of science (see Note 39) Fig. 3
6.43 Chua, L.O. Yoon, S. Dogaru, R.: A nonlinear dynamics perspective of Wolfram's new kind of science (see Note 39) Fig. 22, Fig. 24
6.44 Müller-Schloer, C. (Ed.): Schwerpunktthema: Organic Computing – Systemforschung zwischen Technik und Naturwissenschaften. In: it Information Technology **4** (2005)
6.45 Horn, P., 2001: Autonomic Computing: IBM's Perspective on the State of Information Technology. IBM: New York (2002); Kephart, J.O., Chess, D.M.: The Vision of Autonomic Computing. In: IEEE Computer Society **1** (2003) 41–50
6.46 Small, M.: Applied Nonlinear Time Series Analysis. Applications in Physics, Physiology and Finance. World Scientific: Singapore (2005)
6.47 Pfeifer, R., Scheier C.: Understanding Intelligence. M.I.T Press: Cambridge MA (2001) 280
6.48 For chapter 6.4 compare Mainzer, K.: Philosophical foundations of neurobionics. In: Bothe, H.W. Samii, M. Eckmiller, R. (eds.): Neurobionics. An Interdisciplinary Approach to Substitute Impaired Functions of the Human Nervous System. North-Holland: Amsterdam (1993) 31–47
6.49 Cf. Samii, M. (ed.): Peripheral Nerve Lesions. Springer: Berlin (1990)
6.50 Eckmiller, R. (ed.): Neurotechnologie-Report. Bundesministerium für Bildung, Wissenschaft, Forschung und Technologie: Bonn (1995); Stein, R.B. Peckham, P.H. (eds.): Neuroprostheses: Replacing Motor Function After Disease or Disability. Oxford University Press: New York (1991)
6.51 Cosman, E.R.: Computer-assisted technologies for neurosurgical procedures. In: Bothe, H.W. Samii, M. Eckmiller, R. (eds.): Neurobionics (see Note 48) 365
6.52 Chua, L.O.: CNN: A Paradigm for Complexity (see Note 24) 227 (Fig. 86)
6.53 Chua, L.O. Roska, T.: Cellular Neural Networks and Visual Computing (see Note 27) 321 (Fig. 16.1)
6.54 Elger, C.E. Mormann, F. Kreuz, T. Andrzejak, R.G. Rieke, C. Sowa, R. Florin, S. David, P. Lehnertz, K.: Characterizing the spatio-temporal dynamics of the epileptogenic process with nonlinear EEG analyses. In: Tetzlaff, R. (ed.): Cellular Neural Networks and Their Applications. Proc. 7th IEEE International CNN Workshop. World Scientific: Singapore (2002) 238
6.55 Foley, J.D. Dam, A. van: Fundamentals of Interactive Computer Graphics. Addison-Wesley: Amsterdam (1982); Foley, J.D.: Neuartige Schnittstellen zwischen Mensch und Computer. Spektrum des Wissenschaft **12** (1987)
6.56 Gibson, W.: Neuromancer. Grafton: London (1986) 67
6.57 Putnam, H.: Reason, Truth and History. Cambridge University Press: Cambridge (1981) 6
6.58 Langton, C.G. (ed.): Artificial Life, Addison Wesley: Redwood City (1989)
6.59 Lindenmayer, A. Rozenberg, G. (eds.): Automata, Languages, Development. North-Holland: Amsterdam (1976); Lindenmayer, A.: Models for multicellular development: Characterization, inference and complexity of L-systems. In: Kelemenová, A. Kelemen, J. (eds.): Trends, Techniques and Problems in Theoretical Computer Science, Springer: Berlin (1987) 138–168

6.60 Langton, C.G.: Self-reproduction in cellular automata. Physica **D 10** (1984) 135–144; Artificial Life, in: Langton [5.58] Plates 1–8

6.61 Langton, C.G.: Life at the edge of chaos. In: Langton, C.G. Taylor, C. Farmer, J.D. Rasmussen, S. (eds.): Artificial Life II, Addison-Wesley: Redwood City (1992) 41–91

6.62 Rechenberg, I.: Evolutionsstrategie: Optimierung technischer Systeme nach Prinzipien der biologischen Evolution, Frommann-Holzboog: Stuttgart (1973)

6.63 Holland, J.H.: Adaption in Natural and Artificial Systems. The MIT Press: Cambridge, MA (1992, 1st ed. 1975)

6.64 Mitchell, M. Forrest, S.: Genetic Algorithms and Artificial Life. Artificial Life **1** (1994) 267–289

6.65 Forrest, S. Javornik, B. Smith, R. Perelson, A.: Using genetic algorithms to explore pattern recognition in the immune system. Evolutionary Computation **1** (1993) 191–211

6.66 Highland, H.J. (ed.): Computer Virus Handbook. Elsevier Advanced Technology: Oxford (1990); Spafford, E.H.: Computer viruses as artificial life. Artificial Life **1** (1994) 249–265

6.67 Taylor, C. Jefferson, D.: Artificial life as a tool for biology inquiry. Artificial Life **1** (1994) 1–13

6.68 Pfeifer, R. Scheier, C.: Understanding Intelligence, MIT Press: Cambridge Mass. (2001)

6.69 Morán, F. Moreno, A. Merelo, J.J. Chacón, P. (eds.): Advances in Artificial Life, Springer: Berlin (1995)

6.70 Mainzer, K.: Gehirn; Computer, Komplexität. Springer: Berlin 1997; Mainzer, K.: KI – Künstliche Intelligenz. Grundlagen intelligenter Systeme, Wissenschaftliche Buchgesellschaft: Darmstadt (2003)

6.71 Bellman, K.L.: Self-Conscious Modeling. In: it Information Technology **4** (2005) 188–194

6.72 Mainzer, K.: Computer, künstliche Intelligenz und Theory of Mind: Modelle des Menschlichen? In: Förstl, H. (ed.): Theory of Mind. Neurobiologie und Psychologie sozialen Verhaltens. Springer: Berlin (2007) 79–90

Chapter 7

7.1 Quesnay: Œuvres économiques et philosophiques. Francfort (1988)

7.2 Figure 7.3a shows a "Tableau économique" from the "Eléments de la philosophie rurale" (1767) of Marquis of Mirabeau; cf. Meek, R.L.: The interpretation of the "Tableau économique", Economica **27** (1960) 322–347

7.3 Serviere, G. de: Recueil d'ouvrages curieux de mathématique et de méchanique, ou description du cabinet de Monsieur Grollier de Serviere, avec des figures en taille-douce. Seconde édition. Lyon (1733) (Fig. 38); Rieter, H.: Quesnays Tableau Economique als Uhren-Analogie. In: Scherf, H. (ed.): Studien zur Entwicklung der ökonomischen Theorie IX. Duncker & Humblot: Berlin (1990) 73

7.4 Smith, A.: The Theory of Moral Sentiments (1759). Ed. D.D. Raphael/A.L. Macfie. Oxford (1979)

7.5 Cf. Skinner, A.S.: Adam Smith, science and the role of imagination. In: Todd, W.B. (ed.): Hume and the Enlightenment. Edinburgh University Press: Edinburgh (1974) 164–188

7.6 Smith, A.: An Inquiry into the Nature and Causes of the Wealth of Nations. Ed. E. Cannan. The University of Chicago Press: Chicago (1976) vol. 1, 478

7.7 Smith, A.: (see Note 6), vol. 1, 18

7.8 Mayr, O.: Adam Smith und das Konzept der Regelung. In: Troitzsch/Wohlauf, G. (eds.): Technikgeschichte. Frankfurt (1980) 245

7.9 Cf. Marchi, N.B. de: The empirical content and longevity of Ricardian economics. Economica **37** (1970) 258–259; Sowell, T.: Classical Economics Reconsidered. Princeton University Press: Princeton (1974) 118–120; Blaug, M.: The Methodology of Economics. Cambridge University Press: Cambridge (1980)

7.10 Mill, J.S.: On the definition of political economy. In: Collected Works. Essays on Economy and Society. Ed. J.M. Robson. University of Toronto Press: Toronto (1967) vol. 4, 330; Hicks, J.: From classical to post-classical: The work of J.S. Mill. In: Hicks, J.: Classics and Moderns. Collected Essays on Economic Theory. Vol. III. Basil Blackwell: Oxford (1983) 60–70

7.11 Georgescu-Roegen, N.: The Entropy Law and Economic Progress. Harvard University Press: Cambridge, MA (1971)

7.12 Walras, L.: Principe d'une théorie mathématique de l'échange. Journal des Economistes **33** (1874) 1–21; Hicks, J.: Léon Walras. In: Hicks, J.: (see Note 10) 85–95

7.13 Keynes, J.M.: The Collected Writings of John Maynard Keynes. Vol. XIV. The General Theory and After. Ed. D. Moggridge. Macmillan: London (1973) 296

7.14 Neumann, J. von/Morgenstern, O.: Theory of Games and Economic Behavior. Princeton University Press: Princeton, NJ (1943)

7.15 Cf. Stegmüller, W.: Probleme und Resultate der Wissenschaftstheorie und Analytischen Philosophie Bd. 1. Springer: Berlin (1969) 391

7.16 Cf. Shubik, M. (ed.): Game Theory and Related Approaches to Social Behavior. John Wiley: New York/London/Sidney (1964); Nash, J.F.: Equilibrium points in n-person games. Proceedings of National Academy of Science (USA) **36** (1950) 48–49

7.17 For a survey compare Anderson, P.W./Arrow, K.J./Pines, D. (eds.): The Economy as an Evolving Complex System. Santa Fé Institute Studies in the Sciences of Complexity vol. 3. Addison-Wesley: Redwood City (1988); Barnett, W.A./Geweke, J./Shell, K. (eds.): Economic Complexity. Chaos, Sunspots, Bubbles, and Nonlinearity. Cambridge University Press: Cambridge (1989)

7.18 Kalecki, M.: Theory of Economic Dynamics. Unwin University Books: London (1954); Samuelson, P.A.: Foundations of Economic Analysis. Harvard University Press: Cambridge, MA (1947)

7.19 Goodwin, R.M.: Chaotic Economic Dynamics. Clarendon Press: Oxford (1990) 19

7.20 Rössler, O.E.: Chaos und Endophysik. In: Mainzer, K./Schirmacher, W. (eds.): Quanten, Chaos und Dämonen. Erkenntnistheoretische Aspekte der modernen Physik. B.I. Wissenschaftsverlag: Mannheim (1993)

7.21 Goodwin, R.M.: Chaotic Economic Dynamics (see Note 19) 113

7.22 Lorenz, H.-W.: Nonlinear Dynamical Economics and Chaotic Motion. Springer: Berlin (1989); Goodwin, R.W.: A growth cycle. In: Hunt, E.K./Schwartz, J.G. (eds.): A Critique of Economic Theory. Penguin: Harmondsworth (1969) 442–449

7.23 This idea stems back to Samuelson, P.A.: Generalized predator-prey oscillations in ecological and economic equilibrium. In: The Collected Scientific Papers of P.A. Samuelson. Ed. R.C. Merton. Vol III. MIT Press: Cambridge, MA (1972) 487–490

7.24 Cf. Lorenz, H.-W.: Nonlinear Dynamical Economics and Chaotic Motion. (See Note 27) 157; Goodwin, R.M.: Chaotic Economic Dynamics. (See Note 19) 117

7.25 Cf. Farmer, J.D./Sidorowich, J.J.: Predicting Chaotic Time Series. Mineo: Los Alamos National Laboratory (1987)

7.26 In his "History of Economical Analysis", J.A. Schumpeter had already discussed the possibility of multiple equilibrium points. But like Poincaré in the physical sciences, he was afraid of chaotic scenarios in economics: Schumpeter, J.A.: Geschichte der ökonomischen Analyse. Vandenhoeck & Ruprecht: Göttingen (1965)

7.27 Arthur, W.B.: Positive feedbacks in the economy. Scientific American, February (1990) 92–99

7.28 Weidlich, W.: Sociodynamics: A Systematic Approach to Mathematical Modeling in the Social Sciences. Taylor & Francis: London (2002). Chapt. 3.2.3

7.29 Weidlich, W.: Sociodynamics: A Systematic Approach to Mathematical Modeling in the Social Sciences (see Note 28) 224 (Fig. 7.1)

7.30 Arthur, W.B./Ermoliew, J.M./Kaniowski, J.M.: Path-dependent processes and the emergence of macro-structure. European Journal of Operational Research **30** (1987) 294–303; Helpman, E./Krugman, P.: Market Structure and Foreign Trade. MIT Press: Cambridge, MA (1985)

7.31 Simon, H.: Administrative Behavior: A Study of Decision-Making Processes in Administrative Organizations, 2nd edn. Macmillian: New York (1957) 198

7.32 Sterman, J.D.: Business Dynamics: Systems Thinking and Modeling for a Complex World. McGraw-Hill: Boston (2000) 597–629

7.33 Wolfram, S.: A New Kind of Science (see Note 5.52) 432

7.34 Bachelier, L.: Théorie de la spéculation. Dissertation. In : Annales Scientifiques de l'Ecole Normale Supérieure **17** (1900) 21–86

7.35 Mainzer, K.: Der kreative Zufall. Wie das Neue in die Welt kommt. C.H. Beck: München (2007) 192

7.36 Fama, E.F.: The distribution of daily differences of stockprices: A test of Mandelbrot's stable Paretian hypothesis. Ph.D. dissertation, University of Chicago Graduate School of Business (1964)

7.37 Markowitz, H.H.: Portfolio Selection: Efficient Diversification of Investments. Yale University Press: New Haven CT (1959)

7.38 Sharpe, W.F.: Capital asset prices: A theory of market equilibrium under conditions of risk. In: Journal of Finance **19** (1964) 425–442

7.39 The expected return-beta equation is $E(r_i) = r_f + \beta_i(E(r_M) - r_f)$. E stands for the mathematical expectation operator. The equation indicates that the expected return r on a security i equals the sum of two numbers. The first of these numbers is the risk-free rate that you would expect to get from something safe like a Treasury bill. The second is Sharpe's beta times the market premium. That is, however much better than the Treasury rate you expect the market M to perform. Each stock has its own beta, or degree to which its price movements correlate to that of the market overall.

7.40 Black, F., Scholes, M.: The pricing of options and corporate liabilities. In: Journal of Political Economy **81** (1973) 637–654

7.41 The price of a call option to buy a stock at a specific price and time is $C_o = S_o N(d_1) - Xe^{-rT} N(d_2)$. Here, C_o is the price of the call option. S_o is the current stock price. X is the exercise price – the price at which the option contract allows you to buy the stock. R is the risk-free interest rate, and T is the time to maturity. The other two functions $N(d_1)$ and $N(d_2)$ are the probabilities of a random number d that follows a bell-curve distribution being less than the quantities $d_1 = [\ln(S_o/X) + (r + \sigma^2/2)T]/\sigma\sqrt{T}$ and $d_2 = d_1 - \sigma\sqrt{T}$, where σ is the standard deviation of the stock price and ln is the natural logarithm.

7.42 Hull, J.C.: Options, Futures, and other Derivatives. Prentice-Hall: Upper Saddle River NJ (1997)
7.43 Merton, R.C.: Theory of Rational Option Pricing. In: Bell J. Eco. Management Sci **4** (1973) 141–183
7.44 Mantegna, R.N., Stanley, H.E.: An Introduction to Econophysics. Correlations and Complexity in Finance. Cambridge University Press: Cambridge (2000) 120
7.45 McCauley, J.L.: Dynamics of Markets. Econophysics and Finance. Cambridge University Press: Cambridge (2004) 4
7.46 Mandelbrot, B.B., Hudson, R.L.: The (Mis)Behavior of Markets. A Fractal View of Risk, Ruin, and Reward. Basic Books: New York (2004) 89
7.47 Mandelbrot, B.B., Hudson, R.L.: The (Mis)Behavior of Markets (see Note 46) 93
7.48 Mandelbrot, B.B., Hudson, R.L.: The (Mis)Behavior of Markets (see Note 46) 112–113
7.49 Kolmogorov, A.N.: The local structure of turbulence in incompressible viscous fluid for very large Reynolds numbers. In: Dokl. Akad. Nauk. SSSR **30** (1941) 9–13 (reprinted in: Proc. R. Soc. Lond. A **434** (1991) 9–13)
7.50 McCauley, J.L.: Dynamics of Markets (see Note 45) chapter 7
7.51 Pareto, V.: Manuel d'économie politique. V. Giard and E. Brière: Paris (1909)
7.52 Evertsz, C.J.G., Hendrych, R., Singer, P., Peitgen, H.-O.: Fraktale Geometrie von Börsenzeitreihen: Neue Perspektiven ökonomischer Zeitreihen. In: Mainzer, K. (ed.): Komplexe Systeme und Nichtlineare Dynamik in Natur und Gesellschaft. Springer: Berlin (1999) 404
7.53 Mandelbrot, B.B.: The Fractal Geometry of Nature. W.H. Freeman & Co.: New York (1982)
7.54 Mandelbrot, B.B., Hudson, R.L.: The (Mis)Behavior of Markets (see Note 46) 118
7.55 Mantegna, R.N., Stanley, H.E.: An Introduction to Econophysics (see Note 7.44) 62
7.56 Mandelbrot, B.B.: The variation of certain speculative prices. In: Journal of Business **36** (1963) 394–419
7.57 Mandelbrot, B.B., Hudson, R.L.: The (Mis)Behavior of Markets (see Note 46) 171
7.58 Google Inc.: Google search for "Arche Noah". http://www.google.com. Cited 2nd May 2007
7.59 Museo Arcivescovile: The dream of the pharaoh. Museo Arcivescovile: Ravenna
7.60 Hurst, H.E.: Long-term storage capacity of reservoirs. In: Transactions of the American Society of Civil Engineers **116** (1951) 770–799, 800–808
7.61 Mandelbrot, B.B., Hudson, R.L.: The (Mis)Behavior of Markets (see Note 46) 188
7.62 Mandelbrot, B.B., Hudson, R.L.: The (Mis)Behavior of Markets (see Note 46) 194
7.63 Feder, J.: Fractals. Plenum: New York (1988)
7.64 Mandelbrot, B.B., Hudson, R.L.: The (Mis)Behavior of Markets (see Note 46) 215
7.65 Mandelbrot, B.B.: Multifractals and 1/f Noise: Wild Self-Affinity in Physics. Springer: New York (1999); Mandelbrot, B.B.: Heavy tails in finance for independent or multifractal price increments. In: Rachev, S.T. (ed.), Handbooks in Finance 30: Heavy Tailed Distributions in Finance. Elsevier: Amsterdam (2003) 1–34; Turiel, A., Pérez-Vicente, C.J.: Multifractal geometry in stock market time series. In: Physica **A 322** (2003) 629–649
7.66 Malthus, R.T.: Population: The First Essay (1798). Univ. Mich. Press: Ann Arbor (1959)
7.67 West, B.J., Deering, B.: The Lure of Modern Science. Fractal Thinking. World Scientific: Singapore (1995) 281

7.68 Montrol, E.W., Badger, W.W.: Introduction to Quantitative Aspects of Social Phenomena. Gordon and Breach: New York (1974)
7.69 Mainzer, K.: Symmetry and Complexity (see Note 5.44) Chapters 2–3
7.70 McCauley, J.L.: Dynamics of Markets (see Note 45) Chapter 1

Chapter 8

8.1 Ehrenberg, V.: Der Staat der Griechen. Zürich (1965); Meier, C.: Die Entstehung des Politischen bei den Griechen. Frankfurt (1983)
8.2 Platonis Opera, 5 vols. Ed. J. Burnet. Oxford (1899–1906) (especially Politeia, Politikos); Cf. Cross, R.C./Woozley, A.D.: Plato's Republic. A Philosophical Commentary. London (1964)
8.3 Aristotle: Ethica Nicomachae, Politica (Oxford Classical Texts); Atheniesium Res Publica. Ed. F.G. Kenyon. Oxford (1920); Höffe, O.: Praktische Philosophie – Das Model des Aristoteles. München (1971); Barker, L.E.: The Political Thought of Plato and Aristotle. New York (1959)
8.4 Hobbes, T.: The Elements of Law Natural and Politic (1640). Ed. F. Tönnies. London (1889); Leviathan, or the Matter, Form and Power of a Commonwealth, Ecclesiasticial and Civil (1651). Ed. C.B. Macpherson. Harmondsworth (1968); Gough, J.W.: The Social Contract. A Critical Study of its Development. Oxford (1957); Höffe, O. (ed.): Thomas Hobbes: Anthropologie und Staatsphilosophie. Freiburg (1981)
8.5 Locke, J.: Two Treatises on Government (1689). Ed. P. Laslett. Cambridge (1970). Book II: Essay concerning the true original, and end of civil government
8.6 Hume, D.: Essays and Treatise on Several Subjects. In two Volumes. I. Containing Essays, Moral, Political, and Literary. A New Edition. London (1784)
8.7 H.-G. Gadamer: Hegels Dialektik. Fünf hermeneutische Studien, Tübingen (1971)
8.8 G. Gurvitch: Dialectique et Sociologie. Flammarion: Paris (1962)
8.9 Flood, R.L.: Complexity: A definition by construction of a conceptual framework. Systems Research **4**(3) (1987) 177–185; Flood, R.L./Carson, E.R.: Dealing with Complexity. An Introduction to the Theory and Application of Systems Science. Plenum Press: New York (1988); An early inquiry of complexity in social sciences was written by Hayek, F.A. von: The Theory of Complex Phenomena. J.C.B. Mohr: Tübingen (1972)
8.10 For a survey compare Cambel, A.B./Fritsch, B./Keller, J.U. (eds.): Dissipative Strukturen in integrierten Systemen. Nomos Verlagsgesellschaft: Baden-Baden (1989); Tainter, J.: The Collapse of Complex Societies. Cambridge University Press (1988); Dumouchel, P./Dupuy, J.-P. (eds.): L'Auto-Organisation de la Physique au Politique. Editions Du Seuil: Paris (1983)
8.11 Malthus, T.R.: An Essay on the Principle of Population. London (1798)
8.12 Spencer, H.: First Principles. D. Appleton & Co.: New York (1896) 527; Carneiro, R.L.: Structure, function, and equilibrium in the evolutionism of Herbert Spencer. Journal of Anthropological Research **29** (1973) 77–95
8.13 For a review compare, for instance, Carneiro, R.L.: Successive reequilibrations as the mechanism of cultural evolution. In: Schieve, W.C./Allen, P.M. (eds.): Self-Organization and Dissipative Structures. Applications in the Physical and Social Sciences. University of Texas Press: Austin (1982) 110–115
8.14 Allen, P.M.: Self-organization in the urban system. In: Schieve, W.C./Allen, P.M. (eds.): Self-Organization and Dissipative Structures. (See Note 13) 142–146

8.15 Durkheim, E.: De la Division du Travail Social. Paris (1893); Les Règles de la Méthode Sociologique. Paris (1895)

8.16 Weidlich, W.: Das Modellierungskonzept der Synergetik für dynamisch sozio-ökonomische Prozesse. In: Mainzer, K./ Schirmacher, W. (eds.): Quanten, Chaos und Dämonen. (See Note 26) Figs. 6.12–16; Weidlich, W.: Stability and cyclicity in social systems. In: Cambel, A.B. Fritsch, B./Keller, J.U. (eds.): Dissipative Strukturen in Integrierten Systemen. (See Note 10) 193–222

8.17 For a survey compare Scott, W.R.: Organizations: Rational, Natural, and Open Systems. Prentice-Hall: Englewood Cliffs, NJ (1992): Taylor, M.: The Possibility of Cooperation. Cambridge University Press: Cambridge (1987)

8.18 Bott, E.: Urban families: Conjugal roles and social networks. Human Relations **8** (1955) 345–383; Knipscheer, C.P.M./Antonucci, T.C. (eds.): Social Network Research: Substantive Issues and Methodological Questions. Swets and Zeitlinger: Amsterdam (1990)

8.19 Schelling. T.C.: Micromotives and Macrobehavior. W.W. Norton & Co.: New York (1978)

8.20 Glance, N.S./Huberman, B.A.; Organizational fluidity and sustainable cooperation, dynamics of computation group: Xerox Palo Alto Research Center: March 12 (1993) (Fig. 3)

8.21 See Note 20, Fig. 4

8.22 Dawkins, R.: The Selfish Gene. Oxford University Press: New York (1976)

8.23 Popper, K.R.: Evolutionary epistemology. In: Miller, D. (ed.): Popper Selections. Princeton University Press: Princeton, NJ (1985) 78–86

8.24 Miller, M.S./Drexler, K.E.: Comparative ecology: A computational perspective. In: Huberman, B.A. (ed.): The Ecology of Computation. North-Holland: Amsterdam (1988) 51–76

8.25 Compare Luhmann, N.: The self-reproduction of the law and its limits. In: Teubner, G. (ed.): Dilemmas of Law in the Welfare State. Berlin (1986) 111–127; Mayntz, R.: Differenzierung und Verselbständigung. Zur Entwicklung gesellschaftlicher Teilsysteme. Frankfurt (1988)

8.26 Minsky, M.: The Society of Mind. Simon & Schuster: New York (1985) 314

8.27 For a survey compare Kellermayr, K.H.: Lokale Computernetze LAN. Technologische Grundlagen, Architektur, Übersicht und Anwendungsbereiche. Springer: Wien (1986); Sharp, J.A.: Verteilte und parallele Computernetze. Architektur und Programmierung. VCH-Verlagsgesellschaft: Weinheim (1989); Barz, H.W.: Kommunikation und Computernetze. Konzepte, Protokolle und Standards. Carl Hanser: München (1991)

8.28 Cerf, V.G.: Netztechnik, Spektrum der Wissenschaft Nov. **11** (1991) 76

8.29 Halabi, B.: Internet-Routing-Architekturen: Grundlagen, Design und Implementierung. Carl Hanser Verlag: München (1998) 122 (Fig. 5.8); Calvert, K.M./Doar, M.B./Zegura, E.W.: Modeling Internet topology. IEEE Transactions Communication 1 (1997) 160

8.30 Fukuda, K./Takayasu, H./Takayasu, M.: Spatial and temporal behavior of congestion in Internet traffic. Fractals 7 1 (1999) 67

8.31 Mainzer, K./Büchs, M./Kundish, D./Pyrka, P.: Physical and virtual mobility: analogies between traffic and virtual highways. In: Mayinger, F. (ed.), Mobility and Traffic in the 21st Century. Springer: Berlin (2001) 243–318

8.32 Takayasu, M./Fukuda, K./Takayasu, H.: Application of statistical physics to the Internet traffics. Physica A (1999) 3 (Fig. 1)

8.33 Takayasu, M./Takayasu, H./Sato, T.: Critical behaviors and 1/f noise in information traffic. Physica A 233 (1996) 825 (Fig. 1)

8.34 Wang, X.F.: Complex networks: topology, dynamics and synchronization. International Journal of Bifurcation and Chaos in Applied Sciences and Engineering 12 5 (2002) 885–916

8.35 Crestani, F./Pasi, G. (eds.): Soft Computing in Information Retrieval: Techniques and Applications. Physica Verlag (Springer): Heidelberg (2000); Zadeh, L.A.: From computing with numbers to computing with words – from manipulations of measurements to manipulation of perceptions. IEEE Transactions on Circuits and Systems-I: Fundamental Theory and Applications 45 1 (1999)

8.36 Mainzer, K.: KI – Künstliche Intelligenz. Grundlagen intelligenter Systeme. Wissenschaftliche Buchgesellschaft: Darmstadt (2003)

8.37 Boughanem, M./Chrisment, C./Mothe, J./Soule-Dupuy, C./Tamine, L.: Connectionist and genetic approaches for information retrieval. In: Crestani, F./Pasi, G. (eds.): Soft Computing in Information Retrieval (see Note 35) 179 (Fig. 1)

8.38 Merkl, D./Rauber, A.: Document classification with unsupervised artificial neural networks. In: Crestani, F./Pasi, G. (eds.): Soft Computing in Information Retrieval (see Note 35) 102–121

8.39 Cho, S.B.: Artificial life technology for adaptive information processing. In: Kasabov, N. (ed.): Future Directions for Intelligent Systems and Information Sciences. Physica-Verlag (Springer): Heidelberg (2000) 24

8.40 Huberman, B.A./Adamic, L.A.: Growth dynamics of the World Wide Web. Nature 401 (1999) 131

8.41 Minsky, M.: The Society of Mind. (See Note 26)

8.42 Hegel, G.W.F.: Phänomenologie des Geistes. (1807) In: Werke in zwanzig Bänden, vol. 3. Suhrkamp: Frankfurt (1972)

8.43 Ackhoff, R., Emery, F.: On Purposeful Systems. Aldine Atherton: Chicago IL (1972)

8.44 Balke, W.-T., Mainzer, K.: Knowledge representation and the embodied mind: Towards a philosophy and technology of personalized informatics. In: Althoff, K.-D. et al. (eds.): Professional Knowledge Management. Lecture Notes in Artificial Intelligence 3782. Springer: Berlin (2005) 586–597

8.45 Holland, S., Kießling, W.: Situated preferences and preference repositories for personalized database applications. In: Int. Conf. on Conceptual Modeling (ER'04). Shanghai, China (2004)

8.46 Kaplan, C., Fenwick, J., Chen, J.: Adaptive hypertext navigation based on user goals and context. In: Int. J. User Modeling and User-Adapted Interaction. **3(3)** (1993)

8.47 Wagner, M., Balke, W.-T., Kießling, W.: An XML-based multimedia middleware for mobile online auctions. In: Enterprise Information Systems III. Kluwer, The Netherlands (2002)

8.48 Dreyfus, H.L.: What Computer's Can't Do – The Limits of Artificial Intelligence. Harper & Row: New York (1979)

8.49 Balke, W.-T.: A Roadmap to personalized information systems by cognitive expansion of queries. In: Int. Workshop on Content-Knowledge Management and Mass Personalization of E-Services (EnCKompass 2002). Paris (2002)

8.50 Minker, J.: An overview of cooperative answering in databases. In: Int. Conf. on Flexible Query Answering Systems (FQAS'98). Lecture Notes in Computer Science 1495. Springer: Berlin (1998)

8.51 Kießling, W.: Foundations of preferences in database systems. In: Int. Conf. on Very Large Data Bases (VLDB'02). Hong Kong (2002)

8.52 Balke, W.-T., Wagner, M.: Cooperative discovery for user-centered web service provisioning. In: Int. Conf. on Web Services (ICWS'03). Las Vegas NV (2003); Balke, W.-T., Wagner, M.: Towards personalized selection of web services. In: Int. World Wide Web Conference (WWW'03). Budapest, Hungary (2003)

8.53 Balke, W.-T., Mainzer, K.: Knowledge representation and the embodied mind: Towards a philosophy and technology of personalized informatics. In: Althoff, K.-D. et al. (eds.): Professional Knowledge Management. Lecture Notes in Artificial Intelligence 3782. Springer: Berlin (2005) 595

8.54 Weiser, M.: The computer for the 21st century. Scientific American 9 (1991) 66

8.55 Gershenfeld, N.: When Things Start to Think. Henry Holt: New York (1999)

8.56 Norman, D.A.: The Invisible Computer. Cambridge University Press: Cambridge (1998)

8.57 Hofmann, P.E.H. et al.: Evolutionäre E/E-Architektur. Vision einer neuartigen Elektronik-Architektur für Fahrzeuge, DaimlerChrysler: Esslingen (2002)

8.58 Weiser, M.: The computer for the 21st century. In: Scientific American **265(3)** (1991) 94–104

8.59 Müller, G., Torsten, E., Kreutzer, M.: Telematik- und Kommunuikationssysteme in der vernetzten Wirtschaft. Oldenbourg: München (2004)

8.60 Müller, G., et al. (eds.): Geduldige Technologie für ungeduldige Patienten: Führt Ubiquitous Computing zu mehr Selbstbestimmung? Total vernetzt – Szenarien einer informatisierten Welt. Springer: Berlin (2003); Sackmann, S., Eymann, T., Müller, G.: EMIKA – Real-time controlled mobile information systems in health care applications. In: Proceedings of the Second Conference on Mobile Computing in Medicine. GI-Edition Lecture Notes in Informatics. Springer: Berlin (2002); Scheer, A.-W., Chen, R., Zimmermann, V.: Prozessmanagement im Krankenhaus. Krankenhausmanagement Schriften zur Unternehmensführung. SzU: Wiesbaden (1996)

8.61 Paramythis, A., Loidl-Reisinger, S.: Adaptive learning and e-learning standards. In: Electronic Journal on e-Learning **2 (1)** (2004) 181–194

8.62 Sampson, D., Karagiannidis, C., Cardinali, F.: An architecture for web-based e-learning promoting re-usable adaptive educational e-content. In: Educational Technology & Society **5(4)** (2002) 1–14

8.63 Tomaschek, N.: Systemic Coaching. A Target-Oriented Approach to Consulting. Carl-Auer: Heidelberg (2006)

8.64 Nonaka, I., Toyama, R., Byosière, P.: A theory of organizational knowledge creation: Understanding the dynamic process of creating knowledge. In: Dierkes, M., Berthoin Antal, A., Child, J., Nonaka, I. (eds.): Handbook of Organizational Learning & Knowledge. Oxford University Press: Oxford (2001) 491–517

8.65 Hedberg, B., Holmqvist, M.: Learning in imaginary organizations. In: Dierkes, M., Berthoin Antal, A., Child, J., Nonaka, I. (eds.): Handbook of Organizational Learning & Knowledge. Oxford University Press: Oxford (2001) 733–752

8.66 Dierkes, M., Berthoin Antal, A., Child, J., Nonaka, I. (eds.): Handbook of Organizational Learning & Knowledge (see Note 65)

8.67 Mainzer, K.: Leben in der Wissensgesellschaft. In: Deutscher, Hochschulverband (ed.): Glanzlichter der Wissenschaft. Lucius und Lucius: Stuttgart (2000) 133–142; Mainzer, K.: Computernetze und Virtuelle Realität. Leben in der Wissensgesellschaft. Springer: Berlin (1999)

Chapter 9

9.1 Goodmann, N.: Fact, Fiction and Forecast. London (1954); Lenk, H.: Erklärung, Prognose, Planung. Rombach: Freiburg (1972)

9.2 Montgomery, D.C./Johnson, L.A.: Forecasting and Time Series Data Analysis and Theory. Holt, Rinehart and Winston: New York (1975)

9.3 Abraham, B./Ledolter, J.: Statistical Methods of Forecasting. Wiley: New York (1983); Pindyek, R./Rubinfeld, D.: Econometric Models and Economic Forecasts. McGraw-Hill: New York (1980)

9.4 Makridakis, S./Wheelwright, S.C.: Forecasting Methods for Management. Wiley: New York (1989); Makridakis, S./Wheelwright, S.C./Mc Gree, V.E.: Forecasting: Methods and Applications. Wiley: New York (1989)

9.5 Kravtsov, Y. A. (ed.): Limits of Predictability. Springer: Berlin (1993) 82; Sarder, Z./Ravetz, J.R. (eds.): Complexity: fad or future? In: Futures. J. Forecasting, Planning, and Policy **26** 6 (1994)

9.6 Mayntz, R.: The Influence of natural science theories on contemporary social science. In: Meinolf, D./Biervert, B. (eds.): European Social Science in Transition. Assessment and Outlook. Westview Press: Boulder, Colorado (1992) 27–79

9.7 Merton, R.K.: The Sociology of Science. Theoretical and Empirical Investigations. Chicago (1973); Sorokin, P.A.: Social and Cultural Dynamics. New York (1962); Price, D. de Solla: Science since Babylon. New Haven (1961); Goffman, W.: An epidemic process in an open population. In: Nature **205** (1965), 831–832); Mathematical approach to the spread of scientific ideas – the history of mast cell research. In: Nature **212** (1966), 449–452; Berg, J./Wagner-Döbler, R.: A multidimensional analysis of scientific dynamics. In: Scientometrics **3** (1996), 321–346

9.8 Bruckner, E.W./Ebeling, W./Scharnhorst, A.: The application of evolution models in scientometrics. In: Scientometrics **18** (1990), 33

9.9 Small, H./Sweeny/Greenlee, E.: Clustering the science citation index using co-citations. II. Mapping science. In: Scientometrics **8** (1985) 321–340; Callon, M./Courtial, J.-P./Turner, W.A./Bauin, S.: From translation to problematic networks – An introduction to co-word analyses. In: Social Science Information **22** (1983), 191–235

9.10 Mayntz, R.: Soziale Diskontinuitäen: Erscheinungsformen und Ursachen. In: Hierholzer, K./Wittmann, H.-G. (eds.): Phasensprünge und Stetigkeit in der natürlichen und kulturellen Welt. Wiss. Verlagsgesellschaft: Stuttgart (1988) 15–37

9.11 Mainzer, K.: Die Würde des Menschen ist unantastbar. Zur Herausforderung von Artikel 1I Grundgesetz durch den technisch-wissenschaftlichen Fortschritt. Jahrbuch der Universität 1989. Augsburg (1990) 211–221

9.12 For a historical survey compare Schwemmer, O.: Ethik. In: Mittelstraß, J. (ed.): Enzyklopädie Philosophie und Wissenschaftstheorie. Vol. 1. B.I. Wissenschaftsverlag: Mannheim/Wien/Zürich (1980) 592–599; Abelson, R./Nielsen, K.: History of ethics. In: Edwards (ed.): The Encyclopedia of Philosophy. Vol. III. New York (1967) 81–117

9.13 Irwin, T.: Plato's Moral Theory. The Early and Middle Dialogues. Oxford (1977)

9.14 Kenny, A.: The Aristotelean Ethics. A Study of the Relationship between the Eudemian and Nicomachean Ethics of Aristotle. Oxford (1978)

9.15 Compare Raphael, D.D. (ed.): British Moralists (1650–1800) I–II. Oxford (1969)

9.16 Paton, H.J.: The Categorical Imperative. A Study in Kant's Moral Philosophy. London (1947); Williams, B.A.D.: The Concept of the Categorical Imperative. A Study of the Place of the Categorical Imperative in Kant's Ethical Theory. Oxford (1968)

9.17 Mill, J.S.: Utilitarianism. London (1861); Rescher, N.: Distributive Justice. A Constructive Critique of the Utilitarian Theory of Distribution. Indianapolis (1966)

9.18 Rawls, J.: A Theory of Justice. London (1972)

9.19 Simon, J.: Friedrich Nietzsche. In: Höffe, O. (ed.): Klassiker der Philosophie II. München (1981) 203–224, 482–484

9.20 Heidegger, M.: Die Technik und die Kehre. Pfullingen (1962); Loscerbo, J.: Being and Technology. A Study in the Philosophy of Martin Heidegger. The Hague (1981)

9.21 Jonas, H.: Das Prinzip Verantwortung. Versuch einer Ethik für die technologische Zivilisation. Suhrkamp: Frankfurt (1979)

9.22 Lenk, H.: Zwischen Wissenschaft und Ethik. Suhrkamp: Frankfurt (1992); Mittelstraß, J.: Leonardo Welt. Über Wissenschaft, Forschung und Verantwortung. Suhrkamp: Frankfurt (1992)

9.23 Chua, A.: World on Fire, Doubleday: New York (2003)

9.24 Kant, I.: Zum ewigen Frieden (1795), in: Kants gesammelte Schriften, ed. Königlich Preußische Akademie der Wissenschaften Bd. VIII, Berlin (1912/23)

9.25 Wolfram, S.: A New Kind of Science (see Note 5.47) 750

9.26 Kant, I.: Kritik der reinen Vernunft (1787), B 833

Subject Index

activator 108, 109
adaptive e-learning 405
adiabatic elimination 65
agent
  mobile 397
  stationary 397
algorithmic complexity 8, 9, 179, 183, 192, 193, 206, 225, 374
algorithmic information content 193
algorithmic information theory 194
allometric equation 119
Alzheimer's disease 69, 310
analytical geometry 135
AND gate 211
aperiodicity 80, 81
arbitrage 343
Aristotelian dynamics 25
ars inveniendi 181, 228
ars iudicandi 181, 228
artificial intelligence (AI) 8, 10, 11, 134, 138, 142, 179, 180, 227, 229, 242, 243, 260, 301, 334
artificial life 10, 227, 300, 301, 305, 308, 406
associative memory 7, 174
associative network 131, 140, 141
astronomy 20, 30
atom 21
attractor 4, 6, 7, 12, 13, 17, 40, 56, 59–61, 78, 82, 83, 89, 121, 143, 144, 166–169, 171, 205, 255, 281, 283, 286, 303, 338, 367, 368, 376, 377, 383, 399, 425, 429, 438
  chaotic 7, 80, 81, 255, 259, 305, 325
  fixed point 7, 41, 46, 59, 60, 114, 202, 255, 322, 422
  periodic 7, 107, 255
  quasiperiodic 7, 223, 255
auditive cortex 150
augmented reality 411
autocatalysis 94, 100, 108
autonomy 397
axon 132

background knowledge 309
backpropagation 10, 154, 155, 247, 251, 252, 396
backward chaining 234
BACON program 240
balance 372
Baldwin effect 306
bell curve 339
Belousov–Zhabotinsky (BZ) reaction 4, 57, 177, 291, 305
Bénard convection 152
Bénard experiment 55, 61
bifurcation 57, 59, 60, 98, 110, 117, 222, 311, 375, 422, 423, 425
Big Bang 70, 100
biochemistry 7, 8, 163
Bionic Eye 293
black hole 217
black noise 198, 199
Black–Scholes formula 342–345
Boltzmann distribution 145
Boltzmann machine 10, 145, 146, 252
Boolean cube 278–281, 283
Boolean function 103, 221, 239, 245, 265, 266, 277, 278, 284
Boolean network 6, 103
Born–Oppenheimer procedure 96
bounded rationality 14, 309, 334–336, 399
boundedness 80, 81
BPN (biological pulse-processing) network 289
Bradford-distribution 425

brain 123, 132, 155, 258, 438
Brownian motion 61, 93, 94, 339, 340, 343, 344, 346–348, 352–355, 357, 359
Buckminsterfullerene 73, 77, 222
business cycle 325, 376
butterfly effect 2, 82, 164, 311, 320, 367

C^{++} 409
Capital Asset Pricing Model (CAPM) 340, 342
catalysis 100
Cauchy distribution 202
causa efficiens 23
causa finalis 23
causa formalis 23
causa materialis 23
causality 2, 129, 168, 432
causality of freedom 432
cell assemblies 7, 8
cell assembly 161, 162, 165, 177, 196
cellular automaton (CA) 10, 11, 179, 217, 219, 220, 223, 277–279, 284, 302–305, 337, 338, 374, 392, 393, 424
cellular neural network (CNN) 10, 227, 261, 272, 278, 392, 393
cellular nonlinear network 271
central limit theorem 9, 202, 352
central nervous system (CNS) 8, 133, 134, 137, 174, 175, 195, 290
cerebellum 151
chaos 63, 78, 111, 179, 200, 217, 262, 311, 321, 322, 370, 375, 391
chaos theory 5, 338, 362
chaotic dynamics 78, 82, 304
chaotic pattern 64
chaotic regime 58, 59
Chinese philosophy 20, 29
Chinese room 172, 173
Chomsky grammar 301
Church's thesis 186–188, 228, 235
cluster 74
CNN Universal Machine (CNN-UM) 11, 273–275, 277, 292, 293
cognitive psychology 157
cognitive science 129, 156, 163, 165, 287, 308, 405, 410
coin tossing 9, 200, 339, 357
collapse of the wave-packet 208
communication network 390, 391
communism 373
competitive learning 147
complex system 1, 14, 15, 152, 157, 161, 163–165, 255, 259, 260
$1/f$ complexity 194
complexity class 304
complexity theory 188, 189, 191, 211
computability 8, 11, 179, 183, 186–188, 410, 424, 438
computational complexity 193, 196, 197, 216, 308, 365
computational dynamics 345
computational ecology 390
computational irreducibility 224, 225, 335, 424
computational neuroscience 290, 292
computational system 14
computational time 188–191, 211, 389
computer experiment 179, 304, 306, 309, 328, 337
computer virus 307, 308
congestion 393
connectionism 139
consciousness 7, 8, 123, 124, 155, 159, 161, 162, 176, 208, 299, 309, 310, 372, 373, 413
consistency proof 191
constitution 372, 433
control parameter 4, 58, 63, 105, 109, 112, 167, 169, 257, 304, 305, 331, 349, 393, 422
Conway's game of Life 226
correlation coefficient 78
correlation dimension 82
correlation integral 82
correspondence principle 51–53
cortex 174
Cosmological Principle 70
crystallography 21
cyberspace 298
cyborg 296

DALTON 240
data glove 297
data traffic 395
database 231, 399, 401, 404, 413
DAX 350, 351
decision theory 335
Delphi method 430

democracy 372, 436
DENDRAL 230, 235, 242
dendrite 132
derivatives 342, 343
determinism 80
deterministic chaos 4, 44, 50, 54, 223, 427
dialectics 373, 436
diffusion 109, 116, 167, 338
diffusion-reaction equation 271
dissipative structure 376, 387
dissipative system 55, 57
distribution function 330
DNA 301
Doppler effect 150, 151
Dow Jones 344, 346–348
dynamic complexity 84
dynamical system 32, 33, 38, 44, 46, 47, 54, 57, 262

e-commerce 399
ecology 112
economic complexity 373
economic equilibrium 321
economics 316, 317, 321, 327, 328, 346, 369, 375, 419, 433, 435
econophysics 311, 362, 364, 365
ecosystem 112, 113
EDGE CNN 263–267
EEG 175, 295
efficient markets hypothesis 340
electrodynamics 36, 44
elementary particle physics 22
ELIZA 300
embodied
  artificial intelligence 227
  artificial life 227
  mind 174, 401
  robotics 285
emergence 4, 5, 7, 9, 11, 14, 17, 61, 65, 67, 68, 87, 105, 106, 116, 123, 124, 152, 155, 156, 158, 176, 179, 200, 208, 261, 271, 285, 286, 309–311, 368, 373, 375, 383, 417, 423
  controlled 309
energy landscape 143, 144, 169, 170
entangled quantum state 2, 207, 214
entanglement 216
entelechy 5, 182
entropy 62, 92–94, 97, 98, 195, 224, 304
Entscheidungsproblem 192
epicycle-deferent technique 25, 26, 28
epistemology 47, 128, 157, 166, 287, 372, 417, 431, 438
EPR (Einstein–Podolsky–Rosen) experiment 52, 207, 214, 215
EPR correlation 208
equilibrium 19, 22, 38, 60, 92, 156, 221, 223, 252, 317, 320, 339, 371, 372, 375, 377, 433
equilibrium economics 318
equilibrium point 39
equilibrium thermodynamics 98, 252, 257, 311
ethics 431–433
Euclidean geometry 30, 34
EURISKO 239
evolution 5, 98, 117, 130, 151, 155, 245, 309, 321, 327, 328, 368, 373, 375
evolutionary architecture (EvoArch) 407–409
expert system 179, 180, 230–234, 260, 285, 296
exponential curve 363

feedforward network 244, 252
ferromagnet 62, 142, 144, 332
financial market 12, 340, 350
financial theory 311, 338
finite-difference equation 79
fitness 306, 395
fixed point 4, 59, 78, 83, 143, 197, 222, 305, 326, 376, 383, 386, 427
flicker noise 199
fluctuation 117, 142, 152, 163, 164, 198, 199, 205, 251, 255, 257, 329, 331, 336, 346, 350, 352, 383, 392, 394, 435
fluid dynamics 63, 305, 425
forecast 252, 417
  long-term 321
forward chaining 234
Fourier analysis 9, 197
Fourier transformation 55
fractal 12, 17, 77, 86, 110, 111, 120, 199, 352–354, 356
fractal attractor 223
fractal dimension 56, 57, 82, 84, 85
fractal geometry 360
fractality 9, 84, 119, 200

freedom 430, 433
future contract 342
fuzzy logic 296, 394

Galilean invariance 35, 37, 365
Galilean transformation 35, 61
game-of-life CA 273
gate 211, 212
  classical 212
  quantum 212
gauge groups 67
Gaussian attractor 204
Gaussian bell curve 9, 200, 205, 340, 341, 344, 346
Gaussian distribution 9, 202, 355
Gaussian probability attractor 350
General Problem Solver (GPS) 229
general relativity 65, 67, 69, 70, 209, 217
GENERATE_AND_TEST 235, 237
GENESYS program 308
genetic algorithm 305, 307, 394
genetic information 195
genotype 306
Gestalt psychology 252
GLAUBER 240
global village 399
globalization 367, 435, 436
GPS (Global Position System) 405, 409
grand unification 67
grandmother neuron 161, 162
graph theory 189
Greatest Good 439

hallucination 293, 298
halting problem 192
Hamilton's problem 190, 191
Hamiltonian dynamics 45, 46
Hamiltonian equation 2
Hamiltonian function 44, 48, 53, 207
Hamiltonian operator 52, 53, 207
Hamiltonian system 44
hand calculating machine 181, 183
Hansen–Samuelson model 322, 324
harmonic oscillator 37, 48, 52, 79
harmony 19
Hartree–Fock procedure 96
health 6
heart 110–112, 121
Hebb-like rule 140, 142, 161, 225
Hebb-synapse 139, 140
Hebb-type learning 252
hedging 345
heuristic knowledge 231
hidden unit 152, 246, 248
high energy physics 67
Hilbert curve 85
Hilbert space 2, 52, 207
Hodgekin–Huxley equation 177
homo oeconomicus 316, 317, 321, 334, 345
Hopfield system 10, 144, 159, 257
Horowitz–Maldacena model 217
HTML 392
Hurst exponent 358–360
Hurst parameter 198
hydrodynamical law 61
hypercycle 100, 103
hysteresis 258

immune system 307
incompleteness 9, 191, 193
incompressibility 193
ε_0-induction 191
industrial revolution 6
inertial system 35–37
inflationary universe 67, 70
information content 194
information entropy 9, 195
information flood 393, 395, 401
information flow 13, 84, 179, 196, 392
information retrieval (IR) 390, 394, 397
information theory 9, 194, 195, 200
inhibitor 108
instability point 64
intentionality 7, 8, 165, 166, 173, 174
Internet 367, 391–395, 405
invariance 362, 364, 365
investment bubble 360
invisible hand 11
irregularity 222
irreversibility 65, 224
iustitia commutativa 369
iustitia distributiva 369

Java 409
Joseph effect 12, 198, 311, 356–360

Kalecki model 322
KAM theorem 47–49
KEKADA 240, 241
knowledge based economy 411
knowledge based system 240, 243, 421
knowledge representation 229, 232, 285, 309
knowledge society 398, 415
Königsberg river problem 189
Kolmogorov–Sinai (KS) entropy 84, 179, 196, 197, 223, 295, 417
Kondratieff-cycle 414

L-system 301
Laplacean demon 2, 17, 47, 369, 378, 417
laser 64, 65, 259, 291
Lausanne School 316, 317, 350
learning algorithm 10, 296, 395
Leviathan 371, 372, 399, 432
Lévy-stable distribution 205, 354, 355, 360, 365
lex inertiae 33
life science 71
limit cycle 6, 40–42, 47, 78, 143, 197, 269, 322, 326, 385, 421, 422
limit point 47
limit set 39, 40
limit theorem 202
linear differential equation 54
linear dynamics 2, 3
linear separation 246
linearity 207, 317, 420
linearly non-separable rule 279, 284
linearly separable rule 279, 283, 284
Liouville's theorem 45, 46, 55
LISP 193, 229, 236, 238, 239, 243
local activity principle 261, 272
logistic curve 58, 323, 363, 375, 388
logistic map 57, 59, 60, 322, 422, 425, 427
logos 3, 19
long-term dependency 358, 359
long-term memory 198
long-term potentiation (LTP) 175
Lorentz transformation 37
Lorenz attractor 79, 80
Lotka–Volterra equation 421, 425
Lundberg–Metzler model 324
Lyapunov exponent 83, 84, 295, 422

M-theory 17
macroscopic pattern 62, 63, 175, 176
macrostate 93, 172, 176, 195, 315
macrovariable 330
management 391
many-body problem 43, 44, 49, 175
many-layered network 138
market equilibrium 338
market fluctuation 356
master equation 13, 61, 65, 330, 331, 380–382, 426
materials science 17, 71
mathesis universalis 181–183, 187, 228
maximax utility criterion 319
maximin utility criterion 318
Maxwell's demon 87, 97, 99
McCulloch–Pitts neuron 218
MCP (McCulloch & Pitts) network 219, 243, 289
measurement process 2, 207
meme 389
META-DENDRAL 237
metabolism 94, 118, 119, 159
metamorphosis 91, 106
meteorology 6, 321
Mexican hat 148
microreversibility 216
microstate 93, 99, 172, 195
migration 13, 382, 383, 426
migration dynamics 384, 385, 387
mind-body problem 7, 123, 124, 163
minimax risk rule 319
Minkowskian geometry 37
mobile learning 412
mobility 397
Modern Portfolio Theory (MPT) 340
Möbius strip 324
monadology 90, 127, 128, 300, 308
morality 433
morphogenesis 105, 106, 108–110
mother neuron 161
MPT 342
MRI 291
multi-layered network 147, 152, 247, 395, 397
multifractal 12, 17, 77, 86, 360, 361
mutation 91, 104, 306, 307, 395
MYCIN 230, 237

nanomachine 72
nanorobot 77
nanostructure 71–73, 75
Navier–Stokes equation 349, 365
Necker cube 253–255
neocortex 148, 176
NETalk 248, 250, 251
neural computer 251, 260
neural network 7, 10, 142, 148, 162, 243, 249, 253–255, 260, 290, 423
neurobionics 227, 285, 287, 288, 294, 296
neurocomputer 258
neuron 132
neurosurgery 11, 288, 290
neurotransmitter 273
Newtonian equation 2
Newtonian physics 5, 11, 311, 367, 417
Newtonian space-time 35, 37
no-cloning theorem 213
Noah effect 12, 311, 355, 359, 360
noise 198, 335
$1/f^b$ noise 179
non-equilibrium thermodynamics 97, 177, 329
non-Gaussian attractor 205
non-Gaussian probability density function 350
non-locality 52
nonlinear causality 6
nonlinear complex system 54
nonlinear data analysis 78
nonlinear differential equation 54, 55, 262, 266, 270
nonlinear dynamical system 262, 282
nonlinear dynamics 1, 4–6, 8, 14, 77, 111, 121, 123, 165, 168, 179, 198, 266, 271, 273, 281, 286, 329, 333, 336, 338, 350, 367, 379, 390, 410
nonlinear feedback 142, 176
nonlinear network 154
nonlinear time series analysis 295
nonlinearity 2, 4, 54, 60, 98, 99, 301, 322, 324, 327, 380, 399, 436, 439
normal distribution 339, 340, 342, 347
NOT gate 211
NP-complete problem 191
NP-problem 190, 191

Ockham's razor 14, 173, 209, 327, 375, 434
ontology 285, 401, 403, 404, 434
option
  American 343
  European 342
OR gate 212, 245
order parameter 4, 6–8, 12, 14, 65, 109, 115, 123, 169, 175–177, 255, 257, 258, 286, 305, 307, 330–332, 393, 424, 428, 435, 436
organic computing 285, 286
oscillation 63, 113, 427
oscillator 326

P-problem 191
paradox of quantum information 217
parity problem 245
partial differential equation 262, 271
pattern formation 7, 11, 17, 104, 179, 223, 255–257, 271, 291, 337, 393
pattern recognition 7, 11, 123, 143, 227, 229, 255, 256, 261, 271, 291, 307
Pauli principle 96
perceptron 244, 245, 272
periodic oscillation 59, 60, 322
periodic pattern 221
periodic time series 269, 270
periodic trajectory 114
periodicity 59, 81, 198
personalization 400, 403, 404
personalized information service 411
personalized information system 13, 367, 390, 400, 403, 411, 412
personalized learning 413
PET (positron emission tomography) 176
phase portrait 33, 39–43, 84, 85, 113–115, 135, 143, 226, 334, 383, 384, 435
phase space 45, 48, 56, 57, 79–81, 270, 283, 365, 421
phase state 93
phase transition 3, 4, 12, 17, 60–62, 65–67, 104, 105, 112, 123, 143–145, 163, 164, 169, 216, 222, 252, 255, 304, 311, 350, 367, 368, 372, 373, 376, 379, 387, 388, 393, 395, 425, 436
philosophy 7, 77, 238, 243, 415, 429
philosophy of mind 127, 158, 405

physicalism 12
physiocratic school 311–314, 375
physiology 120, 121
pink noise 198–200
Platonic solids 75
Poincaré map 48, 49, 57, 270
Poisson distribution 53
polis 368, 399, 432
political economy 315
portfolio 341, 342, 344, 345, 347
power law 6, 87, 117–120, 205, 345, 358, 360, 362
power law distribution 9, 350, 355, 393
power spectrum 84, 394
predator 113, 375, 421
predictability 196, 337, 421
prediction 420
preference 400–404, 410
prey 113, 115, 375, 421
principle of uncertainty 70, 96
probabilistic attractor 9, 179
probabilistic distribution 3, 61, 65, 194, 195, 383, 384
probability 194, 200, 339
probability density function 9, 202, 205, 350, 353, 354
profile 400, 402–404, 410
program-controlled computer 173
PROLOG 243

qualia 177
quantum bit (qubit) 195, 211
quantum chaos 53, 54
quantum complexity 179, 206
quantum computer 9, 179, 206, 210, 211, 216, 217
quantum dote 75
quantum effect 210
quantum field theory 3
quantum fluctuation 69, 70
quantum information 9, 195, 206, 212–214, 216
quantum mechanics 2, 3, 9, 45, 50, 52, 70, 96, 208, 212, 215, 321
quantum parallelism 216
quantum randomness 216
quantum state 52, 207, 211
quantum system 2, 9, 52, 208
quantum teleportation 214, 216
quantum universe 69
quantum vacuum 67

Rössler attractor 324
random 100, 104, 105
random noise 78
random variable 195, 202, 204, 205
random walk 201, 338, 339, 352
randomness 10, 12, 14, 179, 193, 217, 223, 224, 262, 311, 321, 335, 336, 338, 424, 438, 439
rationality 318, 320
reaction 109
recursive function 186, 187
red noise 198
reduction of complexity 65
register machine 183–185, 187
regular body 21
relativistic space-time 37
responsibility 430, 431, 435
reversibility 3, 96, 212, 224
reversible cellular automaton 225
Reynolds number 63, 349
RM-computable function 184, 185
RMI (Remote Method Invocation) 407
RNA 72
robotic 227, 285, 295, 308
router 392
RPC (Remote Procedure Call) 407

scala naturae 89
scale invariance 351
scaling law 119
Schrödinger equation 2, 52, 53
science fiction 295, 298
second law of thermodynamics 5, 87, 92–94, 97, 195, 224, 225
selection 65, 117, 119, 395
self-adaption 409, 410
self-assembly 74
self-configuration 406, 410
self-consciousness 7, 8, 123, 158, 173, 178
self-control 310
self-diagnosis 406
self-healing 409
self-organization 3, 5, 7, 9, 10, 62, 64, 67, 73, 75, 76, 87, 91, 96, 104, 105, 139, 141–143, 146, 149, 151, 173, 176,

179, 180, 200, 222, 243, 252, 255, 286, 288, 292, 308, 321, 332, 333, 373, 374, 386, 391, 406, 408, 431
conservative 64, 76
dissipative 64, 76, 87, 104
self-referential state 260
self-referentiality 299, 398
self-reflection 310
self-replication 72
self-reproduction 99–101, 243, 321, 426, 427
self-similarity 6, 9, 85, 117, 118, 121, 200, 351, 352, 354, 396
sensitivity 80
sensorimotor coordination 136, 137, 151
separatrix 40, 41, 422
sequential space 102
SHRDLU 230
SIBYL 421
Sierpinski curve 85
slaving principle 4, 14, 64, 256
social dynamics 364, 387, 388
social force 363
social science 374, 375, 417, 431
socio-economic dynamics 380
sociobiology 115, 329, 390, 396
socioconfiguration 13, 330, 383
sociodynamics 365
sociology 14, 380, 384, 410
solid-state physics 72
soul 124, 126
special relativity 45, 50
$1/f$ spectrum 9, 199
speed of light 70
spin glass 143, 332
STAHL 240
state space 30, 33, 38, 40–42, 45, 106, 136, 143, 144, 156, 171, 325
stochastic differential equation 61, 343
stochastic dynamics 362, 426
stochastic equation 61, 365
stochastic process 200–202, 205, 311, 338
non-Gaussian stable 205
stochastic symmetry breaking 61, 62
stock 340, 341, 350, 423
stock quotation 253
Stoics 29, 432
strange attractor 56, 57, 270, 324–326, 383, 393, 396, 421
superorganism 405
superposition 2, 209, 211, 212, 320
superposition principle 57, 207, 317, 374
superstring theory 17, 67
supramolecular chemistry 302
supramolecule 71, 73
survival of the fittest 100
swarm intelligence 390, 396, 405
symbiosis 6, 115
symmetry 18, 19, 25, 26, 60, 67, 73, 109, 216, 321, 364, 365
symmetry breaking 4, 17, 18, 60, 61, 65, 66, 163, 164, 332, 333, 386
synapse 132, 133
synaptic weight 262, 263
synaptic weight space 138, 139
synergetic computer 10, 243, 256–259
synergetics 4, 7, 12, 13, 255, 259, 271, 331, 332, 380, 381

Tableau économique 312–314
tacit knowledge 413
Taoism 29
teleology 5, 89, 127, 182
tensor transformation 136, 156
theology 70
theory of games 318
Theory of Mind (ToM) 8, 176, 310
thermal equilibrium 3, 5, 62, 64, 76, 94, 98, 123, 142, 145, 159, 163, 259, 305, 321, 376
thermodynamic equilibrium 330
thermodynamics 3, 63, 93, 97, 156, 316, 330, 337, 344, 376
three-body problem 374
three-layer network 138, 139
threshold 142, 162, 245, 259, 262, 263, 286, 376, 385, 395
THRESHOLD CNN 265
time scales 360, 396
time series 5, 79, 81, 84, 85, 121, 325, 328, 351, 359, 394, 418
time series analysis 17, 77, 78, 82, 334
trading time 361, 362
trajectory 33, 38, 39, 42, 46, 47, 53, 81, 83, 259, 324, 325
travelling salesman problem 191
truth table 266–268, 281

turbulence 9, 55, 111, 200, 206, 345, 347, 348, 362, 365
Turing machine 185–188, 190, 192, 210, 218, 219, 225
  deterministic 190
  non-deterministic 190
twin paradox 37
two-layer network 138

ubiquitous computing 13, 367, 405, 410
uncertainty relation 51, 53, 69, 134, 213, 217, 317, 318
undecidability 9, 193, 225, 302, 335
unification 66
uniform distribution 202
unitary operator 212
universal Turing machine 8, 77, 185, 187, 227, 273, 279, 302
unsolvability 192
urban dynamics 380
urban system 377–379
utilitarianism 433
utility 318–320, 345, 402

variance 201, 202, 205, 341
vector field 33, 326
vector space 135, 139
Verhulst equation 363
virtual reality 11, 298
visual cortex 133
volatility 361
von-Neumann computer 218, 219, 391
vortex 63, 385
vortex point 39, 46

weather 321, 322
Web service 402
white noise 199, 200
Widrow–Hoff rule 153, 154
Wigner distribution 53
World Wide Web 13, 199, 391, 393, 397, 399

XOR gate 245–247

Name Index

Aigeas 418
Albertus Magnus 89, 125
Alcmaeon of Croton 124
Allen, P. 12, 377, 378, 381
Anaxagoras 20, 124
Anaximander 18, 88
Anaximenes 18, 19
Apollonius of Perga 25, 26
Aristotle 4, 5, 17, 23–25, 29, 88, 89, 98, 124, 125, 182, 227, 315, 368, 369, 432
Arnold, V.I. 2, 47
Aspect, A. 2, 207
Aurelius, M. 89
Avicenna 125

Babbage, C. 182
Bachelier, L. 12, 311, 338–340, 344, 353, 354, 357, 358, 360, 365
Barlow, H.B. 161
Bénard, A. 55
Bentham, J. 433
Bergson, H.L. 96, 155
Berkeley, G. 298
Birkhoff, G. 54
Black, F. 340, 342, 365
Bohr, H. 28
Bohr, N. 2, 45, 51, 96, 208
Bois-Reymond, E. du 130
Boltzmann, L. 5, 62, 87, 92–95, 97, 98, 113, 195
Bondi, H. 69
Bonnet 133
Borelli, G.A. 90
Boyle, R. 240
Brentano, F. 166
Broglie, L. de 50
Brown, J.H. 119
Brown, R. 339

Cantor, G. 192
Carnap, R. 218, 237
Carnot, S. 92
Chaitin, G.J. 193
Chomsky, N. 301
Chua, L.O. 11, 261, 275
Church, A. 179, 186
Churchland, P. 157
Clausius, R. 92
Conway, J. 220
Copernicus, N. 17, 27, 28, 30
Curie, P. 65

Darwin, C. 5, 87, 100, 113, 305, 434
Democritus 7, 21, 88, 125
Descartes, R. 7, 20, 43, 90, 125–127, 129, 135, 227, 299, 312, 370, 432
Drexler, E. 71, 72
Driesch, A. 96
Duhem, P. 97
Durkheim, E. 380
Dyson, F. 101, 103

Eccles, J. 7
Eigen, M. 5, 98–101, 302, 306
Einstein, A. 17, 29, 35, 50, 65, 67, 69, 92, 97, 216, 298, 338, 339, 346, 350
Empedocles 20, 21, 29
Enquist, B.J. 119
Euclid 88
Euler, L. 189
Everett, H. 208, 209
Exner, S. 95

Feigenbaum, M.J. 57, 230
Feynman, R. 71, 72, 77
Fischer, E. 73, 74
Flohr, H. 161
Fuller, R.B. 73

Galen, C. 89
Galilei, G. 24, 30, 32, 33, 43, 89, 90, 370, 371
Gentzen, G. 191
Gerisch, G. 5
Gibson, W. 298
Gierer, A. 108, 109
Glance, N.S. 386
Gödel, K. 9, 180, 183, 191
Goethe, J.W. 96, 106, 163
Goffman, W. 425
Gold, T. 69
Goodwin, R.M. 326
Gorman, R.P. 248
Grossmann, S. 57

Haeckel, E. 92
Haken, H. 4, 7, 64, 97, 169, 258, 271
Hamilton, W. 44, 189
Harrod, R. 318
Hartle, J.B. 69
Harvey, W. 372
Hawking, S.W. 67, 69, 217
Hebb, D.O. 131, 139, 140, 161, 196, 243
Hegel, G.W.F. 12, 372, 373, 433, 436
Heidegger, M. 434
Heiles, C. 48
Heisenberg, W. 2, 19, 22, 51, 208, 317
Heitler, W.H. 96
Helmholtz, H. von 43, 129, 130
Hempel, C.G. 318
Hénon, M. 48
Heraclitus of Ephesus 4, 17, 19, 21, 23, 65, 69, 124
Herbrand, J. 229
Hilbert, D. 183, 191–193, 218
Hinton, G.E. 145
Hippocrates 124, 287
Hobbes, T. 1, 367, 368, 370, 371, 432
Hoff, M.E. 153
Holland, J. 306
Hopfield, J. 142, 143, 261
Hoyle, F. 69
Huberman, B.A. 386
Hume, D. 128–130, 140, 298, 367, 372, 432
Hurst, H.E. 357
Huygens, C. 40, 42, 90

James, W. 130, 132
Jonas, H. 434
Joy, B. 77

Kant, I. 5, 47, 67, 87, 117, 129, 132, 155, 372, 432, 433, 436–438
Kauffman, S. 6, 103
Kelso, J.A. 169
Kepler, J. 25, 27–29, 51, 365
Keynes, J.M. 318
Köhler, W. 155, 156
Kohonen, T. 147–149, 156, 161
Kolmogorov, A.N. 2, 47, 193, 200, 201, 349
Krebs, H. 240, 241
Kuhn, T.S. 420, 429

Lamettrie, J.O. de 1, 7, 90, 129
Lange, L. 35
Langton, C.G. 301–303
Laplace, P.-S. 155
Leibniz, G.W. 7–9, 17, 32, 33, 37, 43, 46, 49, 90, 97, 127, 128, 133, 155, 179–182, 187, 191, 192, 206, 216, 228, 301, 346
Lévy, P. 204, 350
Linde, A. 70
Lindenmayer, A. 301
Liouville, J. 46
Locke, J. 128, 367, 372, 432
London, F. 96
Lorenz, E. 6, 54, 55, 78, 322, 421, 423
Loschmidt, J. 93
Lotka, A.J. 6, 113, 326, 375, 425
Lovelock, J. 113

Mach, E. 93
Malebranche, N. 127
Malsburg, C. von der 161
Malthus, T.R. 316, 362, 364, 375
Mandelbrot, B.B. 352, 354, 355
Markowitz, H.M. 340, 341, 344, 365
Marshall, A. 346
Marx, K. 117, 373
Maxwell, J.C. 36, 44, 46, 87, 97, 98
McCarthy, J. 187
McClelland, J.L. 252
McCulloch, W.S. 218
Meinhardt, H. 5, 108, 109

Mendel, J.G. 92
Merton, R. 425
Mill, J.S. 316, 433
Minsky, M. 183, 185, 229, 230, 244, 245, 272
Monod, J. 5, 87, 95, 97, 98
Monroe, M. 298
Morgenstern, O. 318, 320
Moser, J.K. 2, 47
Moses, J. 229
Müller, A. 74
Müller, J. 129, 130

Navier, C.L. 349
Neumann, J. von 10, 11, 185, 218, 219, 261, 273, 302, 318, 320, 321
Newell, A. 187, 229
Newton, I. 2, 5, 17, 29, 32–35, 37, 43, 49, 65, 87, 89, 97, 312–315, 363, 364,370
Nicolis, G. 115
Nietzsche, F. 433

Ohm, G.S. 240
Oresme, N. 31
Orwell, G. 298, 369

Papert, S. 244, 245
Pappos 235
Pareto, V. 316, 350
Parmenides of Elea 19, 23
Pascal, B. 180
Penrose, R. 3, 7, 67, 69, 209
Pitts, H.W. 218
Planck, M. 50
Plato 7, 19, 21, 22, 25, 26, 30, 88, 125, 368, 369, 432
Poincaré, H. 2, 17, 33, 47, 54, 55, 93, 98, 224, 322, 427
Polya, G. 237
Popper, K. 389, 429
Post, E. 185
Prigogine, I. 92, 97, 98, 104, 258, 271, 381
Ptolemy 25, 26, 28
Putnam, H. 298, 299
Pythia 417, 430

Quesnay, F. 311

Rashevsky, N. 106
Rawls, J. 433
Rayleigh, Lord 43
Rechenberg, I. 305
Ricardo, D. 316
Robinson, J.A. 229
Roosevelt, F.D. 376
Rosenberg, C.R. 248, 249
Rosenblatt, F. 243, 244, 261, 272
Rumelhart, D.E. 252
Russell, B. 218, 229

Samuelson, P.A. 340
Savage, C.W. 319
Schelling 96
Schickard, W. 180
Scholes, M.S. 340, 342, 365
Schrödinger, E. 2, 51, 96, 97, 104, 208, 215, 317
Schumpeter, J.A. 323, 414
Searle, J. 166, 171–173
Sejnowski, T.J. 145, 248, 249
Shannon, C.E. 179, 194, 195, 414
Sharpe, W.F. 340, 341, 344, 365
Shaw, G.L. 229
Sibyl of Cumae 421
Simon, H.A. 14, 187, 229, 240, 241, 334
Singer, W. 161
Smalley, R.E. 77
Smith, A. 2, 11, 182, 311, 312, 314–317, 345, 372, 373, 414, 432
Snow, C.P. 12
Socrates 432
Sorokin, P. 425
Spencer, H. 5, 87, 98, 219, 376, 389
Spinoza, B. 127, 432

Takens, F. 82
Tarski, A. 228
Thales of Miletus 18, 87, 88
Thomae, S. 57
Thompson, W. (Lord Kelvin) 97
Thoreau, H. 112
Turing, A.M. 5, 9, 180, 185, 192, 193, 218, 228, 242, 299, 300

Verhulst, P.-F. 363, 375
Virchow, R. 123
Volterra, V. 6, 113, 326, 375

Walras, L. 316, 317
Weidlich, W. 13, 380
Weizenbaum, J. 300
West, G.B. 119
Whitehead, A.N. 96, 229
Widrow, B. 153
Wiener, N. 421
Wigner, E. 2, 208
Wolfram, S. 10, 179, 221, 225, 279

Yang, L. 11, 261, 275

Zermelo, E. 93
Zuse, K. 185